# Magnetic Resonance Imaging
## Theory and Practice

Springer
*Berlin*
*Heidelberg*
*New York*
*Hong Kong*
*London*
*Milan*
*Paris*
*Tokyo*

Marinus T. Vlaardingerbroek
Jacques A. den Boer

# Magnetic Resonance Imaging

## Theory and Practice

With a Historical Introduction
by André Luiten

Third Edition
with 168 Figures and 57 Image Sets

Springer

Dr. Ir. Marinus T. Vlaardingerbroek
Haagbeuklaan 7
6711 NK Ede (Gld), The Netherlands
e-mail: R. Vlaardingerbroek@tue.nl

Dr. Ir. Jacques A. den Boer
Zweerslaan 3
5691 GN Son, The Netherlands
e-mail: denboer@iae.nl

*Cover Figure:* 1.5 T whole body MR system with short cylindrical magnet and online image display, supporting interventional radiology.

ISBN 978-3-642-07823-1

Library of Congress Cataloging-in-Publication Data applied for.

Die Deutsche Bibliothek – CIP-Einheitsaufnahme

Vlaardingerbroek, Marinus T.: Magnetic resonance imaging: theory and practice/
Marinus T. Vlaardingerbroek; Jaques A, den Broer. With a historical introd. by André Luiten. –
3.ed. – Berlin; Heidelberg; New York; Hong Kong; London; Milan; Paris; Tokyo: Springer, 2002
(Physics and astronomy online library)



Springer-Verlag is a part of Springer Science+Business Media
springeronline.com


Printed in Germany



Cover design: *design & production* GmbH, Heidelberg

Printed on acid-free paper

# Foreword

When retired it is a blessing if one has not become too tired by the strain of one's professional career. In the case of our retired engineer and scientist Rinus Vlaardingerbroek, however, this is not only a blessing for him personally, but also a blessing for us in the field of Magnetic Resonance Imaging as he has chosen the theory of MRI to be the work-out exercise to keep himself in intellectual top condition. An exercise which has worked out very well and which has resulted in the consolidated and accessible form of the work of reference now in front of you.

This work has become all the more lively and alive by illustrations with live images which have been added and analysed by clinical scientist Jacques den Boer.

We at Philips Medical Systems feel proud of our comakership with the authors in their writing of this book. It demonstrates the value we share with them, which is "to achieve clinical superiority in MRI by quality and imagination".

During their careers Rinus Vlaardingerbroek and Jacques den Boer have made many contributions to the superiority of Philips MRI Systems. They have now bestowed us with a treasure offering benefits to the MRI community at large and thereby to health care in general: a much needed non-diffuse textbook to help further advance the diffusion of MRI.

Freek Knoet
Director of Magnetic Resonance
Philips Medical Systems

# From the Prefaces of the First and Second Editions

In this textbook we have undertaken the task of developing a coherent theoretical description of MRI which can serve as a background for thorough understanding of recent and future developments. Although we start with the basic theory, the textbook is not meant for making a first acquaintance with MRI. For this goal we refer the reader to other textbooks, of which some are mentioned in Chap. 1.

It is interesting to note here that many of the building blocks of the theory that we need for our task were already available in the papers on NMR published long before the invention of magnetic resonance imaging in 1972 – for example, in the early works of Bloch, Purcell, Ernst, Hahn, Hinshaw, and many others. Andre Luiten, who was already active in the development of MRI during "the first hour", has provided us with his notes on that period and thereby added a much needed historic perspective to the subject of our book.

This textbook also presents a short global description of the MR system and its components, as far as this knowledge is necessary for understanding the application capabilities of the system. The design task itself requires much more detail and is beyond the scope of this textbook.

Each theoretical chapter is followed by a number of image sets. They were specially acquired for the purpose of demonstrating the effects resulting from the MR physics, the system design, and the properties of the sequences under consideration. The images were not taken for medical purposes: they were usually taken from healthy volunteers. However, many problems that are met in practice are illustrated in the image sets and are extensively discussed in the captions.

The image sets in this book were all generated on Philips Gyroscan systems. This choice means that the images shown were obtained using the particular acquisition methods available on that system type. No guarantee can be given of the equivalence of these methods with others that have the same names, but are implemented using MR systems of different make. Many scan methods are named according to usage within Philips. To facilitate comparison, we have tried to list the brand names of related scan methods.

In an appendix we propose a systematic nomenclature for the imaging sequences. This is done jointly with Prof. E.M. Haacke, one of the authors of another physics book on MRI, see [8] of Chap. 1.

In the second edition we added an index, we replaced many figures, and, of course, we corrected the errors that we were made aware of.

As one might expect, we did not think that the basic chapters of our book needed much work, apart from some minor improvements. Extensions were felt necessary in the chapters on Contrast and Signal-to-Noise Ratio and Motion and Flow. A major extension to the theoretical part of our book is an alternative theoretical description of MRI in Chap. 8. Following Prof. J. Hennig (who also kindly read our manuscript) we named this theoretical description the "Theory of Configurations", and we show that this theory is well suited to describing the properties of the conventional sequences as well as those of multi-pulse sequences.

Finally we added a number of image sets, mainly in connection with the chapter on Motion and Flow. Furthermore, a new image set explaining SENSE, a new way to combine information from receive coil arrays to reduce scan time, was added.

July 1995 and April 1999

M.T. Vlaardingerbroek
J.A. den Boer

# Preface to the Third Edition

In the three years since the second edition of this book, the development of MR imaging has continued at a rapid pace. The need for an overview of its mathematical principles has remained, and influenced the decision to write this third edition. Our report of these principles is still valid, as it should be, but the availability of stronger gradients in new MR systems has been instrumental in focusing the development of imaging sequences onto new areas, such as balanced FFE and advanced diffusion imaging. Such areas are now also addressed in the text and the images of this book. We are grateful for the reactions of readers who helped us with some remaining errors. We hope that this new edition will reach the students of MRI, the brains that will eventually determine the future of this fascinating imaging technique.

Ede, Son, The Netherlands
July 2002

Marinus T. Vlaardingerbroek
Jacques A. den Boer

# Acknowledgements

This book evolved from the education that one of us (MTV) received from his coworkers during the period that he acted as project leader for (mainly) 1.5 T MRI systems. After a long career in other fields of physics and engineering (plasma physics, microwave devices and subassemblies, lasers, etc.) and industrial management, he joined the MR development group with practically no knowledge of system design in general and MRI in particular. With much patience, colleagues undertook the task of teaching their project leader, and this education lies at the root of the theoretical part of this textbook. It is in a way a modest compilation of the broad knowledge at all levels of MRI system design, system testing, and (clinical) application of the MR department. To mention all names here would be unwieldy but the friendly lessons of all colleagues are highly appreciated.

The writing of this textbook was further supported by a course on system design, which we organized within the development group. Together with a number of colleagues who specialized in the different disciplines, we prepared notes for this course. We (the present authors) were allowed to use these notes for the preparation of this textbook. We acknowledge the lecturers of this course, who were also willing to criticize our text. They are: M. Duijvestijn, C. Ham, W.v. Groningen, P. Wardenier, J. den Boef, F. Verschuren, L. Hofland, P. Luyten, B. Pronk, H. Tuithof, and G.v. Yperen. Many discussions with J. Groen, P.v.d. Meulen, M. Fuderer, R. de Boer, M. Kouwenhoven, J.v. Eggermond, A. Mehlkopf, and many others were very stimulating.

Part of the internal Philips course was later also presented at the Institut für Hochfrequenztechnik of the Rheinisch Westfälische Technische Hochschule (Technical University) in Aachen, Germany. We also thank the students for teaching us how to explain difficult concepts such as $\vec{k}$ space.

One of us (JAdB) undertook the task of designing and collecting image sets for the purpose of illustrating a number of essential problems in the interpretation of MR images of human anatomy. The text to those images was read carefully by J.v.d. Heuvel of the Philips MR Application department. All MR images presented in this textbook are with the courtesy of Philips Medical Systems.

Most of the images were produced especially for this book on a 1.5 T, S15 ACS (Advanced Clinical System) installed at the hospital "Medisch Spectrum Twente" in Enschede, The Netherlands. The system was kindly made available for this purpose by the management of this hospital. We wish to thank the operators of this system: Dinja Ahuis-Wormgoor, Tienka Dozeman, Annie Huisman-Brouwer, Richard van der Plas, and Francis Welhuis for their skilled and enthusiastic work. Some other images, in part at other field strengths, were put at our disposal by K. Jansen and A. Rodenburg of Philips Medical Systems.

During the preparation of the second edition we again had the support of many of our colleagues at Philips Medical Systems. We especially mention P. Folkers and J. Smink, but realize that there were also useful discussions with the other colleagues at PMS, mentioned earlier. Also the cooperation with Prof. P. Wijn. Dr. D. Kaandorp and Ir. S. Nijsten of the Technical University of Eindhoven is much appreciated. We are grateful to Prof. J. van Engelshoven of the Academic Hospital in Maastricht for the opportunity to use the MR system for the generation of part of the new image sets. Very important for the work described in Chap. 8 was the cooperation between one of us (MTV) and Prof. T.G. Noll and his coworkers Dr. A.R. Brenner and Dipl.-Ing. J. Kürsch in preparing a follow-up course to the MRI course at the Technical University of Aachen, Germany. Through this cooperation we were made aware of the "Theory of Configurations", which also appeared to be a useful addition to the standard theory of conventional multi-pulse sequences.

In the third edition the new image sets were obtained by active support of G. van Yperen, F. Visser and R. Springorum of Philips Medical Systems.

The continuous support and encouragement of the management of Philips Medical Systems are highly appreciated. Especially we thank F. Knoet and M. Duijvestijn for their active support.

We have enjoyed our cooperation and we apologize to our dear Annie and Tineke, life long companions and wives, for the long period during which we invested such a large part of our attention to the writing of this book.

Marinus T. Vlaardingerbroek<br>Jacques den Boer

# Contents

# List of Image Sets

# Magnetic Resonance Imaging: A Historical Introduction

By A.L. Luiten

The discovery and development of magnetic resonance imaging is one of the most spectacular and successful events in the history of medical imaging. However, there is a time gap of almost thirty years between the discovery of nuclear magnetic resonance simultaneously and independently by Bloch [1] and by Purcell [2] in 1946 and the first imaging experiments in the 1970s by Lauterbur and by Damadian.

NMR became a very important technique for non-destructive chemical analyses since the discovery by Proctor and Yu in 1950 of the chemical shift effect in NMR spectra. Nuclei of the same type had different resonance frequencies depending on the chemical composition caused by the typical field screening effected by the electrons in the molecule. In those days, Gabillard [3] was already experimenting with NMR signals of samples in inhomogeneous magnetic fields and suggested the possibility of localizing resonating nuclei by using inhomogeneous magnetic fields with a linear field strength gradient.

Interest in the medical diagnostic possibilities of NMR began in 1971 with the study by Damadian [4] of the differences in relaxation times $T_1$ and $T_2$ between different tissues and between normal and cancerous tissue. In spite of the lack of uniformity in the results and in the physical explanations of various experimenters, the subject raised immediate interest at cancer research centers. In 1972, Damadian [7] patented a device that could selectively measure the resonance signals of localized human tissue samples in situ by using a "single-point" technique, which he named FONAR (field focussing NMR) [6]. In the center of the magnet a typical symmetric field inhomogeneity is created that has a sort of 3-dimensionally saddle-shaped field distribution. Then only in the vicinity of the "saddle point" is the field sufficiently homogeneous to raise a measurable resonance signal. This device could not only perform localized measurement of resonance signals and relaxation parameters but also produce an image in a point-by-point scan when the examined person was moved. In this way the first whole-body human chest scan was obtained in July 1977 in a 0.05 T home-made superconducting magnet with a scan time of 4.5 hours and a resolution of the order of a cm. The technique was further developed to a scanning system with a permanent magnet of 0.3 T and became the first commercial MR scanner.

In the early 1970s research on NMR imaging with different approaches began in a number of research groups in the US and the UK. The first of these was the group under Lauterbur at the Stony Brook University (NY). Lauterbur had the unique opportunity in the summer of 1971 of observing the work of a graduate student, at the Johns Hopkins Medical School, who was repeating and confirming the tissue relaxation experiments performed by Damadian one year earlier. Although he initially discounted Damadian's findings he now became convinced of the potential of NMR in disease detection and he came up with a method of spatially localizing tissue NMR signals using Gabillard's idea of applying linear field strength gradients, and displaying the signals in a pictorial map.

In 1973 Lauterbur [5] was the first to present a 2-dimensional NMR image of a water-filled structured object. The image was reconstructed from a number of NMR measurements each obtained in the presence of a linear field gradient with a different direction. Using the obtained spectra as the one-dimensional image projections perpendicular to the applied gradient direction he could reconstruct the NMR image in a similar projection-reconstruction procedure as used in CT scanners. Lauterbur [8] called his technique "Zeugmatography". This name includes the Greek word zeugma, meaning "that which joins together", because of the joint use of a static magnetic field and high-frequency RF fields in the imaging process. Other investigators introduced other names for their techniques: spin mapping, spin imaging and NMR imaging. Eventually the name magnetic resonance imaging (MRI) became generally accepted; the term "nuclear" was left out as this could suggest some relation with radioactivity and nuclear medicine techniques.

Also in the early 1970s, Mansfield and Hinshaw started their research with independent groups at the University of Nottingham, both within the department headed by Professor Andrew. At the same time Hutchison [14] started at the University of Aberdeen in Professor Mallard's department. They were soon followed by research groups in laboratories at other universities and medical equipment companies.

In 1974 Hinshaw introduced the "sensitive point" technique [9, 10], another sequential point scanning technique using three alternating gradients fields, which suppress all NMR signals from the whole object except from the point where all gradient fields are zero. Later he applied two alternating gradients and one constant gradient, the "multiple sensitive point" technique, with a frequency encoding along the sensitive line by the stable gradient. Scanning was carried out by parallel shifting the sensitive line in a selected plane. In 1977 Hinshaw published a surprisingly detailed cross-sectional image of the human wrist using this technique [11]. The excitation technique then used was a small-flip-angle steady-state-free-precession (SSFP) technique, which was later abandoned after the introduction of the spin-echo (SE) technique. Some ten years later the SSFP technique became important again in the fast-imaging techniques. The first medical images were obtained by the Nottingham group using this sensitive line technique [12, 13]. Better results were obtained later in a sensitive plane technique by applying only an alternating

Z gradient in combination with Lauterbur's projection reconstruction (PR) technique, with a field gradient rotated in a number of directions during detection.

An important improvement introduced by Garroway, Grannell and Mansfield [15] in 1974 was the selective slice excitation technique that is generally used today, in which a field gradient perpendicular to the selected plane is applied during a tailored excitation pulse. Through a combination of excitation pulses and orthogonal gradient pulses a line-scan technique was developed by the Mansfield group in Nottingham [16] and used for the first human whole-body scan in 1978 [17]. The very early imaging experiments used FID signals for image detection, but owing to the poor homogeneity of the magnets, $T_2^*$ was as low as a few ms and images of acceptable clinical quality could not be obtained.

The breakthrough in the planar techniques occurred with the application of the 2-dimensional Fourier imaging technique (2DFT) first reported in 1975 by Kumar, Welti and Ernst [18]. This imaging procedure is based on the 2D Fourier spectroscopy, which was invented much earlier by Ernst. This imaging method was modified by Edelstein and Hutchison [19], from the Aberdeen group, with some practical changes in the applied gradient pulse technique, and it was published as spin warp imaging in 1980. The 2DFT method uses a 2D Fourier transformation which requires only two gradient directions as orthogonal coordinates. The imaging method eliminates the detail unsharpness due to magnet field inhomogeneity that was the problem with the back-projection technique. Now even with magnets of relatively poor field homogeneity and gradient linearity acceptable clinical images could be produced. Since then the projection-reconstruction technique as applied by Lauterbur has been generally abandoned and has only recently been seen in certain very fast imaging methods.

The Fourier reconstruction technique also enabled 3-dimensional volume scanning and reconstruction (3DFT), which produced all the possible cross-sectional images of a selected volume in one scanning procedure. However, the long scanning time of this technique made it clinically unpractical. Furthermore the multi-slice technique introduced by Crooks [20] at that time, which produced a series of parallel shifted images within the scan time of a single image, offered a much better solution for the problem of volume scanning. The 3DFT had to wait for some ten years unil the introduction of the fast imaging techniques before it regained practical interest.

The conference held in Winston-Salem (USA) in 1981 was one of the first international conferences devoted entirely to NMR imaging and spectroscopy for medical purposes as it brought together practically all research groups active at that time. The group photograph taken at this conference shows all speakers, with an indication of those who were particularly involved in the initial physical or clinical research on MR imaging, many of whom are mentioned in this book.

This conference also led to the birth to two independent societies: the Society of Magnetic Resonance in Medicine (SMRM) and the Society of Magnetic Resonance Imaging (SMRI). Both competing societies in fact covered the same field, albeit that in the beginning the SMRM was somewhat more oriented to physcial science and the SMRI to education and clinical applications. After more than ten years of practically parallel activities both societies merged to form the present International Society of Magnetic Resonance Imaging (ISMRM).

Many physical NMR phenomena that initially created problems or artifacts in imaging results later became important principles for new imaging possibilities. The understanding of the cause of flow artifacts resulted in the birth of MR angiography. Movement and flow studies have been carried out since the early experiments by Singer [21] in 1959. However, it was not until 1984 that the utility of flow direction and velocity using phase effects for the imaging of motion and flow was published by Van Dijk [22] and Bryant [23]. Later, with use of the then available image reformatting techniques, the observed flow enhancement and phase effects became the basis of the successful "time-of-flight angiography" described by Rossnick et al. in 1986 [24] and Groen et al. in 1988 [25] and the "phase-contrast angiography" (PCA) in 1989 (Dumoulin et al.) [26].

In the mid-1980s there was a passionate debate on the subject of the optimum field strength of MR imagers between advocates of high field (1.5 T) and low field (0.5 T) systems. The controversy faded away when the initial advantage of better S/N ratio of 1.5 T imagers was later reduced by the development of improved signal detection techniques, which enabled low-field systems to produce images of excellent quality. It has led to the present wide range of imagers with different field strengths and performance levels matching the different clinical needs.

A very useful tool for the understanding and characterization of the numerous imaging techniques, introduced by Twieg [27] in 1983, was the k-trajectory formulation of the MR imaging process, which has now become a fundamental concept in the analyses of imaging techniques. A variety of new or alternative imaging sequences could be conceived by simply varying the itinerary in k space.

Improvements in signal detection, fast data handling and gradient technology, advanced understanding of spin systems, pulse sequences and artifact suppression have eventually eliminated the major problem of the scan time. The first reports on fast scanning using small excitation angles and gradient echoes were published in 1986 by Van der Meulen et al. [28] and Haase et al. [29]. Simultaneously, Hennig et al. [30] proposed a fast scanning method based on the use of multiple spin echoes with different phase encoding. Since then the development of fast spin echo and gradient echo techniques has resulted in a wide range of "turbo" sequences for high-speed and high-resolution imaging that have outdated the "classical" SE sequences.

The most advanced idea in very fast imaging originates from Mansfield [31] and dates back to the early days of MR imaging. The "echo-planar" technique he proposed, which for a long time was seen as an intelligent but impracticable idea, had to wait for more than a decade before technical system limitations no longer inhibited its realization and clinical application. The EPI technique and its derivatives have again paved the way for a number of new clinical imaging applications, such as diffusion and perfusion imaging, functional brain imaging, etc.

Since the first publications the field of MR imaging has kept on growing at an enormous rate and with surprising diversity and its maturity is not yet within sight.

1. Bill Moore (Nottingham, 2. André Luiten (Eindhoven), 3. Bill Edelstein (Aberdeen), 4. Frank Smith (Aberdeen), 5. Ian Young (London), 6. Raymond Damadian (New York), 7. Paul Bottomley (Nottingham), 8. Paul Lauterbur (Stoney Brook), 9. Larry Crooks (San Francisco), 10. Graham Bydder (London), 11. John Gore (London), 12. Brian Worthington (Nottingham), 13. Waldo Hinshaw (Nottingham), 14. Peter Mansfield (Nottingham), 15. David Hoult (Bethesda), 16. Jim Hutchison (Aberdeen)

Speakers on NMR imaging at the Conference on NMR in Medicine in Winston-Salem 1981. Details see opposite page

# 1. MRI and Its Hardware

## 1.1 Introduction

There are several textbooks on magnetic resonance, giving an excellent treatment of the basic concepts of MR physics, see for example [1, 2]. So in this chapter we shall restrict ourselves to summarizing these basic physics concepts and we refer the reader to other books for qualitative descriptions [3–7] and to the detailed technical treatment in [8]. Since magnetic resonance imaging (MRI) is the topic of this textbook, there will be no discussion on MR spectroscopy and, furthermore, only imaging on the basis of proton spin will be considered. Taking these restrictions into account in Sect. 1.1 we summarize the basic physics concepts of magnetization, precession, excitation, and relaxation. The signal obtained from precessing spins is the free induction decay, which will be turned into a spin echo, as is described in Sect. 1.2. Then the methods to determine the position of certain spins in the object on the basis of gradient fields will be discussed. As a first example of an imaging method the "Spin-Echo" measuring sequence is described. This Spin-Echo sequence is the oldest imaging method used [9], it is followed by an overwhelming amount of newer methods [10], the underlying theory of which is the topic of this book.

The knowledge of an example of a measuring sequence brings us to a position to explain what functions an MRI system must be able to perform, so a global architecture of an MRI system can be sketched. Following this we shall briefly discuss the properties of the main components of the system as far as they are relevant for the user. For system design a much more detailed knowledge is necessary, but this is beyond the scope of this book since this depends heavily on the system philosophy and the applied hardware. For more information on the hardware we refer the reader to [11].

### 1.1.1 Spin and Magnetization

Living tissue can be considered to consist of 60–80% water in which macro-molecules are suspended. Both water and macro-molecules have protons as one of their constituents and protons have spin $^1/_2$, see Chap. 4 in [6]. In a magnetic field most protons align along the magnetic field lines. Excitation means that the total magnetization vector formed by the spins is rotated away from the direction of the main magnetic field. The water is either free or bound to the surface of the macro-molecule, and it is its interaction with the macro-molecules that determines the relaxation properties (the speed with which the equilibrium situation is restored after excitation) of a certain tissue. The signal level in an MRI experiment is determined by the proton density and the relaxation properties of the tissue. The proton spins of a macro-molecule, however, are not visible in an MR experiment, because of their short relaxation time, so only the spins of the protons of free water generate the actual signal of a tissue. In Chap. 6 a more detailed treatment of these phenomena will be given.

A correct description of what happens when tissue is immersed in a magnetic field relies on quantum mechanics. Fortunately all the theory necessary for MRI can be based on a simple classical model in which the protons can be considered as small magnets. These small magnets align along the magnetic field, either parallel or anti-parallel. The resulting magnetic moment is called "magnetization". This magnetization has a magnitude, depending on the number of protons per cubic centimeter, and a direction and is therefore characterized by a vector. In equilibrium, the magnitude of the magnetization, $M$, is proportional to the magnetic field strength $H$, according to $M = \chi_p H$, where $\chi_p$ is the susceptibility of the nuclear spins. The magnetic flux $B$ is given by $B = \mu_0 H(1 + \chi_p)$.

### 1.1.2 Precession: Rotating System of Reference

When the magnetization has another direction than the main magnetic field a torque is exerted on the magnetization, which is perpendicular to both the magnetic field and the magnetization. This results in a precession motion, as described in Sect. 2.2, in which the magnetization vector rotates around the main magnetic field, $B_0$. This is shown in Fig. 1.1. We follow the convention that the $z$ axis of our system of reference is always chosen along the main magnetic field. This rotation, named "precession" has an angular velocity, $\omega_0$ (the Larmor frequency) given by the Larmor equation (see Sect. 2.2)

$$\omega_0 = \gamma |B_0|. \tag{1.1}$$

The proportionality constant, $\gamma$, is the gyromagnetic ratio. It is characteristic for the nucleus considered. In the case of protons, the only nucleus considered in imaging, we have $\gamma\!\!\!/ = \gamma/2\pi = 42.6 \times 10^6$ hertz/tesla (Hz/T).

A very simple but important concept in MRI is the rotating frame of reference $(x', y', z)$, which rotates at an angular velocity equal to $\omega_0$ radians per second around the $z$ axis. In this system of reference the magnetization vector does not move as long as the magnetic field is exactly equal to $B_0$. It is as if the effective magnetic field in the rotating frame is equal to zero. As we shall see later, all the information necessary for forming an image is contained in the motion of the $x', y'$ component of the magnetization vector

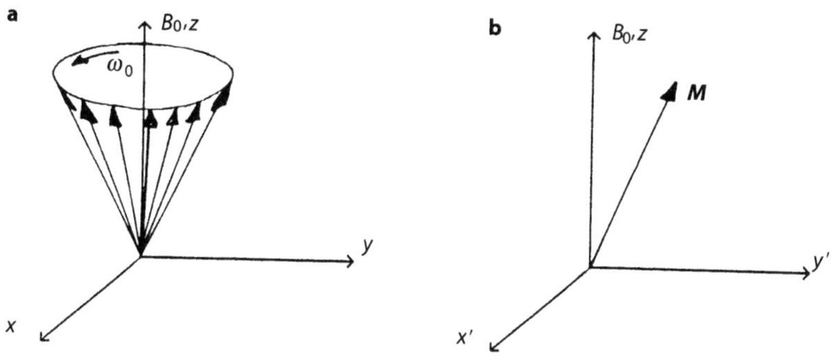


**Fig. 1.1.** **a** Precession around the main magnetic field in the laboratory frame of reference and **b** absence of precession in the rotating frame of reference

due to deviations from the main magnetic field as a result of extra gradient fields or RF fields (or unwanted deviations which result in image artifacts).

### 1.1.3 Rotation: Excitation by RF Pulses

As was said earlier, the magnetization aligns along the magnetic field when no extra fields are present. Now a method must be found to give the magnetization vector another direction so that precession can start. To do this we consider the rotating frame of reference and excite a magnetic field, $B_1$, along (for example) the $x'$ axis. As this now is the only active magnetic field in the rotating frame of reference, the magnetization vector will start a precession motion around the $x'$ axis (in the $z', y'$ plane, when we start from equilibrium). The precession velocity, $\omega_1$, is given by (1.1):

$$\omega_1 = \gamma B_1. \tag{1.1a}$$

After a time $t_{90} = \pi/(2\omega_1)$, the magnetization vector is rotated by 90°, which means that if it started from equilibrium along the $z$ axis it is along the $y'$ axis at $t = t_{90}$. This rotation is called "excitation" or, sometimes, "nutation".

From the known values of $\omega_1 t = \pi/2$ and $\gamma$, it can be calculated that, if we assume a realistic time for $t_{90}$ of 1 ms, the value of $B_1$ is 6$\mu$T. The index expresses the fact that the magnetic field $B_1$ along the $x'$ axis in reality rotates with $\omega_0$ in the laboratory system. Such a rotating magnetic field can be excited with a linearly polarized high-frequency field, which can be decomposed into two circularly polarized fields rotating in opposite directions, as shown in Fig. 1.2. One of these fields must rotate with $\omega_0$ to stay along the $y'$ axis (the other rotates in the opposite direction and has no interaction with the spins, and therefore the energy necessary for this field is actually lost). So we need high-frequency (RF) energy for the excitation of the spins. We took the example of a 90° excitation pulse, but 180° pulses, which invert the magnetization, are also possible. In an equal time they need twice the am-

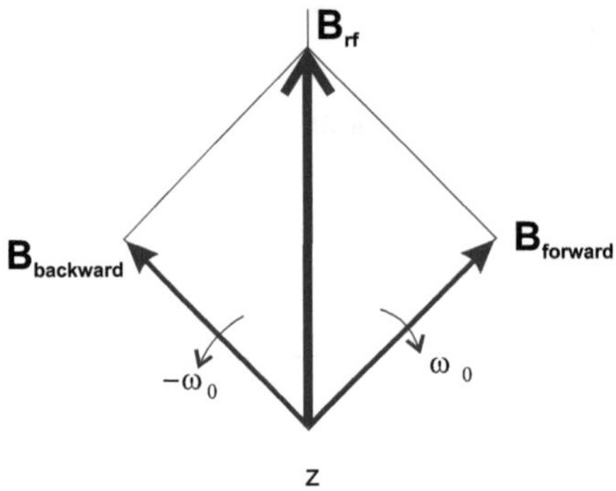


**Fig. 1.2.** Decomposition of a linearly polarized field in two circularly polarized fields

plitude and therefore four times the RF energy. In modern sequences pulses with flip angles different from 90° or 180° are also applied.

The question may be raised: "How can the small RF magnetic fields ($6\mu$T) superimposed on the much bigger main magnetic field (1 T) cause the magnetization vector to rotate over substantial angles". This question can be answered by looking from the $+z$ direction down to the transverse $x, y$ plane, as shown in Fig. 1.3. In the equilibrium situation $B_1 = 0$ the magnetization vector lies along the $z$ axis, which points toward us and shows up in the origin of Fig. 1.3. The total magnetic field including $B_1$ is slightly away from the origin and rotates along the small circle, around $(x, y) = (0, 0)$, which passes through $y = 1, 2$, due to the transverse RF field. Assume that at $t = 0$ the magnetic field vector is in point $y = 1$. The magnetization starts to precess around the total magnetic field and reaches point $y = 3$. In the meantime the magnetic field vector moves to point $y = 2$. So now the distance of the magnetization vector from the magnetic field vector is equal to the distance between the points $y = 3$ and $y = 2$. Further precession brings the magnetization vector to point 8, however, in the mean time the magnetic-field vector moves to point 1 again. Precession now brings the magnetization vector in point 11, etc. Actually the process is continuous, but this explanation gives the correct idea. In Fig. 1.4 the picture of excitation is shown when viewed

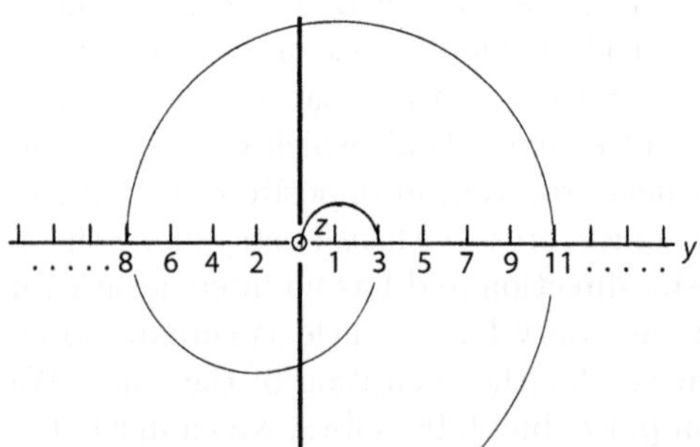


**Fig. 1.3.** The magnetization vector viewed from a direction parallel to the static magnet field when a transverse RF field with frequency $\omega_0$ is applied

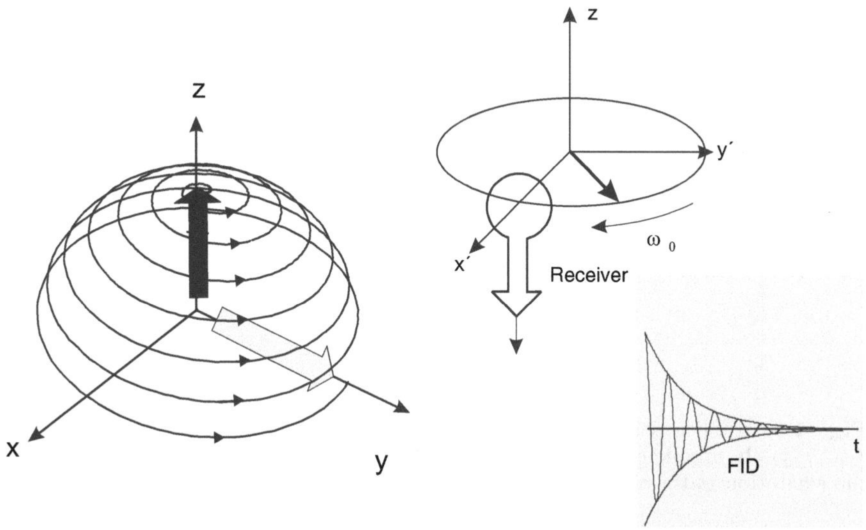


**Fig. 1.4.** Excitation of the magnetization when viewed in the laboratory frame of reference

from another angle, from which the magnetization can be seen to spiral down into the transverse plane. After the 90° excitation pulse, the magnetization remains in the transverse plane, where it rotates with the gyromagnetic frequency $\omega_0$.

When a current loop, which is sensitive for transverse magnetization, is placed in the vicinity of this rotating magnetization the changing magnetic flux will induce a current in this loop (Fig. 1.4). This current is the MR signal, which is amplified for further processing. The signal dies out after some time due to relaxation. This is further described in Sects. 1.1.5, 2.2, and 6.3.

### 1.1.4 Excitation of a Selected Slice: Gradient Field

The "hard" excitation pulse described before hits all spins in the object, irrespective of their position. For localization one must find a method to excite only the spins in a certain slice. This can be done by superimposing on the homogeneous main magnetic field a small linear position-dependent magnetic field, called the gradient field. We follow the convention that the $z$ axis of our reference system is always along the main magnetic field. The gradient field is assumed to have only a $z$ component, and this $z$ component is linearly dependent on the position (according to Maxwell's laws this is not possible, but for low-gradient fields it is a good approximation; for high-gradient fields see Sect. 3.3.2.3). For ease of argument we choose the position dependence along the $z$ axis. The gradient field $\delta B_{G,z}$ now has the form:

$$\delta B_{G,z}(z) = G_z z. \tag{1.2}$$

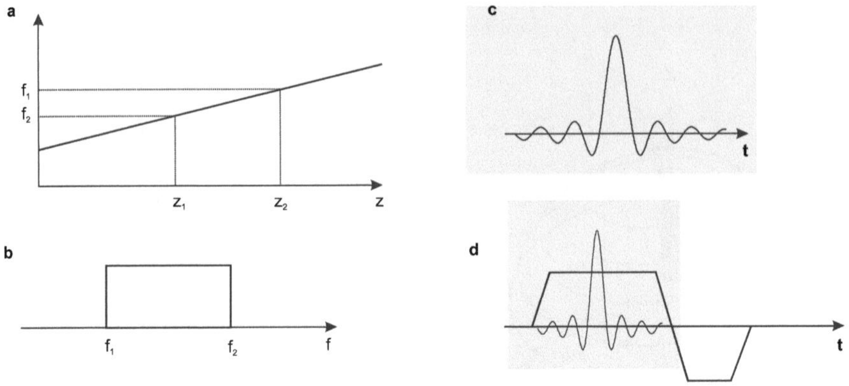


**Fig. 1.5.** **a** Slice selection with a gradient field defining the frequency band $f_1, \ldots, f_2$. **b** Frequency spectrum of the RF pulse. **c** Envelope of amplitude of the excitation pulse versus time. **d** Excitation pulse with refocussing gradient

Since the Larmor frequency depends on the total magnetic field $B_0 + G_z \cdot z$, according to (1.1) this frequency depends on $z$, as shown in Fig. 1.5. If we wish to excite the transverse slice, defined by $z_1 < z < z_2$, we need an RF signal containing the frequencies given by $f_1 < f < f_2$. So a frequency spectrum is needed which ideally has a constant amplitude for $f_1 < f < f_2$ and which has zero amplitude outside this region, as shown in Fig. 1.5b. The Fourier transform [12, 13] of this spectrum in the time domain shows a harmonic signal at the Larmor frequency, which is amplitude modulated with the well-known "sinc" (sin $t/t$) function as shown in Fig. 1.5c. This signal must be presented to the transmitting coil to excite a slice. A problem is, however, that this signal has infinite duration, so in practical situations it has to be cut off. The greater the number of side lobes of the sinc function that are cut off, the more the frequency spectrum deviates from the ideal form of Fig. 1.5b and the more the slice profile also deviates.

When the selection gradient is switched off, we still do not have the excitation profile that we want. Only in the central plane of the slice the Larmor frequency matches the rotation velocity of the rotation frame of reference, so in this plane the $z$-magnetic field is compensated for. To the right of the center there is an uncompensated positive gradient field (in the situation drawn in Fig. 1.5a) and therefore, apart from the desired rotation of the magnetization vector from the $z$ axis into the $(x'–y')$ plane there is also a precession of the magnetization vector from the $z$ axis due to the field of the selection gradient. To the left of the central plane the (negative) uncompensated gradient field rotates the magnetization vector in the opposite direction. Therefore the spins are not aligned across the slice and the total magnetization vector of the slice is decreased. This can be corrected by switching on a reverse-gradient field for a short time, as shown in Fig. 1.5d. During this "refocussing", the field gradient causes opposite precession to correct for the unwanted precession around the $z'$-axis through the selection gradient.

We will discuss the excitation in more detail in Sect. 2.3. For now it is sufficient to realize that a slice has been excited (in our example a transverse slice) and that the signals coming from the spins in this slice must be measured in such a way that their position is determined and an image can be generated.

### 1.1.5 Free Induction Decay (FID)

After excitation with a $90^\circ_x$ pulse, which causes rotation around the $x$ axis, the spins in a slice are precessing in the transverse $x, y$ plane (or are all aligned along the $y'$ axis in the rotating frame). If there is a current loop in a plane parallel to the $x, z$ plane the changing magnetic flux in the loop will induce a current of frequency $\omega_0$ in this loop, as shown in Fig. 1.6. This current will die out after some time and the measured signal is called the free induction decay.

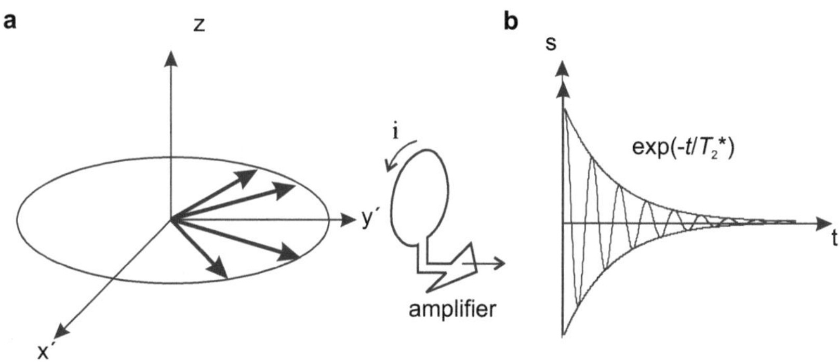


**Fig. 1.6. a** Dephasing of the magnetization vectors in the rotating frame. Measurement of the FID. **b** Decay of the signal

The decay can be explained as follows. Directly following the excitation pulse all spins precess in the same phase with their magnetic moments parallel, adding up to the maximum induced signal. After some time they start to loose their phase coherence due to two effects. First, other spins and the molecular magnetic fields, due to the macro-molecules in the tissue, $\delta B_m$ perturb the total magnetic field experienced by the different free water protons during their fast random (diffusion) motion through these fields. This causes the precession frequency to change per proton in a random way, resulting in dephasing of the proton spins and thus in a decay of the total magnetization. This is called "$T_2$ decay" and the decay time $T_2$ is 40–200 ms, depending on the tissue considered. Second, the "macroscopic" magnetic field in the tissue, $\delta B_s$, which is constant over distances longer than the diffusion length of the free protons, is not exactly homogeneous, due to susceptibility variations in the tissue considered (for example, air–tissue interfaces or changes in their component parts). The total vector sum of the magnetic moments within a volume element considered (several cubic mm) decreases and so the total magnetization and the induced signal decrease. The total decay caused by both effects is called "$T_2^*$ decay" and is two to three times as fast as pure $T_2$ decay. The decay mechanisms are further discussed in more detail in Sect. 6.3.

Actually the FID is not measured in imaging applications (as opposed to spectroscopy applications), since the measurements can start only some time after the effective excitation time (which is the center of the excitation pulse at the maximum of the main RF lobe), so that the $T_2^*$ process already has partly destroyed the signal. This problem can be avoided by using an RF spin echo, which was proposed in 1950 by Hahn [9].

## 1.2 Spin Echo

Throughout this book we shall use as our reference system a Cartesian systemwith its $z$-axis along the magnetic field. For ease of argument we shall always deal with transverse slices so that the selection gradient is always in the $z$ direction (the "selection direction"). The read-out direction (Sect. 1.2.1) is always taken along the $x$ axis, and the phase-encode direction (Sect. 1.2.2) is taken along the $y$ axis. Slices with other angulation and orientation are related to the transverse slice by straightforward transformation, so that the restriction to transverse slices does not restrict the validity of the arguments.

After excitation, the precessing spins causing the FID are dephased by inhomogeneities in the magnetic field. In the example discussed here, we will assume that the excited slice is parallel to the $(x, y)$ plane and that the inhomogeneity of the field is such that a constant gradient exists in the $x$ direction:

$$\delta B_{G,z}(x) = G_x x, \tag{1.3}$$

In this gradient field the $z$ component of the magnetic field increases linearly as a function of $x$. Spins at position $x = 0$ ($\delta B_{G,z}(0) = 0$) have a precession velocity nearly equal to the rotation velocity of the rotating frame of reference, since they experience only the field $B_0 + \delta B_m + \delta B_s$, where $B_0 \gg \delta B_m + \delta B_s$. Therefore their magnetization vectors barely move in the rotating frame and dispersion is due only to the molecular and susceptibility fields, $\delta B_m$ and $\delta B_s$. For $x \neq 0$ the gradient field soon dominates all other fields and the magnetization vectors of volume elements outside the isocenter ($x = 0$) rotate with respect to the rotating frame at an angular velocity approximately equal to

$$\delta\omega = \gamma \delta B_{g,x} = \gamma G_x x, \tag{1.3a}$$

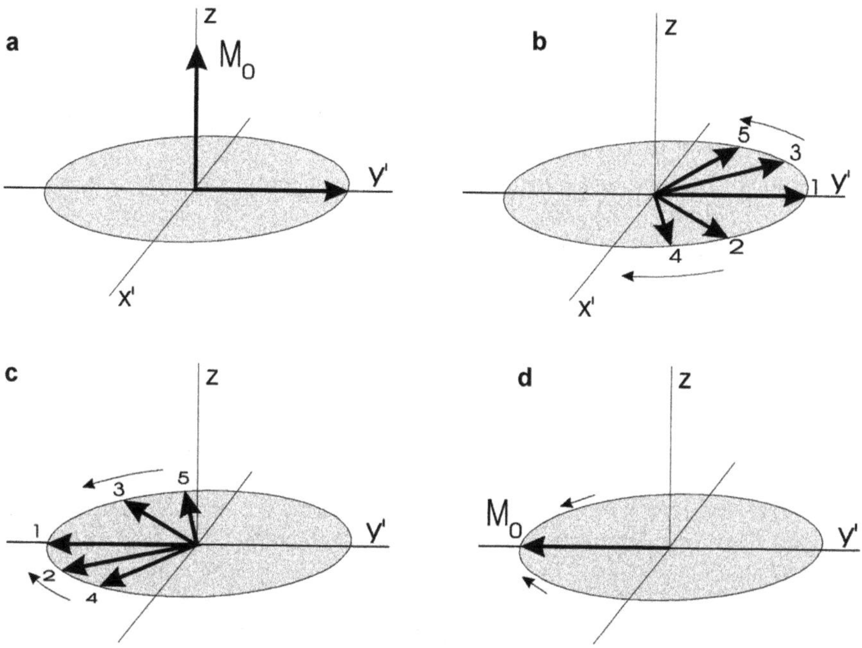


**Fig. 1.7.** **a** Excitation with $90^\circ_x$ pulse. **b** Dephasing by $x$ gradient field. **c** Rotation with 180° refocussing pulse. **d** Rephasing by gradient field When $t = \mathrm{TE}$ all spins are in phase

So, the spins dephase in the rotating frame depending on their position.

Following this dephasing, at $t = \mathrm{TE}/2$ all magnetic moments are rotated over 180° around the $x'$ axis by an RF pulse called "refocussing pulse". In the magnetic field, which is inhomogeneous according to (1.3), the spins continue to move in the same direction, depending on their position along the $x$ axis. This is shown in Fig. 1.7c and d for five points. Point 1 is at $x' = 0$. The entire set of spins now starts to rephase, and at the echo time TE rephasing is complete. The spins of all points are now grouped along the negative $y'$ axis. When $t > \mathrm{TE}$, the spins will dephase again and the vector sum of their magnetization will decrease again. The signal induced in the receiver by this magnetization is called the "spin echo". This echo can be sampled during its build-up and its decay and the information will be more complete than when a FID is sampled.

The important feature of the spin echo is its insensitivity for magnet field inhomogeneities, whatever their distribution. For any inhomogeneity $\delta B_0$ which is present locally in the object, the excess phase of the local magnetization vector in the $(x', y')$ plane built up before the refocussing pulse is reversed by this pulse and subsequently also refocused at the spin echo time. The spin echo that is so formed does not depend on the local inhomogeneity as long as it is stationnary during the echo time. Such an inhomogeneity could be caused by a deviating susceptibility or by the design limits of the magnet. Inhomogeneities caused by fields that are not constant during the echo time of course are not refocused in the spin echo. An example of such a field is the molecular field $\delta B_m$. So, the signal observed at the spin echo will have experienced the true $T_2$, decay only.

We must now see how information on the position of the spins within the slice can be obtained so that the measurement of the RF spin echo can be used for image formation.

### 1.2.1 Determination of Position in the Read-Out Direction

In imaging experiments, a gradient as described in (1.3) is added deliberately to the main magnetic field. The method to achieve such a linear gradient field (which is a field in the $z$ direction that increases linearly in the $x$ direction) is described in Sect. 1.3.3. As explained above, rephasing of the signal will take place after the refocussing pulse and a spin echo is formed. During rephasing (and dephasing after the echo) the signal induced in the receiver coil by the total magnetization is measured by taking over time a sufficiently large number of samples. The signal contains many frequencies because the frequency depends on the position of the spins in the gradient field, see (1.3a) and Fig. 1.8. The shape of the signal, sampled in the time domain, can be seen as the result of the interference of the position-dependent contributions from the various parts of the sample, that all have frequencies within a certain frequency band. The original frequencies and their amplitudes can be obtained from this sampled temporal shape of the signal by Fourier analysis [12]. The frequency characterizes the position of the spins in the direction of the frequency-encoding (read-out) gradient (also called the measuring gradient); the amplitude is proportional to the magnetization at this position. This is, however, not enough to make an image, since nothing is known about the $y$ direction yet.

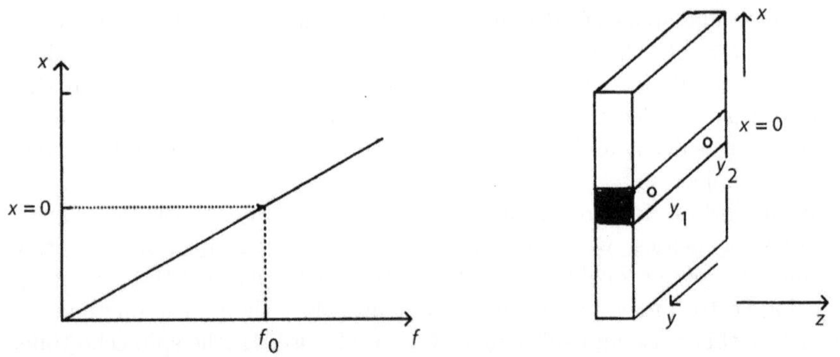


**Fig. 1.8.** Gyromagnetic frequency as a function of the position in the read out direction

### 1.2.2 Determination of Position in the Phase-Encode Direction

The direction perpendicular to the read-out direction is called the phase-encode (or preparation or evolution) direction. In our discussions this direction is usually called the $y$ direction. A detailed discussion on how to obtain information on the magnetization distribution in the $y$ direction is given in Sect. 2.4. Here a simple explanation will be given for the case of our example. Consider the points $(0,\ y_1)$ and $(0,\ y_2)$, shown in Fig. 1.8, and assume that only these two points generate a signal. We first measure the signal as described in the previous section. We find from this measurement the sum of the signals ($S_{\text{tot}}$) from these two points:

$$S_{1,\text{tot}} = S(0,\ y_1) + S(0,\ y_2). \tag{1.4}$$

After the measurement we first wait for equilibrium and then excite again. During the period between excitation and refocussing, not only the dephasing gradient in the $x$ direction, but also a gradient field in the $y$ or phase-encode direction, $G_y$, the phase-encode gradient, is applied during a time $T_y$. The signals from the points $(0,\ y_1)$ and $(0,\ y_2)$ will now have different phases, because they experienced different gradient fields. We assume that their phase difference is 180° by satisfying the equation

$$\varphi_2 - \varphi_1 = \gamma G_y (y_1 - y_2) T_y = \pi. \tag{1.5}$$

Now we find for the total signal

$$S_{2,\text{tot}} = S(0,\ y_1) - S(0,\ y_2). \tag{1.6}$$

Equations (1.4) and (1.6) now form two equations with two unknowns, so $S(0,\ y_1)$ and $S(0,\ y_2)$ can be calculated from $S_{1,\text{tot}}$ and $S_{2,\text{tot}}$. It will be clear that for three objects we need three measurements.

In practice, however, we wish to distinguish $N$ points ($N$ is usually a power of two, for example 128 or 256), which means that we must repeat this measurement $N$ times. This will be further discussed in Sect. 2.4.

### 1.2.3 Measuring Sequence

The example we have described so far constitutes a valid method for obtaining an MR image. The method is called Spin Echo (SE) because of the use of the refocussing pulse and sampling during the rise and fall of the RF spin echo. The measuring sequence (or "sequence", for short) of events that is played off in the Spin Echo imaging method is shown in Fig. 1.9. After each measurement the magnetization is left behind in a disturbed state and we must wait some time for the relaxation to restore the equilibrium situation, the repetition time TR. Each measurement takes something of the order of 10–30 ms and typical wait times are of the order of seconds. During the waiting periods one can measure other slices in the object. This is "multiple-slice acquisition" (as opposed to 3D acquisition, to be discussed later) and is

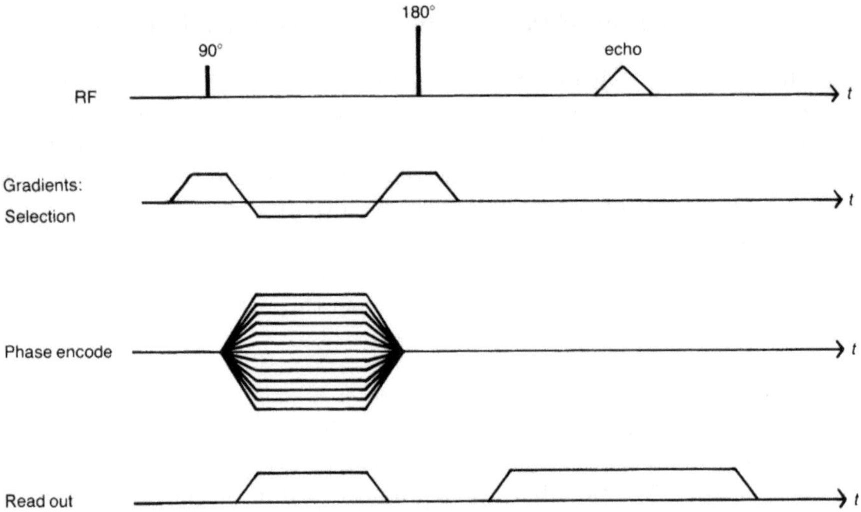


**Fig. 1.9.** Measuring sequence of the Spin Echo Imaging method

an important acquisition method. For now it is important that we understand that an MRI system must be able to generate sequences of RF and gradient waveforms with high precision, which brings us to a position from which to describe a very global system architecture.

The Spin Echo sequence is just one example of the many imaging methods that are possible [10]. Actually it was the only method used in the oldest MRI systems, because of its insensitivity to the inhomogeneities of the magnetic field in early systems. With the more homogeneous modern magnets many other imaging methods became possible and will be described in the following chapters.

### 1.2.4 Object Slice: Voxels and Image Pixels

As the last part of our introduction to the fundamental ideas of MRI we shall describe how the MR image is built up. An object slice is considered with a thickness $d$. The area of the slice that is imaged is the "field of view" (FOV). When not explicitly stated differently the FOV is taken to be square. The position in the slice is described by two dimensions: the read-out ($x$) direction and the perpendicular phase-encode ($y$) direction. The object slice is divided into voxels (volume elements) with dimensions equal to the field of view divided by $N$ in the transverse directions and $d$ in the longitudinal direction as sketched in Fig. 1.10. So the voxel volume is $(\mathrm{FOV}/N)^2 d$. All magnetization within a voxel is summed to yield the total magnetization vector of the voxel.

Usually the length of the magnetization vector of each voxel is displayed as the brightness of a point in the image. This point is called a pixel (picture

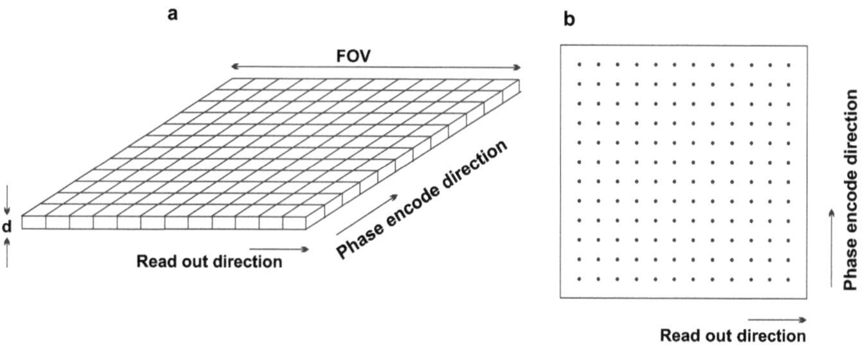


**Fig. 1.10.** **a** Object slice divided into voxels. **b** Image formation with pixels

element). All pixels together form the image on the monitor or the film. There are many techniques to influence the appearance of the image (grey scale) or to perform operations such as "zoom", "rotate", "compare", "filter", etc. These techniques are brought together under the common name "viewing". For "viewing" we refer the reader to the relevant literature [14].

## 1.3 System Architecture

In the previous section the Spin Echo imaging method has been described in some detail as an example of the many other methods possible with an MRI system, which are discussed in later chapters. This brings us to the functions that an MRI system must be able to perform and to a sketch of the global system architecture. A global picture of what the system is doing helps in applying the system properly. For the design of the system a much deeper knowledge of system architecture is required and this is certainly not the aim in this text. During the discussion of the imaging methods in the following chapters, there will be opportunities to mention the main requirements imposed by these imaging methods on the system.

So let us start with the actions of the user – see Figs. 1.11a (schematic) and 1.11b (usual). At first the patient is positioned on the "patient support", which brings the patient into the magnetic field in such a way that the anatomy to be imaged is in the region where the field is homogenecus. In most cases the magnet has a circle cylindrical bore and its centre, which is also the center of the sphere in which the magnetic field is sufficiently homogeneous (better than, say, 5 ppm), is called the "isocenter" and is shown in Fig. 1.11a. The magnet is situated in an RF-screened room to avoid spurious signals from the surroundings, both from foreign transmitters and from the spurious signals generated in the driving electronics of the system itself. The user now wishes the system to perform a certain imaging task. Of course first the patient data must be entered into the administration section. Now the

a

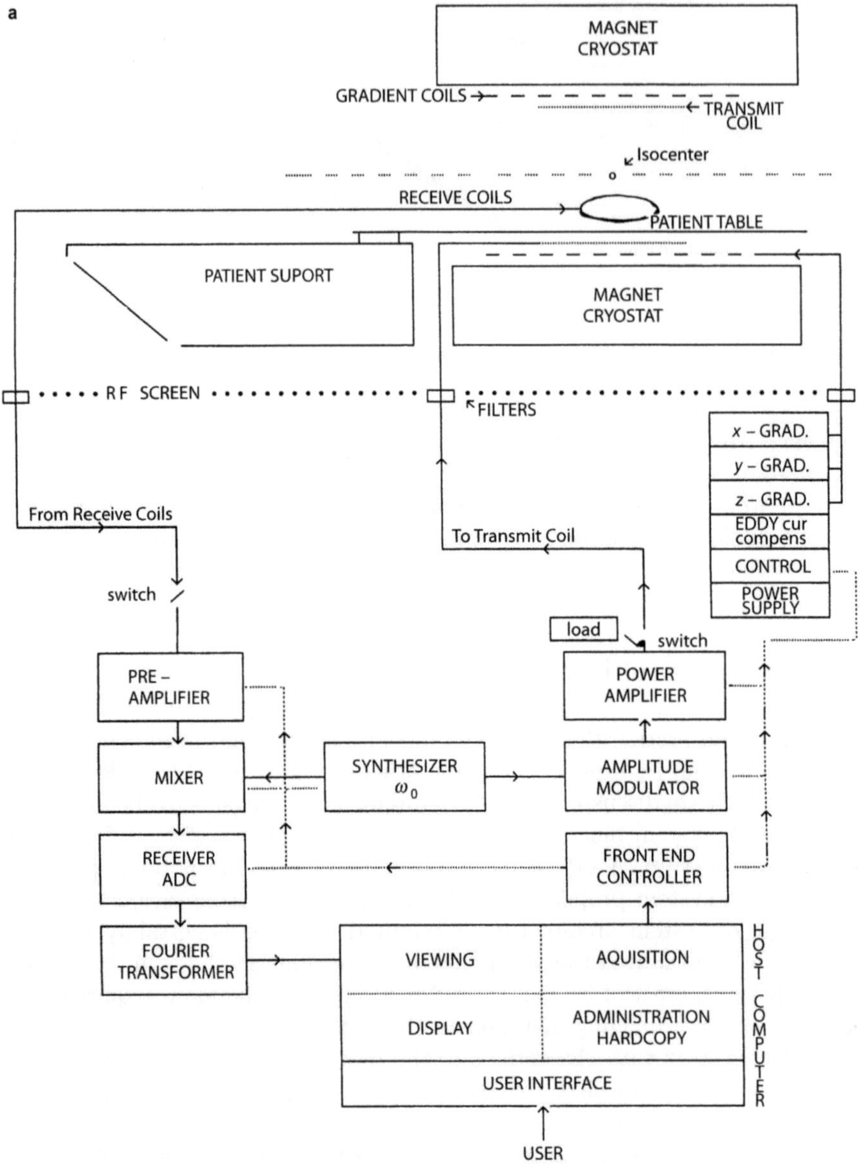


**Fig. 1.11a.** Global system achitecture. For description see text. - - - - control lines, ——— signal lines, ADC = analog-digital converter

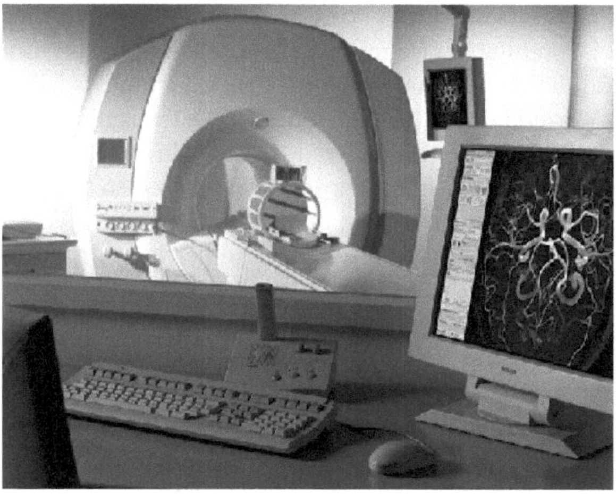

**Fig. 1.11b.** View of an MR system from the position of the operator, outside the RF screened room. The window in the RF screen has a fine copper mesh that does not block the sight

system must be "told" the required scan, including the geometrical parameters, the imaging method, and the sequence timing via the user interface. In most cases only a name is sufficient (when the scan definition is present in the memory of the host computer as a preset procedure). For new imaging methods the gradient field strength as a function of time for each direction and the RF waveform must be entered into the system and written into the memory of the host computer, so that the system can perform this scan. The "language" in which this happens is specific for the system used. This information is then "downloaded" to the "spectrometer", which consists of the "Front-end controller" (controlling the magnet, gradients, RF transmitter and receiver, RF coil switches, etc.) and the data acquisition around the receiver (the receiver switches and the ADC).

After downloading, the spectrometer has many tasks before the sequence can start. At first the synthesizer frequency must be tuned to the gyromagnetic frequency. The RF receiving coils must be tuned. The RF power and the receiver gain must be adjusted to the measurement to come, etc. The details of this "scan initialization phase" depend very much on the actual design philosophy used and will not be further described in this book.

After this "initialization phase", the Spin Echo sequence can start with switching on the selection gradient (the $z$ gradient in our example). All other components (except of course the magnet) are switched off. When the selection gradient reaches its required value (after some time, the rise time, the minimum of which is imposed by the inherent gradient coil inductance and the maximum voltage of the gradient amplifier, see Sect. 1.3.3.1), the RF power amplifier is activated and its output is switched from a matched load (its default position) to the transmit coil. The input signal of the power amplifier comes from the synthesizer, which generates a harmonic signal with frequency $\omega_0$. This signal is modulated (we restrict the discussion to the simple case of amplitude modulation) in the waveform generator and fed into the power amplifier. The AM modulated RF signal drives the transmit coil so that the desired RF magnetic field, $B_{RF}$, is emitted. During this time the receive coils must be detuned and the pre-amplifier must be blocked so that the large transmit signal cannot burn out the pre-amplifier and receiver. When the RF pulse is ready the output line of the power amplifier is switched to the matched load again. This switch is necessary because the output noise of a power amplifier with zero input signal is so high that it could mask the MR signal to be measured. Also the selection gradient is now switched off, but this again takes some time due to the coil inductance and the gradient-amplifier voltage. The result of these actions is that the spins in a single slice are excited.

Now the dephasing gradient and the phase-encode gradient are switched on to the strength required by the spectrometer (see Fig. 1.11a). They can be applied simultaneously, since (in the absence of RF power) their effects are linear and therefore additive. When these gradient pulses have their required surface (the integral of the gradient over time, which determines the phase angle in the rotating frame) they are switched off and the refocussing pulse can now be applied. This 180° RF pulse is usually slice selective and applies to the same slice as the excitation pulse. Its shape is similar to that of the excitation pulse but with twice the amplitude and therefore four times the power. In single-slice methods, slice selectivity of the refocussing pulse may be left out, making a pulse of shorter duration possible.

After the refocussing pulse the read-out gradient is switched on. During this gradient pulse the receiver coil is in a tuned state and the receiver is activated. The magnetization is measured during the time the read-out gradient is switched on and the ADC samples the received signal at certain points in time, separated by the sampling time $t_s$. After a wait time, the spectrometer will proceed to the next logical element of the imaging sequence, until the imaging scan is completed. The samples are sent to the array processor, the circuit that performs the fast Fourier transform, and the image results from this transformation, which is sent to the viewing section. Here the user can display the images on the monitor and/or can reproduce them on a hard-copy camera.

Usually after this first scan more scans of the same anatomical area follow. These scans are taken of slices positioned on the basis of this first image. The orientation and the off-center distance of the slices addressed in such scans can be selected by the user. Usually the system supports the user by offering a graphical interactive mode, allowing adjustment of the desired slices from their cross section on the first scan, which is now available as a survey. The adjustments for oblique orientations of the slices can be realized by a combined drive of all gradient coils for each of the logical functions (select, phase encode, or read out). Off-center positioning of the slices can be realized by adjustment of the synthesized frequency. The rotations and translations needed are translated into these adjustments by the host computer using simple matrix transforms.

This introduction gives an initial idea of what an MR imaging system must be able to do and of the system architecture. Now we shall discuss some properties of the front-end hardware, which are important to the user of the system. The discussion is not meant for those who wish to work on system design, they must refer to the specialized literature. Also a discussion of the back end is restricted to mentioning its functions, more is not necessary for the understanding of imaging sequences. Furthermore the back end is very dependent on the design philosophy of the different manufacturers.

### 1.3.1 Magnets

A constant magnetic field is generated by either a dc electric current (electromagnet) or a permanent magnetic (ferromagnetic) material. Let us first concentrate on the most frequently applied type, the electromagnet built up of a series of solenoid coils.

The relation between an electric current and the magnetic field associated with it is given by Biot and Savart's law:

$$\mathrm{d}\vec{B} = \frac{\mu_0 I}{4\pi r^3}\vec{r} \times \mathrm{d}\vec{l}, \tag{1.7}$$

where $I$ is the electric current. $\mathrm{d}\vec{l}$ describes the length and the direction of the current element considered, $\vec{r}$ is the vector between the current element $\mathrm{d}\vec{l}$ and the position where the magnetic field is measured, and "$\times$" means the vector product. By integrating over all current elements $\mathrm{d}\vec{l}$ one can find the magnetic field as a function of the overall current geometry. This – of course – results in complicated computer programs and optimization procedures used for designing magnets. For MR imaging a sufficiently large region is needed in which the field is homogeneous within a certain specification. We shall not discuss the design problems; in this text we shall only try to build up a general understanding.

We therefore first consider an infinitely long solenoid and consider the magnetic field $B_0$, within this solenoid, which is ideally homogeneous; see Fig. 1.12a. This magnetic flux, $\Phi$, within the solenoid is proportional to the number of turns per meter, $N_\mathrm{m}$, and the current $I$ through the wire:

$$B_0 = \mu_0 I N_\mathrm{m}. \tag{1.8}$$

Practical systems are – of course – of finite lengths and their homogeneity is deteriorated (they also generate a magnetic field outside the solenoid). This is easily seen by looking at the field of a single current loop (very short magnet) which can be calculated by integrating (1.7); see Fig. 1.12b. Such a loop generates a very inhomogeneous field. So for a magnet of finite length an

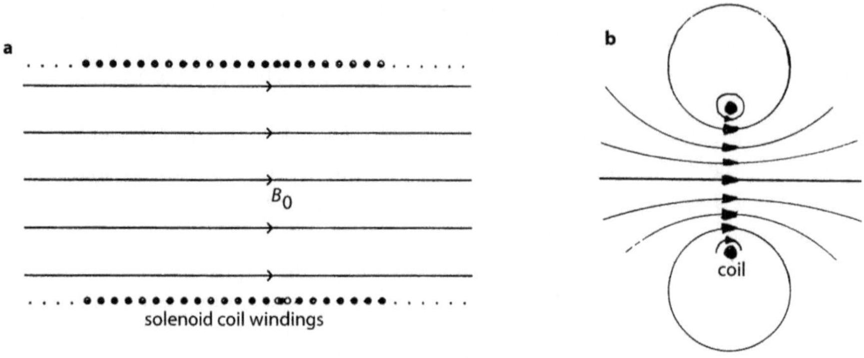


**Fig. 1.12.** **a** Homogeneous field in an infinite solenoid. **b** Magnetic field of a single loop

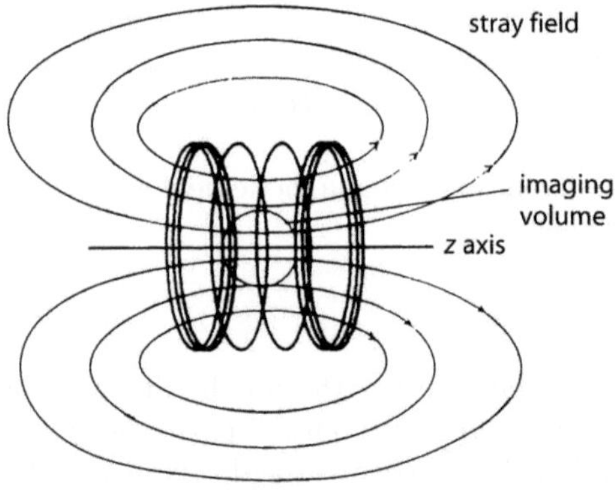


**Fig. 1.13.** Magnet coils with homogeneity sphere and stray field

intermediate situation results, namely a number of coils of which the dimensions and the distance are chosen so as to optimize the radius of the sphere of homogeneity (within which the homogeneity is better than, say, 5 ppm). A practical example is the four-coil magnet shown in Fig. 1.13. A disadvantage of this geometry is that there is a large stray field outside the magnet, which is undesirable due to its interaction with other measuring systems, magnetic materials, and data storage media, and also due to the extensive safety precautions required to prevent, for example, interaction with pace makers. Therefore modern magnets are shielded by either an iron shield which conducts the stray field away (passive shielding) or by extra coils that have a larger diameter than the original windings (active shielding) and that carry opposite current. Passive shielding has the disadvantage that many tonnes of iron are necessary around the coils to absorb and return a sufficiently large part of the stray magnetic flux. For a 1.5 T magnet this shield may have a weight up to 20 tonnes. Since both the passive and active shielding decrease the field inside the primary coils, the current in the primary coils must be increased in comparison to a non-shielded magnet with the same strength.

#### 1.3.1.1. Superconducting Magnets

Copper-wired electromagnets for whole-body systems with a field above 0.3 T impose unwieldy requirements on the total power and the stability of the current supply. Therefore modern MR systems frequently rely on superconductive magnets, which require (after activation) no current supply. A sketch of such a magnet is shown in Fig. 1.14. It has six coils to excite the field and two coils at larger diameter to compensate the stray field. The wire is made of niobium titanium alloy filaments embedded in copper and is superconductive below a certain maximum temperature ($< 12\,\mathrm{K}$). For high current densities the temperature must be lower than this maximum value. To reach the required temperature the coils are immersed in liquid helium (4.3 K). Apart from the low temperature, a current density below a certain maximum value and a sufficiently low magnetic field everywhere in the wire are also required, see [11]. These requirements impose important design criteria, since violation of any of them would make some part of the wire resistive and heat generation would take place, heating the surrounding wires and making them resistive, etc. This is typically an unstable process leading ultimately to the dissipation of all stored magnetic energy and the subsequent boiling off of the helium. This process, called "quenching", is a very violent process and constructional precautions are needed for safety, such as an over-pressure safety valve and venting channels for gaseous helium. With correct magnet design a quench is very improbable in practice and if it takes place it is a controlled process.

In order to reduce heat conduction from the outside world to the helium vessel, the vessel is surrounded by two radiation shields (or vessels), which are kept at 15 and 60 K, respectively, by a cryocooler for example. The three vessels are suspended in the outer vacuum vessel by thin, long rods with high heat resistance, to reduce heat conduction. These fragile rods make transport of the magnet on air-suspended vehicles necsessary.

Of course in the long run some heat still leaks into the magnet, which causes helium evaporation. Another problem is that the magnet current decreases slowly, due to non-zero ohmic losses. This means that the gyromagnetic frequency also decreases slowly and the synthesizer must be tuned to ever lower frequencies in the "initialization phase". This, of course, cannot go on forever, since the frequency can come outside the tuning range of the electronic components, such as RF coils. As a result the magnet current must be brought back to its original value, which is often combined with helium replenishment and maintenance of the "cold head". The length of the intervals between these "refills" is an important design criterion.

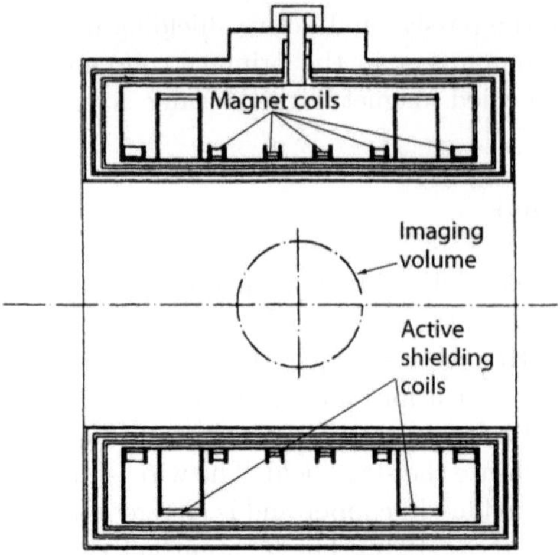


**Fig. 1.14.** Superconducting magnet in its cryostat

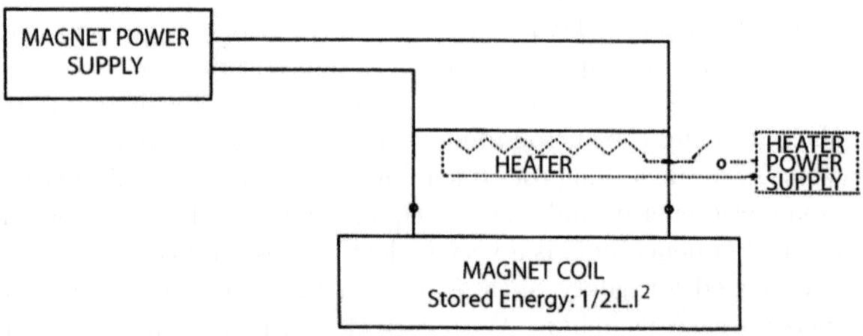


**Fig. 1.15.** Activating the magnet

The power supply of the magnet is only in use when the magnet is (re)energized. This is accomplished with the circuit shown in Fig. 1.15. During normal operation the magnet coils are short-circuited with a superconductive wire and the current does not "see" the magnet power supply. However when the magnet is energized this wire is heated and made resistive so that the coil wires are directly attached to the power supply voltage, which can add current to the coil. Note that this is a slow process since the coil resists changes in the current according to

$$U = L\frac{\mathrm{d}I}{\mathrm{d}t} \tag{1.8a}$$

where $U$ is the voltage of the power supply and $L$ is the (very large) inductance of the coil assembly. For reasonable power supply voltage, $U$, the rate of change of the current, $\mathrm{d}I/\mathrm{d}t$, is low and thus energizing a magnet takes a long time.

The same type of superconducting magnets can be built also with niobium tin wire embedded in copper. Although this material is more expensive, it has the advantage that the critical temperature is 18 K and it can be cooled completely with cryocoolers so that He filling is no longer necessary.

### 1.3.1.2. Other Magnet Types

So far we have discussed the magnet type that is most frequently used in practical systems with field strengths above 0.3 T. It is, however, possible to design other magnets. Although it is tempting to go into more depth we shall restrict ourselves to only mentioning the several other types.

**a. Resistive Electromagnets.** An electromagnet as shown in Fig. 1.13 is sometimes used for MR applications. In order do reduce the magnetic "resistance" of the return magnetic flux path an iron yoke is applied in most cases. This makes the efficiency (tesla/amp) of the magnet up to four times as high as in a magnet with only air coils. Two such types of electromagnets are shown in Fig. 1.16. The magnet with two (or more) return yokes is called an "H magnet" (the form of the hole), the magnet with one return yoke is called a C-arm magnet. A problem with resistive magnets is that the current supply must be very stable. Also the temperature of the magnet must be very constant, since the temperature coefficient of the coils and the permeability of iron, $\mu_0$, causes the magnetic field to change with 2000 ppm per K. For slow changes of either the temperature or the current the gyromagnetic frequency can be measured from time to time and subsequently the synthesizer is tuned to the instantaneous frequency.

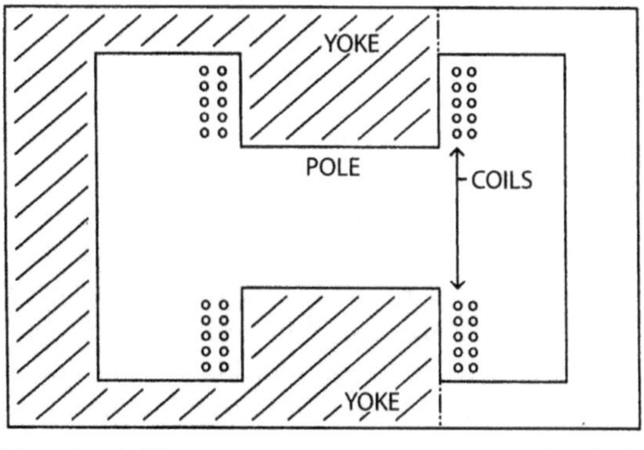


**Fig. 1.16.** Electromagnet with iron yoke. The dashed yoke makes a C-arm magnet, the full yoke an H magnet

A C-arm magnet (see Fig. 1.16) has the advantage that the patient can be approached from the side. This advantage is somewhat less manifest than expected, because the diameter of the poles in practice can not be made smaller than 2.5 times the diameter of the homogeneity sphere. The gap length must be larger than 1.5 times the diameter of the homogeneity sphere. For a (rather small) homogeneity sphere with a diameter of 35 cm, the gap length is at least 52 cm and the pole diameter 92 cm, around which the coils must still be positioned. This optimum geometry is obtained by using pole pieces with a special field-correcting shape or by shaping the incoming magnetic flux through the choice of an appropriate geometry of the windings.

**b. Permanent Magnets.** The second alternative class of magnets discussed here is the group of permanent magnets. The applied permanent magnetic material is ferromagnetic and is characterized by a *B-H* curve (relation between magnetic flux and magnetic field strength) as shown in Fig. 1.17. In the fourth (upper left) quadrant the flux and the field strength have opposite directions, which means that the material can act as a source of magnetic flux. To see this, let us consider a C-arm magnet, as shown in Fig. 1.18, in which the coils are replaced by a layer of permanent magnetic material behind the pole faces. We assume the field to be homogeneous over the cross-section area $A$. This is not actually true, but we aim here at a crude estimate. From Maxwell's laws we know that for the closed path $l$:

$$\oint H_l \cdot \mathrm{d}l = 0 \quad \text{and} \quad B_0 A = \text{Constant}. \tag{1.9}$$

The second equation discribes the continuity of the total magnetic flux. Writing now the index m for the material, g for the gap, and i for the length of the path in iron, we can easily evaluate the integral of (1.9) to yield $H_\mathrm{m} l_\mathrm{m} + H_\mathrm{g} l_\mathrm{g} + H_\mathrm{i} l_\mathrm{i} = 0$. The total flux is also constant, so $B_\mathrm{g} A_\mathrm{g} = B_\mathrm{m} A_\mathrm{m} = B_\mathrm{i} A_\mathrm{i}$, where $A_\mathrm{g}$ is the area of of the pole, $A_\mathrm{m}$ the cross section of the ferromagnetic material, and $A_\mathrm{i}$ that of the iron. Since $H_\mathrm{i} = B_\mathrm{i}/\mu_0\mu_\mathrm{i}$, where $\mu_\mathrm{i}$ is very large, we can disregard the term $H_\mathrm{i} l_\mathrm{i}$. Combining both results yields

$$B_\mathrm{m} H_\mathrm{m} V_\mathrm{m} = \mu_0^{-1} B_0^2 V_\mathrm{g}, \tag{1.10}$$

where $V$ means volume. $B_\mathrm{m} H_\mathrm{m}$ is the total magnetic energy present per unit volume in the permanent magnet, which is a fixed property of this material. Therefore when we wish to increase the magnetic field in the gap by a factor of two we need to increase the amount of permanent magnetic material by a factor of four. When we increase the linear dimension of the homogeneity sphere by 1.2 we need about 1.7 times the amount of material. The permanent magnet material used most frequently is neodinium boron iron and is rather expensive. The magnets made of this material are restricted to a field strength of about 0.25 T by their weight and price (which should be lower than the price of a superconducting magnet). Their temperature must be very constant, since the field changes with temperature at about 1000 ppm per K.

Another type of magnets based on permanent magnetic material is the cylindrical magnet. Examples are shown in Fig. 1.19. Zijlstra [15] has shown

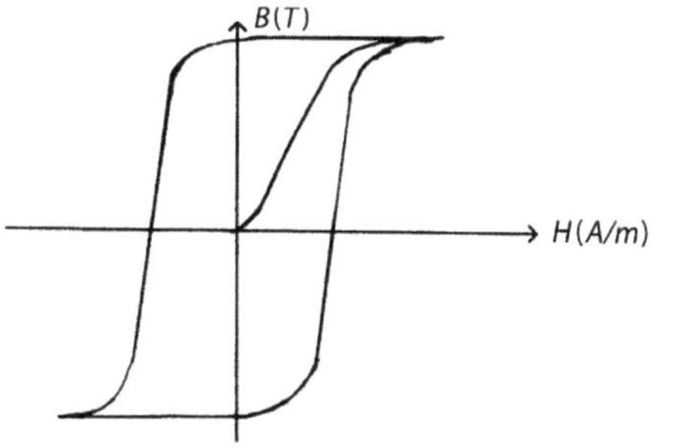


**Fig. 1.17.** Magnetization curve for ferroelectric material

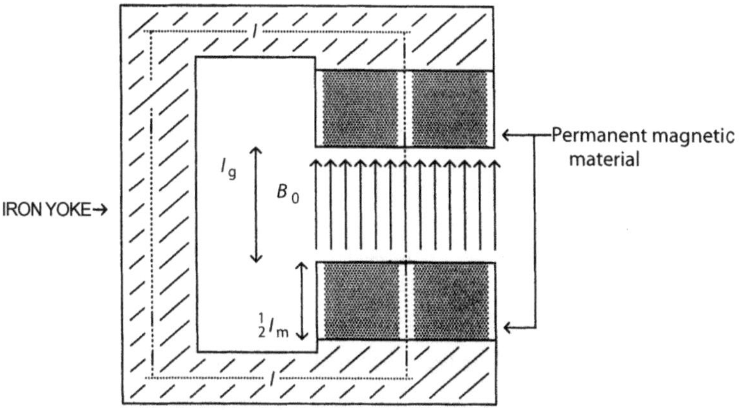


**Fig. 1.18.** Integration path $l$ for a permanent magnet with iron yoke

for the magnet in Fig. 1.19a, that when the angle of the magnetization in a certain point is twice the angle between the position vector and the vertical, we have an ideal homogeneous vertical field (note that the field direction is transverse and differs from that in a superconducting magnet; this has consequences for the RF coil design). Also in this case the finite length of the magnet deteriorates the homogeneity of the magnetic field.

The type of cylindrical magnet shown in Fig. 1.19b is called the prism magnet. This configuration can also be proved to have a homogeneous field (for infinite length [15]). The weight of prism magnets is high: for a 0.15 T magnet with a gap height of 0.7 m and a length of 1.05 m the total weight is 3400 kg, due to the contribution of the (very expensive) permanent magnetic material, NdBFe (33%), and the iron yoke (60%).

Concluding, one can say that for field strengths above 0.3 T all practical achievable magnets are superconducting. Below 0.3 T, electromagnets are feasible and probably cheaper, and below about 0.25 T permanent magnets are the choice (the price and the weight of the material are important parameters). Yoke magnets and cylindrical permanent magnets have transverse

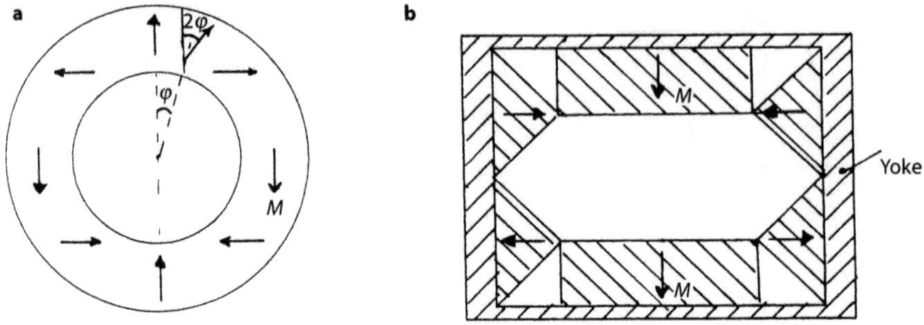


**Fig. 1.19.** **a** Circle cylindrical magnet. **b** Prism magnet

fields, which is of consequence for the design of the RF coils. Also, low-field systems impose design requirements on the RF coils that differ from those in high-field systems (see Sect. 6.4.3).

The design requirements become completely different, when dedicated systems are being considered. When a system for extremities is designed the magnet can be smaller, so the constraints on price and weight are much relaxed. On the other hand, in a system for interventional imaging, in which the patient should be reached while lying in the magnet, accessibility of the patient is very demanding on the magnet design.

### 1.3.2 Deviations from the Homogeneous Magnetic Field

As already stated, there is no magnet with a completely homogeneous field. Deviations from the homogeneous field in a spherical volume can be described by spherical harmonics. The idea is similar to Fourier analysis: just as a signal can be decomposed into its temporal harmonic functions or a spatial pattern can be decomposed into spatial harmonics. In spherical coordinates we use spherical harmonics for the same purpose. It is useful to have some understanding of these harmonics as a background for understanding what may happen in actual systems. For a complete treatment we refer the reader to the literature [11, 16].

In stationary fields ($\mathrm{d}/\mathrm{d}t = 0$) and current-free regions Maxwell's equations reduce to ($\nabla = \vec{i}\mathrm{d}/\mathrm{d}x + \vec{j}\mathrm{d}/\mathrm{d}y + \vec{k}\mathrm{d}/\mathrm{d}z$)

$$\nabla \times \vec{B} = 0 \quad \text{and} \quad \nabla \cdot \vec{B} = 0. \tag{1.11}$$

Both expressions can be combined by applying the rotation operation to the first one:

$$\nabla \times (\nabla \times \vec{B}) = \nabla(\nabla \cdot \vec{B}) - \nabla^2 \vec{B} = \Delta \vec{B} = 0, \tag{1.12}$$

where $\Delta = \nabla^2$. The Laplace operator $\Delta$ works on all components of $\vec{B}$ separately so we can restrict ourselves to solving $\Delta B_z = 0$. (We assume that the field is oriented in the $z$ direction. The transverse components will then be small and their influence on the gyromagnetic frequency is relatively small: they only rotate the total magnetic field a little, almost without changing its overall strength.) This equation can be written in spherical coordinates, which are defined in Fig. 1.20. Its general solution is found to be, after separation of the variables ($m \leq n$),

$$B_z = \sum_n \sum_m C_{nm} r^n P_{nm}(\cos\Theta)\cos m\varphi + S_{nm} r^n P_{nm}(\cos\Theta)\sin m\varphi, \quad (1.13)$$

where $P_{nm}$ are the associated Legendre functions [16] and $C_{nm}$ and $S_{nm}$ are coefficients of the series expansion. We restrict ourselves to writing only the lower-order Legendre functions explicitly:

$$\begin{aligned}
&P_{00} = 1, \\
&P_{10} = \cos\Theta, && P_{11} = \sin\Theta, \\
&P_{20} = \frac{1}{2}(3\cos^2\Theta - 1), && P_{21} = 3\sin\Theta\cos\Theta, && P_{22} = 3\sin^2\Theta, \\
&P_{30} = \frac{1}{2}(5\cos^3\Theta - 3\cos\Theta), && P_{31} = \frac{3}{2}\sin\Theta(5\cos^2\Theta - 1), && P_{32} = 15\sin^2\Theta\cos\Theta, \\
& && && P_{33} = 15\sin^2\Theta, \text{etc.}
\end{aligned}$$

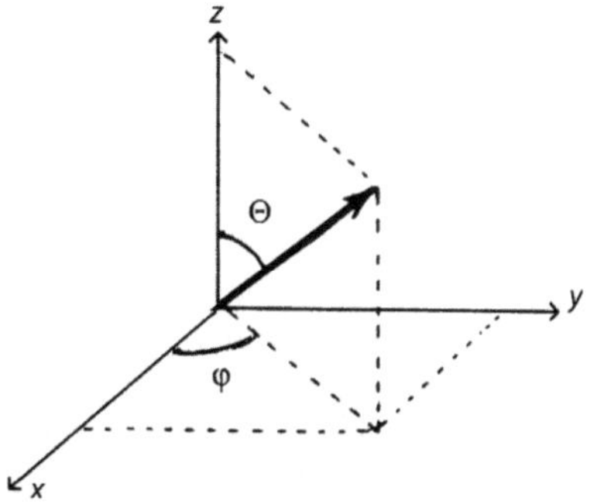


**Fig. 1.20.** Definition of spherical coordinates

The functions quickly become more complicated for higher orders; for our purpose it is not necessary to consider them in detail. Therefore only the general definition is given [16]:

$$P_{nm}(\cos\Theta) = \frac{r^n}{2^n n!}(1 - \cos^2\Theta)^{\frac{m}{2}} \frac{\mathrm{d}^{n+m}(1 - \cos^2\Theta)^n}{\mathrm{d}(\cos\Theta)^{n+m}}. \quad (1.14)$$

We now consider the field on the surface of a sphere around the isocenter, so $r^n$ is constant. Let us first take $n = 0$ and $m = 0$. Then only the term with coefficient $C_{00}$ exists and this yields a constant field on the sphere. We now take the solution of (1.13) for $n = 1$ and $m = 0$. We find $B_{z10} = C_{10} \mathrm{r} \cos\Theta = C_{10} z$. This clearly is the gradient field in the $z$ direction. For $n = 1$ and $m = 1$

we find from (1.13) that $B_{z11} = C_{11} r \sin\Theta \cos\varphi + S_{11} r \sin\Theta \sin\varphi$. The first term clearly is the $x$ gradient and the second term is the $y$ gradient (see Fig. 1.20). The higher-order terms describe more complicated (and mostly unwanted) field forms.

We can now look at the field on the surface of a sphere with unit radius for each separate term in (1.13) in the same way as we look at a cartesian map of the Earth: the surface of a sphere is imaged onto a rectangular plane with cartesian coordinates. We take the length circles as characterized by $\Theta$ and the width circles as characterized by $\varphi$. Then we can look at the areas where a certain term, given by $n$ and $m$ yields a positive or a negative contribution to the $B_z$ field. We shall do that for only the lower-order terms and show the result in Fig. 1.21. This exercise shows that when the field on the surface of a sphere around the isocenter is known, one can decompose this field into its spherical harmonics so as to find the coefficients $C_{nm}$ and $S_{nm}$. Then through (1.13) the field within this sphere is also known. So correction of the field at the surface of the sphere is sufficient to correct the field inside the sphere. This is what is done in practice. The sphere for which correction is obtained within the required homogeneity is called the sphere of homogeneity. Correction can then be made either with extra coils of the appropriate form [11], so as to excite only one of the spherical harmonics, or by adding iron to the magnet system outside the homogeneity sphere to compensate for the field errors. This is called shimming and is actually a complicated process which requires both computer programming and practical skill. The results are astonishing: magnets with homogeneity of less than 3 ppm within a 50 cm sphere are no exception any more, giving much freedom in the design of scan methods.

Finally we shall make some remarks on magnet design. To do that, we apply (1.13) to a rotational symmetric magnet, which is also mirror symmetric with respect to $z = 0$. The first symmetry requires all coefficients $C_{nm}$ and $S_{nm}$, with $m > 0$ to be zero. So only $C_{n0}$ terms exist. The second symmetry requires $C_{n0}$ to be zero for odd $n$. Now from the description of the superconducting magnet in Sect. 1.3.1.1 we know that we have six primary coils, which means that the first five terms can be compensated for by the coil geometry and the first term that causes deviation from the homogeneous field is the term with $C_{12.0}$. Its field then behaves as $z^{12}$, so it remains small for small $z$, but above a certain value it very quickly takes off to large values, so we have a field that is homogeneous around the isocentre, but outside the region of homogeneity it quickly increases to unacceptable values. Examples of the distortion of images due to inhomogeneous magnetic field are shown in image set II-6.

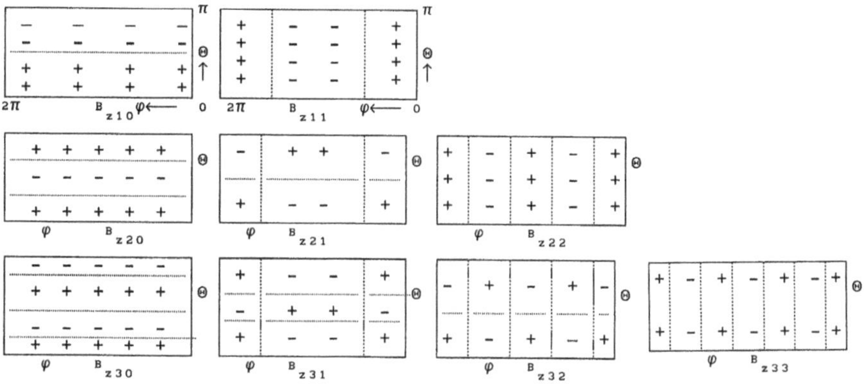


**Fig. 1.21.** Only the $\cos m\varphi$ terms are shown. The corresponding sine terms are 90° shifted in the $\varphi$ direction. In the $\varphi$ direction we see $2m$ zeros. In the $\Theta$ direction we have $2(n-m+1)$ zeros for $0 < \Theta < 2\pi$ (displayed is $0 < \Theta < \pi$)

### 1.3.3 The Gradient Chain

The gradient fields that are added to the main magnetic field to position encode the signal emitted by the proton spins in the object are generated with (resistive) gradient coils. Gradient fields are described by the equation

$$B_{G,z} = \frac{\mathrm{d}B_z}{\mathrm{d}x}x + \frac{\mathrm{d}B_z}{\mathrm{d}y}y + \frac{\mathrm{d}B_z}{\mathrm{d}z}z = G_x x + G_y y + G_z z = \vec{G} \cdot \vec{r}. \tag{1.15}$$

The dimension of $G$ is T/m, but the gradient-field values are usually given in mT/m. It should be noted that according to Maxwell's equations, fields as described by (1.15) are in principle impossible: there are always transverse components. However, for low-gradient fields, (1.15) is a good approximation because the (small) transverse components have a negligible influence on the total length of the magnetic field vector (and thus the gyromagnetic frequency). A situation in which this approximation is not allowed is the case of very high gradient fields (which is discussed in Sect. 3.3.2.3), or for low main magnetic fields.

The three different terms in (1.15) are excited with three different coils. The last term, the $z$ gradient, is in the direction of the main field and is excited with a pair of coils with Helmholtz geometry and with opposite currents, as shown in Fig. 1.22. The $z$-gradient strength depends linearly on the current: $G_z = CI$, where $C$ is the sensitivity of the gradient coil. The region in which the magnetic field increases linearly with the distance from the isocenter ($B_{G,z} = G_z z$) should be at least as large as the homogeneity sphere of the magnet (see image set II-2). This can be realized by using coils with several turns in which the density increases linearly with this distance (Fig. 1.22b). The application of more turns, however, also leads to an increase in the wire length and in the self-inductance of the total coil. From elementary circuit

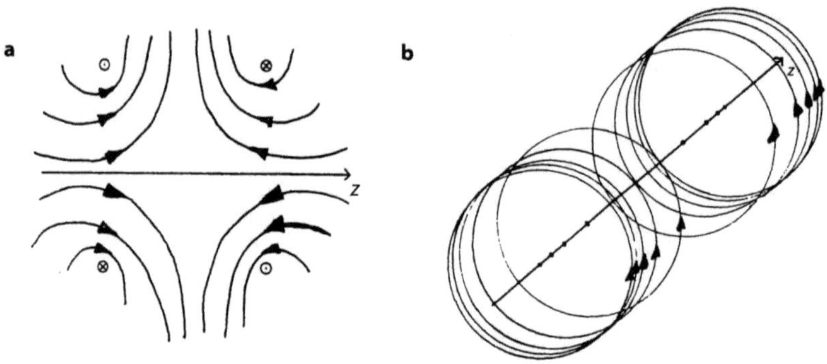


**Fig. 1.22. a** A Maxwell coil pair as a $z$-gradient coil. **b** A $z$-gradient coil with large linearity region

theory we know that the voltage $U$ across a coil is proportional to the rate of change of the current $I$ according to

$$U = L\mathrm{d}I/\mathrm{d}t + I\,R, \tag{1.16}$$

where $L$ is the self-inductance and $R$ the (very low) resistance of the conductor. When a steep current ramp is required, so a large $\mathrm{d}I/\mathrm{d}t$ (low "rise time"), we must choose $L$ to be as low as possible (the voltage $U$ is restricted by practical design problems). However, this implies a small coil with high non-linearity. So in gradient-coil design we have a trade-off between the coil self-inductance and the linearity of the gradient field [17].

The $y$ gradient is rotated by 90° around the magnet ($z$) axis with respect to the $x$-gradient coil, so in a rotational symmetric geometry only one of the two gradients needs to be discussed. The geometry of the $x$-gradient coil is shown in Fig. 1.23a. Its field in the $(x, z)$ plane near the isocenter is shown in Fig. 1.23b. In comparison with the ideal gradient field we see that along the $z$ axis, with increasing distance from the isocenter, the field is lower than the ideal field (the field lines diverge) and that along the direction $x = z$ the field increases more than proportional. The deviation from an ideal field (with equidistant lines of constant gradient field: $B_{G,x} = G_x x$) is shown in Fig. 1.23c, which clearly shows that a fifth-order field is superimposed onto the ideal $x$-gradient field ($2(n - m - 1) = 10$, with $m = 1$ see Fig. 1.21). This higher-order mode can again be counteracted (and the linearity region increased) by increasing the dimensions of the coils (so that the return current path is farther away). This, however, will increase the self-induction, and we find again a trade-off between the self-inductance and the dimensions of the linearity region, just as in the case of the $z$-gradient coil.

The free space within the gradient coil is a critical parameter in the design of an MRI system (claustrophobia), because the power necessary for energizing the coil is proportional to the fifth power of its radius, $r$. This can be

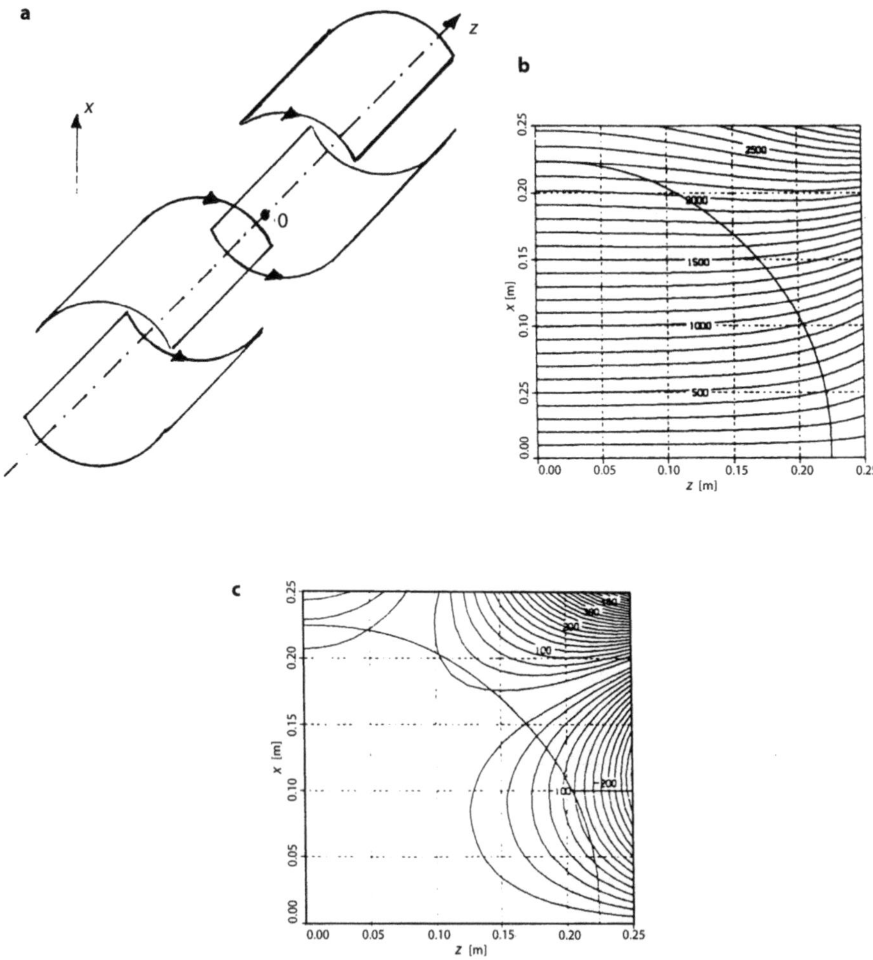


**Fig. 1.23.** **a** An $x$-gradient coil. **b** The actual field in the $(x, z)$ plane. **c** The deviation of the ideal field shows 10 lobes for $0 < \Theta < 2\pi$

understood by writing the equation for the stored energy of, for example, the $x$ gradient:

$$E_G = \frac{1}{2} \int_v \vec{B}_G \cdot \vec{H}_G \mathrm{d}V = \frac{1}{2\mu_0} \int_v G_x^2 x^2 \mathrm{d}x \mathrm{d}y \mathrm{d}z, \tag{1.17}$$

where $V$ is the volume. The integration in the $x$ direction yields a factor proportional to $r^3$. The $y$ integration yields a factor proportional to $r$ and the $z$ dimensions of a gradient coil always scale with $r$, so the $z$ integration also yields a factor proportional to $r$. The integral in (1.17) is therefore proportional to $r^5$. Decreasing $R$ by 12%, for example, means that the stored energy decreases by almost a factor of 2. We return to this point in the next section.

The choice of the coil properties depends very much on the manufacturer of the system, some "ball-park" figures are: $L = 200\,\mu$H and the coil sensitivity $G/I = C = 0.033(\mathrm{mT/m})\,\mathrm{A}^{-1}$. Many other features of the coils are of importance in practice (for example: differential linearity, current (re)distribution, technology, and price), but they are decided upon during the development process, which is beyond the scope of this textbook. For open types of magnets (H or C-arm magnets) the gradient coils have other (but related) geometry, so as to match the geometry of the magnet and the patient space, and are mostly mounted in a plane just before the magnet pole faces.

#### 1.3.3.1. Gradient Power Supply and Rise Time

The relation between the voltage on the gradient coil, $U$, and the current, $I$, is given by (1.16). Figure 1.24 shows the voltage waveform necessary to realize a trapezoidal gradient field. We distinguish three regions: (a) the increasing-field region, (b) the constant-field region, and (c) the decreasing-field region.

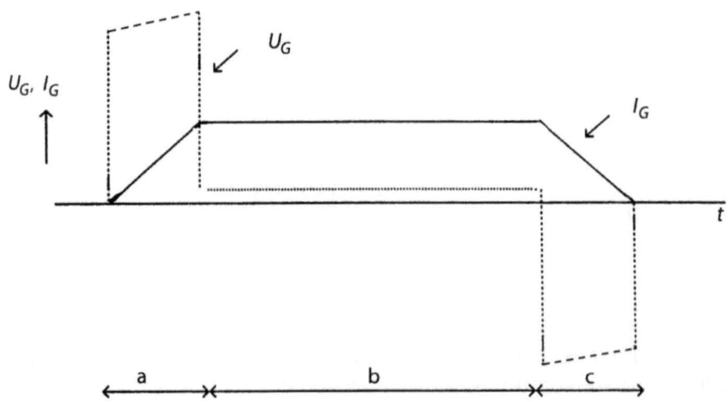


**Fig. 1.24.** Voltage and current waveform for generating a trapezoidal gradient field

During the increasing-field region (a) a high voltage is needed for a fast current rise (the term $IR$ may be neglected in this region). When the current rise is linear we can write $U = LI_{\mathrm{max}}/\tau$, where $I_{\mathrm{max}}$ is the maximum current the current supply can deliver and $\tau$ is the minimum time necessary to reach this current. The peak power to be delivered at the end of the current rise is $P = UI = LI_{\mathrm{max}}^2/\tau$. This peak power is an important design parameter, which to a large extend determines the price of the gradient amplifier. For a gradient coil with $L = 200\,\mu$H, and a sensitivity $C^{-1} = 30\,\mathrm{A(mT/m)}^{-1}$, we need for a gradient field of 12 mT/m and a rise time of 0.6 ms a peak power of 43 kW (neglecting the ohmic voltage drop and the losses in the supply wires).

The energy delivered to the coil during the rise time is the energy stored in the gradient field. Its maximum value is easily estimated from

$$E_{G,\max} = UI_{\max} \int_0^{\tau} \frac{t}{\tau} \mathrm{d}t = \frac{1}{2} L I_{\max}^2, \tag{1.18}$$

which is the well-known result for stored energy in a coil and is equivalent with (1.17).

In the constant-current region (b) the term $\mathrm{d}I/\mathrm{d}t$ is zero so only the ohmic term is of relevance, describing the bulk of the power dissipated in the coil. Of course there is also dissipation during the rise time, for which we can correct by adding a time $\tau$ to the length of region (b). The resistance of a gradient coil is of the order of tens of milli-ohms and dissipation becomes important in modern high-gradient systems.

In the decreasing-current region the voltage must be large and negative, since $\mathrm{d}I/\mathrm{d}t$ is negative. The energy stored in the gradient coil is now returned to the power supply. It depends on the technology of the gradient power supply whether this energy is dissipated or stored for re-use. Modern gradient power supplies use pulse-width-modulated amplifiers, which rely on mosfet or IGBT switches in which the dissipation is very low and which can store the energy returning from a gradient coil in a large capacitor. For a detailed description of such an amplifier see [18]. The requirements on linearity and delay time are very tight and impose severe demands on the design of the amplifier.

Gradient coils for H or C-arm magnets have, of course, different geometry and can be realized in the planes parallel to and just before the pole shoes. In their design, possible problems with hysteresis in the iron pole shoes must be taken into account. Eddy currents in the pole shoes cause similar problems as in the case of superconducting magnets.

#### 1.3.3.2. Eddy Currents

In a superconductive MRI system the gradient coils are surrounded by the metal inner bore of the magnet cryostate, the radiation shields, the helium vessel, and the magnet coil formers. The material used is mainly aluminum, a conducting material. When the current in the gradient coil is changed, the gradient field changes and eddy currents are induced in the surrounding conductors, which intercept the stray field of the gradient coils. The fields caused by these eddy currents are opposing the original field of the gradient coil so as to decrease the rate of change of this field. These eddy currents die out due to the resistance of the surrounding conductors. However, the time constants of this process are of the order of one to hundrets of ms (depending on the resistance of the shield in which they are induced) and the desired fast rise of the gradient field is suppressed: the magnetic field approaches the

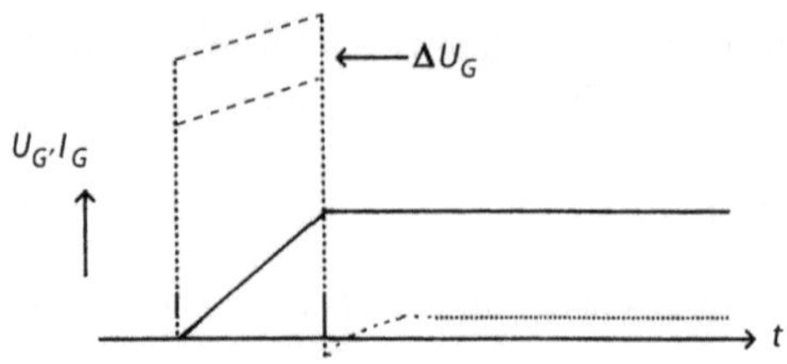


**Fig. 1.25.** Current compensation of eddy currents

desired value with a time constant determined by the decay times of the eddy currents.

One way to cope with this problem is to drive the coil with a current overshoot which dies out with the same time constant as the eddy current. The actual resulting total field inside the coil then has the desired time dependence as shown in Fig. 1.25. However, this method only compensates for those eddy-current fields that have the same spatial characteristics as the original gradient field (only the linear fields). As we shall see later, one stringent requirement in MRI is that at every moment during a scan the integral $\int G(t)\mathrm{d}t$ must be accurately known (see Sect. 2.4). Every inaccuracy in the integral leads to artifacts, especially when the unwanted eddy-current fields are nonlinear, which correspond to unwanted (and frequently unknown) position-dependent contributions to the gradient field.

One can visualize the situation induced in the inner bore by changes in the gradient current somewhat by using the concept of a mirror current, where the mirror is, for example, the inner bore of the magnet. The gradient coil is mirrored at the other side of the inner surface of the bore by an imaginary coil with an opposite current. The field shape of this imaginary coil is actually the shape of the field of the eddy current (a similar situation areises with the other shields in the magnet). So the coil sensitivity decreases due to the presence of the inner-bore conductor during the time the eddy currents exist. Therefore the current overshoot is necessary. The imaginary mirror coil has a larger radius than the gradient coil, but the same length, so the dimensions of the coil are no longer optimal and another spherical harmonic content may exist. This means that by compensating for the linear components by current overshoot, extra higher-order components can arise.

The field of the gradient coil outside the inner bore of the magnet is almost zero, due to the shielding effect of the eddy-current pattern on this conductive cylinder. This idea can be used to design a shielded gradient coil (just as for an actively shielded magnet): when a second layer of windings around the gradient coil is designed with a current pattern resembling as much as possible the eddy-current pattern of a conducting cylinder with the same radius, the field outside this layer is compensated for and therefore there are no eddy currents in the inner bore. Most modern systems have these "shielded gradient coils" and are to a certain extend free from eddy currents. An example is shown in Fig. 1.26. The actual severe requirements for the design of shielded gradient coils result from the need to reduce the amount of the "rest eddy currents" to a very low level [19]. This design will not be discussed here, there are various rest eddy-current field components with different time constants and different spherical harmonics content (see Sect. 1.3.2).

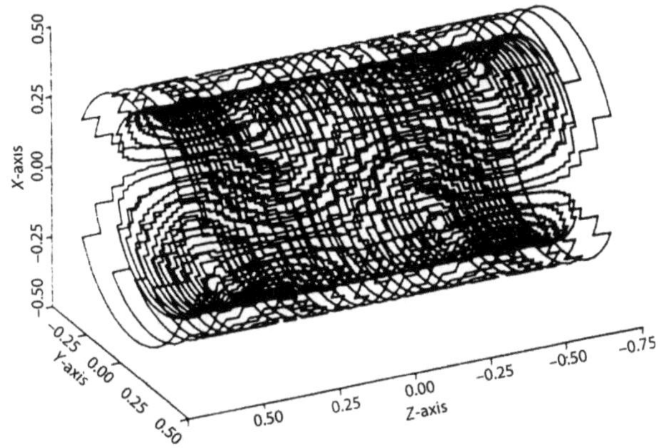


**Fig. 1.26.** Shielded $x$-gradient coil: fluent pattern

### 1.3.4 The RF Chain

The RF chain consists of the transmitting part and the receiving part. As discussed in Sect. 1.3.1, the transmitting chain starts with the frequency synthesizer, followed by the amplitude modulator (we do not consider frequency-modulated RF pulses here). The signal from the amplitude modulator is fed into the power amplifier. This amplifier is capable of delivering a peak RF power of 5 to 23 kW at $\omega_0$, which is about 21 MHz at 0.5 T and 64 MHz at 1.5 T. The relation between the peak power and the excited field is discussed in Sect. 6.4.2. Apart from the peak power, the average power that can be delivered is also of importance for several imaging sequences (such as Turbo Spin Echo, where many 180° pulses are applied in a short time; see Sect. 3.2). The actual requirements imposed on the RF power amplifier (linearity, harmonic content, distortion, etc) will not be discussed here.

When the power amplifier is active it is connected to the transmit coil, which is a large coil with a large region with a (nearly) homogeneous RF magnetic field. This is desirable for homogeneous slice excitation. The technology and build-up of this coil will be discussed in Sect. 1.3.4.1. During transmission all "receive-only" coils are open circuited to prevent the transmitting power from entering the receiving system. When the power amplifier is not active, its output is disconnected in order to prevend its output noise from interfering with the very small signals in the system. Also during reception the transmitting coil is open circuited with a switch to avoid mutual coupling with the receive coils.

The receive chain starts with the receive coil, which is tuned to the signal frequency (or detuned or open circuited when no signal is to be received). In the coil, a first-amplifier stage is integrated to avoid signal loss due to the (long) cable between the coil and the receiver (see Sect. 6.5.2). Then the signal is again amplified and subsequently digitized in an analog-to-digital converter. The signal is in a form that can be fed into the Fourier transformer, which must be a very fast circuit to make the images available within a second after the scan is completed. We shall describe the receiver and the ADC in somewhat more detail in Sect. 1.3.4.2.

#### 1.3.4.1. RF Coils

In this section the fundamental principles of RF coil design are treated. The actual realization of an RF coil is, notwithstanding the fundamental knowledge, more an art than a well-controlled action. This is due to the parasitic impedances and couplings that are always present in an actual circuit, especially when its dimensions are large, as in our case. We start again with (1.7), Boit and Savart's law, which relates the electric current through a wire to its magnetic field. In principle this law is only correct for dc fields; however, the frequencies of nuclear spins are sufficiently low to allow this approximation. Actually the dimensions of the circuit considered are of importance: when they are small in comparison with the wavelength the low-frequency approximation is allowed. For 63 MHz the wavelength in air is 4.7 m (actually 1/4 of a wavelength, 1.17 m is the important parameter here).

The simplest coil is the surface coil, which has the form of a circle or a rectangle in a plane (see Fig. 1.27b). This coil is still widely used, especially since it can be made flexible. The field of this coil is easily calculated from Biot and Savart's law, as is shown in Fig. 1.27a. Along the line through the center of the circle and perpendicular to the plane of the coil we find for the coil sensitivity

$$\beta(y) = \frac{B(y)}{I} = \frac{\mu_0 R^2}{2(R^2 + y^2)^{3/2}}. \tag{1.19}$$

From this equation we see that for $y = 0$ (and $x = z = 0$, in the center of the coil) the field is $\mu_0 I/(2R)$; so small coils yield a larger field in the centre than do large coils. For increasing $z$, however, the field decays faster for small coils than for large coils. This effect is shown in Fig. 1.28. The field is apparently very inhomogeneous and rather inadequate for exciting a slice, for which we need a homogeneous amplitude for a constant flip angle in the slice. Therefore in imaging, surface coils are not used as transmitting coils (we do not speak in this book about spectroscopy, where very high fields are needed so that small

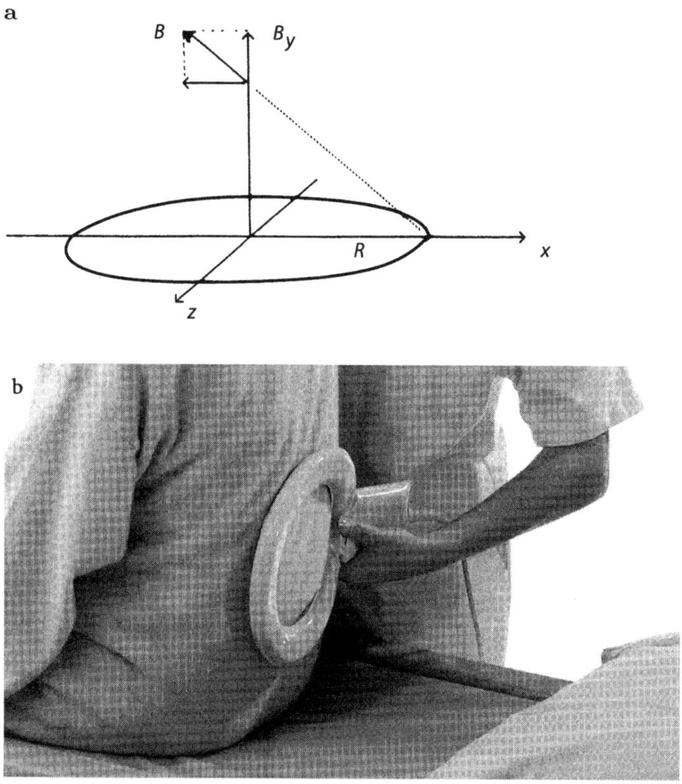


**Fig. 1.27.** Circle cylindrical RF coil. **a** Schematic. **b** Visual

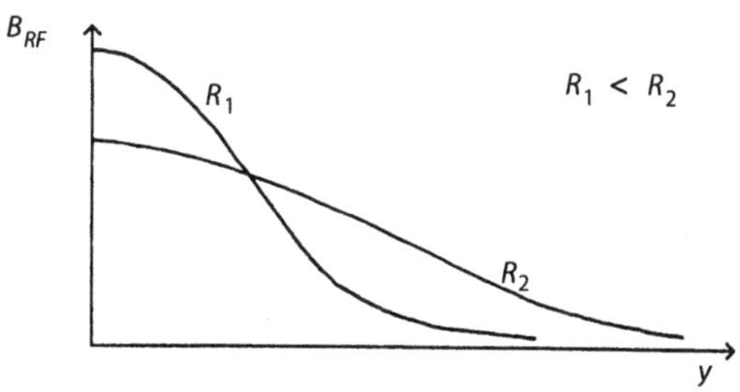


**Fig. 1.28.** RF magnetic field along the center line of a circular coil

inhomogeneous transmit coils must be used in connection with adiabatic RF pulses [20]).

However, for reception one uses the smallest possible coil compatible with the required FOV and depth below the surface. The reason is that the noise in an MRI system is mainly due to losses (eddy currents) in the patient, which couple into the receiving coil and thus manifest themselves as a damping. With a small coil only the noise from a small volume in the patient is received

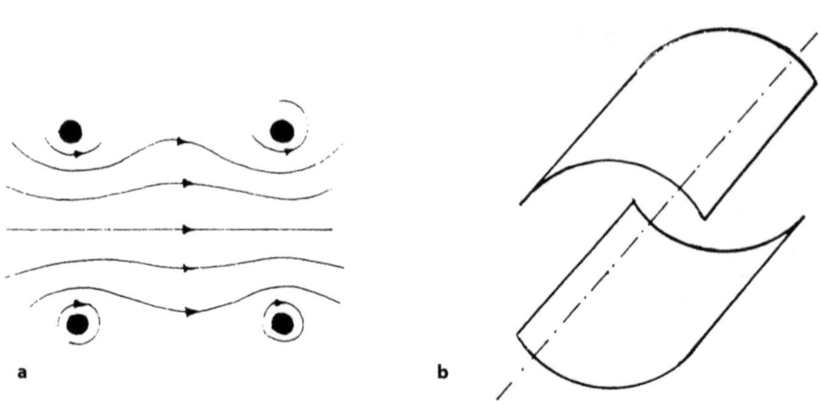


**Fig. 1.29.** **a** Helmholtz coil pair. **b** Saddle coil

and therefore the noise power is small. This point is treated in detail in Chap. 6. The coils discussed so far are referred to as "surface coils". The eddy currents in the patient also cause an left-right inhomogeneity in the images obtained with surface coils. This inhomogeneity is explained in [21].

The homogeneity of the magnetic field is much improved when two coils with radius $R$ at a distance $D$ are used, with the currents in the same direction (a "Helmholtz pair"). The field on the axis of the system can be read of from (1.19) to be (Fig. 1.29a).

$$\beta(y) = \frac{\mu_0 R^2}{2} \left\{ \frac{1}{\{R^2 + (D+y)^2\}^{3/2}} + \frac{1}{\{R^2 + (D-y)^2\}^{3/2}} \right\}. \tag{1.20}$$

In the center of the Helmholtz coil pair the sensitivity is $\beta_0 = \mu_0 R^2 (R^2 + D^2)^{-3/2}$.

A coil frequently used in early MRI systems is the saddle coil. In essence it is a Helmholtz pair, one coil above and one coil below the patient, so that a transverse RF field is excited. The saddle coil fits into the magnet bore, so $D = 2R$, the coil planes are curved (see Fig. 1.29b). For a homogeneous field the length of the coil is about equal to the diameter of the bore, so $D = 2R$ and the coil sensitivity in the center can be approximated by $\beta_0 \cong \mu_0 (11R)^{-1}$ (see (1.20)). An essential disadvantage of this coil is the fact that the field deviates from homogeneity just at the place of the shoulders of the patient, causing severe artifacts when off-center images (shoulder images) are required.

This problem is avoided by the "birdcage coil" design. The idea is that when there is a cylinder in the $z$ direction in which the current can flow only in this longitudinal direction and when this current has a $\varphi$ dependence like $I_z = I_0 \cos\varphi$, the field within this cylinder is transverse and homogeneous. This current distribution can be approximated by a system of rods that conduct the current in a longitudinal direction and which are supported by two

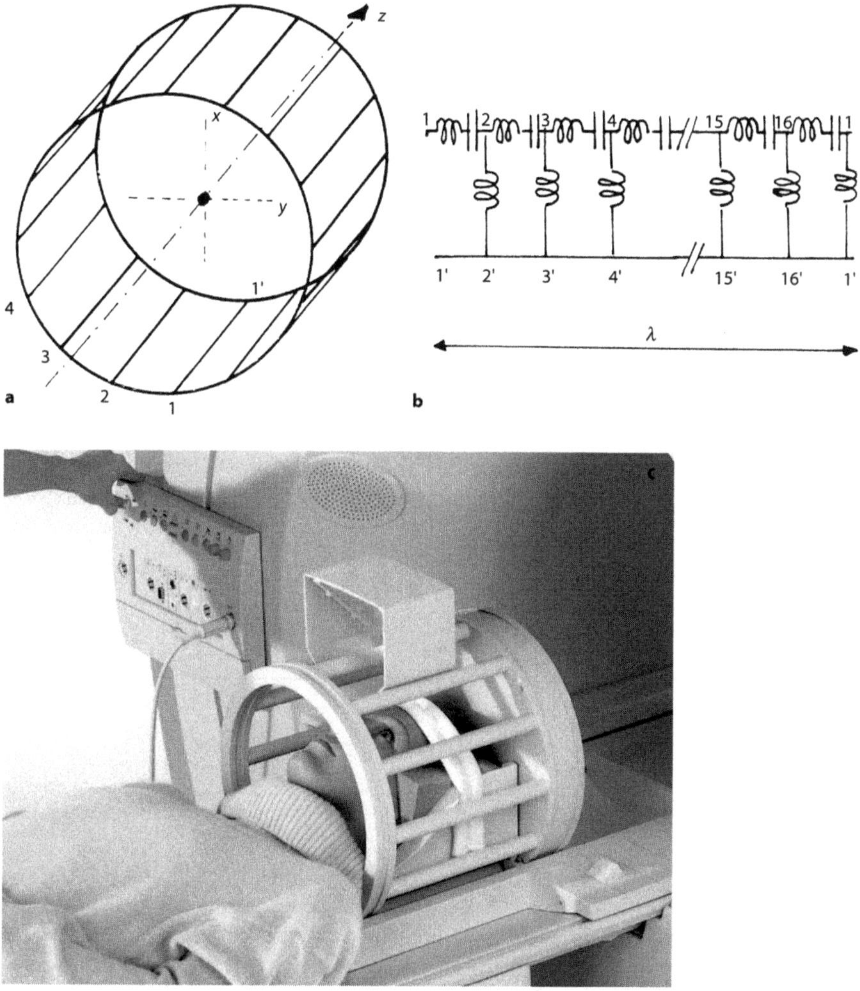


**Fig. 1.30.** **a** Birdcage coil, schematic. **b** Slow-wave structure. **c** Birdcage head coil

rings (see Fig. 1.30a). These rods together with the circle segments between them serve as inductances in a slow-wave structure (Fig. 1.30b) with the wave propagation along the circumference. With the appropriate capacitances the wavelength along this slow-wave structure can be chosen such that it is just equal to the circumference of the birdcage coil. In that case the requirement $I_z = I_0 \cos\varphi$ is satisfied. The slow-wave structure can be a high-pass filter, a low-pass filter, or a band-pass filter. In Fig. 1.30 a high-pass filter is shown. The wavelength can be found by the standard methods of the long-line theory. The practical realization is not that simple, because the actual impedances of the conductors and the components are difficult to measure at 63 MHz. So there is much trial and error, especially because parasitic modes may obscure the behavior (for example an $I_z = I_0 \cos 2\phi$ mode with nearly the same frequency). By exciting rod 1 we obtain a linearly polarized RF magnetic field in the $y$ direction (see Fig. 1.30a). The coil type is therefore called a "linear birdcage coil".

When a linear birdcage coil contains as the object to be imaged a conductive cylinder, the "quadrupolar" intensity modulation is a well-known image artifact. This artifact is caused by induced RF currents in the conducting medium and also occurs in patients. It appears as diametrically located bright and dark image areas [22] and is unacceptable at 1.5 T (64 MHz) in medical diagnostics.

Actually, a birdcage coil is rotationally symmetric, so when rod 5 (see Fig. 1.30a) is also excited, but shifted by 90° in phase with respect to rod 1, we also have a vertical RF magnetic field, which adds to the horizontal field to form a circularly polarized field. The quadrupole artifact is now much less manifest. The "quadrature birdcage coil" is one of the most frequently applied RF coils in MRI. The circuit for energizing the coil is shown in Fig. 1.31.

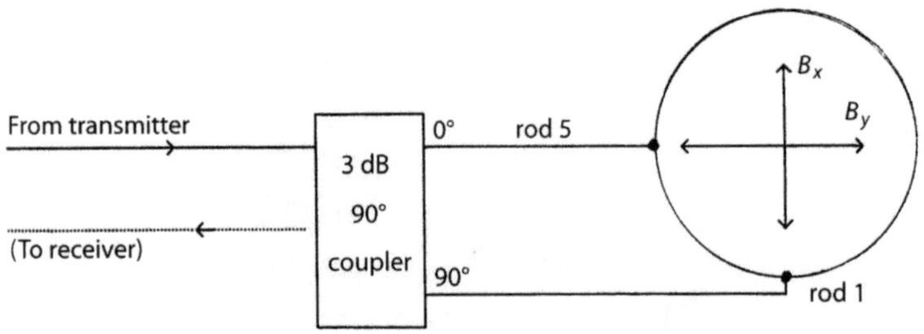


**Fig. 1.31.** Circuit for quadrature (receiving) coil

In most modern MR imaging systems the transmit coil is a birdcage coil, which is fitted just inside the gradient coil (see Fig. 1.11, to allow the maximum possible patient space). In Sect. 2.3.2.1 the amount of power for a certain excitation or echo pulse is calculated to be of the order of kilowatts. A linear transmitting coil is not favorable with respect to power consumption, because a linear field can be decomposed into two circularly polarized fields and only one of the two rotates synchronous with the rotating frame of reference. The other component, rotating in the opposite direction, only costs energy and can be avoided. In practice, especially at high fields, the transmit coil is always a quadrature coil. This quadrature coil suppresses the counter-rotating field and is (almost) twice as efficient as a linear coil.

Apart from the transmitting coil many receiving coils can also be designed as a birdcage type, for example a knee coil, a head coil, a neck coil, etc. A quadrature version of these coils gives in principle a factor 2 (3 dB) improvement in the signal-to-noise power ratio, since the receiving birdcage coil is made sensitive only for one circular polarized mode (so the noise power in the

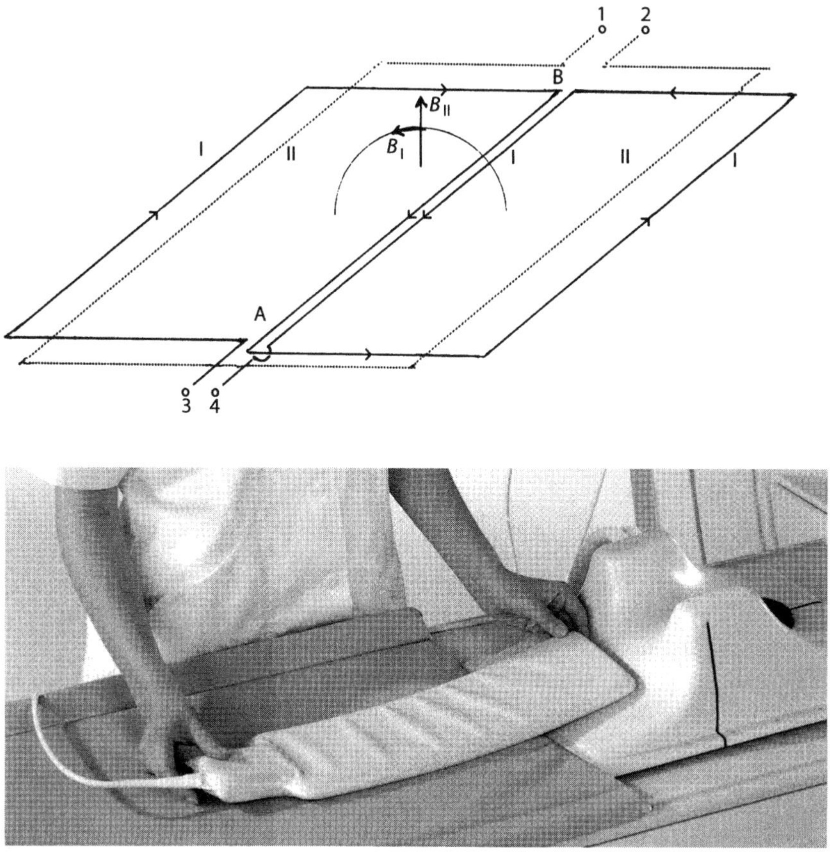


**Fig. 1.32.** Butterfly coil. Coils I and II are 90° shifted in phase. The region with circular polarization is above the line A–B

opposite rotating mode is not collected). Note that since the displayed image is based on amplitudes, a 3 dB improvement in the signal-to-noise power ratio means a 41% improvement in the image. The circuit is the same as that of Fig. 1.31, but works in the opposite direction (from coil to receiver). A full discussion of the influence of coil properties on the signal-to-noise ratio is given in Chap. 6.

The principle of quadrature reception would also be of benefit for the planar coils, like the circular coils discussed above. As has been said, these coils are chosen as small as the required field of view and penetration depth allow, to avoid too much noise entering the receiver. To also have the advantage of quadrature reception, the "butterfly coil" as depicted in Fig. 1.32 is applied. These coils can also be flexible and are used, for example, for wrap-around body coils and spine coils.

Finally, multi-channel coils (sometimes called Phased Array or Synergy coils) will be briefly mentioned [23]. Such coils are of advantage in cases where a large FOV (field of view) and a relatively small penetration depth are needed (for example, in the spine where a maximum length, 50 cm, is necessary with a penetration depth of 16 cm). A single coil for this application cannot be successfully optimized, because in that case a large coil is needed; the received noise will be high and thus the signal-to-noise ratio will be low. To improve this situation several small coils can be used in parallel, which all "see" a small part of the FOV, but, because of their small dimensions, they do so with a good signal-to-noise ratio. The "sub" images of these coils are combined during the reconstruction process into one image with large FOV and a good signal-to-noise ratio. The details of the circuit (multi-receiver or multiplexing) will not be discussed here.

#### 1.3.4.2. The Receiver

The last part of the RF chain to be discussed is the receiver. Although the receiver gain is set automatically during the preparation phase, it is good to learn some general properties of the receiving circuit, which must be kept in mind during the application of an MRI system.

The MR signal, as received by the receiving coil, contains frequencies in a frequency band of $\omega_0 \pm \delta\omega$. This frequency band depends on the field of view (FOV) and the gradient strength of the read-out gradient: $\delta\omega = \gamma G_{\mathrm{m}}$ FOV/2. This MR signal is mixed with the signal of the frequency synthesizer with frequency $\omega_0$ into the frequency band around $\omega = 0$. The frequencies in this band are just the rotating velocities of the magnetization vectors in the rotating frame of reference (there are other strategies of reception, but we restrict ourselves to the oldest method, which is sufficient for our purpose of understanding the use of the system). Therefore there are "positive" and "negative" frequencies, which must be distinguished because they come from positions on opposite sides of the isocentre. We shall first discuss the mixing process.

The received signal, $\exp j[(\omega + \delta\omega)t + \varphi]$, is mixed with $\cos(\omega_0 t)$ and also with $\sin(\omega_0 t)$ in a quadrature detector, as shown in Fig. 1.33. The output voltages now contain all the desired information because both the real and the imaginary parts (or the amplitude and phase) are known. So negative frequencies can also be distinguished. The signals are further amplified in a low-frequency amplifier.

Ultimately both amplified analog signals will be converted into a digital signal by AD converters (ADC). The maximum of the signal in a scan will be amplified up to the full scale of the ADC. This gain optimization is one of the steps that is automatically done during the preparation phase. The AD conversion process adds noise to the signal. From the requirement that this additional noise is limited below a certain percentage of the overall receiver

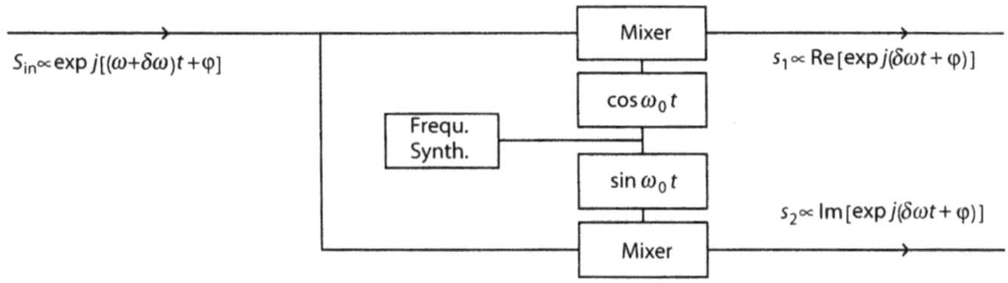


**Fig. 1.33.** Quadrature detector

noise one can calculate the dynamic range of the ADC. This will be shown below.

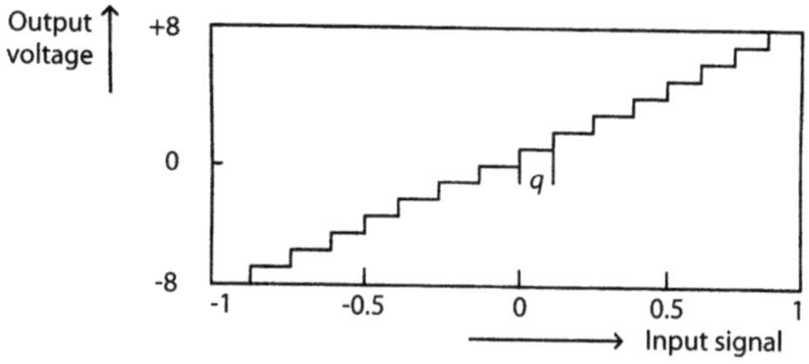


**Fig. 1.34.** Output vs input signal of a 4-bits analog-digital converter

We assume that the RMS value of the maximum amplitude of the input signal is $A$ and that the superimposed RMS noise amplitude, $a$, is $aq$, where $q$ is the quantization step (see Fig. 1.34). When the maximum signal matches the full scale of the ADC we have

$$2^n \cdot q = 2A\sqrt{2}, \tag{1.21}$$

where $n$ is the number of bits of the ADC. The output signal is too low for an input signal in the right half of the quantization step and too high for the left half. When $a > 1$ all possibilities of the error signal occur evenly distributed, so the additional rms noise voltage in the output signal is found to be

$$U_\mathrm{N} = \sqrt{\frac{1}{q}\int_{-q/2}^{q/2} x^2 \cdot \mathrm{d}x} = \sqrt{\frac{q^2}{12}}. \tag{1.22}$$

When the sampling frequency of the ADC is $f_\mathrm{m}$, the frequency band of the input signal must be restricted to $f_\mathrm{m}/2$ (see for an explanation the discussion of the Nyquist theorem in Sect. 2.4.1). The RMS noise voltage per Hz, $U_\mathrm{ND}$, can now be calculated, accounting for the fact that we must work with spectral density of $|U_\mathrm{N}|^2$, to be

$$U_{\mathrm{ND}} = \sqrt{\frac{q^2/12}{f_{\mathrm{m}}/2}} = \sqrt{\frac{q^2}{6f_{\mathrm{m}}}}. \tag{1.23}$$

With this knowledge the noise figure of the ADC, $F_{\mathrm{ADC}}$, can be calculated from its formal definition. When $S$ is the RMS value of the signal amplitude and $N$ is the RMS value of the noise amplitude, we find

$$F_{\mathrm{ADC}} = \frac{\{S/N\}^2_{\mathrm{in}}}{\{S/N\}^2_{\mathrm{out}}} = \frac{N^2_{\mathrm{out}}}{N^2_{\mathrm{in}}} = \frac{(aq)^2 + q^2/6f_{\mathrm{m}}}{(aq)^2} = 1 + \frac{1}{6f_{\mathrm{m}}a^2}. \tag{1.24}$$

Note that $S_{\mathrm{out}} = S_{\mathrm{in}}$. The dynamic range, $R_{\mathrm{ADC}}$, can be found by dividing the maximum RMS signal amplitude by the RMS input-noise voltage

$$R_{\mathrm{ADC}} = \left\{\frac{2^n}{2a\sqrt{2}}\right\}^2. \tag{1.25}$$

The value of $a$ can now be eliminated from (1.24). Furthermore the ADC is characterized by a figure of merit, FOM, defined by

$$10^{\mathrm{FOM}} = f_{\mathrm{m}}2^{2n}. \tag{1.26}$$

Using (1.24, 1.25) and (1.26) we find for the ADC noise figure

$$F_{\mathrm{ADC}} - 1 = \frac{8R_{\mathrm{ADC}}}{6 \times 10^{\mathrm{FOM}}}. \tag{1.27}$$

From this equation we read that a given FOM and a required $F_{\mathrm{ADC}}$ determines the magnitude of $R_{\mathrm{ADC}}$. If we require the ADC to add 1% to the noise voltage ($F_{\mathrm{ADC}} = 1.02$) and FOM $= 14$, which is a practical value, we find $R_{\mathrm{ADC}} = 15 \times 10^{11}$.

The MRI signals, however, have a dynamic range of about $8 \times 10^{15}$ at 1.5 T and $10^{15}$ at 0.5 T. The dynamic ranges are measured, but can also be estimated from the theory of the signal-to-noise ratio, as given in Chap. 6, by considering the maximum signal to come from a single "pixel", which is as big as the complete excited volume. Since the maximum signal strength is proportional to this excited volume it is therefore highest in 3-dimensional scans. The minimum signal is the available thermal noise ($kT\delta f$) at the temperature of the patient (see Sect. 6.4.1).

To accommodate this large dynamic range, which is too large for the ADC, as calculated above, in the receiver, it is necessary to have an attenuator in the receiving chain. For 1.5 T its maximum value, $L$, is about 5000 (37 dB) as is shown above. This attenuator increases the overall noise figure of the receiver, especially for 3D scans with high signal levels, but also in several other cases. It is important to know when such scans are applied. The place of this attenuator in the receiver is the next design decision (see Fig. 1.35).

We finally make some remarks on the theoretical ideas useful for the design of the receiver. With the equation of Friis the overall noise figure, $F_0$, of the receiver can be written as

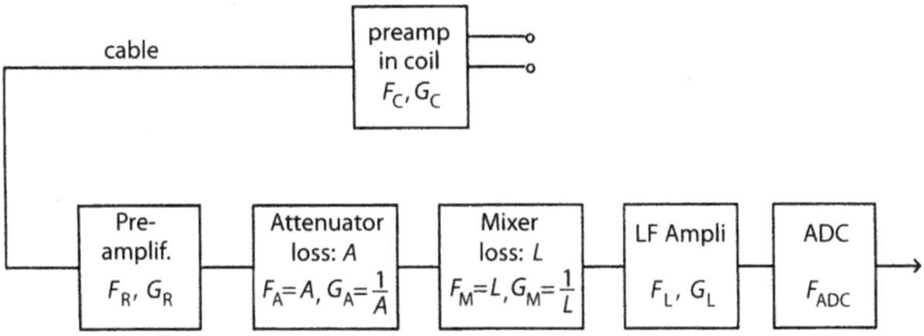


**Fig. 1.35.** Functional scheme of the receiver

$$F_0 = F_C + \frac{F_R - 1}{G_C} + \frac{F_A - 1}{G_C G_R} + \frac{F_M - 1}{G_C G_R G_A} + \frac{F_L - 1}{G_C G_R G_A G_M} + \frac{F_{ADC} - 1}{G_C G_R G_A G_M G_L}, \tag{1.28}$$

with the noise figures ($F$) and the gains ($G$) as defined in Fig. 1.35. In Sect. 6.5.2 it is shown that a passive lossy four pole with a (power) loss factor $L$ can be considered as a four pole with a noise figure $F = L$ and with a gain $G = 1/L$. We have used this theorem for both the attenuator and the mixer (in the latter case it may be considered as a good approximation). It is clear that $F_A$ can be very high when a large attenuation is required. At the same time $G_A$ is very low, so several terms in the Friis formula become large, increasing the overall noise figure. To limit this effect the attenuator is most frequently placed behind the RF amplifier, where the signal is already sufficiently amplified. It cannot be placed later in the chain, because a too-large signal (before the attenuator) would cause linearity problems. Here we have another fundamental trade-off in system design. It is not the task of this textbook to discuss system design in any depth, so we leave the subject by stating that some of the main design considerations of the receiving chain have been mentioned.

### 1.3.5 Physiological Signals

The tissue to be imaged is not always stationary. There is blood flow, cardiac motion, respiratory motion, and peristaltic motion. This motion causes artifacts when the displacement of tissue during an imaging sequence is appreciable. Methods to study blood flow, or to avoid artifacts due to blood flow, are described in Chap. 7. In the case of cardiac motion it is important to acquire the measurements for one specific image at the same heart phase during different heart beats, so as to obtain a stroboscopic effect. Also a cine of the cardiac motion is required. In that case one must take very fast images at a certain heart phase. Many methods, which are a mixture of these two extremes, are proposed. In all cases we need to synchronize the system to the heart rhythm. A similar situation occurs with the respiratory motion when we wish to acquire the most important measurements for an image during the time when the motion is small. The sensors providing us with the relevant information on motion are the electrocardiogram, the peripheral pulse sensor, the respiratory sensor.

**1. Electrocardiogram.** The subtle signals due to the depolarization and repolarization waves in the myocard, which cause contraction and release of the heart, are recorded in an electrocardiogram (ECG) [24, 25]. The R top in the ECG (see Fig. 1.36) is used for triggering. The ECG must be detected in violent electromagnetic surroundings, due to gradient switching, which induces large peaks in the ECG and due to induced RF power during the excitation pulses. Measures must be taken to make the spurious signals at least significantly smaller than the QRS peak to avoid wrong triggering. Furthermore the blood flow in the aorta causes an extra signal component in the electrocardiogram, which is large between the QRS wave and the T wave and which can also cause false triggering. This is due to the Hall effect in the aorta bow, where the blood travels perpendicular to the magnetic field. So the ECG triggering on the R top must be done very carefully and the ECG signal cannot be used as a patient monitoring signal during imaging.

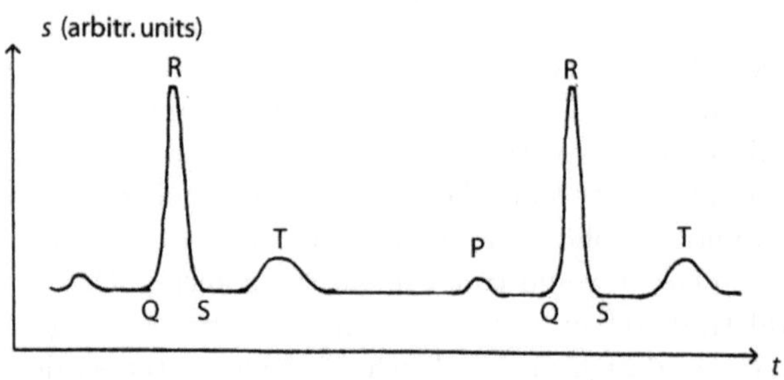


**Fig. 1.36.** The electrocardiogram

**2. Peripheral Pulse Sensor.** This is a sensor which measures the peripheral blood flow, for example in the finger tip. This blood flow can be detected by measuring the reflection or transmission of near infra-red light in the finger tip. The absorption depends on the amount of blood present. This amount of blood is – of course – correlated with the heart beat and can also be used for triggering, so as to avoid artifacts due to blood flow or of Cerebral-Spinal Fluid (CSF).

**3. Respiratory Sensor.** There are several methods to monitor motion due to respiration. For MRI the bellow sensor is mostly used. A non-elastic strap, interrupted by a bellow is wrapped around the chest of the patient. The volume of this bellow will depend on the respiratory motion of the chest. The pressure in the bellow is measured with a solid-state pressure transducer and is related to the respiration of the patient. From the resulting signal one can find periods during which the motion is minimal (after exhalation). The acquisition can be restricted to these periods to avoid motion artifacts. This is of course time consuming. Therefore methods are developed in which only the measurements determining the appearance of the image are taken during these periods (see Sect. 7.2.1), the remaining acquisitions can be taken during the periods with motion (high $k$ values; see Sect. 2.4.1).

Respiration can also be detected by measuring the impedance changes of the chest or with the reflection of ultrasound waves, but the vicinity of RF coils inhibit these methods.

Finally we must mention recently developed sensors, which are used in connection with forced motion (for example of the knees) and cine studies of this motion. Also in the case of dynamic studies with contrast agents it can be important to know the exact timing of the administration of the agent.

### 1.3.6 The Back End

As indicated in the introduction, a detailed description of the back end is outside the scope of this textbook. Therefore we shall restrict ourselves to a short overview of the main components and functions in order to give the reader some idea.

As can be seen from Fig. 1.11a, the back end performs the following functions.

1. User interface and display; communication with the user concerning the measuring method to be performed and display of the results. Normally the user interface is implemented as a computer program on a work station. But this can in principle also be a PC or a dedicated control device.
2. Acquisition and control. Real-time control of all the front-end functions: gradients, RF pulses, frequencies, phases, ADC, and front-end switches via the "front-end controller". The final result is always a series of ADC data. For most commercial MRI systems this data-acquisition system is implemented on a separate computer that controls a set of hardware devices for timing and control. The division between hardware and software, which mainly determines the flexibility of the system, is very vendor dependent.
3. Fourier transformer. The ADC data produced by the data acquisition system must be Fourier transformed to obtain viewable images. This transform can either be executed by dedicated hardware or by general-purpose computer hardware. For most systems the former solution has been chosen in which the reconstruction function is implemented on a separate (very powerful) computer. Invariably the Fast Fourier Transform (FFT) algorithm is used because of its high computational efficiency. It requires a matrix of $2^m \times 2^n$ data points. The logical consequence is that the matrix of MR data acquired usually has that size. (See however Sects. 3.6.1 and 6.7.)

4. Viewing and processing. There are many manipulations and processing steps possible on the images, before they are presented to the user [14]. Because a lot of user interaction is necessary during viewing and processing, these functions are integrated in the user-interface part of the system. For the actual processing either the host work station or a dedicated viewing processor (accelerator) is used.
5. Administration and storage. Another function of the main computer is the storage (and administration) of all the acquired images. This may be a short-term storage on disk or a long-term storage on other devices such as tapes, CDs, optical disks, etc. Finally there are also functions to transfer images and patient data to other scanners, work stations, hospital information systems, etc., making viewing away from the MRI system itself possible (teleradiology). The connection to these other systems must be specified and is at present standardized [26].
6. Hard copy. The final results of most scanners are at this moment not displayed on a screen, but are printed on laser hard-copy films. Therefore most imagers are equipped with a connection to such devices and with the functions to transfer the images and to print them with the administrative data.

# 2. Conventional Imaging Methods

## 2.1 Introduction

In this chapter we shall describe, more formally than in Chap. 1, excitation, precession, and relaxation on the basis of the Bloch equation. These are the elements that we need in order to discuss the conventional scan methods called Spin Echo (SE) and Field Echo (FE). For better understanding, and also as a preparation for the discussion of the modern fast and ultra-fast methods, we introduce the concept of $k$ space. The fundamental artifacts of SE and FE will be treated.

## 2.2 The Bloch Equation

Magnetic resonance is a process that should be treated in terms of quantum mechanics. Fortunately the coupling of the nuclear spins mutually and with the surrounding matter is weak, which allows a classical treatment on the basis of the Bloch equation extended with terms describing the relaxation in a phenomenological way [1]. There are some phenomena in MR for which a quantum mechanical treatment is necessary. These phenomena are beyond the scope of our discussions. The general Bloch equation will not be deduced here, it can be found in many textbooks as

$$\frac{\mathrm{d}\vec{M}}{\mathrm{d}t} = \gamma(\vec{M} \times \vec{B}), \tag{2.1}$$

where the vector $\vec{M}$ denotes the magnetization vector, $\vec{B}$ denotes the magnetic fieldstrength, and $\gamma$ is the gyromagnetic ratio, which for protons (spin 1/2) is $2\pi(42.6 \times 10^6)$ rad/(sec T). The vector product $\vec{M} \times \vec{B}$ is again a vector that is perpendicular to both $\vec{B}$ and $\vec{M}$ with magnitude $|\vec{M}|\,|\vec{B}| \sin\alpha$, where $\alpha$ is the angle between $\vec{M}$ and $\vec{B}$.

The Bloch equation tells us that the vector $\mathrm{d}\vec{M}/dt$ is always orientated perpendicular to the plane of $\vec{B}$ and $\vec{M}$, so the point of $\vec{M}$ moves in a circular path (see Fig. 2.1), thus giving rise to precession. In a stationary situation we may state that $\vec{M}$ rotates around $\vec{B}$ with an angular speed of $\omega$. So if $\vec{M}$ is not aligned with $\vec{B}$, $\mathrm{d}\vec{M}/\mathrm{d}t = \omega|\vec{M}| \sin\alpha$. The right-hand side of (2.1) is

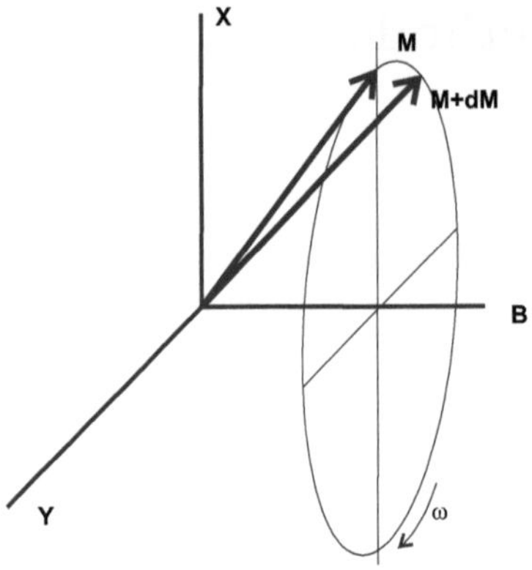


**Fig. 2.1.** Path of the magnetization vector

$\gamma|\vec{M}|\,|\vec{B}|\sin\alpha$, so it follows directly that the precession occurs at a specific rate determined by the magnetic field:

$$\omega = \gamma|\vec{B}|. \tag{2.2}$$

This is the Larmor equation, where $\omega$ is called the "Larmor frequency".

In practical MRI systems we deal with three different magnetic fields, adding up to $\vec{B}$. The first is the main magnetic field in the $z$ direction, $\vec{B}_0 + \vec{\delta}B$, produced by the magnet ($\vec{\delta}B$ describes the field inhomogeneity). The second is the gradient field $\vec{k}(\vec{G}\cdot\vec{r})$, which has ideally only a $z$ component and is linearly dependent on the position ($\vec{r}$ is the position vector, $\vec{G}$ is the proportionality constant, and $\vec{k}$ is the unit vector in the $z$ direction). The third field is the magnetic component $\vec{B}_1$ of the forward-rotating RF field excited by the RF system. Therefore we can rewrite (2.1) as

$$\left\{\frac{\mathrm{d}\vec{M}}{\mathrm{d}t}\right\}_{B_0} + \left\{\frac{\mathrm{d}\vec{M}}{\mathrm{d}t}\right\}_{\delta B,G,B_1} = \gamma\left\{\vec{M}\times\left[\vec{B}_0 + \delta\mathrm{B} + \vec{k}(\vec{G}\cdot\vec{r}) + \vec{B}_1\right]\right\}, \tag{2.3}$$

where the first term on the left-hand side describes the precession due to $B_0$, the homogeneous part of the main magnetic field, and the second term describes the precession due to the field inhomogeneity, $\vec{\delta}B$, the gradient field $\vec{G}\cdot\vec{r}$, and the RF magnetic field, $\vec{B}_1$. Note that the gradient field is always in the direction of $\vec{B}_0$. We now introduce a new system of reference that rotates around the $\vec{B}_0$ direction with an angular velocity

$$\omega_0 = \gamma|\vec{B}_0|. \tag{2.4}$$

In this rotating reference system $(x', y', z)$ the precession due to $B_0$ is not seen, so only precession due to $\delta\vec{B}$, $\vec{G}\cdot\vec{r}$, and $\vec{B}_1$ remains:

$$\left\{\frac{\mathrm{d}\vec{M}}{\mathrm{d}t}\right\}_{\delta B,G,B_1} = \gamma\{\vec{M}\times[\delta\vec{B} + \vec{k}(\vec{G}\cdot\vec{r}) + \vec{B}_1]\}. \tag{2.5}$$

For the time being we shall neglect the term $\delta\vec{B}$; it will become important later in the last part of this chapter in the discussion of artifacts. As before we shall take $\vec{B}_0$ as being along the $z$ direction. Furthermore only the $x'$ and $y'$ components of the RF magnetic field are of interest. The $z$ component can (in general) be neglected in comparison with the main magnetic field $B_1 \cong 10^{-5}B_0$. Using this, (2.5) can be formulated as a vector equation or a matrix equation:

$$\begin{aligned} \left\{\frac{\mathrm{d}\vec{M}}{\mathrm{d}t}\right\}_{G,B_1} &= \gamma \begin{vmatrix} \vec{i} & \vec{j} & \vec{k} \\ M_{x'} & M_{y'} & M_z \\ B_{1x'} & B_{1y'} & \vec{G}\cdot\vec{r} \end{vmatrix} \\ &= \gamma \begin{pmatrix} 0 & \vec{G}\cdot\vec{r} & -B_{1y'} \\ -\vec{G}\cdot\vec{r} & 0 & B_{1x'} \\ B_{1y'} & -B_{1x'} & 0 \end{pmatrix} \begin{pmatrix} M_{x'} \\ M_{y'} \\ M_z \end{pmatrix}, \end{aligned} \tag{2.6}$$

where $\vec{i}$, $\vec{j}$, and $\vec{k}$ are unit vectors in the $x'$, $y'$, and $z$ directions, respectively. The straight bars mean "determinant". The last form is the matrix formulation of the Bloch equation in the rotating frame of reference. Equation (2.6) describes the motion of the magnetization vector in the rotating frame of reference, $(x', y', z)$. In the equilibrium situation $M_{x'}$ and $M_{y'}$ are zero, so when $\vec{B}_1$ is also zero (no excitation field), there is no motion $\mathrm{d}\vec{M}/\mathrm{d}t$ and the magnetization vector is parallel to the main magnetic field (has only a longitudinal component).

In a non-equilibrium situation two processes exist that drive the magnetization back to equilibrium. One of these destroys the overall transverse magnetization and the other brings the longitudinal magnetization back into the equilibrium situation. These relaxation processes must be introduced in the Bloch equation. For a very elaborate and elegant description of the physical and biological backgrounds of these relaxation effects read the survey paper by Fullerton [2]. A summary is given in Sect. 6.3. Here we present only the mathematics.

The spin–spin relaxation, with characteristic decay time $T_2$, causes the transverse magnetization, $M_\mathrm{T}$, which is perpendicular to the main magnetic field, to relax back to zero by dephasing the individual spins. According to the notation used above $M_\mathrm{T} = \vec{i}M_{x'} + \vec{j}M_{y'}$. We shall assume that this happens in a simple exponential manner (which is not necessarily the case):

$$\frac{\mathrm{d}M_\mathrm{T}(t)}{\mathrm{d}t} = -\frac{M_\mathrm{T}(t)}{T_2}, \quad \text{or} \quad M_\mathrm{T}(t) = M_\mathrm{T}(0)\exp\left(\frac{-t}{T_2}\right). \tag{2.7}$$

The longitudinal component of the magnetization, $M_z$, relaxes back to its equilibrium value $M_0$, due to spin–lattice relaxation with a characteristic time $T_1$. This results in a slightly different equation:

$$\frac{\mathrm{d}M_z(t)}{\mathrm{d}t} = -\frac{M_z(t) - M_0}{T_1}, \quad \text{or}$$

$$M_z(t) - M_0 = \{M_z(0) - M_0\} \exp\left(\frac{-t}{T_1}\right). \tag{2.8}$$

When we combine (2.7) and (2.8) with (2.6) we find

$$\left\{\frac{\mathrm{d}\vec{M}}{\mathrm{d}t}\right\}_{G,B_1} = \begin{pmatrix} -1/T_2 & \gamma\vec{G}\cdot\vec{r} & -\gamma B_{1y'} \\ -\gamma\vec{G}\cdot\vec{r} & -1/T_2 & \gamma B_{1x'} \\ \gamma B_{1y'} & -\gamma B_{1x'} & -1/T_2 \end{pmatrix} \begin{pmatrix} M_{x'} \\ M_{y'} \\ M_z \end{pmatrix} + \begin{pmatrix} 0 \\ 0 \\ M_0/T_1 \end{pmatrix}. \tag{2.9}$$

This equation is the basis for most of the theories concerning MRI, both for the design of the imaging method and for RF pulse design.

### 2.2.1 Precession

Precession is the rotation of the magnetization vector around the direction of the main magnetic field ($z$ axis). In the absence of RF fields ($B_1 = 0$) the magnetization is described by

$$\left\{\frac{\mathrm{d}\vec{M}}{\mathrm{d}t}\right\}_{G} = \begin{pmatrix} -1/T_2 & \gamma\vec{G}\cdot\vec{r} & 0 \\ -\gamma\vec{G}\cdot\vec{r} & -1/T_2 & 0 \\ 0 & 0 & -1/T_1 \end{pmatrix} \begin{pmatrix} M_{x'} \\ M_{y'} \\ M_z \end{pmatrix} + \begin{pmatrix} 0 \\ 0 \\ M_0/T1 \end{pmatrix}$$

which can be written in an easier form by using complex notation: $M_\mathrm{T} = M_{x'} + jM_{y'}$, with $j = \sqrt{-1}$. Integration of the equation thus obtained yields

$$M_\mathrm{T}(t) = M_\mathrm{T}(0) \exp\left(-j\gamma\vec{r}\cdot \int \vec{G}(t)\mathrm{d}t\right) \exp\left(-\frac{t}{T_2}\right), \tag{2.10a}$$

(the first exponential describing the precession in the $(x', y')$ plane, the second exponential describing the relaxation) and

$$M_z(t) - M_0 = [M_z(0) - M_0] \exp\left(-\frac{t}{T_1}\right), \tag{2.10b}$$

which is, of course, equal to (2.8). Throughout our treatment we will use the word precession for the motion of the magnetization vector due to the main magnetic and gradient fields only. Rotation (or "nutation") is the motion of the magnetization vector when RF fields are also present.

## 2.3 Excitation

The RF field $B_1$, applied for excitation of the magnetization, is in modern systems always given in the form of short RF pulses: during excitation the effect of $T_1$ and $T_2$ relaxation can be neglected ($T_1 \cong 1000\,\mathrm{ms}$ and $T_2 \cong 100\,\mathrm{ms}$) because the relaxation times are usually very long compared with the length of the RF pulse ($< 6\,\mathrm{ms}$).

### 2.3.1 Non-selective Pulse

We first describe a very simple pulse model in which the effect of the gradient field is zero (or can be neglected). This model can be used for non-slice-selective pulses. The Bloch equation (2.9) then reduces to

$$\left\{\frac{\mathrm{d}\vec{M}}{\mathrm{d}t}\right\}_{G,B_1} = \begin{pmatrix} 0 & 0 & 0 \\ 0 & 0 & \gamma B_{1x'} \\ 0 & -\gamma B_{1x'} & 0 \end{pmatrix} \begin{pmatrix} M_{x'} \\ M_{y'} \\ M_z \end{pmatrix}, \tag{2.11}$$

where the RF field $B_1$ is taken in the $x'$ direction. Equation (2.11) means two linear coupled differential equations, and upon elimination of $M_z$ from these equations we find

$$\frac{\mathrm{d}^2 M_{y'}}{\mathrm{d}t^2} = -\gamma^2 B_{1x'}^2 M_{y'},$$

$$\text{so} \quad M_{y'} = A\sin(\gamma B_{1x'}t) + B\cos(\gamma B_{1x'}t), \tag{2.12}$$

where $A$ and $B$ are complex constants depending on the boundary conditions. Taking for $t = 0$, $M_{y'} = M_{y'}(0)$ and $M_z = M_z(0)$, one easily finds

$$\begin{pmatrix} M_{x'}(t) \\ M_{y'}(t) \\ M_z(t) \end{pmatrix} = \begin{pmatrix} 1 & 0 & 0 \\ 0 & \cos\omega t & \sin\omega t \\ 0 & -\sin\omega t & \cos\omega t \end{pmatrix} \begin{pmatrix} M_{x'}(0) \\ M_{y'}(0) \\ M_z(0) \end{pmatrix}, \tag{2.13}$$

where $\omega = \gamma \cdot B_{1x'}$, and $t$ is the duration of the $B_1$ field. The matrix describes a rotation around the $x'$ and can be deduced also on the basis of simple geometrical arguments. A rotation around the $y'$ axis can be described in a similar way and yields a matrix similar to (2.13) after interchanging the first and second column and the first and second row. For a 90° pulse we have $\omega t = \pi/2$ and for a 180° pulse we have $\omega t = \pi$. For example, if we assume a 180° pulse of 1/2 ms duration, it follows that the RF field strength needed is 23.5 $\mu$T. The power needed for this excitation pulse will be discussed later.

Note that precession can be described by a similar matrix when the gradient is constant, but now the value 1 appears as the 3,3 element in the matrix. The angular frequency now becomes $\Omega = \gamma\vec{G} \cdot \vec{r}$ and $t$ is the duration of the gradient field.

### 2.3.2 Slice-Selective RF Pulses

In the case of RF pulses used for slice selection, the RF pulse is given in combination with a gradient field (selection gradient). Now we cannot possibly make the assumption made in the previous section that $G = 0$, since at the boundary of the slice the effect of the selection gradient is comparable with that of the RF field. Although the general description of RF pulses is beyond the scope of this book, we shall deduce some essential properties of RF pulses by describing another limiting case, but now including the gradient field. We shall treat here the case of small flip angles [3], and assume that the flip angle is so small that $M_z(t)$ remains constant ($\cong M_0$). In this case, for a mathematical description the first two equations of (2.9) suffice, since at small flip angles $M_z$ can be assumed to be constant. Surprisingly this approximation appears (when compared with realistic mathematical simulations) to be correct up to flip angles of 30° and to give reasonable results even up to 90° [4]. Both of the remaining equations in (2.9) combine to give

$$\frac{\mathrm{d}M_\mathrm{T}}{\mathrm{d}t} = -j\gamma\left(\vec{G}\cdot\vec{r}\right)M_T + j\gamma B_1 M_0, \tag{2.14}$$

where $M_\mathrm{T} = M_{x'} + jM_{y'}$, $B_1 = B_{1x'} + jB_{1y'}$, and $j = \sqrt{-1}$. The general solution of this nonlinear differential equation reads

$$M_\mathrm{T}(t,\vec{r}) = A(t)\exp\left(-j\gamma\vec{r}\cdot\int_{t_1}^{t}\vec{G}(t')\mathrm{d}t'\right),$$

where $t_1$ is the time at which the pulse starts. Introducing this solution into (2.14), solving for $A(t)$ using $M(t=t_1,\vec{r}) = M_0$, and substituting the result yields

$$M_\mathrm{T}(T/2,\vec{r}) = j\gamma M_0\int_{-T/2}^{T/2}B_1(t)\exp\left(-j\gamma\vec{r}\cdot\int_{t}^{T/2}\vec{G}(t')\mathrm{d}t'\right)\mathrm{d}t. \tag{2.15}$$

Here we have assumed that the RF pulse starts at $t = -T/2$ and lasts $T$ s. Note that the integral in the exponential runs from $t$ to $T/2$, so it is zero at the end of the pulse. Also the exponent is not constant during the RF pulse, it is dependent on position and time.

We now first restrict ourselves to the case of a constant gradient in the $z$ direction. Equation (2.15) then reduces to

$$M_\mathrm{T}(T/2,z) = j\gamma M_0\exp\left(-j\gamma zG_z\frac{T}{2}\right)\int_{-T/2}^{T/2}B_1(t)\exp(j\gamma zG_z t)\mathrm{d}t. \tag{2.16}$$

One can conclude two facts from this equation. In the first place we see that the slice profile, $M_\mathrm{T}(z)$, is equal to the Fourier integral of $B_1(t)$. Second, the direction of $M_\mathrm{T}(z)$ in the $(x',y')$ plane depends on $z$, as shown by the exponential outside the integral. This latter fact means that there is a phase dispersion of the transverse magnetization over the slice thickness and this dispersion can in fact be sufficient for signal to be lost. Correction of this phase dispersion is necessary for a proper result of slice selection. This is done by introducing a gradient of reverse direction and of half the time length of the pulse, $T/2$, after the pulse. The result is (see also (2.10))

$$M_\mathrm{T}(T,z) = jM_0\int_{-k_T}^{k_T}\frac{B_1(k)}{G_z}\exp(jkz)\mathrm{d}k, \tag{2.17}$$

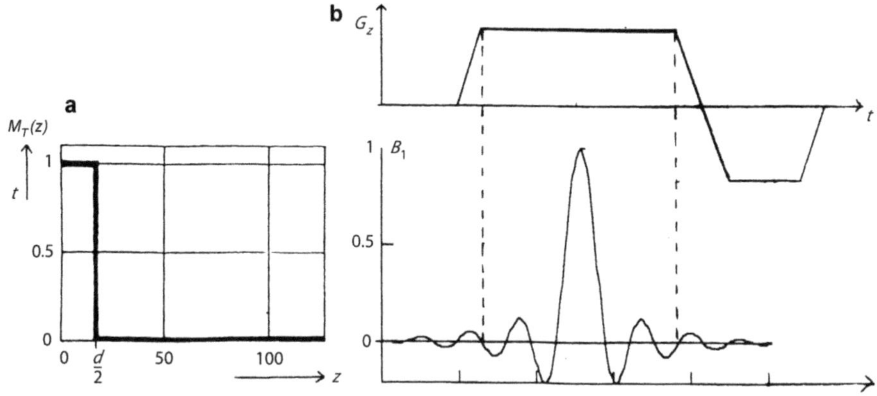


**Fig. 2.2.** Slice profile (**a**) and selection gradient with RF waveform (**b**) as Fourier transform; side lobes outside the duration of the selection gradient are cut off, yielding a non-ideal slice profile

where $k = \gamma G_z t$ and $k_T = \gamma G_z \frac{T}{2}$. Now only the Fourier integral remains and $M_\mathrm{T}(z)$ has a constant direction in the $(x', y')$ plane for all $z$ across the selected slice ($M_\mathrm{T}$ is always perpendicular to $B_1$). Ideally the transverse magnetization follows a block function in space: $M_\mathrm{T}(z) = M_0 \sin\alpha$ for $|z| < d/2$ ($d$ is the slice thickness), otherwise $M_\mathrm{T} = 0$. Its Fourier transform is known to be (see Fig. 2.2)

$$B_1(t) = jG_z d \frac{\sin kd/2}{kd/2} \sin\alpha. \tag{2.18}$$

This sinc-shaped RF pulse is used in practical MR systems, together with a suitable gradient $G_z$, followed by a reverse gradient lobe (see Fig. 2.2). The duration of the main lobe of the pulse, $\tau$, is found from $k\,d/2 = \pm\pi$, or $\gamma G_z d\tau = 4\pi$. It is easier to work with $\gamma\!\!\!- = \gamma/2\pi$ and with the half width of the main lobe $\tau' = \tau/2$, so

$$\gamma\!\!\!- G_z d\tau' = 1, \tag{2.19}$$

which is a well-known relation. For $d = 3\,\mathrm{mm}$, $\gamma\!\!\!- = 42.6 \times 10^6\,\mathrm{Hz/T}$, $G_z = 10\,\mathrm{mT/m}$ we obtain $\tau' = 0.78\,\mathrm{ms}$. $\tau'$ is called the effective length of the selective RF pulse. It is equal to the length of a block pulse with an RF field strength $B_1(0)$ that would result in the same flip angle if there were no gradient field. For a 90° pulse of length $\tau' = 0.78\,\mathrm{ms}$, $B_1(0) = 7.4\,\mu\mathrm{T}$.

The sinc pulse in principle extends from $-\infty < t < \infty$ and has an infinite number of side lobes on both sides of the main lobe. Truncation is therefore necessary to restrict the time duration of the pulse. The penalty paid is that the slice profile deteriorates, but depending on the requirements of the sequence this trade-off must be made. In practice, for a very good slice profile one needs the main lobe and at least two side lobes on each side, resulting in our example in a pulse with a duration of $6 \times 0.78\,\mathrm{ms} = 4.7\,\mathrm{ms}$. This time is too long for fast imaging and here a compromise between the quality of the slice profile and the length of the pulse is made. For example, one could use the main lobe and only a single side lobe preceding the main lobe, resulting in a short RF pulse of 2.3 ms, but in a slice profile which is far from ideal. The consequences of imperfect slice profiles are studied in image sets II-11 and II-12.

The abrupt truncation is often smoothed by multiplying the sinc function (or the large tip-angle pulse derived from a numerical calculation) with a Gauss function $\exp(-at^2)$.

An important note must be made at the end of this discussion: although in the mathematical description the selection gradient was assumed to be in the $z$ direction, it is in practice not necessary to let this direction coincide with the direction of the main magnetic field, the $z$ axis. The direction of the selection gradient can be freely chosen.

### 2.3.3 Other RF Pulses

The RF pulses described in Sect. 2.3.2 are formally only correct for small flip angles. However, it has been shown [4] that for larger flip angles an analytical theory is also possible (formally up to about 30°), which is still reasonably correct up to 90°. For flip angles larger than 90° an analytical solution is no longer possible and the RF pulses used in practice often have a more complex and non-intuitive numerical design, using successive approximations and resulting in a large variety of RF pulses with robust performance [5]. In the design of such RF pulses one can try to improve the performance of the pulse for such aspects as the slice profile, phase coherence, robustness for deviations in $\vec{B}_1$, etc. The modern way of designing pulses relies on an efficient forward calculation of the RF pulse shape and has been proposed by Shinnar and la Roux. A practical survey of their theory is presented in [6].

It is sometimes necessary to restrict the power of an RF pulse designed with a constant selection gradient as described above; for example, to protect against a too-large power deposition in the patient. This can be done by allowing the selection gradient to change in time so as to form a variable rate selective excitation (VERSE) pulse [7]. Looking at (2.17), we note that if during the pulse the gradient field strength is lowered by a factor of $\alpha(< 1)$, while at the same time the RF field strength is lowered by the same factor, so that their quotient remains constant, the only essential difference in the integral comes from the fact that $k$ is also lowered by the same factor. This can be compensated for by replacing $t$ by $\alpha^{-1}t$, which means that the time is stretched by a factor $\alpha^{-1}(> 1)$; see Fig. 2.3. So when an RF pulse requires too much power, in view of safety requirements, one can always lower the peak amplitude and the average power of the RF field at the cost of time. The interesting fact is that the peak power is reduced quadratically and the time is increased linearly so the dissipated power is decreased linearly with the time increase in a VERSE pulse. If a pulse design with a constant gradient

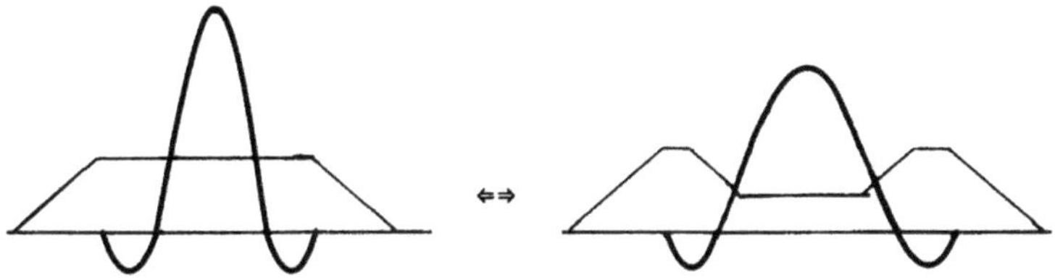

**Fig. 2.3.** Gradient and RF waveform for pulses with identical slice selection

field is available it can always be scaled to a VERSE pulse. It is expected that this argument can also be used for large flip angles.

Although most RF pulses in practical use are spatially selective in one direction only, this is by no means a principal limitation. An example of an RF pulse that can be used to excite or saturate a cylindrical region instead of a slice will be given in Sect. 3.7. That particular design is possibly useful in the imaging of the coronary arteries for suppressing the signal of the aorta blood or to reduce the field of view. Its description also gives an idea of what is required from the gradient system for two-dimensional pulses. We will, however, postpone our discussion until the concept of "$\vec{k}$ space" is introduced in connection with imaging and close our treatment on excitation temporarily (until Sect. 3.7).

### 2.3.4 Power Dissipation in an RF Pulse

In order to find the power delivered by the RF pulse we use the formal definition of the quality of the transmit coil (loaded $Q$ of the coil with patient):

$$Q = \frac{2\pi \,\text{Stored energy}}{\text{Dissipation per cycle}} = \frac{2\pi E_{\text{st}}}{P_{\text{c}}}. \tag{2.20}$$

The stored energy in a coil is $E_{\text{st}} = 1/2 \int \vec{B}(\vec{r}) \cdot \vec{H}(\vec{r})\text{d}\vec{r}$. We now assume a fictitious coil with a homogeneous RF field, $B_{\text{RF}}$, in a volume $V_{\text{eff}}$. The value of $B_{\text{RF}}$ in this fictitious coil is taken to equal the magnetic field strength in the active part of the actual coil. We now equate the stored energy of the actual coil with that of the fictitious coil ($E_{\text{st}} = 1/2 B_{\text{RF}}^2 V_{\text{eff}}/\mu_0$) and find a value for $V_{\text{eff}}$, the effective volume of the actual coil. All coils are characterized (for this type of calculations on power and also on noise) by their quality $Q$ and the effective volume $V_{\text{eff}}$. The dissipation per second is $P_{\text{diss}}\tau' = \omega/2\pi\tau' P_{\text{c}}$, where $\omega_0$ is the angular frequency of the RF field; so for the effective time of a pulse, the dissipation during one single pulse is $P_{\text{pulse}} = P_{\text{diss}}\tau' = \omega/2\pi\tau' P_{\text{c}}$. For (2.20) we can therefore write

$$Q = \frac{\omega_0 B_{\text{RF}}^2 V_{\text{eff}}}{2\mu_0 P_{\text{diss}}}, \quad \text{or} \quad P_{\text{diss}} = \frac{\omega_0 B_{\text{RF}}^2 V_{\text{eff}}}{2\mu_0 Q}. \tag{2.21}$$

As a practical example, we assume that we have an 180° pulse, which means that $\gamma B_{\text{RF}}\tau' = \pi$. Taking the example of Sect. 2.3.2 ($\tau' = 0.78\,\text{ms}$), this yields $B_{RF} = 15.5\,\mu T$. Note that this is the $B_{RF}$ field in the rotating system of reference. To find the linear field that can produce such a rotating field we need twice the field strength, so $B_L = 2B_{RF} = 31\,\mu T$. So to find the power needed we must introduce $B_1$ instead of $B_{RF}$ into (2.21). This yields, for a coil with quality factor $Q$ equal to 40 and an effective volume of $1\,m^3$ (body coil) at 1.5 T,

$$P_{diss} = \frac{4\pi^2 \omega_0 V_{eff}}{2\mu_0 \gamma^2 Q \tau'^2} = 3.8\,kW, \tag{2.22}$$

where $P_{diss}$ is the peak power the RF amplifier must be able to deliver to the coil. Note that there are always losses in the circuitry (say 1 dB), so the real power to be delivered by the amplifier is 4.8 kW. The average power to be delivered by the power amplifier is $P_{pulse}N = P_{diss}\tau' N$ ($N$ is the number of pulses per second). For a turbo spin-echo sequence (described in Sect. 3.2), with an 180° echo pulse every 12 ms, this means an average power of 322 W (including circuit losses). For a 5 kW peak power amplifier with 5% duty cycle, we find 250 W, which is not enough for this sequence (we have to increase the pulse distance to 15 ms).

As has been explained in Sect. 1.3.4.1, one can also use quadrature coils which only excite the wanted circularly polarized field. In this case for the pulse described above, two linear fields with an amplitude of $15.5\,\mu T$ and with 90° phase shift are needed. The power of each of these two fields is a quarter of the power of the linear field, so the total peak power needed is 1.9 kW excluding losses. The average power needed for the turbo spin-echo sequence, mentioned above, now is 161 W, which is within the possibilities of the power amplifier. In all modern systems the transmit coil is therefore a quadrature coil.

Part of the power is dissipated in the patient. This part can be calculated from the $Q$ of the coil with the patient and the $Q$ of the empty coil $Q_e$. The relation is

$$P_{patient} = P_{diss}(1 - Q/Q_e).$$

In many practical cases, more than half of the RF power is dissipated in the patient. This RF dissipation has to be controlled for safety reasons. Present safety rules, for example IEC, impose a limit to $P_{patient}$ [8]. Since modern systems are capable of surpassing these limits, they are equipped with automatic controls to keep the dissipated power in compliance with the legal limits.

## 2.4 The Spin-Echo Imaging Sequence

We shall now describe the well-known Spin-Echo sequence [9] in detail, using an approximation that is commonly used but seldom stated explicitly. Although we have seen that the durations of the RF pulses, necessary for the

Spin-Echo sequence, are several milliseconds, for the spin system we assume the rotation is around the $x'$ axis and occurs instantaneously at the moment of maximum RF field. This rotation is then described by the rotation matrix given in (2.13). For ease of argument for the moment we also neglect the relaxation; its effect can easily be added afterwards. During our treatment we shall meet concepts such as sampling, the Fourier transform, $\vec{k}$ space, etc, which are much more general than within the context of spin-echo; we generalize in a later chapter.

Let us first describe what happens in a Spin-Echo sequence. As usual we describe the course of the magnetization vectors in the rotating frame of reference, as shown in Fig. 2.4. At first the longitudinal magnetization (along the $B_0$ field) is rotated by a 90° pulse into the $(x', y')$ plane. Then precession occurs under the influence of a gradient field $G_x x$, where $x$ is the read-out direction (also called the measuring direction or frequency-encoding direction), and $G_y y$, where $y$ is the phase-encoding direction (sometimes called the preparation or evolution direction). Note that this precession is position-dependent. This is shown for $G_y = 0$ and for several $x$ positions (numbered 1 to 5) in Fig. 2.4. This precession is described by (2.10a). A 180° pulse subsequently rotates the magnetization around the $x'$ (or $y'$) axis, after which the measuring gradient is switched on again. The precession continues in the same direction until (for zero phase-encode gradient) the spins meet at the $-y'$ axis to form the maximum signal, the "echo top", after a time equal to the time difference between the 90° and 180° pulses. (When a phase-encoding gradient wave is added to the sequence the spins from an element d$y$ meet at another direction having a deviation from the $-y'$ axis that depends on the strength of this gradient.) The spin-echo measuring sequence is shown in Fig. 2.5, in which the gradient waveforms are shown as a function of time. The duration of the RF pulses is neglected (see first paragraph of this section). We shall use that approximation to describe mathematically the spin response during the spin-echo sequence.

The total transverse magnetization of an excited slice (perpendicular to the $z$ axis, the direction of the magnetic field) can be written as:

$$M_{\mathrm{T}}(t) = \iint_{\text{slice}} m(x,y) \exp\left(-j\int \omega(x,y,t)\mathrm{d}t\right) \mathrm{d}x\mathrm{d}y, \tag{2.23}$$

where $m(x, y)$ is the distribution of the magnetization over the slice at the time just after the excitation. For the moment it is assumed that an ideal situation exists, so there is no decay of the magnetization due to $T_2$ relaxation. Also, field inhomogeneity $\delta B$ is still neglected. The only deviation from $B_0$ comes from the presence of the gradients. The signal received is proportional to the transverse magnetization. Since we have assumed that the RF pulses have zero duration in this model, we can follow the time evolution of the transverse magnetization by substituting (see (2.10)):

$$\omega(x,y,t) = \gamma\vec{r}\cdot\vec{G}(t).$$

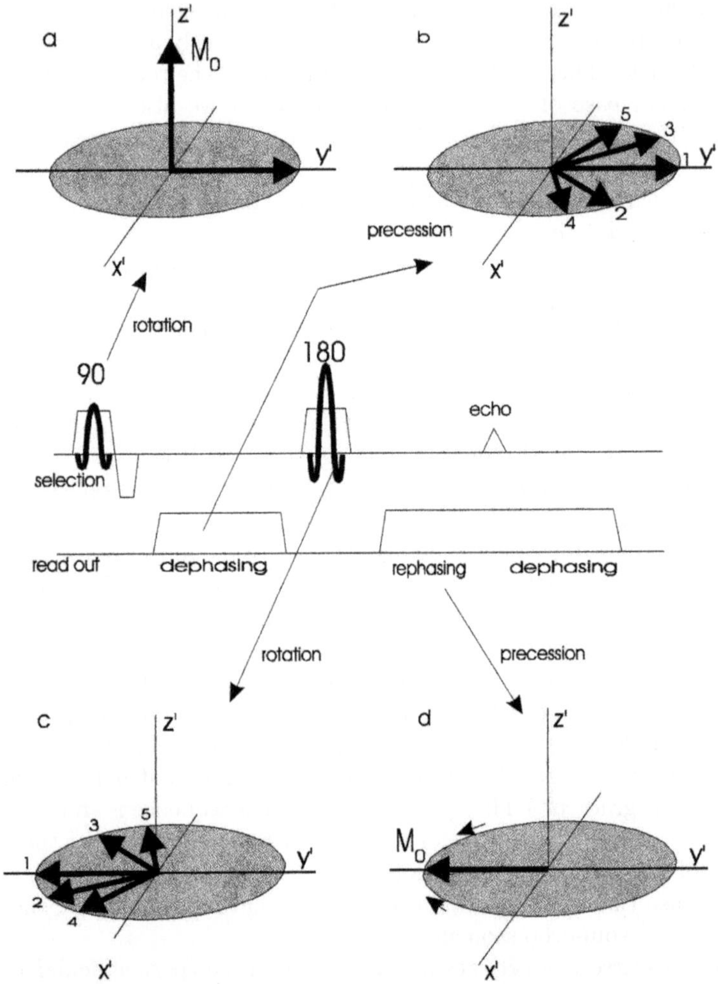


**Fig. 2.4.** Spin-Echo sequence. **a** Rotation. **b** Precession, dephasing. **c** Rotation. **d** Precession, rephasing

We shall consider the evolution of the exponent in (2.23) for the gradients as depicted in Fig. 2.5. So we can write for the exponent (the phase $\theta(t)$)

$$\begin{aligned}\theta(\mathrm{TE}) &= \int_0^{\mathrm{TE}} \omega(x,y,t)\mathrm{d}t \\ &= -\gamma\left\{z\int_0^{\mathrm{TE}/2} G_z(t)\mathrm{d}t + y\int_0^{\mathrm{TE}/2} G_{yn}(t)\mathrm{d}t + x\int_0^{\mathrm{TE}/2} G_x(t)\mathrm{d}t\right\} \\ &\quad +\gamma z\int_{\mathrm{TE}/2}^{\mathrm{TE}} G_z(t)\mathrm{d}t + \gamma x\int_{\mathrm{TE}/2}^{\mathrm{TE}} G_x(t)\mathrm{d}t, \qquad (2.24)\end{aligned}$$

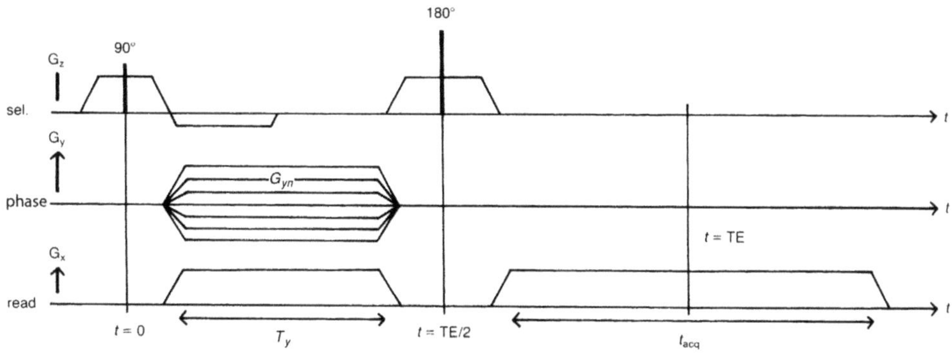


**Fig. 2.5.** Waveforms of a Spin-Echo sequence

where the minus sign is due to the 180° pulse. Each of the integrals describes the area under the gradient-vs-time curve. It is seen from Fig. 2.5 that the terms with the $z$ gradients disappear (remember that the refocussing gradient lobe, associated with the slice selection pulse, has half the surface of the selection gradient; see (2.16)). The term with the $y$ gradient remains and the $x$-gradient waves are adjusted so that the area under the $G_x(t)$ curve is zero at $t = \mathrm{TE}$. This moment is chosen because it is the moment at which the spin-echo pulse refocusses the dephasing of static-field inhomogeneities. Introducing a relative time, measured with respect to TE, $t' = t - \mathrm{TE}$, we find for (2.24)

$$\int_0^t \omega(x,y,t)\mathrm{d}t = \gamma G_{yn} T_y y + \gamma G_x t' x = k_y y + k_x x, \tag{2.25}$$

where the assumption has been made that $G_x$ and $G_y$ have constant strength during the integrations. Clearly we defined

$$k_y = \gamma G_{yn} T_y = \gamma \int_0^{t'} G_{yn}(t)\mathrm{d}t$$

and

$$k_x = \gamma G_x t' = \gamma \int_0^{t'} G_x(t)\mathrm{d}t. \tag{2.26}$$

The equation for the total transverse magnetization in a slice (which for simplicity we took to be an $(x, y)$ plane) now reads:

$$M_{\mathrm{T}}(t') = \iint_{xy} m(x,y) \exp(-j(k_x x + k_y y))\mathrm{d}x\mathrm{d}y. \tag{2.27}$$

This is the fundamental equation for MR imaging (2D acquisition) and in the next section we shall describe how to use it. Notice that $k_x$ and $k_y$ are functions of $t$ and have the dimension $m^{-1}$. Again it must be noted that although we used, for ease of argument, $x$, $y$, and $z$ for the read-out, phase-encode, and selection directions respectively, in practice these directions do not have to coincide with the main directions of the system.

### 2.4.1 The $\vec{k}$ Plane

Equation (2.27) expresses a function in $t$ in a double integral, having the form of a 2-dimensional Fourier transform. At each moment in time, $t$, we know the values of $k_x(t)$ and $k_y(t)$, according to (2.26), so we can write $M_\mathrm{T}$ in the form of $M_\mathrm{T}(k_x, k_y)$. Now (2.27) defines the Fourier pair $m(x, y)$ and $M(k_x, k_y)$, which images the function $m(x, y)$ in the $x, y$ plane on a function $M_\mathrm{T}(k_x, k_y)$ in the $k_x$, $k_y$ plane [10]. As we see in (2.27) we measure a function of $t$, so we artificially have to arrange that we measure it at all necessary $k_x$ and $k_y$ values (which lie for a Fourier transform on a rectangular grid in the $\vec{k}$ plane) and can perform the Fourier transformation

$$m(x,y) = \frac{1}{2\pi} \int_{k_x} \int_{k_y} M_\mathrm{T}(k_x, k_y) \exp(+j(k_x x + k_y y)) \mathrm{d}k_x \mathrm{d}k_y. \qquad (2.28)$$

This transformation yields the required distribution of the magnetization in the selected slice at the time of measurement. Actually with the spin-echo method, described in the previous section, all the values of $M_\mathrm{T}(k_x, k_y)$ necessary for reconstruction are acquired. Let us see what happens with the $k_x$ and $k_y$ values in the sequence shown in Fig. 2.5. Between $t = 0$ and $t = \mathrm{TE}/2$ both $k_x$ and $k_y$ increase linearly (see (2.26)) and $\vec{k}$ moves from O to A in Fig. 2.6. Since the 180° pulse inverts the sign of the phase, so the signs of $k_x$ and $k_y$, the point reached when $t = \mathrm{TE}/2$ is mirrored through the origin (point B in the figure). Then the read-out gradient is switched on and a line with constant $k_y$ (line BC) is described in the $k$ plane. During the time that the $\vec{k}$ vector moves from B to C, $t_\mathrm{acq}$, the data is acquired for constant $k_y$. The acquired data for a line in the $\vec{k}$ plane with constant $k_y$ is called a "profile". The spin system is now given time to relax back to the equilibrium situation and the sequence is repeated, but now with another phase-encode gradient, so that another horizontal line in the $\vec{k}$ plane is acquired. This is repeated until the full $\vec{k}$ plane is scanned ("raw data").

Let us think about the physical meaning of the $k$ values. The exponential in (2.28) is periodic, since $\exp(jk_x x)$ repeats itself when $k_x x$ increases by $2\pi$. This means that the $k$ value describes a wavelength: $\lambda = 2\pi/k$. This is the wavelength of the spatial harmonics in which the real object is decomposed. The highest $k$ value, $k_{x,\mathrm{max}}$ (see Fig. 2.6), is given by

$$k_{x,\mathrm{max}} = \gamma G_x t_\mathrm{acq}/2 = \gamma G_x t_\mathrm{s} N/2, \qquad (2.29)$$

where $N$ is the number of samples per profile. The highest $k$ value determines the smallest wavelength and thus the resolution. The largest wavelength occurring in the $x$ direction is equal to the field of view (FOV):

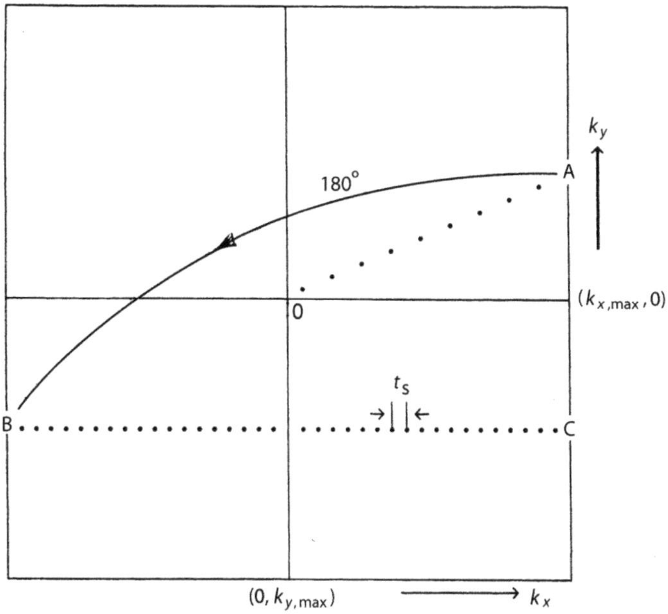


**Fig. 2.6.** $\vec{k}$-plane trajectory of a spin echo. Variable $t_s$ is the sampling time and $N$ the number of samples per profile

$$\lambda_{\max} = 2\pi k_{x,\min}^{-1} = \frac{2\pi}{\gamma G_x t_s} = \mathrm{FOV}, \tag{2.30}$$

as can be expected. The smallest wavelength is therefore $\lambda_{\min} = 2\,\mathrm{FOV}/N$. The resolution is $1/\lambda_{\min}$.

### 2.4.1.1. Discrete Sampling

In the preceding paragraph the transverse magnetization is described in the rotating frame of reference. The physical phenomenon that shows the same behavior is the voltage at the output of the receiver chain, after demodulation with the gyromagnetic frequency. This signal is sampled at a finite sampling rate, so that per profile a discrete set of $N$ samples is obtained.

We shall now discuss the consequences of the discrete sampling. First we consider the discrete sampling during acquisition in the read-out (frequency-encoding) direction. The signal $M_{\mathrm{T}}(k_x, k_y = \text{const})$, where $k_x = \gamma G_x t$, is sampled (acquired) during the time $t_{\mathrm{acq}}$, in which $\vec{k}$ travels (see Fig. 2.6) from B to C. Let us describe what this means: since we only sample the signal at discrete times it is as if the signal $M_{\mathrm{T}}(t)$ is multiplied by a series of $\delta$ functions with distance $t_{\mathrm{s}}$, which stands for the sampling time. So

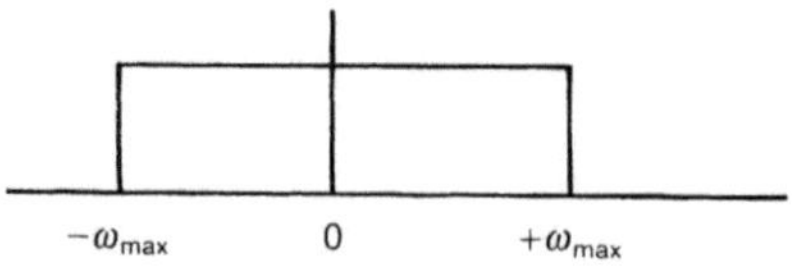


**Fig. 2.7.** Frequency spectrum of the received signal

$$M'_{\mathrm{T}}(t) = \sum_{-\infty}^{\infty} M_{\mathrm{T}}(t)\delta(t - nt_{\mathrm{s}}), \tag{2.31}$$

where, according to the definition of the $\delta$ function,

$$\int_{-\infty}^{\infty} M_{\mathrm{T}}(t)\delta(t - nt_{\mathrm{s}})\mathrm{d}t = M_{\mathrm{T}}(nt_{\mathrm{s}}).$$

Now it is known that the Fourier transform of a series of $\delta$ functions in the time domain is another series of $\delta$ functions in the frequency domain. The Fourier pair is (the reader can check this by considering the sequence of $\delta$ functions as a periodic repetition of a single $\delta$ pulse and by finding the Fourier series of this periodic function)

$$\sum_{n=-\infty}^{\infty} \delta(t - nt_{\mathrm{s}}) \circ\!\!\overset{\mathfrak{F}}{-\!\!-\!\!-}\!\!\circ \frac{2\pi}{t_{\mathrm{s}}} \sum_{m=-\infty}^{\infty} \delta\left(\omega - \frac{2\pi m}{t_{\mathrm{s}}}\right). \tag{2.31a}$$

This means that the frequency spectrum of a series of $\delta$ functions consists of all harmonics of $2\pi/t_{\mathrm{s}}$, all having equal amplitude (note also the term with $m = 0$). The frequency spectrum of the signal $M_{\mathrm{T}}(\omega)$ (see Fig. 2.7) contains frequencies (in the case of quadrature detection) between zero and $\omega_{\mathrm{max}} = \gamma G_x\,\mathrm{FOV}/2$, where FOV stands for field of view. Due to the discrete sampling in the time domain it will be convolved with the harmonics of the sampling frequency:

$$M_{\mathrm{T}}(t) \sum_{n=-\infty}^{\infty} \delta(t - nt_{\mathrm{s}}) \circ\!\!\overset{\mathfrak{F}}{-\!\!-\!\!-}\!\!\circ m_{\mathrm{T}}(\omega)\frac{2\pi}{t_{\mathrm{s}}} \otimes \sum_{m=-\infty}^{\infty} \delta\left(\omega - \frac{2\pi m}{t_{\mathrm{s}}}\right), \tag{2.31b}$$

where $\otimes$ means convolution. Figure 2.8 shows the resulting spectrum. When $\omega$ becomes larger than $\pi/t_{\mathrm{s}}$ these different frequency bands will overlap. Since $x$ runs from $-\mathrm{FOV}/2$ to $+\mathrm{FOV}/2$ when $\omega$ runs from $-\omega_{\mathrm{max}}$ to $+\omega_{\mathrm{max}}$ we see that in the image, $m(x, y)$, the left-hand and right-hand side start to overlap (aliasing) as shown in Fig. 2.9, see also image set II-7. To avoid aliasing we must always have

$$\omega_{\mathrm{max}} = \gamma G_x\,\mathrm{FOV}/2 \leq \frac{\pi}{t_{\mathrm{s}}}.$$

Or, when we use a sampling frequency as low as possible ($\leq$ becomes $=$), we can write this in another useful way (writing $\omega = 2\pi\nu$):

$$\gamma G_x\,\mathrm{FOV} = \frac{1}{t_{\mathrm{s}}}, \quad \text{or} \quad 2\nu_{\mathrm{max}} = \frac{1}{t_{\mathrm{s}}} = \frac{N}{t_{\mathrm{acq}}} = \nu_{\mathrm{s}} \tag{2.32}$$

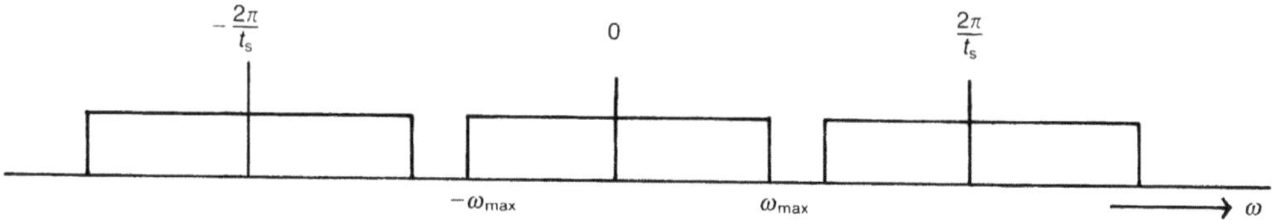


**Fig. 2.8.** Convolution of the frequency band of Fig. 2.7 with $\delta(\omega - 2\pi m/t_s)$

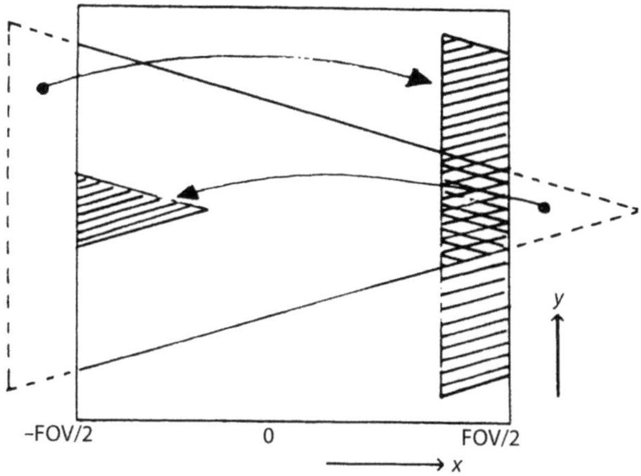


**Fig. 2.9.** Image of a triangular object with aliasing

This states that the sampling frequency must be at least twice the maximum signal frequency. It is one of the most important fundamental relations in magnetic resonance imaging and is called "the Nyquist sampling theorem". Aliasing can be avoided by choosing $t_s$ very short (oversampling). Note that the noise in the image also depends on $t_s$, as discussed in Sect. 6.5.

Before we draw conclusions we have to look at the transform in the $y$-direction. We wish to make this transformation fully symmetric with the $x$ transformation. Now we know that during acquisition of a profile ($x$ direction) the maximum frequency of the spins in the rotating frame of reference is half the sampling frequency. So between two sampling points the phase of the spins at the edge of the field of view advances 180°. We apply the same reasoning to the gradient $G_{yn}$ and determine its values for the adjacent profiles $n$ and $n-1$, (acquired $M_T$ on a line BC in Fig. 2.6) in such a way that

$$\gamma \left\{ G_{y,n} - G_{y,(n-1)} \right\} T_y \mathrm{FOV}/2 = \pi . \qquad (2.33)$$

This directly yields the highest value of $k_y$ required for a scan. When an image with matrix $N^2$ is taken, we need in principle $N$ profiles of which $N/2$ profiles are measured with $G_{yn}$ positive and $N/2$ with $G_{yn}$ negative, so:

$$\gamma G_{y,\max} T_y \mathrm{FOV} = \pi N . \qquad (2.34)$$

Actually we now have all information needed to design a Spin-Echo sequence. This will be done in the next section, but before doing so we have to correct one omission in our discussion on imaging, namely the finite length of the acquisition time of a profile.

#### 2.4.1.2. Sampling Point-Spread Function

A last consequence of our measuring method is that the acquisition time, $t_{\mathrm{acq}}$, for one "profile" (the "echo" for a certain value $k_y$) is not unlimited, as is inherently assumed. In reality the acquisition time is finite as shown by $t_{\mathrm{acq}} = Nt_{\mathrm{s}}$ ($t_{\mathrm{s}}$ is sampling time) and this gives rise to blurring described by the "sampling point-spread function". We can describe this mathematically by multiplying the acquired $M_{\mathrm{T}}'(t)$ (see (2.31)) with a unit step function $U(t)$, given by $U(t) = 1$, when $-1/2t_{\mathrm{acq}} < t < 1/2t_{\mathrm{acq}}$, otherwise $U(t) = 0$. The Fourier transform of such a unit step function is

$$U(t_{\mathrm{acq}}) \circ\!\!\xrightarrow{\;\mathfrak{F}\;}\!\!\circ\; t_{\mathrm{acq}} \frac{\sin(\gamma G_x x t_{\mathrm{acq}}/2)}{\gamma G_x x t_{\mathrm{acq}}/2}. \tag{2.35}$$

When we have in the time domain a multiplication with the function $U(t)$, it means that in the object space (frequency domain: $\omega = \gamma G_x x$) we have a convolution with a sinc function. Let us consider a point object (in the form of a $\delta$ function) in the origin, which means that in the $\vec{k}$ plane we have equal signals at every sample point. Due to the finite sampling time, the $\delta$-function (point) object is imaged as a sinc function in the origin of the $(x, y)$ plane. This function has a half maximum width, given by $\gamma G_x \Delta x t_{\mathrm{acq}} = 1.2\pi$, or

$$\Delta x = 1.2\pi/(\gamma G_x t_{\mathrm{acq}}). \tag{2.36}$$

A point-like object is imaged as a sinc function with a certain effective width. The function is called the "sampling point-spread function". Note that $\Delta x$ is *not* the width of a pixel! The width of a pixel is FOV/$N$, which determines the resolution.

#### 2.4.1.3. Thinking in Terms of $\vec{k}$ Space

In this interlude we shall recapitulate what we have learned about $\vec{k}$ space by making some schematic examples. In Fig. 2.10 an overview of SE scanning is shown. The acquisition process yields the sampling matrix $M_{\mathrm{T}}(k_x, k_y)$, according to the rules described in the previous section. When the $\vec{k}$ plane is filled with sampled values of $M_{\mathrm{T}}(k_x, k_y)$, as shown in perspective in Fig. 2.10, fast Fourier transform yields the image $m(x, y)$. Now it is very instructive to do some simple transforms from $\vec{k}$ space to the image or vise versa. We look at the object plane to be imaged, as depicted in Fig. 2.11. First we consider a very small object in point $(-x_2, 0)$. We know that the phase of the signal

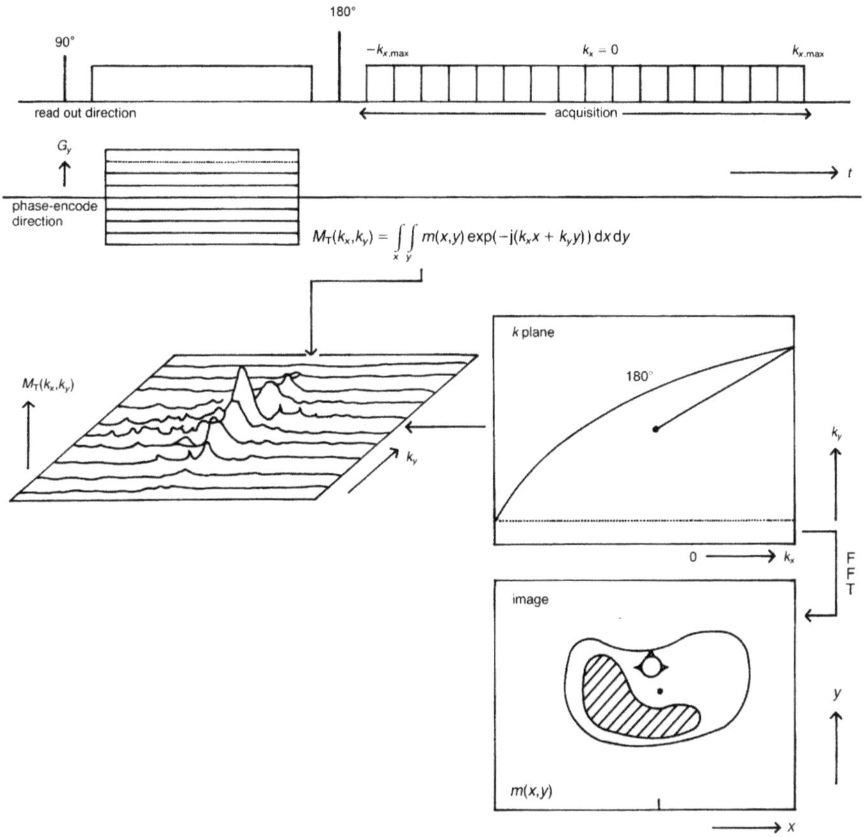


**Fig. 2.10.** Overview of the imaging process for a SE sequence

from this object in the rotating frame of reference is $\gamma G_x x_2 t = k_x x_2$. So the signal is given by

$$M_\mathrm{T}(k_x, k_y) = M_\mathrm{T} \exp(\mathrm{j} k_x x_2) = M_\mathrm{T}[\cos(k_x x_2) + \mathrm{j} \sin(k_x x_2)],$$

where $M_\mathrm{T}$ is the total equilibrium magnetization of the object. The dimension of the object in the $y$ direction is zero and also its $y$ coordinate is zero. Therefore a preparation gradient has no influence and $M_\mathrm{T}(k_x, k_y)$ is constant in the $k_y$ direction. In the $k_x$ direction we see an harmonic function with a wavelength of $2\pi/x_2$. When the object moves to the origin the wavelength becomes longer, because $x$ becomes shorter (see Fig. 2.11b). Finally, when the object is located in the origin the wavelength in $\vec{k}$ space becomes infinite, so $M_\mathrm{T}(k_x, k_y)$ becomes constant, as shown in Fig. 2.11c. The transformation from object space to image space is clearly a 2D Fourier transformation which could be performed in this simple case without the formal mathematical methods.

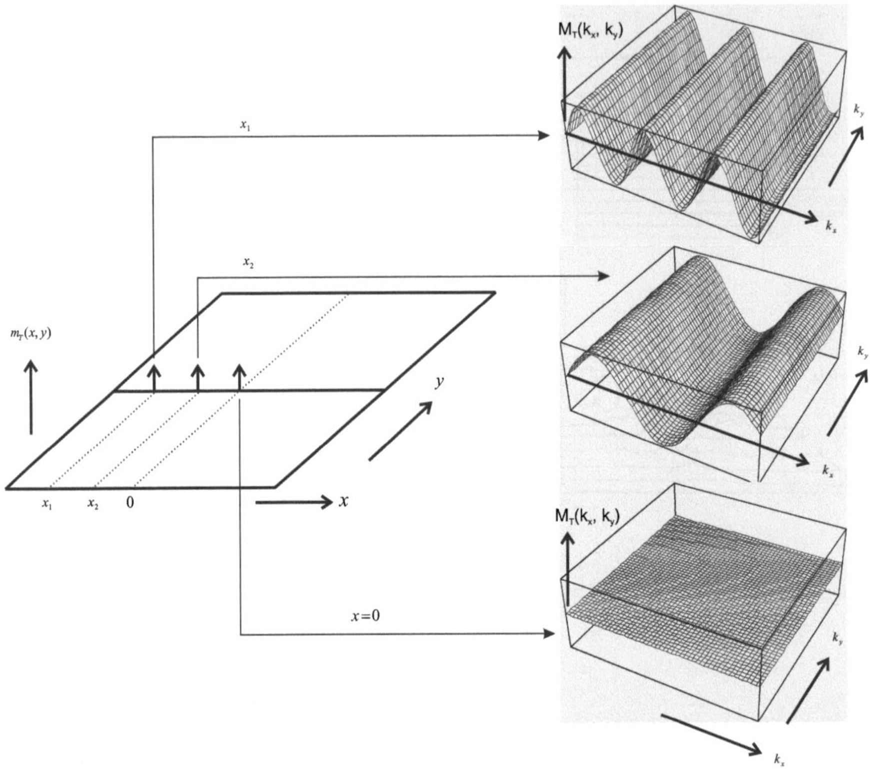


**Fig. 2.11.** Imaging a very small object at several positions along the $x$ axis

So far we have discussed only objects located on the line $y = 0$. When we consider a $\delta$-like object at a point $(x_2, y_2)$ (see Fig. 2.12), we find on the line $k_y = 0$ exactly the same profile as shown for the object in $(x_y, y = 0)$ in Fig. 2.11. However, if we measure another profile in the $k$ plane we must apply a preparation gradient before the refocussing (180°) pulse, so this profile will have a phase shift with respect to the $k_y = 0$ profile and is shifted to the right. As the shift is linearly dependent on $y_2 : \varphi_y = \gamma G_y T_y y_2$ ($G_y$ is the preparation gradient and $T_y$ is its duration), the Fourier transform of the object is again a sine wave as in the case of an object on the $y = 0$ line, but now rotated in a direction with coefficient $\tan\alpha = y/x$ (see Fig. 2.12). For one single object we need only one single preparation gradient, but in practical cases we must distinguish more "objects" (= points) on the line $x_2 =$ constant, so we need more different preparation gradients. For 256 points we need 256 preparation gradients and the measurement must be repeated 256 times to fill all necessary points on a cartesian grid in $\vec{k}$ space. The grid of all $k_x$–$k_y$ values necessary for the image is called "the acquisition matrix", sometimes abbreviated to "matrix".

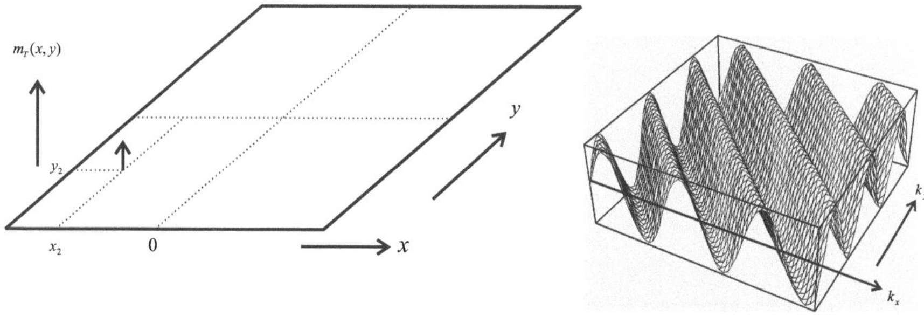


**Fig. 2.12.** The function $M_{\mathrm{T}}(k_x,\, k_y)$ for a $\delta$-like object in $(x_2, y_2)$

Until now we have discussed only $\delta$-function-type objects. More complicated objects can be thought of as being built up from $\delta$ functions. As an exercise, in Fig. 2.13 we looked at objects built up of two, three, five, and nine $\delta$ functions at locations along the $x$ axis. In the $\vec{k}$ space we know that there is no $k_y$ dependence so only the $k_x$ dependence is shown. For two small objects one clearly sees the "beat" of the two frequencies, for nine small objects one can recognize the sinc function, which we have already met in the theory of RF pulses and the description of point spreading.

In Sect. 2.4.1.1 it was shown that the $k$-plane is scanned by acquiring $N_y$ profiles with $N_x$ sampling points. This takes a long time, since for $T_2$ weighted Spin-Echo sequences the equilibrium situation must be restored before a new profile can be acquired. When we assume that the time between the excitation pulses is 1 s, the total scan time for 256 profiles is about 4.5 minutes. So there will always be a request for faster imaging, especially for monitoring dynamic situations.

The simplest way to shorten the scan time is to use the fact that the measured magnetization in the $\vec{k}$ plane is in general located in the centre of the $\vec{k}$ plane (the exception being in the unusual case when very small objects are imaged, such as the examples of Figs. 2.11 and 2.12). That means that the measurement of the outermost profiles can be omitted and the values taken as zero (zero filling). This technique is named "reduced matrix acquisition" and a reduction up to 30% is frequently used. It has, of course, consequences for the resolution and for the signal-to-noise ratio, which is increased due to the fact that also the noise of the outer profiles is made zero (see Sect. 6.7).

Another method to reduce the number of acquired profiles uses the fact that the value of the magnetization in points symmetrical to the origin of the $k$-plane is complex-conjugated. In this case, for example, the profiles from $-N_y/2$ to $+N_y/8$ (see Fig. 6.9) are acquired (to the 62.5% level). The magnetization profiles for $+N_y/8 < k_y < +N_y/2$ are obtained by taking in each sampling point the complex conjugate of the the magnetization measured in points mirrored through the origin of the $\vec{k}$ plane ("half matrix acquisition"). Here we do not sacrifice resolution, but still the signal-to-noise ratio

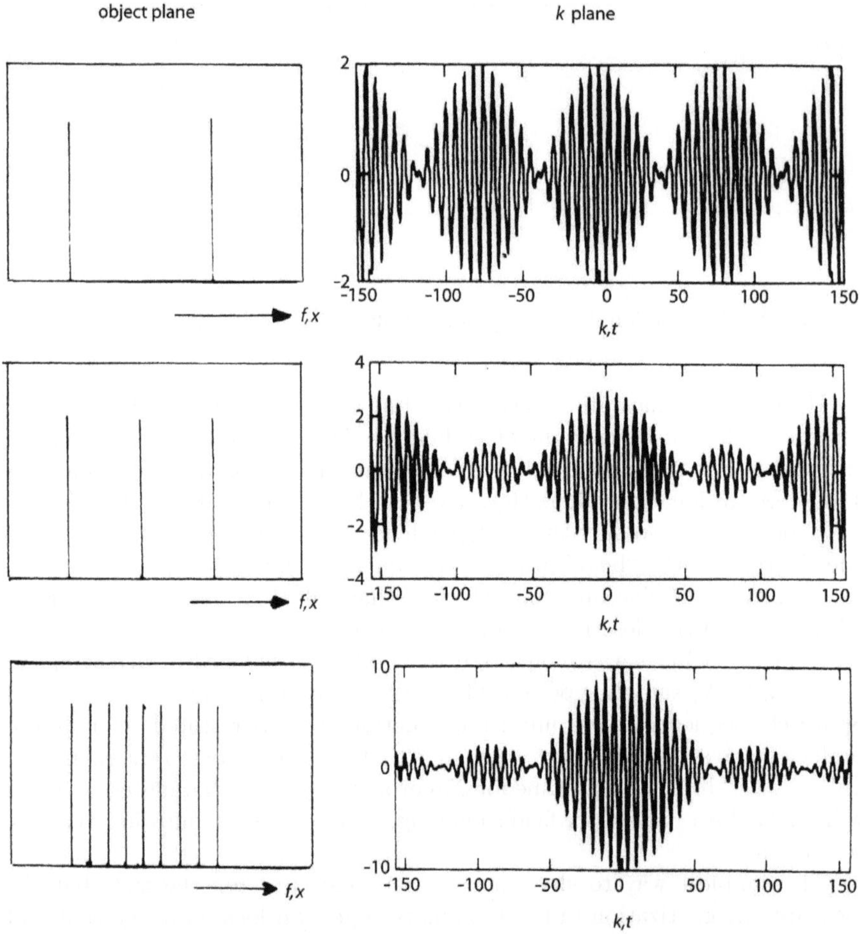


**Fig. 2.13.** One-dimensional Fourier transform of objects composed of $\delta$ functions

is decreased, as will be explained in Sect. 6.7. An example of images obtained with reduced matrix acquisition is given in image set II-13. In the same image set, an example is also given in which, based on prior knowledge of the image, certain $k_y$ profiles are not acquired. The method leads to very acceptable images, although their reconstruction becomes more complicated (see also RIGR [11]). In image set II-14, a recent method to decrease scan time (named SENSE [21]), based on combining the signals from several receiving coils, is described.

There are many other ways to decrease the scan time (or to improve the time resolution in a dynamic scan). An example is "keyhole imaging" [12] used in dynamic imaging after contrast injection. At first all $k_y$ profiles of the image before administration of a contrast agent are measured. After injection of the contrast agent, only the profiles with low $k_y$ values are updated and subsequently completed with the higher $k_y$ profiles of the first image, so as to follow changes in contrast as a function of time.

In cases in which a short echo time is required, we can choose to measure only (for example) 62.5% of each profile – see Fig. 6.10 ("half echo acquisition"). Then the dephasing and the rephasing lobes can be made shorter, leading to a short echo time. The profile is that sampled from $-N_x/8$ to $+N_x/2$ and the unacquired points are again obtained by mirroring, see Sect. 6.7.

### 2.4.2 Contrast in Spin-Echo Sequences

So far the Spin-Echo sequence has been discussed as if there were no relaxation effects. That means that the signal is proportional with $M_{\mathrm{T}}(x,y)$, which describes density contrast. However, during the time delay between the excitation and the echo top, the magnetization has decreased by a factor of $\exp(-\mathrm{TE}/T_2)$ because of $T_2$ decay. Since this factor is a function of $T_2$, it causes $T_2$ weighting of the image. When $\mathrm{TE} \ll T_2$ this weighting is not important since the factor $\exp(-\mathrm{TE}/T_2)$ is close to unity for all tissues. When TE is somewhere in the range of relevant $T_2$ values the factor $\exp(-\mathrm{TE}/T_2)$ depends heavily on the value of $T_2$, so $T_2$ weighting is dominant (see image set II-1).

It is also possible to bring $T_1$ weighting in the image, as is shown in image set II-1. This can be done by choosing TR, the repetition time between the excitations, to be smaller than $T_1$. In that case the the longitudinal magnetization does not relax back to the equilibrium situation, but only to $M_0[1-\exp(-\mathrm{TR}/T_1)]$. Again when TR lies within the range of $T_1$ values of tissue (300–1500 ms) the image becomes $T_1$ weighted as well. For a purely $T_1$ weighted image one must choose TE as small as possible. When TE is not small enough one has the combined effect:

$$M_{\mathrm{T}} = M_0[1-\exp(-\mathrm{TR}/T_1)]\exp(-\mathrm{TE}/T_2).$$

Note that for $T_1 \ll \mathrm{TR}$ and $T_2 \ll \mathrm{TE}$, this equation reduces to $M_{\mathrm{T}} = M_0$. The signal magnitude depends only on the proton density $M_0$ (proton density contrast).

### 2.4.3 Scan Parameters and System Design

Let us now collect together all the general relations we have found during our previous discussion on excitation and spin-echo imaging. For the effective time $\tau'$ of a slice-selective excitation with a sinc pulse (which is half the time of the main lobe or equal to the time of a side lobe) we found

$$\boxed{\begin{aligned}&\gamma \mathrm{G_z} d\tau' = 1,\\ &\text{or } G_{\mathrm{z}}[\mathrm{mT/m}]d[\mathrm{mm}]\tau'[\mathrm{ms}] = 23.5[(\mathrm{mT/m})\mathrm{mm\,ms}].\end{aligned}} \qquad (2.19')$$

The power (in Watts) necessary for this pulse is

$$P_{\mathrm{diss}} = \frac{\omega B_{\mathrm{RF}}^2 V_{\mathrm{eff}}}{2\mu_0 Q} = 100 \frac{B_0[\mathrm{T}] B_{\mathrm{RF}}^2[\mu\mathrm{T}] V_{\mathrm{eff}}[\mathrm{m}^3]}{Q}, \tag{2.21$'$}$$

where the RF magnetic field in the active-coil region $B_{\mathrm{RF}}$ is found from the effective length of the pulse, $\tau'$, and the flip angle, $\alpha$, expressed in radians (rads) or as part of a total revolution ($= 2\pi$ rads) called "rev":

$$\begin{aligned} &\gamma B_{\mathrm{RF}} \tau' = \alpha[\mathrm{radians}], \\ &\text{or } B_{\mathrm{RF}}[\mu\mathrm{T}] \tau'[\mathrm{ms}] = 23.5\alpha[\mathrm{rev}]. \end{aligned} \tag{2.37}$$

For good slice selection we need at least an RF pulse duration of $4\tau'$.

For acquisition we found

$$\begin{aligned} &\gamma G_x \mathrm{FOV} t_{\mathrm{s}} = 1; \\ &\text{or } G_x[\mathrm{mT/m}]\mathrm{FOV}[\mathrm{mm}] t_{\mathrm{s}}[\mathrm{ms}] = 23.5[(\mathrm{mT/m})\mathrm{mm\,ms}], \end{aligned} \tag{2.32$'$}$$

from which more relations can be deduced. If we consider an image of $N^2$ pixels, then the acquisition time for a profile must be

$$t_{\mathrm{acq}} = N t_{\mathrm{s}}, \tag{2.38}$$

which can be combined with the previous equation, using the fact that FOV/$N$ is the resolution, $R$, to give

$$G_x[\mathrm{mT/m}] R[\mathrm{mm}] t_{\mathrm{acq}}[\mathrm{ms}] = 23.5[(\mathrm{mT/m})\mathrm{mm\,ms}]. \tag{2.38a}$$

The bandwidth, $\Delta\Omega$, of the signal received is

$$\Delta\Omega = \frac{\omega_{\max}}{2\pi} = \frac{1}{t_{\mathrm{s}}}, \tag{2.39}$$

and so the bandwidth per pixel $\Delta\nu$ is $\Delta\Omega/N$, which is equal to $1/t_{\mathrm{acq}}$. For the phase-encode gradient we finally found

$$\gamma G_{y,\max} T_y \mathrm{FOV} = N/2. \tag{2.34$'$}$$

The resolution is $R = \mathrm{FOV}/N$, so (2.34') can be written as

$$G_{y,\max}[\mathrm{mT/m}] T_y[\mathrm{ms}] R[\mathrm{mm}] = 12[(\mathrm{mT/m})\mathrm{ms\,mm}]. \tag{2.40}$$

These relations are sufficient for calculating the demands on essential performance aspects of an MRI system, once a specified spin-echo method is required. That is nice for future scenarios of system design. In normal life the reasoning is the other way round: given an MRI system, can we perform a certain Spin-Echo sequence with it. For future designs one should even take sequences into account, which are expected to become important in a few years. However, many fundamental relations remain valid, only the parameters change as can be seen in the next chapter.

#### 2.4.3.1. Practical Example

We shall take a practical example. Our system has a maximum RF power of 5 kW; the maximum possible gradient is 10 mT/m with a rise time of 1 ms. We now ask: what is the shortest possible echo time TE of a spin-echo sequence at 1.5 T with a slice thickness of 3 mm, a $128^2$ matrix, and a FOV of 128 mm? The sequence is drawn in Fig. 2.5. The answer to the question is fully specified and is found by adding up the different contributions.

**a.** The length of the RF pulse and the selection gradient, counted from the maximum of the RF pulse, which we assume to have one side lobe on each side, is twice the effective pulse time (half maximum width) $\tau'$. The half maximum width is (see (2.19)) $\tau' = 24/(G_z d) = 0.8$ ms. The selection gradient therefore takes 1.6 ms in the interval $0 < t < \mathrm{TE}/2$.
We have to check whether the available power is sufficient for the echo pulse (the excitation pulse requires much less power since it causes a rotation of only 90°). Assuming the quality of the transmitting coil to be 40, we find from (2.37) that the "rotating" RF magnetic field of the "echo" pulse is 15 $\mu$T, for which we need a linear field of 30 $\mu$T. We need, according to (2.21), 1.8 kW. Although this is far below 5 kW, one must realize that a 1 dB power loss in the leads and the components between the power amplifier and the coil is not unrealistic. The amplifier has to deliver 2.5 kW.

**b.** The effective time, $T_y$, of the phase encoding gradient is given by $T_y = 12/(G_{y,\max} R)$, which (since $R = 1$ mm) is 1.2 ms. However, here one must realize that we have a finite risetime of 1 ms. This makes a trapezoid form with a real length of 2.2 mm. This trapezoid can start when the selection gradient starts its downward slope. At the end of the phase-encode gradient the time elapsed since the center of the excitation pulse is $1.6 + 2.2 = 3.8$ ms. Since no RF power is present, the refocussing lobe of the selection gradient can take place in the same time.

**c.** For the echo pulse we need also 1.6 ms in the interval $0 < t < \mathrm{TE}/2$.

**d.** The total time between excitation and the echo pulse is $TE/2 = 5.4$ ms. It follows that the echo time obtainable with our system for the sequence defined is 10.8 ms.

**e.** We wish, in view of the signal-to-noise ratio, to apply the lowest possible acquisition gradient, this means the longest possible acquisition time.

From what has been said earlier we have from $t$ = TE/2 to $t$ = TE a period of 5.4 ms of which (1.6+1) ms will be used for the rest of the selection gradient and its slope. This leaves us 2.8 ms for half the acquisition time (the other half is at $t >$ TE). So we have $t_{\mathrm{acq}} = 5.6$ ms and according to (2.38a) the measurement gradient required is 4.3 mT/m. The gradient surface of the measurement gradient before $T$ = TE is $4.3 \times 2.8 + 2.1 = 14$(mT/m/ms), the term 2.1 (= 4.3/2) is due to the rise time. For $t <$ TE/2 we have the measurement compensation gradient for which we need 3.2 ms. This is 1 ms more than the time needed for the phase encode gradient, so this increases TE by 2 ms, resulting in TE = 12.8 ms.

As stated earlier the equations give all relevant relations for estimating the essential performance aspects of the MRI system, which must be able to perform a certain required spin-echo scan. In later chapters further application of these relations and their modifications for other and faster scan sequences will be discussed.

### 2.4.4 Multiple-Slice Acquisition

In conventional single-slice spin-echo scans, when a single profile is measured after the excitation of a slice (see line BC in Fig. 2.6) for that slice, one has to wait until the spins are in equilibrium again. This waiting time is about $T_1$ s, which is 1 s or more. On the other hand the technical minimum time between excitations is the time needed to collect one profile. It is the echo time plus the second half of the acquisition time and the first half of the selection pulse (see Fig. 2.5). From the previous section it can be found that the profile time is $12.8 + 2.8 + 1 = 15.6$ ms. It is therefore in principle possible to collect the profiles of 64 slices in one TR period (taking some computer overheads into account). The procedure in which more than one slice is measured per TR is called multiple-slice imaging.

In our example we required the minimum echo time; however, when we want a better signal-to-noise ratio it is better to lower the measuring gradient, which means decreasing the acquisition bandwidth (see (2.32) and Sect. 6.5). According to the Nyquist theorem, the noise voltage of a passive system (in this case the receiving coil loaded with the patient and followed by the receiver) is proportional to the square root of its bandwidth, $1/t_{\mathrm{s}}$. So a lower measuring gradient results in an improvement of the signal-to-noise ratio; see (2.32). However, the acquisition time increases (see (2.38a)), and in our example the number of profiles per repetition time (so the number of slices) is reduced. Also the echo time is increased, which results in a smaller signal. Therefore the precise relationship between the signal-to-noise ratio and the scan parameters is a complicated one and will be the subject of Chap. 6.

In Sect. 2.3.2 it was shown that for an ideal pulse many side lobes are needed. However, there is always the trade-off with the echo time, which requires short pulses. Therefore a practical RF pulse is truncated and has only two to four sidelobes and so the slice profile is not ideal either, but is spread out into a wider region. Adjacent slices may therefore influence each other (cross talk) in a multiple-slice imaging method. In image set II-11 the consequences of imperfect slice profiles are shown and discussed.

### 2.4.5 Imaging with Three-Dimensional Encoding

As described by (2.28), the desired magnetization function in a slice is related to the measured magnetization values in the $k_x$–$k_y$ plane. This equation can readily be generalized to the 3-dimensional situation (3D acquisition):

$$\begin{aligned} m(x, y, z) &= \frac{1}{2\pi} \int_{k_x} \int_{k_y} \int_{k_z} M_{\mathrm{T}}(k_x, k_y, k_z) \\ &\quad \exp(\mathrm{j}(k_x x + k_y y + k_z z)) \mathrm{d}k_x \mathrm{d}k_y \mathrm{d}k_z , \end{aligned}$$

where we now need to acquire the measurements in a 3D $\vec{k}$ space. In a manner similar to the scanning of the $k_x, k_y$ plane, shown in Fig. 2.5, we now have to scan a number of phase encode values in the $z$-direction of the $\vec{k}$ space to obtain the measurements necessary for the 3D Fourier Transform. The result of the 3D FFT resolves the space in the $z$-direction into a number of partitions. Since we have to wait for equilibrium after the acquisition of each profile (in a conventional $T_2$ weighted spin-echo scan) the total scan time becomes very long: the number of profiles per plane times the number of planes. So for a $16 \times 256^2$ scan with TR equal to 1 s the total scan time is 68 min. This is of course prohibitive, but the example is illustrative and suggests that we have to invent tricks to shorten this time. One of the tricks is to shorten the TR; we then have to take the resulting $T_1$ weighting for granted. Another possibility is not aquiring all the profiles, as is done for example in a "half-matrix" scan for which only 62.5% of the profiles are sampled (see Sect. 6.7).

The larger number of independent measurements for one single pixel during a 3D scan means that the signal-to-noise ratio is increased by the square root of the number of planes, when compared with a single-slice or multiple-slice measurement. Therefore thinner slices and/or smaller pixels can be measured with an acceptable signal-to-noise ratio in 3D imaging.

The discussion of conventional spin-echo sequences is concluded here. The reason for the elaborate treatment of conventional spin-echo sequences is that much of the basic ideas used can also be applied to most other scan methods. This especially holds for the other "conventional" scan method, the Field-Echo (Gradient-Echo) sequence. Actually all other imaging methods are derived from either the Spin-Echo (SE) or the Field-Echo (FE) methods. Without a good understanding of the concept of $\vec{k}$ space (or plane) their description in the modern literature is difficult to understand. An exception must be made for the Fast Field Echo and turbo-field-echo sequences (see

Chaps. 4 and 5), where the signal is built up from many components excited by several previous excitation pulses. The $\vec{k}$-plane description would only hold for one single component (see Chap. 3).

## 2.5 The Field-Echo Imaging Sequence

Looking again at Fig. 2.6, one might ask, "why the 180° pulse, bringing us from A to B in the $\vec{k}$ space, when one can also sample the signal along the line for which $k_y$ is constant, starting at point A?" The only measure one has to take then is to invert the gradient in the read-out direction. The sequence in which this approach is used is shown in Fig. 2.14 and the path through $\vec{k}$ space is shown in Fig. 2.15.

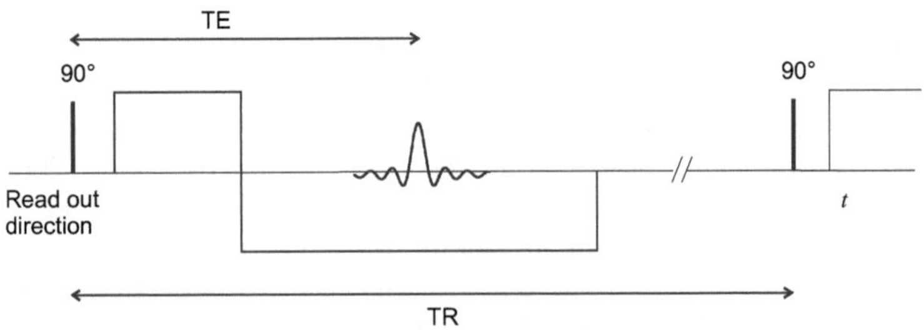


**Fig. 2.14.** Field-Echo sequence; the phase-encode gradient is not shown

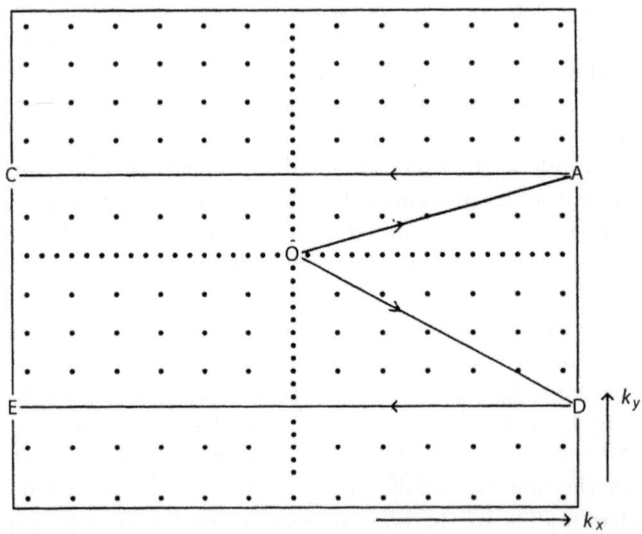


**Fig. 2.15.** $\vec{k}$-plane trajectory for a Field-Echo sequence

This indeed is an important imaging method, called the Field-Echo or Gradient Echo method [13, 14]. It has several features that are different from those of the SE imaging method. To start with, no time is needed for the refocussing pulse, so a shorter echo time is possible. However, the omission of the refocussing pulse also has a disadvantage: the effect of inhomogeneity of the magnetic field is inverted in a Spin-Echo scan due to the 180° pulse, so it is counteracted during the period after the 180° pulse. This is not the case for a FE scan, so imaging problems caused by dephasing can occur more often in this type of scan. Apart from the inhomogeneity of the magnet itself, there is also inhomogeneity due to variations in the susceptibility of the patient due to the air–tissue interfaces near cavities, for example. These variations (for example, near the cavities paranasales) can cause locally enhanced dephasing of the spins within a voxel (see image set II-10). These effects, named off-resonance effects, are superimposed on the $T_2$ decay and the combined effect is described by $T_2^*$ decay (so $T_2^* < T_2$). Therefore in field-echo sequences one must work with short echo times to avoid deterioration of the signal. The sensitivity of FE for dephasing has the consequence that the real image is strongly disturbed. FE images are in practice always modulus images (see image set II-9).

Another distinct feature of FE is the effect that can occur when tissue (water) and fat (having a gyro magnetic precession that differs by 3.3 ppm from that of water due to local molecular shielding of the magnetic field) are contained in the same voxel. In this case the signals from both constituents can compensate each other (are out of phase) at certain echo times, given by

$$3.3 \times 10^{-6} \gamma\!\!\!^{-} B_0 t = (1 + 2n)/2. \tag{2.41}$$

For 0.5 T this time is 7.1 $(1 + 2n)$ ms. For 1.5 T these values are 3 times shorter. For susceptibility changes, the same effect occurs so that at too-large TE values we find black regions in the image.

In Field-Echo scans there is no real reason to restrict the excitation angle to 90°. Other values can be taken, for example smaller values (30°). The interesting aspect of a small flip angle is that the decrease in the longitudinal magnetization is limited. For small flip angles the $\vec{k}$ plane concept remains unimpaired and is now applicable to the transverse component of the magnetization vector.

In the previous discussion we have stated that after acquiring a profile one has to wait for equilibrium before exciting for the next profile. This is not necessarily so, one can also excite the same slice long before equilibrium can be established. The $T_1$ relaxation is then incomplete and depending on the actual $T_1$ value a dynamic equilibrium situation will arise, similarly to what happens in Spin-Echo sequences, as described in Sect. 2.4.2. Together with the short TE, which means almost no $T_2^*$ weighting, we then have a $T_1$ weighted FE sequence with short TR, thus we have a short scan time (see image set II-3). In comparison with the Spin-Echo sequence the Field-Echo sequence is fast and is mainly used for $T_1$ weighted studies, including imaging with 3D encoding.

When we make TR even shorter, comparable with $T_2^*$, a banding structure appears over the image, making it undiagnostic. This is due to the fact that after an excitation there is also transverse magnetization left from several earlier excitations which sometimes adds to and sometimes subtracts from the FID of that last excitation. There are several ways to avoid these bands. Sequences in which such provisions have been taken go by the name of "Fast Field Echo" (FFE). FFE imaging is the subject of a separate study, described in Chap. 4, because these scans cannot be described on the basis of the theory developed in this chapter, where the spin phase was described in terms of the history of the transverse magnetization since the last excitation only.

## 2.6 Artifacts

Artifacts are aspects of the image that are not related to the object, they can be misleading in the diagnosis [15]. Artifacts cannot be avoided, so, if you cannot beat them, at least try to understand them so that you can take them into account. Therefore in this section we shall describe the fundamental artifacts caused by the physics of our "conventional" scan methods, SE and FE. However, we shall set up the mathematical description in such a way that a similar treatment can be used in later chapters for most of the modern (fast) scan methods. Actually this makes the present treatment somewhat more involved than necessary, but later on it pays off.

For a treatment of fundamental imaging artifacts let us go back to (2.27) describing the transverse magnetization as a function of time during the acquisition. This equation describes an ideal measurement, but there are two fundamental effects not taken into account in this description. The first one is the $T_2$ (or $T_2^*$) relaxation and the other one is the "resonance offset" [16]. Resonance offset occurs when the angular velocity of the rotating frame of reference is unequal to the gyromagnetic frequency of the local magnetic field. The angular velocity of the rotating frame is $\omega_0 = \gamma B_0$ (see (2.4)), where $B_0$ is the ideal value of the main magnetic field. However, in reality the main magnetic field is not completely homogeneous and deviations from the required value can play an important role in certain sequences. We denoted the deviations by $\delta B$ in (2.3) and saw that this term remains present in the Bloch equation for the rotating frame (2.5). This introduces the term $\delta B$ into (2.5) and a factor of $\exp(-\mathrm{j}\gamma\delta Bt)$ into (2.10a). $\delta B$ is assumed here to be independent of time, we do not consider eddy currents, which can cause a time-dependent offset making the situation much more complicated. We restrict ourselves to inter-voxel effects.

Coming back now to (2.27) we see that the ideal measurement results are multiplied by a function describing the $T_2$ decay and one describing the resonance offset. The artifacts arising from these effects are fundamental artifacts because the $T_2$ decay cannot be avoided and a purely homogeneous field is impossible in a magnet with finite length. So instead of the ideal value $M_{\mathrm{T}}(t)$ we measure

$$M'_{\mathrm{T}}(t) = M_{\mathrm{T}}(t)\exp(-t/T_2)\exp(-\mathrm{j}\gamma\delta Bt). \tag{2.42}$$

Introducing this into (2.28), we have

$$\begin{aligned} m'(x,y) &= \iint_x M_{\mathrm{T}}(k_x,k_y)\exp(+\mathrm{j}(k_x x + k_y y))\exp(-t/T_2) \\ &\quad \exp(-\mathrm{j}\gamma\delta Bt)\mathrm{d}k_x\mathrm{d}k_y, \end{aligned}$$

where the factor $(2\pi)^2$ is assumed to be part of $m'$.

To understand the artifacts for SE sequences arising in the non-ideal situation, we must replace the time $t$ by an expression in $k_x$ ($k_y$ is constant during the acquisition of a profile). Assuming $t' = t - \mathrm{TE}$ and defining $\gamma G_x t' = k_x$, see (2.26), where $k_x$ runs from $-k_{x,\max}$ to $k_{x,\max}$ (see Fig. 2.6) $t'$ runs from $\mathrm{TE} - 1/2t_{\mathrm{acq}}$ to $\mathrm{TE} + 1/2t_{\mathrm{acq}}$ (see Fig. 2.5). Due to the refocussing pulse in a SE sequence, the phase of the magnetization is reversed at $t = \mathrm{TE}/2$ so that the effect of the magnetic field inhomogeneity is compensated for at the time of the echo, TE. So no term describing the effect of field inhomogeneity over the period $0 < t < \mathrm{TE}$ will occur. For the read-out direction in an SE sequence, (2.42) becomes

$$\begin{aligned} m'(x) &= \exp\left(-\frac{\mathrm{TE}}{T_2}\right)\int\left\{M_{\mathrm{T}}(k_x)\exp\left(\frac{-k_x}{\gamma G_x T_2}\right)\exp\left(\frac{-\mathrm{j}k_x\delta B}{G_x}\right)\right\} \\ &\quad \exp(-\mathrm{j}k_x x)\mathrm{d}k_x. \end{aligned} \tag{2.43}$$

From (2.43) it is concluded that the measured $m'(x)$ is weighted with the term $\exp(-\mathrm{TE}/T_2)$ and is the convolution of Fourier transforms of $M_{\mathrm{T}}(k_x)$ and two exponentials, one describing $T_2$ decay and the other the resonance offset equal to $\gamma\delta B$. Actually, the equation must be extended with a term describing the dephasing of the spins due to that part of $T_2^*$ ($T_2'$ in Sect. 3.3.2.1) that describes dephasing due to field inhomogeneities and which depends on $|t'|$. This term will be omitted in this treatment, it complicates the situation and is small because $\delta B$ and $t'$ are small.

As has been described earlier there is an essential difference between SE and FE: for FE, field inhomogeneity and susceptibility effects are not compensated for at $t = \mathrm{TE}$ by the effect of a refocussing pulse. Therefore for FE the decay is described by (2.43) after replacing $T_2$ by $T_2^*$ and adding a phase term $\exp(-\mathrm{j}\gamma\delta B\mathrm{TE})$ before the integral:

$$\begin{aligned} m'(x) &= \exp\left(-\frac{\mathrm{TE}}{T_2^*}\right)\exp(-\mathrm{j}\gamma\delta B\mathrm{TE}) \\ &\quad \int\left\{M_{\mathrm{T}}(k_x)\exp\left(\frac{-k_x}{\gamma G_x T_2^*}\right)\exp\left(\frac{-\mathrm{j}k_x\delta B}{G_x}\right)\right\}\exp(-\mathrm{j}k_x x)\mathrm{d}k_x. \end{aligned} \tag{2.43a}$$

The consequence of this phase term in an image with aliasing is discussed in image set II-7. In image set II-9 the phase distribution due to inhomogeneous magnetic field is made visible. In FE only the modulus image is used.

Let us first discuss the effect of resonance offset. The position dependence of $\delta B$ can be approximated by

$$\delta B(x) = \alpha + \beta x. \tag{2.44}$$

The first term describes a constant offset and represents the difference in resonance field strength between water and fat, for example. The second term serves as a first approximation to the field changes in inhomogeneous areas, such as at the edge of the homogeneity sphere of the magnet, and near local field changes due to susceptibility effects. Distortion of an image due to susceptibility effects is shown in image set II-8. If $\beta = 0$ the second exponential in (2.43) can be avoided when we introduce $x' = x + \alpha/G_x$. This means that $\delta B = \alpha$ causes a position shift equal to $\alpha/G_x$. For water–fat the shift $\delta B/B_0$ is 3.3 ppm, which means $\alpha = 5\,\mu\mathrm{T}$ at 1.5 T. In a read out gradient of 5 mT/m this means a position shift of 1 mm. For an example of the misregistration of water or fat see image set II-4.

When $\beta \neq 0$ we introduce $x' = x(1 + \beta/G_x) + \alpha/G_x$. The integral of (2.43a) can now for FE be rewritten as (for SE the calculation is similar):

$$m'(x') = \frac{\exp\left(-\frac{\mathrm{TE}}{T_2^*}\right)\exp(-\mathrm{j}\gamma\delta B\mathrm{TE})}{1+\frac{\beta}{G_x}} \int M_{\mathrm{T}}(k_x)\exp\left(\frac{-k_x}{\gamma G_x T_2^*}\right) \exp(-\mathrm{j}k_x x')\mathrm{d}k_x. \tag{2.45}$$

The factor before the integral describes the overall effect of $T_2^*$ decay and resonance offset on the signal strength of the echo. The difference between $m'(x)$ and $m'(x')$, called distortion, is threefold:

- the image is shifted over a distance $(\alpha + \beta x)/G_x$,
- the image is magnified by a factor $(1 + \beta/G_x)$,
- the intensity in the image is multiplied by $(1 + \beta/G_x)^{-1}$.

Finally also the phase of the magnetization depends on $\delta B$. Distortion due to a inhomogeneous magnetic field is shown in image set II-6.

In the read-out direction the value of $G_x$ is normally so high ($mT/m$ versus ppm for $\alpha$ and $\beta$), that the distortion described here is negligible (except for the water–fat shift; see image set II-4). The reason for this detailed treatment of these small effects is that in later discussions of these effects in very fast imaging sequences we shall use the same reasoning and mathematics. As is stated by (2.26), the velocity in $\vec{k}$ space, $\mathrm{d}\vec{k}/dt$, is proportional to the gradient, and in fast sequences we frequently have to deal with low "pseudo gradients", describing the low velocity through $\vec{k}$ space in the phase-encode direction. This will be discussed in connection with Echo Planar Imaging in the next chapter.

Now we shall consider the influence of $T_2^*$ relaxation on the point spread function of FE imaging. The term in (2.45) before the integral, determining the signal strength and phase, can now be omitted:

$$m'(x) = \mathfrak{F}^{-1}\left\{M_{\mathrm{T}}(k_x)\right\} \otimes \mathfrak{F}^{-1}\left\{\exp(-k_x/\gamma G_x T_2^*)\right\}. \tag{2.46}$$

The Fourier transform of the exponential is well known as the complex Lorentz function:

$$\mathfrak{F}^{-1}\left\{\exp(-k_x/\gamma G_x T_2^*)\right\} = \frac{\gamma G_x T_2^*}{1 + \mathrm{j}\gamma G_x T_2^* x}. \tag{2.47}$$

This function has a line width at half maximum given by $\Delta x = 2(\gamma G_x T_2^*)^{-1}$. So the actual image is convolved with this function. But we remember that it is already convolved with the "sampling point-spread function" defined in (2.36), which has a half width equal to $\Delta x = 1.2\pi/\gamma G_x t_{\mathrm{acq}}$. When $T_2^* \gg t_{\mathrm{acq}}$, which is normally the case, we see that the line width due to $T_2^*$ decay (the "blurring") is much smaller than the usual sampling point-spread function. So in the read-out direction we do not have to worry about image degradation by $T_2^*$ decay and when the measuring gradient is large enough, also the water–fat shift and magnetic field inhomogeneity are not very important. This is generally true also for the fast imaging sequences described in the next chapter. So we shall not return to a discussion of image degradation in the read-out direction caused by $T_2^*$ decay and field inhomogeneity. Note that for SE, (2.47) must be changed by replacing $k_x$ by $|k_x|$ which results in a real Lorentz function.

We now have to consider the phase-encode direction. For SE one can see directly that the effect of $T_2$ decay is for each $k_y$ value the same, because each profile is measured in the same time interval around $t$ = TE after the last excitation. Therefore all $k$ values in the profiles are weighted with $\exp(-\mathrm{TE}/T_2)$, which yields the $T_2$ weighting. For SE, the resonance offset is automatically compensated for since the errors that are built up during $0 < t < \mathrm{TE}/2$ are reversed by the 180° pulse and therefore compensated for by the errors during $\mathrm{TE}/2 < t < \mathrm{TE}$. So for spin echo there is no image degradation due to $T_2$ decay and field disturbances in the phase-encode direction. The same arguments hold for FE sequences except that the phase due to resonance offset is not compensated for, but is equal for each profile. The resulting signal is not real but complex. Therefore in FE mostly modulus images are made.

### 2.6.1 Ghosting

We shall finally discuss one further (not fundamental) artifact, which is also very instructive for our later discussions on fast and ultra-fast sequences. In the early days of MRI the gradient-amplifier technology was not yet sufficiently developed to suppress "hum" from the power supply. So the values of the phase-encode gradients were not exactly as required, but there was a variation from profile to profile depending on their timing with respect to the 50 Hz hum. This resulted in a ripple in the phases of the echos (see Fig. 2.16). The amplitude of the ripple is assumed to be $\alpha$ and the periodicity in the $k_y$ direction is $\Lambda\,\mathrm{m}^{-1}$. Since the form is unknown we describe it as a Fourier series. When we introduce this ripple into the equation for the Fourier transform, in the $y$ direction we find

$$m(y) = \int_{k_y} M(k_y)\exp(\mathrm{j}k_y y)\exp\left[\mathrm{j}\alpha\sum\left(c_n\exp\left(\mathrm{j}\frac{2\pi n}{\Lambda}k_y\right)\right)\right]\mathrm{d}k_y. \quad (2.48)$$

Using the fact that $\alpha$ is small so that the exponential can be written as the first two terms of the series expansion, we find

$$m(y)' = m(y) + \mathrm{j}\alpha\sum c_n m\left(y - \frac{2\pi n}{\Lambda}\right). \quad (2.49)$$

The first term describes the image itself; the second term describes a series of shifted images (named "ghosts"; see Fig. 2.16), which are shifted over distances $2\pi n/\Lambda$. Note that the amplitude of the ghost is equal to the amplitude of the phase shift. For a pure harmonic only $c_{\pm 1}$ is equal to 1 and all other $c_n$ are 0, so the ghost level is equal to $\alpha$ and the shift is $2\pi/\Lambda$. In modern systems the artifacts due to insufficiency of the gradient system do not occur any more in conventional SE and FE imaging. However, similar artifacts may arise due to other causes, such as a pulsating blood flow.

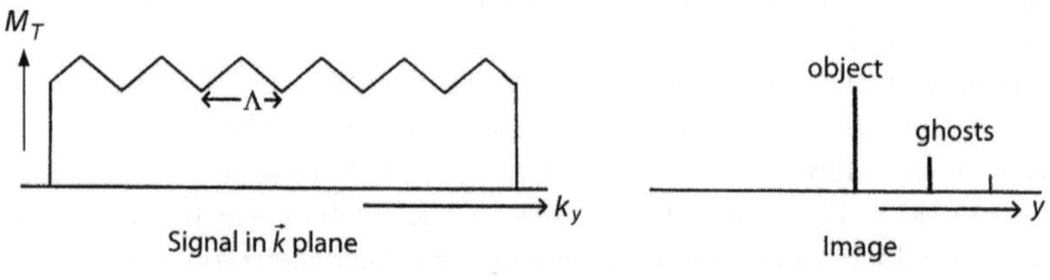


**Fig. 2.16.** Ghost in image due to periodically changing sensitivity

In a SE sequence in diastole the blood in an artery or vein has a low velocity and the 90° pulse hits the same blood as the 180° pulse, causing maximum signal. However, during systole the blood excited by the 90° pulse moves out of the imaged slice before the 180° pulse is applied so the signal is at a minimum. For sequences that are not ECG triggered we have different signals from the artery in the different profiles. In the same way as discussed in connection with (2.49), this yields repeated images of the artery (ghosts) shifted in the phase-encode direction. So when we have a transverse slice through which some vessels cross we see in the SE image of the slice that the arteries repeat themselves in the phase-encode direction. By using ECG triggering we avoid this artifact when each profile is measured in the same heart phase (see image set VII-3).

## 2.7 Magnetization Preparation

Instead of exciting spins starting from their equilibrium situation, $M_0$, one can "prepare" the magnetization state at the excitation pulse with a magnetization preparation pulse (pre-pulse). The aim is to bring a certain "weighting" in the magnetization state, for example a $T_1$ weighting. This means that tissues are distinguished in the image because of differences in their magnetization due to their different $T_1$ relaxation behaviour.

There are, however, more possibilities for bringing the magnetization in a certain state at the start of an imaging sequence by using pre-pulses, so as to carry information on certain other properties, such as $T_2$ relaxation or diffusion, or to exclude a certain tissue from imaging (for example, fat). Pre-pulses are also used to distinguish flowing from stationary tissue ("labelling"). In Sect. 2.7.1, the principle of magnetization preparation will first be explained with the example of a 180° pre-pulse, which inverts the magnetization $M_0$ to $-M_0$. In Sect. 2.7.2, more pre-pulses will be briefly mentioned with reference to later sections in this book.

### 2.7.1 A $T_1$ Preparation Pulse: Inversion Recovery

The inversion recovery method relies on a 180° pre-pulse, which inverts the magnetization vector. The longitudinal magnetization relaxes back to the equilibrium state according to (2.8). When TR is the time between the inversion pulses and TI is the time delay between the inversion pulses and the 90° excitation pulses (inversion time) applied for an imaging sequence, we find for the longitudinal magnetization just before that pulse:

$$M_z(\mathrm{TI}) = M_0 - 2M_0 \exp\left(-\frac{\mathrm{TI}}{T_1}\right) + M_0 \exp\left(-\frac{\mathrm{TR}}{T_1}\right). \tag{2.50}$$

This dependence of $M_z(\mathrm{TI})$ on $T_1$ is shown in Fig. 2.17 for the limiting case where $\mathrm{TR} \gg T_1$ for different tissues, distinguished by their $T_1$ values. An imaging sequence started at $t = \mathrm{TI}$ will therefore show a $T_1$ weighting. When

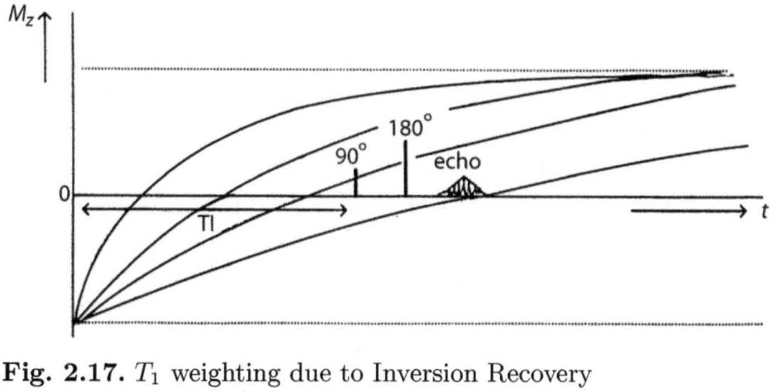


**Fig. 2.17.** $T_1$ weighting due to Inversion Recovery

combined with a Spin-Echo sequence, this scan method is named Inversion Recovery (IR). An inversion pre-pulse, however, can readily be combined with other scan methods, and in the later chapters we shall find several examples – see Sects. 5.3 and 7.5.2.3.1.

A special problem is met in the situation sketched in Fig. 2.17. At the moment of excitation, some tissues still have a negative magnetization vector and some other tissues already have positive magnetization. In an image in which the grey scale depicts the modulus value of the transverse magnetization ("modulus image"), it is possible that two different tissues – with opposite magnetization vectors – obtain the same gray value. This problem can be avoided by using the real value of the magnetization as the basis of the grey scale (a "real image"). A practical example of Inversion Recovery and a discussion of the problems of multi-slice IR imaging is shown in image set II-12.

In the limiting case $\mathrm{TR} \gg T_1$, the longitudinal magnetization crosses zero when $t = t_c = T_1 \ln 2$, and $t_c$ is therefore dependent of the tissue considered. One can choose TI in such a way that $\mathrm{TI} = t_c$ for a certain tissue. The resulting image will not contain a signal of that tissue.

This manipulation can be used for solving a problem that occurs when a slice is imaged that contains both water and fat. Although the protons in water and fat have the same gyromagnetic ratio, fat protons have a molecular surrounding that changes the magnetic field of the fat protons by $\delta_{\mathrm{fw}} = 3.3\,\mathrm{ppm}$. In Sects. 2.5 and 2.6, the consequences of this problem are described: misregistration, phase shift and signal voids, all due to intra-voxel dephasing. This causes problems in the reading of the image. For example, the fat in bone marrow near a joint can shift in an image across the thin cortical bone and over the cartilage (see image set II-4), which results in an undiagnostic orthopaedic image. Also in other cases (for example, in the liver), the fat may obscure important contrast differences in other tissue. Decreasing the water–fat shift by using large read-out gradients is not a real solution, because this decreases the signal-to noise ratio (see Chap. 6).

Since lipids have short $T_1$, after an inversion pulse their longitudinal magnetization vector passes through zero earlier than that of any other tissue. So when at that moment an excitation pulse is applied, the resulting image will not contain a fat signal (fat suppression) and the magnetization of all other tissue is still negative and will show unique $T_1$ weighted contrast. The TI must be chosen with relatively low values, and the method is therefore named: Short-Tau Inversion Recovery (STIR) [17].

In the same way, it is possible to suppress the signal of free water. For example, when a $T_2$ brain image is required the bright signal of CSF can be suppressed when TI is chosen accordingly. Since $T_1$ of CSF is very long (4 s or more), the time at which the longitudinal magnetization of CSF crosses zero, $t_c = \mathrm{TI}$, is relatively large. All other tissues are already completely relaxed to equilibrium at this moment and show the usual contrast (for example, the $T_2$ contrast in a Spin-Echo sequence). This scan method is named Fluid Attenuation by Inversion Recovery (FLAIR). An example of this scan method is given in image set II-9. Inversion pulses are also frequently applied in order to label flowing blood – see Sect. 7.5.2.3.2.

### 2.7.2 Other Types of Magnetization Preparation

In many situations the magnetization preparation is not performed with a single pre-pulse, but by a sequence of pre-pulses and gradient waveforms. For example, instead of inverting the spins in the slice under investigation, it is also possible to invert all spins in the volume, except those of the slice to be imaged. This can be done using two inversion pulses, the first not slice-selective and the second slice-selective. The result is that all spins are inverted except the spins in the slice to be imaged (see Sect. 7.5.2.3.2). In the present section we shall list a number of other methods used for influencing the magnetization state at the start of an imaging sequence.

**1. Saturation Pulse.** In order to avoid the problem that two types of tissue have opposite magnetization vectors and show equal magnetization values in a modulus IR image, we can also apply a 90° (saturation) pre-pulse, followed by a dephasing gradient. In that case, the relaxation occurs according to:

$$M_z(t) = M_0 - M_0 \exp\left(-\frac{t}{T_1}\right), \tag{2.51}$$

which starts for all tissues at zero and negative values of the magnetization do not occur. This method is called "Saturation Recovery". Saturation prepulses are also used in angiographic imaging in order to distinguish arterial from venous blood, because only unsaturated blood yields a signal – see Sect. 7.5.2.1. Thus, when blood is saturated on the venous side of a slice, it will not yield a signal.

In Sect. 2.4.1.1 it has been shown that parts outside the FOV fold back into the image (see Fig. 2.9, aliasing). In the read-out direction this can be solved easily by oversampling, so that a much larger FOV is imaged, from which the unwanted part is omitted. This oversampling does not require extra time, since the sampling frequency can be increased. In the phase-encode direction, however, more profiles mean extra acquisition time. In this case the regions outside the required FOV can be saturated by "regional saturation" (REST) and therefore do not generate a signal. REST can also be used to suppress the signal from moving structures, for example the fat layer of the abdomen, which moves due to respiratory motion.

**2. Spin Locking Pulse Sequences, "$T_{1\rho}$ Pre-pulses".** At first a (for example) $90^\circ_{y'}$ pulse brings the longitudinal equilibrium magnetization in the transverse plane along the $x'$ axis. Then an RF field excites a $B_1$ field along the same axis. When the spins dephase due to the $T_2^*$ effect, they will start a precession motion around the $B_1$ field and not dephase further (they are "locked"). This effect is named "spin locking". After a certain time, the locking pulse ends and the magnetization is flipped back into the longitudinal direction with a $90^{\circ}_{-y'}$ pulse. During the spin-lock period, a special type of $T_1$ relaxation, based on an interaction with slow motions of the macro molecules, is active. This is named $T_{1\rho}$ relaxation. This pre-pulse is sometimes used in angiography – see Sect. 7.5.2.3.2.

**3. $T_2$-weighted Magnetization Preparation.** This preparation pulse sequence is described as a simple spin echo, where at the moment of the echo the (then refocussed) magnetization is brought back into the longitudinal direction using using a $-90°$ pulse. During the time between the $90°$ pulses (the echo time), $T_2$ relaxation takes place just as is the case in regular spin echoes. So the longitudinal magnetization at the end of the preparation sequence is $T_2$-weighted. The echo time can be made longer by using a number of refocussing pulses. A special application of a "$T_2$ prep" sequence is described in Sect. 7.5.2.3.3.

**4. Spectral-Selective Pre-pulse.** A modern method to suppress the signal of lipids is with the help of a spectral selective pulse added just before the excitation pulse of an imaging sequence. The spectral selective $90°$ pulse addresses the lipids only and rotates their magnetization into the transverse plane. This pulse is followed by a dephasing gradient, causing complete saturation of the lipids. The spectral selective pulse is not slice selective. In a multiple-slice acquisition with $n$ slices per TR, the spins of the lipids are influenced by this pulse $n$ times per TR, so that the lipid magnetization before the next SPIR pulse is small.

A problem with this method is that these spectrally selective pulses must saturate the lipids only, and so their bandwidth must be smaller than $\delta_{\mathrm{fw}}\gamma\!\!\!^{-} B_0$, which is about 200 Hz at 1.5 T. The duration of a spectral selective pulse is therefore at least 5 ms. At 0.5 T, a spectral selective pulse lasts more than 15 ms, which is almost prohibitive for incorporation in an actual imaging sequence. The combination of the delay between the SPIR pulse and the excitation pulse and the low value of the magnetization just before the SPIR pulse means that the SPIR pulses have to be given at a flip angle that is considerably larger than $90°$ to cause the desired saturation. This effect is also the reason for the name of this method: Spectral Presaturation by Inversion Recovery, abbreviated to SPIR [18]. An example of the use of SPIR is shown in image set II-5.

**5. Diffusion of Free Water.** An important new development is the measurement of the diffusion of free water in tissue, especially for neurology (see Sect. 7.7). Also in this case, the magnetization at the time of an excitation pulse for an imaging sequence is diffusion-weighted, which can be made visible in an image. A detailed description of diffusion-weighted magnetization preparation is not possible at this stage; we therefore refer to Sect. 7.7.

**6. Magnetization Transfer Pulses.** These rely on the selective excitation of the protons bound to macromolecules. The bound protons have a very short $T_2$ relaxation time due to dipole–dipole interaction (see Sect. 6.3), and so they have a wide frequency spectrum. In contrast, the free-moving protons of the free water have much longer $T_2$, and so a narrower spectrum. A magnetization transfer pre-pulse can be applied at frequencies away from the free proton gyromagnetic frequency so as to address the bound protons only. The result is that the image contrast can be changed since the equilibrium magnetization is lowered and the T1 relaxation time is shortened (see Sect. 6.3.3).

**7. Tagging.** In the study of motion of the cardiac wall the spins in narrow strips are excited by a series of equidistant excitation pulses so as to show a "grid" in the image ("tagging"). The wall motion can be studied by monitoring the motion of these excited strips with dynamic imaging (see Sect. 7.2).

**8. Reset Pulse.** After collecting the echo in a spin echo sequence or the echoes in a turbo spin echo sequence (see Sect. 3.2), it can pay off to create an extra spin echo with an additional refocusing pulse. When a $\pm\pi/2$ pulse is applied at the moment of the extra echo (the sign depends on the number of echoes since the last excitation), the refocused transverse magnetization is converted into positive longitudinal magnetization. The advantage of this method is that the waiting period TR to allow for $T_1$ relaxation can be shortened, so that faster scans are possible. This technique, which is related to the $T_2$-prep technique, see point 3 above, was proposed very early by Becker and was, at that time, called "Driven Equilibrium" [24]. The associated signal enhancement can be large for long $T_2$ tissue, as illustrated in image set III-3.

**9. Black-blood Preparation.** The signal of moving blood can be nulled by preceding the excitation with a pair of inversion pulses, of which the first is non-selective and the second is selective over the image region. The result is that all spins are inverted, except those in the imaging region. Nulling of the blood signal is optimal when the excitation of the imaging region occurs after a delay during which the inverted longitudinal magnetization of the inflowing blood vanishes through relaxation (delay $=$ TI $\ln 2$) [25]. The technique works for 3D as well as for 2D imaging, as long as the blood velocity is fast enough to fill the image region completely during the inversion period. Image set VII-13 shows an example of the use of this method in coronary vessel imaging.

**10. Arterial Spin Labelling.** For perfusion measurement, scan methods can be used that contain inversion prepulses applied in an up-stream zone of the feeding artery. The scan is repeated without these prepulses and the difference signal is a quantitative measure of perfusion. Various subtle difficulties in this technique exist and are discussed in Sect 7.6.2 and in image set VII-17.

# Image Sets
# Chapter 2

## II-1 Contrast in Spin-Echo Images of the Brain: Variation of Repetition Time TR

(1) Spin-Echo (SE) imaging is the classic workhorse among the acquisition methods.

At TRs that are shorter than the $T_1$ values of the tissues of interest and at short TE, the SE image is $T_1$ weighted, that is, the modulation of the image brightness is mainly determined by differences in $T_1$ (see Sects. 2.4.2 and 6.2, (3a)). When TE is kept short, but TR is increased, the $T_1$ weighting is lost. This does not imply a total loss of contrast, as different tissues can show contrast because of differences in their MR visible proton densities (proton-density weighting).

This image set illustrates the transition from $T_1$ weighting to proton-density weighting in the transverse brain. A single-slice technique was used for purpose of demonstration. Subtle effects of the influence on the contrast of the use of multi-slice techniques are demonstrated in image sets II-2, II-11, and VI-4.

Parameters: $B_0 = 1.5\,\mathrm{T}$, FOV $= 230\,\mathrm{mm}$, matrix $= 256 \times 256$, $d = 4\,\mathrm{mm}$, TE $= 15\,\mathrm{ms}$.

| image no. | 1 | 2 | 3 | 4 | 5 | 6 | 7 |
|---|---|---|---|---|---|---|---|
| TR | 63 | 125 | 250 | 500 | 1000 | 2000 | 4000 |
| NSA | 16 | 8 | 4 | 2 | 1 | 1 | 1 |

NSA was adjusted to keep the image time constant (at 4 minutes), except for in images 6 and 7.

(2) The transition from $T_1$ weighting to proton-density weighting is reached by increasing TR from scan to scan by a factor of 2 and is demonstrated in the image contrast. Strong $T_1$ weighting of the brain tissue is visible for TR values of 500 ms and shorter (images 1, 2, 3). At longer TR, $T_2$ weighting cannot dominate the contrast because the echo time used is too short to give a noticeable influence of $T_2$ differences to the contrast. (For $T_2$ values of brain tissue, see image set II-2.)

(3) The signal-to-noise ratio, SNR, (Chap. 6) is measured from the images for grey and white matter and the difference between these values, the contrast-to-noise ratio, CNR, for these tissues is obtained:

| image no. | 1 | 2 | 3 | 4 | 5 | 6 | 7 |
|---|---|---|---|---|---|---|---|
| SNR (white) | 17 | 37 | 52 | 78 | 84 | 108 | 119 |
| SNR (grey) | 12 | 26 | 38 | 57 | 74 | 110 | 136 |
| CNR (white–grey) | 5 | 11 | 14 | 21 | 10 | −2 | −17 |

Given the constant total image times in images 1 to 5, the longest TR gives the best SNR for both grey and white matter. Moreover, a reduction of TR below 500 ms does not pay off for an increase in the $T_1$-weighted contrast-to-noise ratio between these tissues. In images 1, 2, and 3, for which TR =

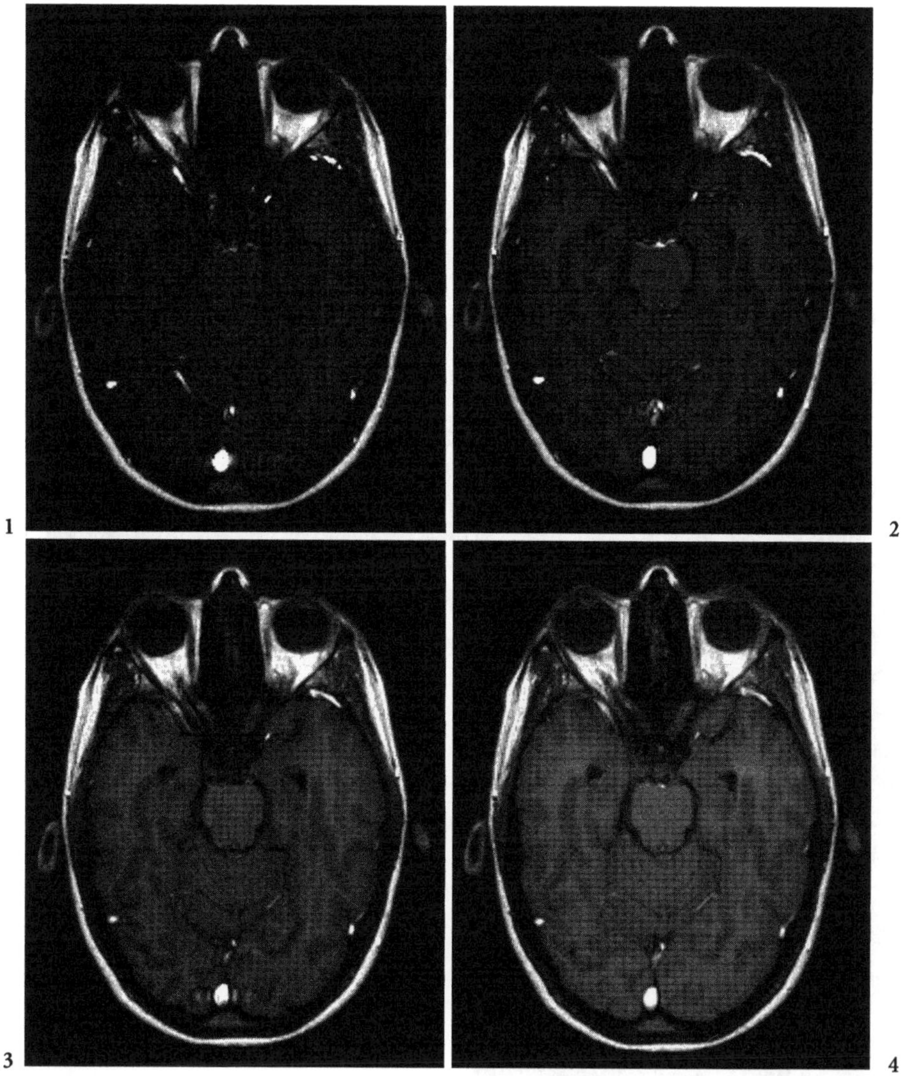

Image Set II-1

63, 125, and 250 ms, respectively, the short TR instead results in a loss of contrast-to-noise ratio at constant imaging time.

(4) In the image with the longest TR, (image 7, 4000 ms), the $T_1$ influence on the image brightness should have disappeared completely for most tissue (see under image set VI-1 for $T_1$ values of brain tissue). In image 7, the contrast between grey and white has reversed in comparison to the short-TR images. This contrast therefore reflects the difference in proton density between these tissues. Although quite visible, the contrast-to-noise ratio in this image is not

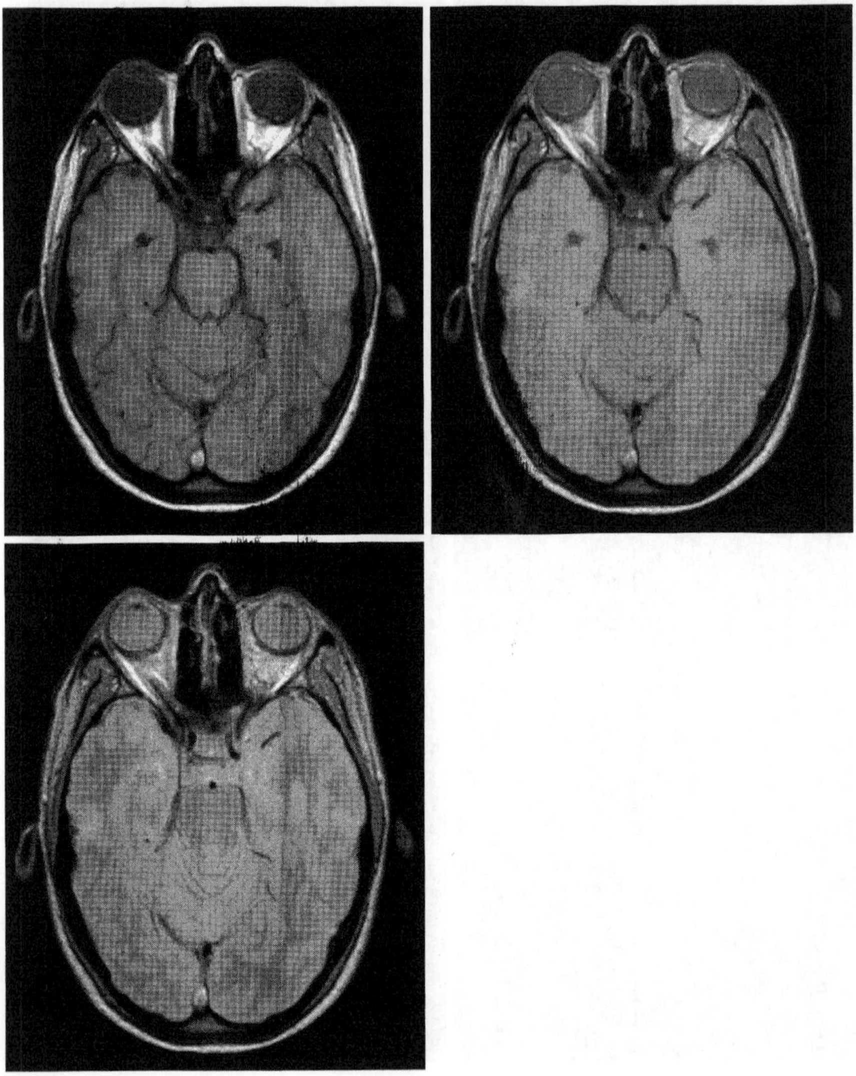

5 6 7

**Image Set II-1.** (*Contd.*)

as clear as in image 4 (TR = 500 ms). Note that even at 4000 ms the CSF in the ventricles still lacks signal. This is due to saturation, which still occurs for CSF because its $T_1$ at 1.5 T has a value exceeding 4000 ms.

(5) In conclusion, the contrast between grey and white matter in SE imaging is best visualized by $T_1$-weighted imaging. For a given total scan time, the best contrast-to-noise ratio in single-slice imaging is obtained at a TR of about 500 ms.

## II-2 Contrast in Spin-Echo Images of the Brain: Influence of TE and of the Spatial Profile of the Refocussing Pulse

(1) In our system, the refocussing pulses used in single-slice Spin-Echo (SE) imaging are not slice selective, whereas in multi-slice SE, slice selectivity is required. As a result, the flip angle (180°) for the refocussing pulse in single-slice SE imaging is more precisely defined over the slice thickness than that in multi-slice imaging. So, where in a single-slice scan the signal loss at TE can be expected to be due to $T_2$ decay only, in multi-slice scans some decay may be the result of non-ideal echo formation.

The image set shows type modulus images (labelled "M") obtained from single-slice (SS) and multi-slice (MS) SE scans with TE of 25 and 100 ms. From both scans, a calculated $T_2$-image (labelled "$T_2$") was obtained as well.

Parameters: $B_0 = 1.5\,\mathrm{T}$; FOV = 230 mm; matrix = $256 \times 256$; $d = 4\,\mathrm{mm}$; TR = 2000 ms; NSA = 1.

| image no. | 1 | 2 | 3 | 4 | 5 | 6 |
|---|---|---|---|---|---|---|
| TE | 25 | 100 | | 25 | 100 | |
| image type | M | M | T2 | M | M | T2 |
| mode | SS | SS | SS | MS | MS | MS |

(2) In the images obtained with the single-slice scan, a train of four equidistant echos was obtained and the images from echos 1 and 4 are shown. In the images obtained with the multi-slice technique, two unevenly spaced echos were obtained and the corresponding images are shown. Although the multi-slice technique allowed the acquisition of more than one slice per TR, to avoid cross talk (see image set II-11) only one slice per TR was imaged.

(3) The images with equal TE (image 1 versus 4 and 2 versus 5) are very similar in contrast, showing that the refocussing pulses of the single-slice technique and the multi-slice technique have a visually almost equivalent performance. In the single-slice images, the ventricles appear somewhat brighter.

(4) The calculated $T_2$ images are obtained by assuming a monoexponential decay. The results for the single-slice and the multi-slice scan techniques can be compared numerically:

| | $T_2$ values (ms) | |
|---|---|---|
| image no. | 3 (SS) | 6 (MS) |
| CSF | 1709 | 703 |
| grey matter | 83 | 76 |
| frontal white matter | 70 | 65 |
| corpus callosum | 71 | 65 |
| subcutaneous fat | 55 | 51 |

For subcutenaous fat, in the literature the values of $T_2$ are above 100 ms. For this tissue, the signal decay is known to deviate from a mono-exponential curve; so that the value of $T_2$ obtained will increase with the value of the echo times used. The estimated values of $T_2$ of grey and white matter are only slightly below the literature values [19]. The laboratory value of $T_2$ of CSF is about 2560 ms [20].

For most tissues, the values of $T_2$ in image 6 (MS) are only marginally below the values in image 3, indicating relatively complete refocussing in the multi-slice scan. For CSF, however, the multi-slice technique clearly gives a too low value of $T_2$. This does indicate imperfect refocussing in multi-slice imaging. This phenomenon can have various causes. Such causes include a systematic error in the flip angle due to the approximate slice profile of the RF pulse, but could also include flow of CSF, deviations in the phase adjustment of the refocussing pulse, or, of course, a combination of such factors.

(5) In conclusion, the slice-selective refocussing pulse of the multi-slice SE sequence has close but not complete similarity to the non-slice-selective pulse of single-slice imaging. In both acquisition methods the loss of signal between the first and second echo is dominated by $T_2$ decay as expected for ideal behaviour.

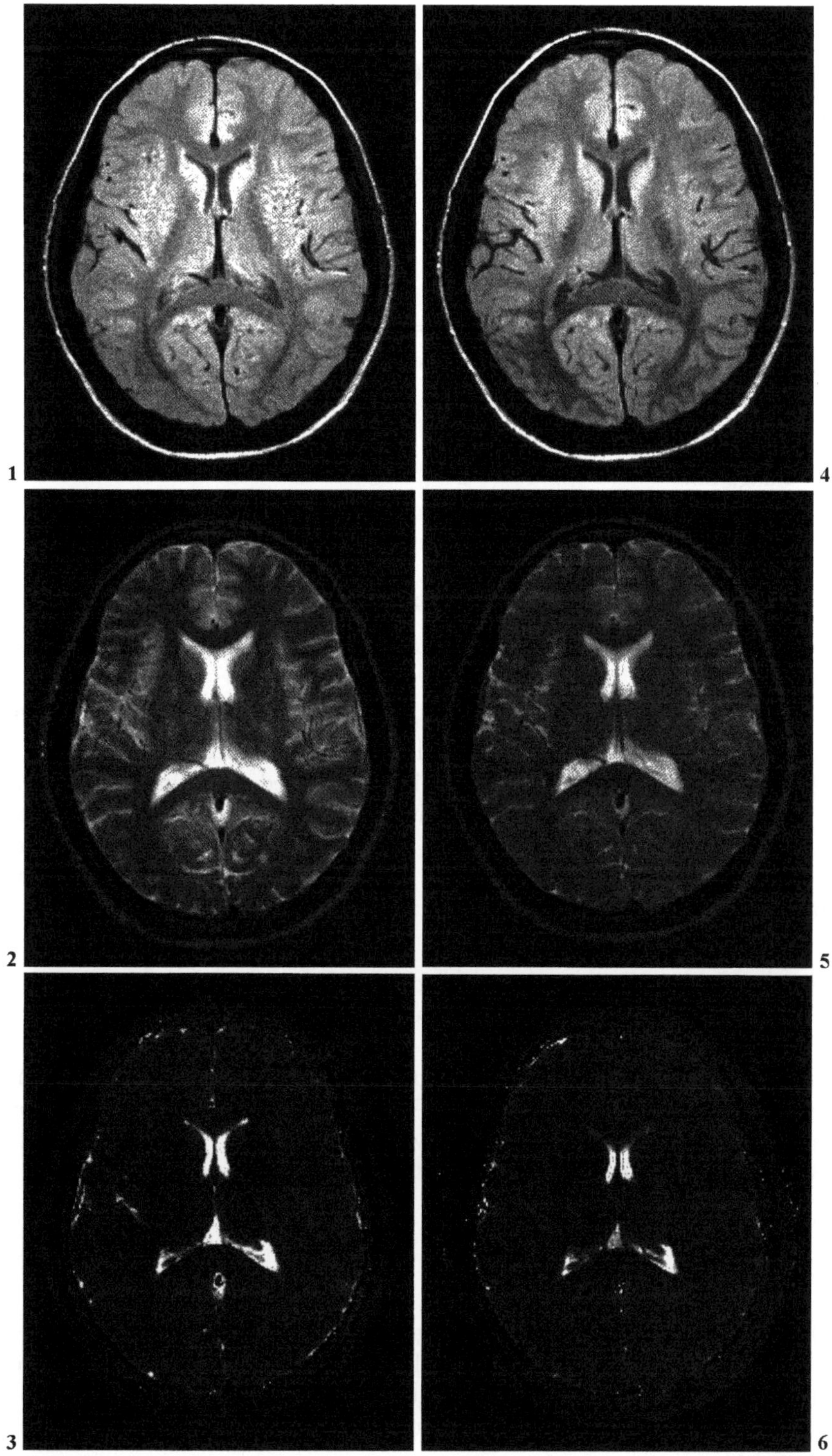


Image Set II-2

## II-3 Contrast in Multi-slice Field-Echo Imaging of the Cervical Spine: Variation of TR and Flip Angle

(1) Field-Echo (FE) imaging with a small flip angle and large TR is used frequently for $T_2$-weighted imaging of the cervical spine, so that the CSF is displayed bright. In this body region, the CSF has a pulsatile motion and SE images sometimes show strong ghosts and flow voids (see image set VII-8). The use of the FE method results in images that are relatively free of these defects. In the present set, FE images are shown with parameters that are approaching the optimum for this type of cervical-spine imaging.

Parameters: $B_0 = 1.5\,\mathrm{T}$; FOV $= 190\,\mathrm{mm}$; matrix $= 256 \times 256$; $d = 4\,\mathrm{mm}$; TE $= 14\,\mathrm{ms}$;

| image no. | 1 | 2 | 3 | 4 | 5 | 6 |
|---|---|---|---|---|---|---|
| TR (ms) | 210 | 210 | 500 | 500 | 800 | 1250 |
| flip angle (degr) | 6 | 20 | 20 | 40 | 40 | 40 |
| NSA | 4 | 4 | 2 | 2 | 2 | 1 |
| scan time | 3′36″ | | 4′18″ | | | 5′25″ |

NSA is adjusted to keep the scan time in the same range; between 3.5 and 5.5 minutes.

(2) The selected value of TE is a compromise; shorter TE would give insufficient contrast between the CSF and the spinal cord, longer TE would make the sequence too sensitive to susceptibility-induced field changes near the vertebral disks. At 14 ms, water and lipids are in phase for the 1.5 T system used. FE images with echo times that bring water and fat in opposed phase are unwanted for imaging of the spine, because they are disturbed by mutual extinction of water and fat in some pixels ("ink line" artifact; see image set IV-3).

(3) Images 1 ($\mathrm{TR}/\alpha = 210/6$); 3 ($\mathrm{TR}/\alpha = 500/20$), and 6 ($\mathrm{TR}/\alpha = 1250/40$) are closely similar in contrast. In these images, the increase in TR balances the increase in the flip angle (see Sect. 2.5 and (6.3g)). For each of them, the flip angle is small enough to avoid significant saturation of the longitudinal magnetization (see Fig. 4.8), so that $T_1$ differences do not influence the contrast and proton density or mild $T_2$ weighting results. This is true even for CSF, giving this tissue a bright aspect in the images.

Comparison of SNR in these images:

| image no. | 1 | 3 | 6 |
|---|---|---|---|
| SNR (CSF) | 17 | 29 | 32 |

(4) FE imaging is useful for imaging of the cervical spine. It can offer a contrast with bright CSF and gives a low artefact level. TR can be adjusted within wide margins. With the advent of Turbo-Spin-Echo imaging, in our institute the clinical interest in the use of FE for this purpose, however, has diminished.

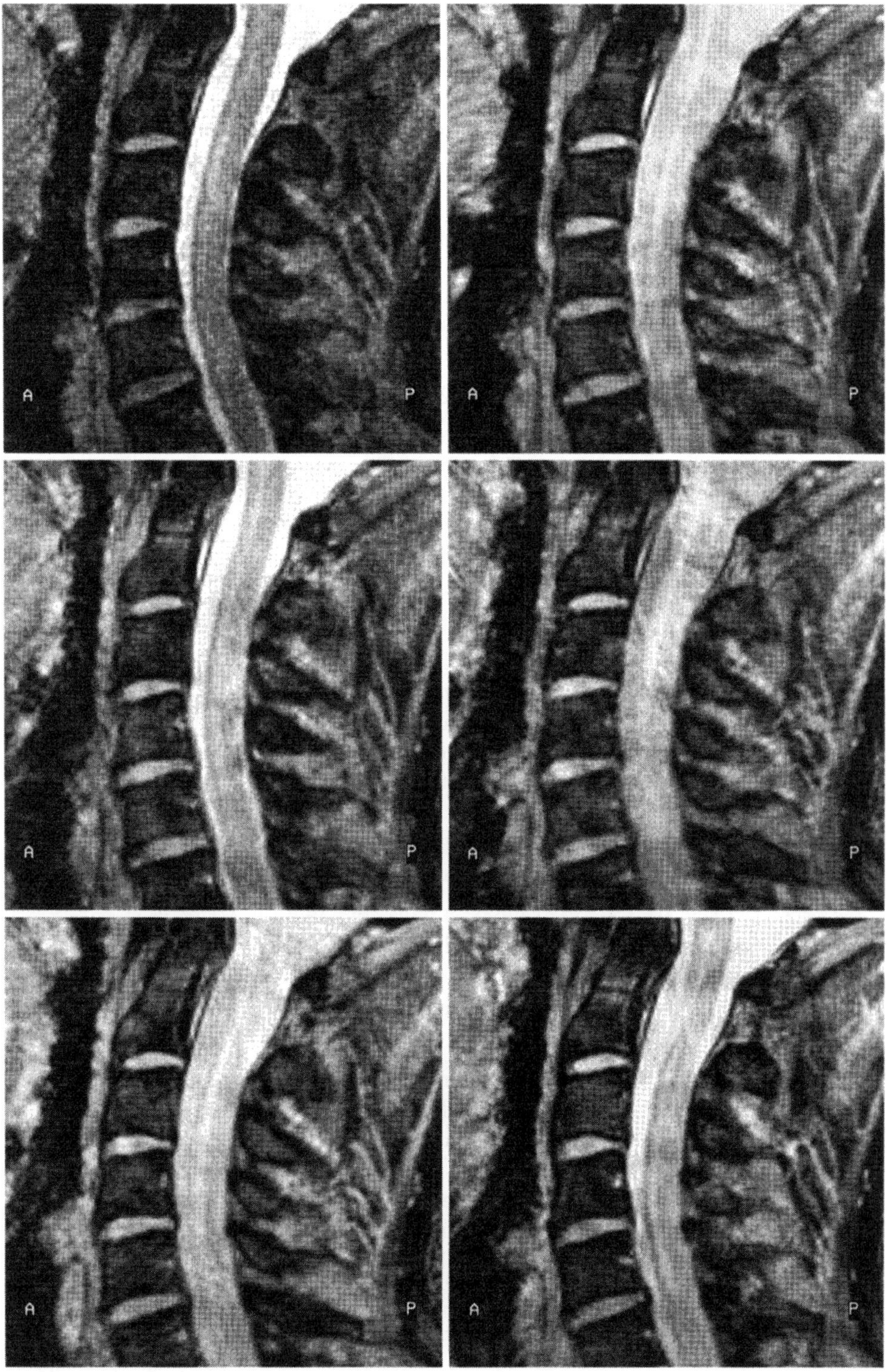


1 2 3 4 5 6

**Image Set II-3**

## II-4 Shift Between Water and Fat in FE

(1) The spatial shift of water and lipids can be an important cause ot difficulties in interpretation of the MR images when tissue containing lipids and water are both present (see Sect. 2.7.1). The relative shift of the lipid image with respect to the water image is in the direction of the read-out gradient as well as the selection gradient. To illustrate this difficulty, this set shows sagittal FE images of the knee, obtained from scans in which the sign of the selection gradient $G_s$ and the sign and direction of the read-out gradient $G_r$ are varied by permutation:

Parameters: $B_0 = 1.5\,\mathrm{T}$; FOV $= 160\,\mathrm{mm}$; matrix $= 256 \times 256$; $d = 3\,\mathrm{mm}$; TR/TE/$\alpha = 500/9.5/90$; NSA $= 2$.

| | | | | | | | | |
|---|---|---|---|---|---|---|---|---|
| image no. | 1 | 2 | 3 | 4 | 5 | 6 | 7 | 8 |
| phase encode gradient | AP | AP | AP | AP | FH | FH | FR | FH |
| $G_r$ | + | + | − | − | + | + | − | − |
| $G_s$ | + | − | + | − | + | − | + | − |

The strength of the read-out gradient was adjusted to correspond to an in-plane water–fat shift of 2 pixels (1.2 mm). The bandwidth of the RF pulses was 1000 Hz, corresponding to a fat shift perpendicular to the slice equal to 0.6 mm; 20% of the slice thickness.

(2) The direction of the in-plane shift of fat versus water can be found by comparison of the images, looking at the orientation of the dark and bright rims that appear in the images between fat- and water-rich tissue. In images 1 to 4, the fat has an in-plane shift in the FH direction. Comparison of images 1 and 2 with images 3 and 4 shows that in our system a positive sign of the direction of the read-out gradient corresponds to a caudal shift of the lipids. Similarly, comparison of images 5 and 6 with images 7 and 8 shows that in these cases a positive sign of the direction of the read-out gradient corresponds to a posterior shift of the lipids.

The direction of the displacement of lipids in other image orientations is a result from the direction and sign of the read-out gradient in those cases. The sign of this gradient in such image orientations can be read from image set II-8.

(3) The shift between fat and water perpendicular to the slice can create rims that are visually quite similar to those from the in-plane water–fat shift. Such rims will occur especially when the plane bordering a fatty tissue crosses the slice at some oblique angle. The width and the brightness of the rims depend on the value of this angle. The interpretation of the "true" anatomy, without shift, is still more difficult than for the in-plane shifts. Reversal of the selection gradient illustrates this reading difficulty (image 1 vs. 2, 3 vs. 4 etc.).

(4) The complexity of reading the knee images is demonstrated by comparison of the images in this set. At all borders between marrow, cortical bone, and

1 2

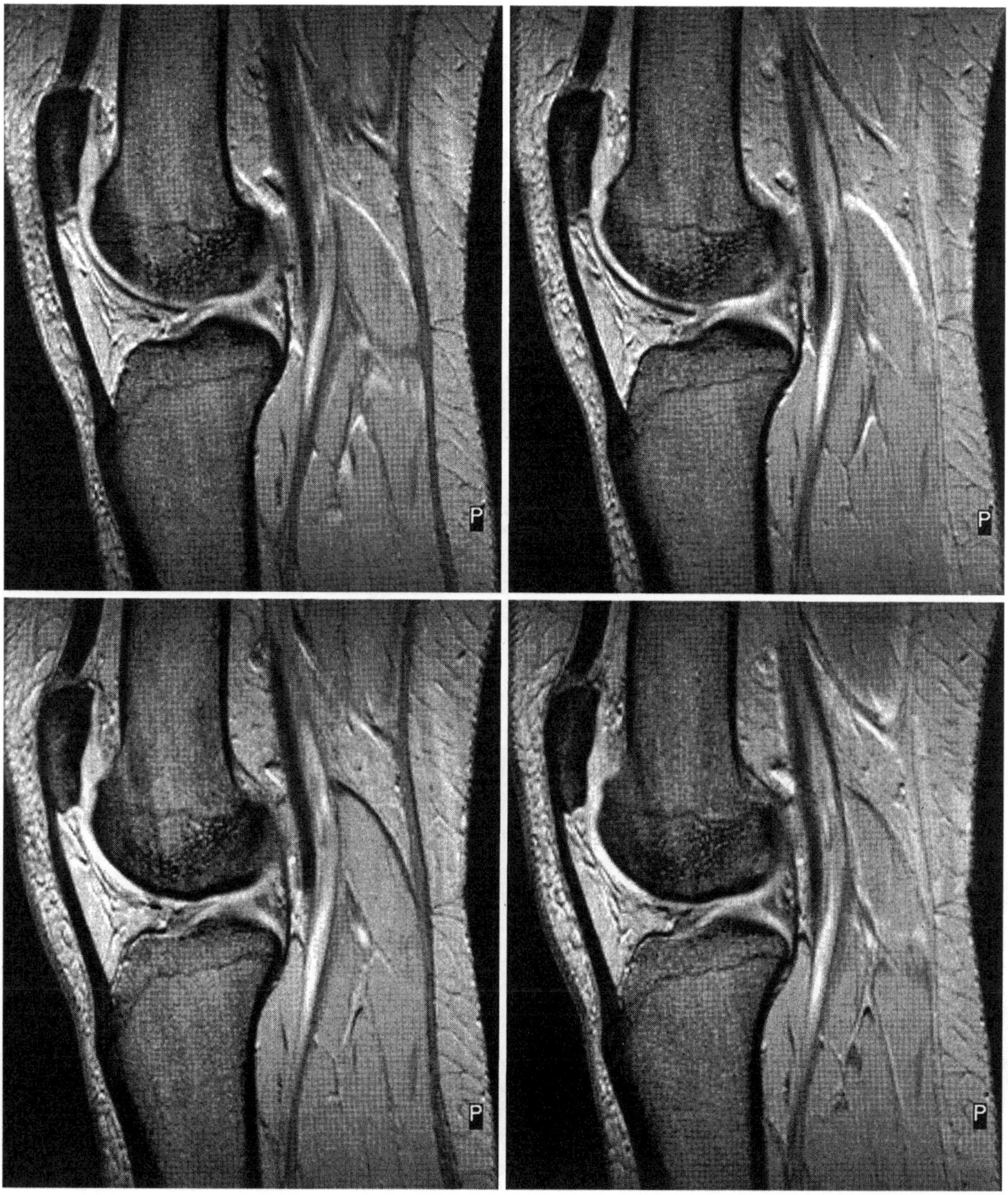

3 4

**Image Set II-4**

cartilage and between muscle and subcutaneous fat, reading difficulties arise from the shift of water and fat in two directions, in-plane and perpendicular to the slice. The true thickness of tissue layers near these borders, such as that

5 6

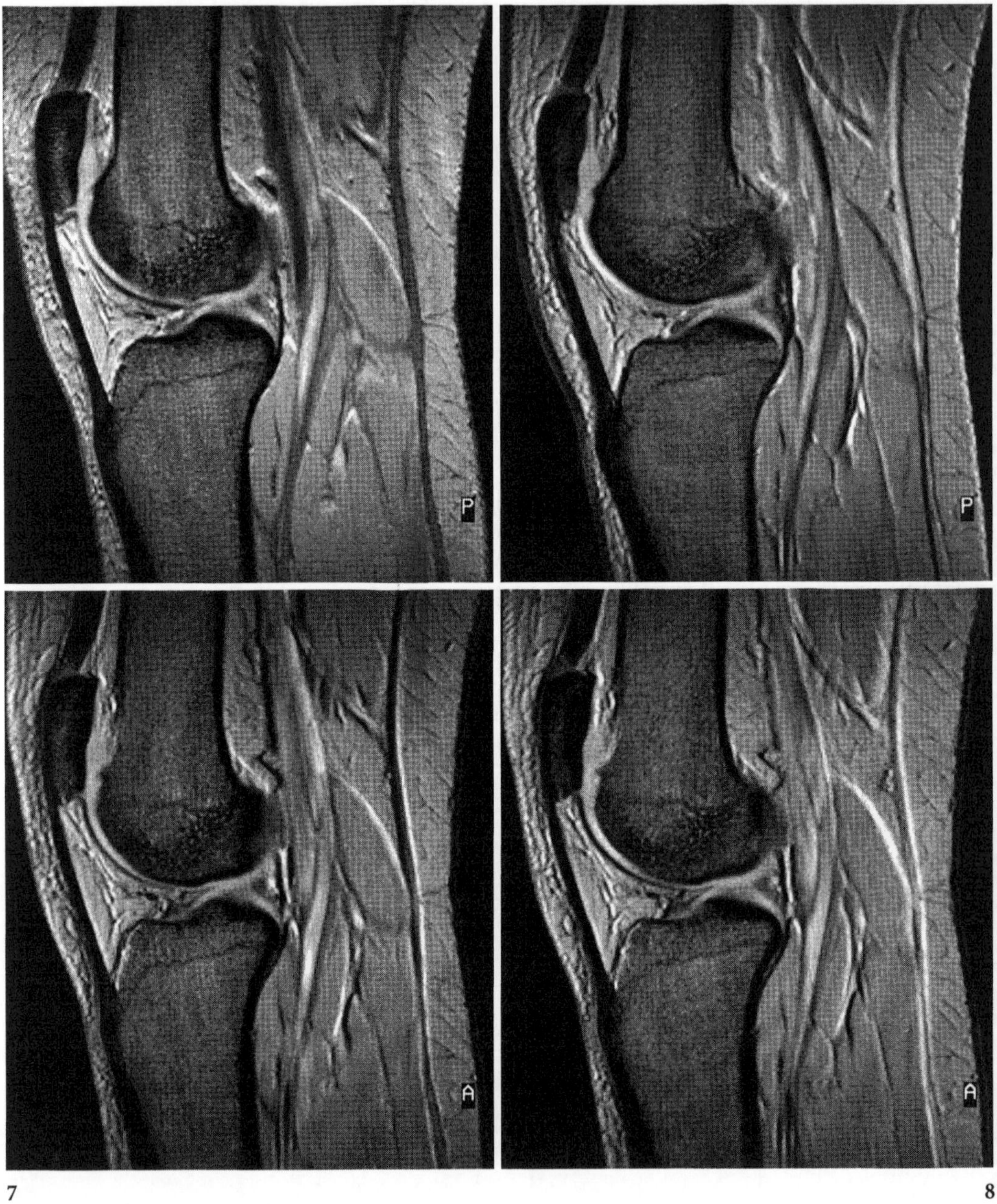


7 8

**Image Set II-4** (*Contd.*)

of cortical bone, cartilage, or tendons are difficult to estimate. The nature of rim-like dark and bright features of the image is difficult to assess. Such difficulties can be avoided when the signal from the fat is suppressed (see image set II-5).

## II-5 Suppression of Fat in Spin-Echo and Field-Echo Imaging

(1) Numerous techniques for the suppression of fat exist. Such techniques can be used to obtain important additional information in cases where fat may obscure the contours of anatomical details or the presence of pathological changes. Two of these techniques are compared in this image set. Short tau inversion recovery (STIR, Sect. 2.7.1) [17] relies on the choice of a suitable inversion time, chosen so that at the moment of excitation the longitudinal magnetization of lipids is zero ("nulling of lipid signal"). In spectral presaturation by inversion recovery (SPIR, Sect. 2.7.1) [18] the inversion pulse addresses the lipids only. So the inversion pulse is spectrally selective. This calls for a long FM-modulated RF pulse, taking about 20 ms at 1.5 T. This pulse has a flip angle that is only slightly larger than 90°, so that at excitation, following shortly after the SPIR pulse, the lipid signal is nulled. The SPIR technique requires a very homogeneous $B_0$ field. As a result it is vulnerable to field inhomogeneities such as occurring in off-centre positions of the image or near irregular air–tissue boundaries. STIR and SPIR techniques can both be applied to Field-Echo (FE) as well as to SE methods. This image set shows a comparison between these approaches:

Parameters: $B_0 = 1.5\,\mathrm{T}$; FOV = 140 mm; matrix = 256 × 256; $d = 4\,\mathrm{mm}$

| image no. | 1 | 2 | 3 | 4 | 5 | 6 |
|---|---|---|---|---|---|---|
| method | STIR + TFE** | SPIR + FE | STIR +SE | SPIR+SE | FE | FE |
| inversion delay | 150 | aut* | 150 | aut* | na | na |
| TR (ms) | 28 | 1200 | 1200 | 1200 | 500 | 500 |
| shot interval** | 1200 | na | na | na | na | na |
| shot length (TR) | 16 | na | na | na | na | na |
| TE (ms) | 14 | 14 | 25 | 25 | 18 | 18 |
| NSA | 2 | 1 | 1 | 1 | 2 | 2 |
| WFS (pixels) | 4.2 | 4.2 | 4.2 | 4.2 | 1 | 6 |

* The delay of the SPIR pulse is small. Its exact value depends on TR and is adjusted automatically by the system.

** The acquisition method TFE is a segmented Turbo-Field-Echo method, introduced in Chap. 5. The shot interval is defined as the time between the start of the segments (shots) in the segmented Turbo-Field-Echo method. All scan times were kept approximately equal (3–5 minutes) by adjustment of NSA. na = not applicable.

(2) The parameters of the scans for images 1 to 4 are aimed at visualization of cartilage. This tissue has a short $T_2$, so that the shortest-available echo time is wanted. On the other side, the absence of signal from lipids can be used to reduce the receiver bandwidth and the image noise (6.19) without the penalty of reading difficulty from the large misregistration between water and fat; (water–fat shift, WFS). This leads to a long acquisition time and

hence a longer TE. In each of the scans, this consideration was used to select a low bandwidth per pixel of 43 Hz. This choice will give a water–fat shift of 4.2 pixels and, in our system, an acquisition time of 23 ms, resulting in echo times of 14 ms (FE) and 25 ms (SE). For the SPIR images (images 2 and 4), local shimming of the $B_0$ field was performed after positioning of the knee so that inhomogeneity in this field due to the presence of the volunteer was minimized.

(3) Noise reduction by the use of a small bandwidth without fat suppression can complicate the reading of the image. To illustrate this, images 5 and 6 are added to this set. These are obtained with FE scans (TR/TE/$\alpha$ = 500/18/90). The scans for these images differ in bandwidth and accordingly in the value of the water–fat shift. Comparison of images 5 and 6 shows the gain in SNR from the use of a small bandwidth, and the associated difficulties in image interpretation at the borders of muscle or cartilage and lipids or bone marrow.

(4) The visualization of cartilage in both SPIR images (images 2 and 4) is about equivalent. In the STIR + TFE scan (image 1), the signal-to-noise ratio is lower than in the STIR + SE scan. This is due to the rather long TFE shots (shot length was 16 TR or 450 ms), so that recovery of magnetization after the shot and before the next inversion pulse lasts only 600 ms. (600 = 1200(shot interval)–450(shot duration)–150(inversion delay)).

(5) In STIR images (images 1 and 3), the signal-to-noise ratio is relatively poor compared with that in SPIR images (images 2 and 4). This is expected from the fact that in the STIR technique the magnetization of all tissue is inverted. When the longitudinal magnetization of the lipids passes through zero, the magnetization of water in most other tissues will be small as well. This is the case when the $T_1$ of such tissue is not sufficiently different from that of the lipids. In the SPIR technique, the magnetization of tissue water is not influenced by the spectrally selective preparation pulse and will be much larger for the same $T_1$.

(6) For the suppression of fat, the SPIR technique appears to give better results than STIR. It can be combined with SE as well as with FE methods. However, the application of the SPIR technique in strongly inhomogeneous $B_0$ fields is impossible because of its spectrally selective character. This may give occasional problems in clinical circumstances; in such cases one could revert to the more robust STIR techniques.

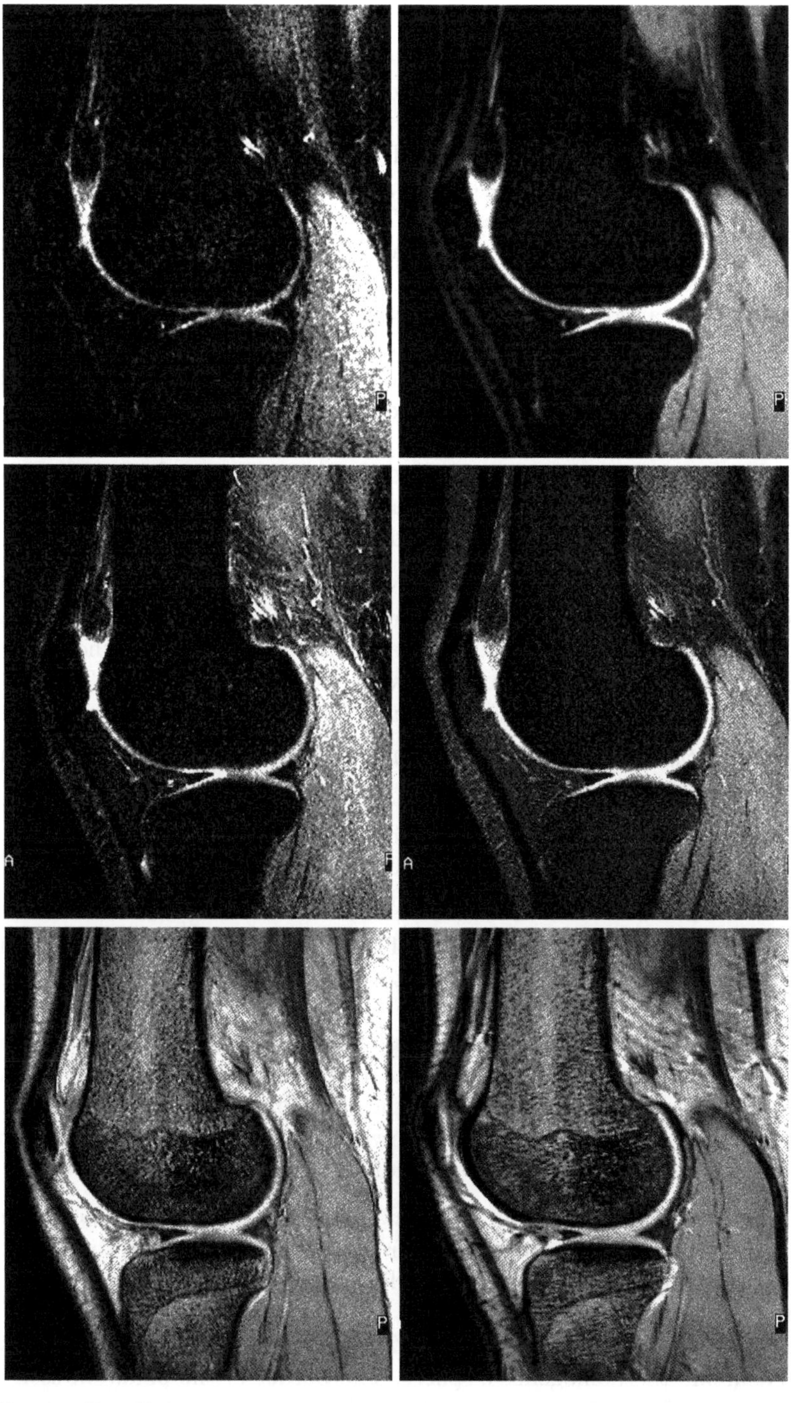

1 2 3 4 5 6

**Image Set II-5**

## II-6 Distortion in a Phantom

(1) Distortion of the MR image is the combined result of inhomogeneity of the main field and non-linearity of the gradient field (see Sect. 2.6). In MR systems, typically the inhomogeneity of the main field $B_0$ is defined at the diameter of a "sphere of homogeneity", within which the inhomogeneity is expressed in ppm of its isocentre value. The $B_0$ inhomogeneity of our system has a value better than 5 ppm within a diameter of 50 cm. As discussed in Sect. 1.3.2, the inhomogeneity grows as a strong power of the distance $\vec{r}$ to the isocentre; in a modern magnet, for example, it grows as $\vec{r}^{12}$. So, distortion due to field inhomogeneity will be observable in particular near the edge of the homogeneity sphere. Distortion from inhomogeneity of the main field will be visible only in the direction of the read-out gradient, in particular when the value of this gradient is low.
Distortion from non-linearity in the gradients will be a less sharp function of $\vec{r}$; it will occur for both the read-out and the phase-encode gradients and it will not depend on the value of the gradient used.

In this image set some transverse images each with a large field of view will be compared. SE images are obtained from a phantom with a square array of equidistant markers that is placed perpendicular to the $z$ axis of the system, at three $z$ positions. The distance between the markers is 5 cm. The "MR visible" diameter of the phantom is 38 cm.

Parameters: $B_0 = 1.5$ T; FOV = 450 mm; matrix = 256 × 256; $d = 8$ mm; TR/TE = 300/20.

| | | | | | | |
|---|---|---|---|---|---|---|
| image no. | 1 | 2 | 3 | 4 | 5 | 6 |
| $z$ position (cm) | 0 | 0 | 10 | 10 | 20 | 20 |
| read-out direction | LR | AP | LR | AP | LR | AP |

(2) In images 1 and 2, the cross section of the phantom is contained entirely in the homogeneity sphere. Very little distortion is visible. The inhomogeneity of the main field and the non-linearity of the gradients appear to be very small for the magnet region corresponding to these images. Close observation shows a slight difference (of the order of 1%) in the horizontal and vertical image size. This difference rotates with the direction of the read-out gradient. It represents a slight misadjustment between the gradient strengths of the read-out gradient and phase-encode gradient.

(3) In images 3 and 4, the slice is 10 cm from the isocentre. The borders of the phantom in this slice are 21.5 cm from the isocentre, close to the edge of the homogeneity sphere. The image distortion still is very low. There is a slight (2%) increase in magnification.
In images 5 and 6, the slice is taken at 20 cm from the isocentre. Part of the slice is outside the homogeneity sphere. The border of the phantom seen in this slice is at 27.6 cm from the isocentre and outside the homogeneity sphere.

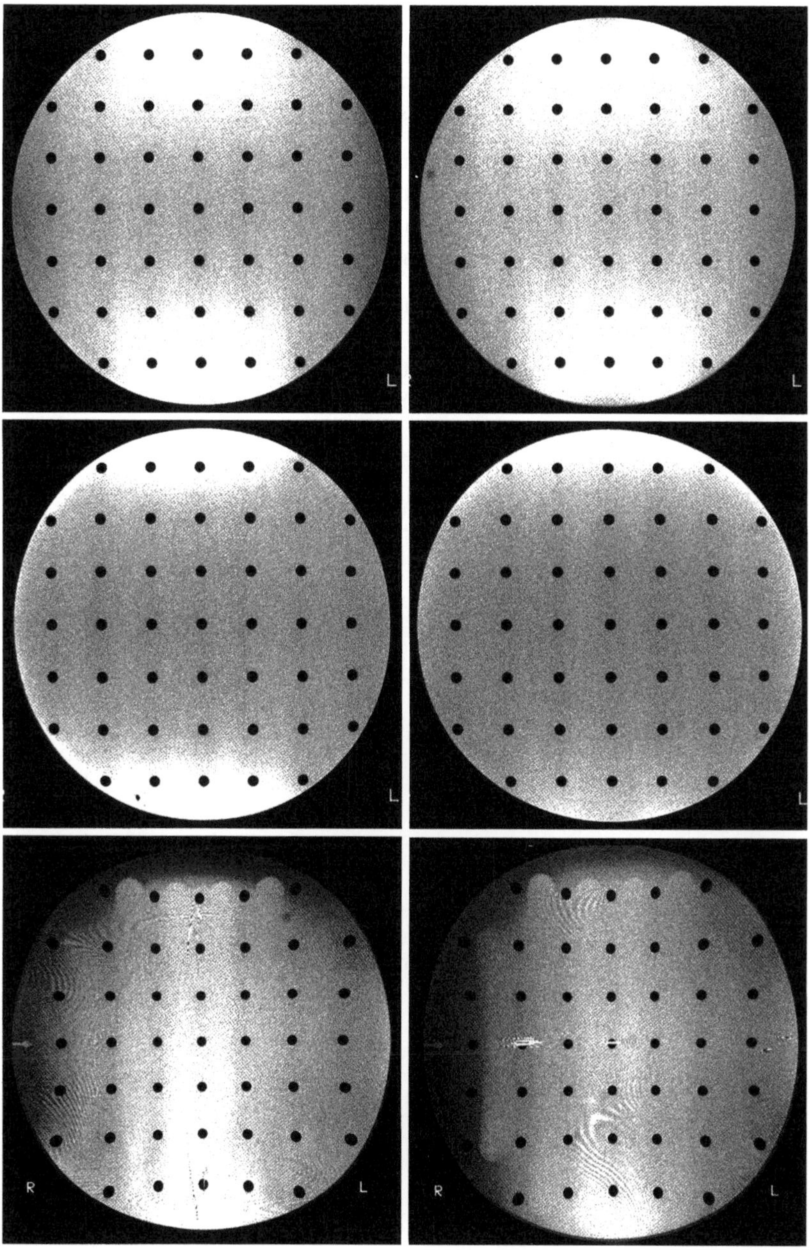


1 2 3 4 5 6

**Image Set II-6**

There is a clearly visible cushion distortion in both images. Everywhere outside the image centre, the magnification has increased. This increase is up to 20% at the border of the phantom. Note that in these images a contribution from a stimulated echo is visible as a sharply delineated bright artifact.

(4) Even at 20 cm off centre, the distorted shape of the phantom in the images is not visibly dependant on the direction of the read-out and phase-encode gradients. This implies that the distortion from the main field is negligible compared with that caused by the gradients.

(5) The specified diameter of the homogeneity sphere appears to be a conservative descriptor of the region that can be used for imaging. A margin of a few cm outside this sphere can contribute to the image when one wants to accept the distortion that will occur in the part of the object within this margin.

One should, however, be careful; distorted images can give rise to complicated aliasing and interference of such aliases with the normal image, as will be discussed with help of image set II-7.

# II-7 Aliasing and Interference in Coronal Abdomen Images: SE and FE

(1) The homogeneous region of the magnet is limited in size (Sect. 1.3.2). Usually a spherical area, the "homogeneity sphere", with a diameter of about 50 cm is specified to be homogeneous within a few ppm. At the borders of the homogeneous area the field deviations increase sharply, which can lead to strong image distortion of eccentric body regions. The distortion pattern near the edge of the homogeneity sphere is demonstrated in image set II-6. Aliasing from strongly distorted regions can occur easily and can be difficult to interpret. In this image set, some cases are compared. The Spin-Echo (SE) and Field-Echo (FE) images are coronal views of the abdomen.

Parameters: $B_0 = 1.5\,\mathrm{T}$; FOV $= 500\,\mathrm{mm(FH)} \times 400\,\mathrm{mm(LR)}$; matrix $= 256 \times 205$; $d = 10\,\mathrm{mm}$; phase-encoding direction is LR. TR/$\alpha$ = 300/90. NSA = 2

| image no. | 1 | 2 | 3 | 4 | 5 |
|---|---|---|---|---|---|
| method | SE | FE | FE | SE | FE |
| TE | 15 | 15 | 3.1 | 15 | 15 |
| phase-encode oversampling | no | no | no | yes | yes |

(2) In the corners of images 1 to 3, aliasing in the phase-encode direction (LR) is visible. Aliasing is the superposition in the normal (central) image of image information obtained outside the field of view (see Fig. 2.9). Aliasing can occur when the field of view is smaller than the object and when no oversampling is used. Oversampling is defined as the provision that avoids aliasing by taking more samples. The oversampled image has a large field of view during acquisition, but not during display. In the images shown, the alias signal originates from a region outside the homogeneity sphere. As a consequence the alias image is strongly distorted. So for instance the hand of the volunteer is distorted into a sharply bordered semicircular shape overlaying the pelvis. When oversampling in the phase-encode direction is used (images 4 and 5, different volunteer), aliasing is prevented.

(3) In the FE images (images 2 and 3), the alias does not simply add to the signal of the central image, but a zebra like interference pattern is generated. The pattern arises from the fact that the aliasing signal is generated outside the homogeneity sphere and is dephased with respect to the normal signal. The dephasing is caused by the difference in the main field strength inside and outside the field of view (see in (2.45) the factor $\exp\{-\mathrm{j}\gamma\delta B\mathrm{TE}\}$). At the interference maxima the alias image and the central image are in phase and at the minima they are in opposed phase. The rapid spatial change of the phase difference is caused by the strong gradient in the main field outside the field of view. (See also image set II-9).

(4) The relation between interference or the zebra pattern and field inhomogeneity is illustrated further by comparing different echo times in the FE scan

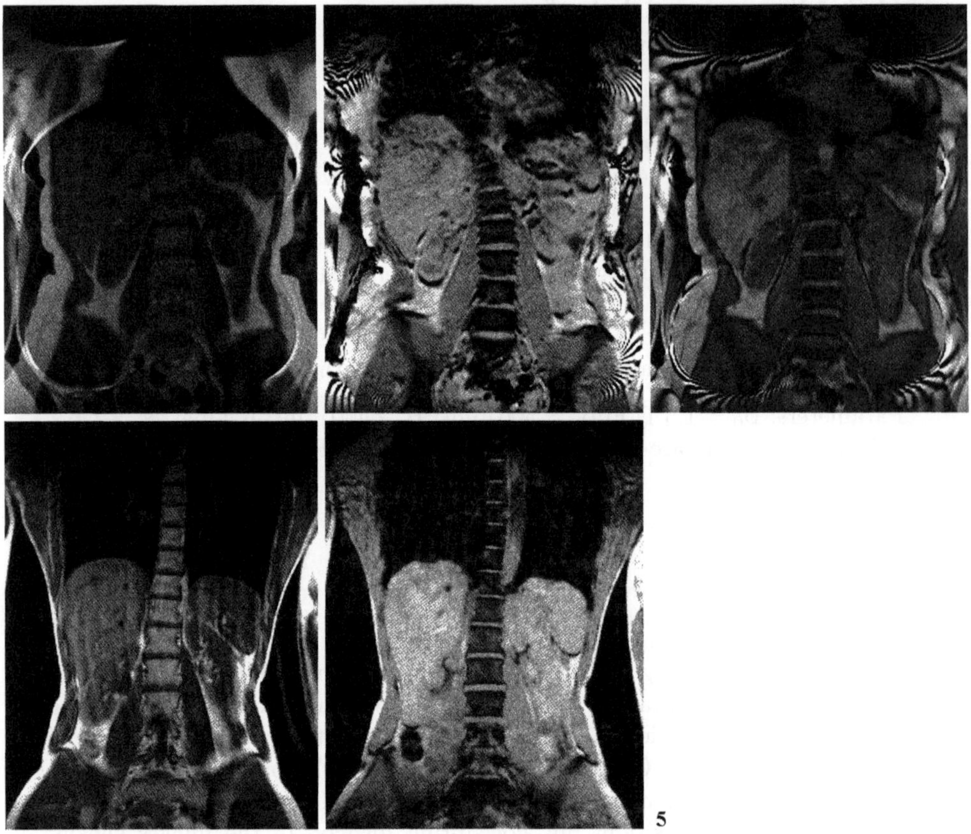
1,2 3 4 5

**Image Set II-7**

(images 2 and 3). In FE imaging, at a given field, the phase is proportional to the echo time (see (2.45)). The distance between the maxima in the zebra pattern in image 3 is much larger than that in image 2. With the long echo time used for this last image, the phase changes so rapidly with position that dephasing within each voxel causes extinction of parts of the alias image, for example in the region of the hands.

(5) The absence of an interference pattern in SE images (image 1) reflects the fact that in the SE method the phase of the signal does not depend on the field inhomogeneity (see (2.43)).

## II-8 Distortion in the Image of a Phantom with Rods of Deviating Susceptibility

(1) Within the homogeneity sphere, the main field will be made inhomogeneous by the presence of the patient, of whom the susceptibility differs from that of air. Especially near air pockets with strongly curved boundary, the gradient in the field value will be large enough to cause visible image distortion. This is illustrated with SE images from a water-filled phantom that contains three sets of round, air filled perspex tubes in three mutually perpendicular directions (see inserted sketch).

Parameters: $B_0 = 1.5\,\mathrm{T}$; FOV $= 300\,\mathrm{mm}$; $d = 5\,\mathrm{mm}$; TR/TE $= 300/25$; NSA $= 1$

| | | | | | | | | |
|---|---|---|---|---|---|---|---|---|
| image no. | 1 | 2 | 3 | 4 | 5 | 6 | 7 | 8 |
| slice angle* | 0 | 0 | 30 | 30 | 60 | 60 | 0 | 0 |
| # tubes cut | 2 | 2 | 4 | 4 | 1 | 1 | 2 | 2 |
| tube angle* | 90 | 90 | 60 | 60 | 30 | 30 | 90 | 90 |
| direction of $G_\mathrm{r}$ | $+x$ | $-z$ | $-y$ | $-z$ | $-y$ | $-x$ | $-z$ | $-z$ |
| value of $G_\mathrm{r}(mT/m)$ | 2.7 | 2.7 | 2.7 | 2.7 | 2.7 | 2.7 | 1.8 | 0.9 |

* The slice angle and tube angle are defined with respect to the direction of the main field ($z$). $x$ and $y$ are the directions perpendicular to this field.

(2) All slices were perpendicular to one set of tubes (tubes cut) and parallel to both other sets. The angle to the $z$ axis of the tubes cut is given and is of course 90° different from the angle between the $z$ axis and the slice. The value of the read-out gradient $G_\mathrm{r}$ cannot be adjusted directly in the user interface of our system, but is the consequence of the choice of bandwidth and field of view.

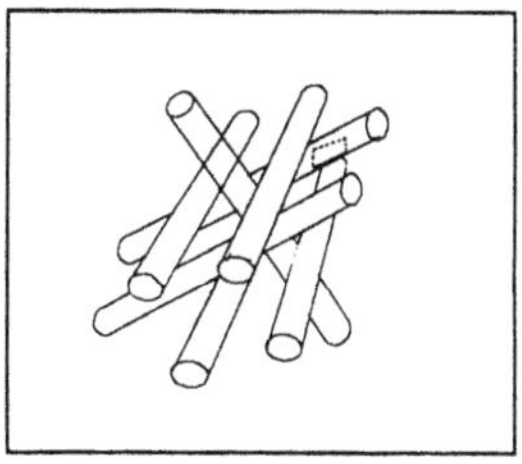

Tubular phantom

(3) The cross section of the rods is seen to be distorted to an arrowhead shape, pointing in the direction of $G_\mathrm{r}$. This shape is the consequence of the field deviation around the tubes. At the tip and at the shaft end of the arrow head the field is increased and the image shifts to the positive direction of $G_\mathrm{r}$. At the wings of the arrowhead, the field is decreased and the image shifts to the negative direction of $G_\mathrm{r}$. The distortion is the largest when the rods are perpendicular to the main field (images 1 and 2, compared with images 3 to 6) and when $G_\mathrm{r}$ has a low value (images 7 and 8 compared with 1 and 2). The distortion is accompanied by modulation of the image brightness. (See Sect. 2.6, below (2.45) for a mathematical description).

(4) In all images, the arrowheads point either to the right side of the image or to the top of the image. This, of course, is the consequence of the choice of the sign of $G_\mathrm{r}$ used by the manufacturer. Note that in image 6 the sign

1 2

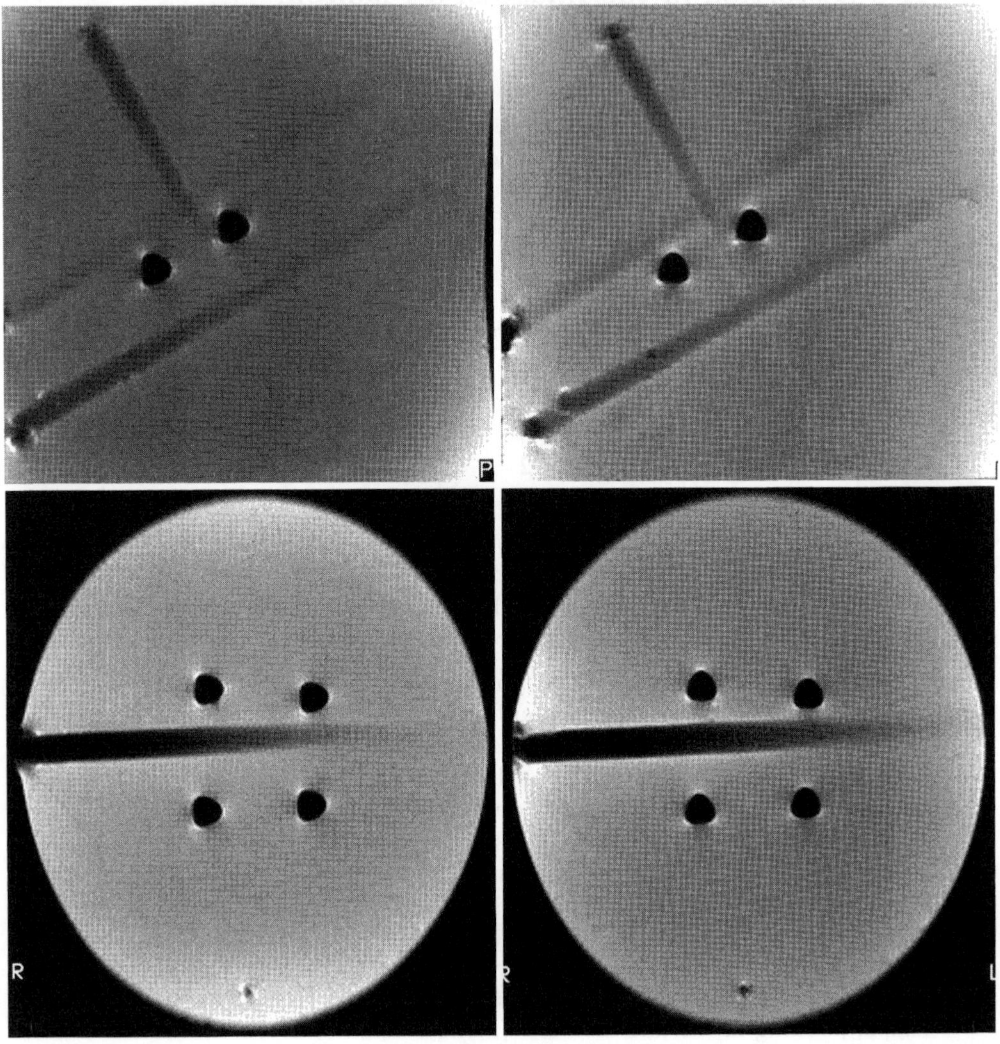


3 4

**Image Set II-8**

of $G_r$ differs from that in image 2. As far as we know, there is no agreement between manufacturers on the choice of the sign of the gradients used in MR imaging.

(5) In image set III-7, other examples are given of image distortion from field deviations near air pockets (nasal sinuses); showing the influence of the field strength on this distortion.

5 6

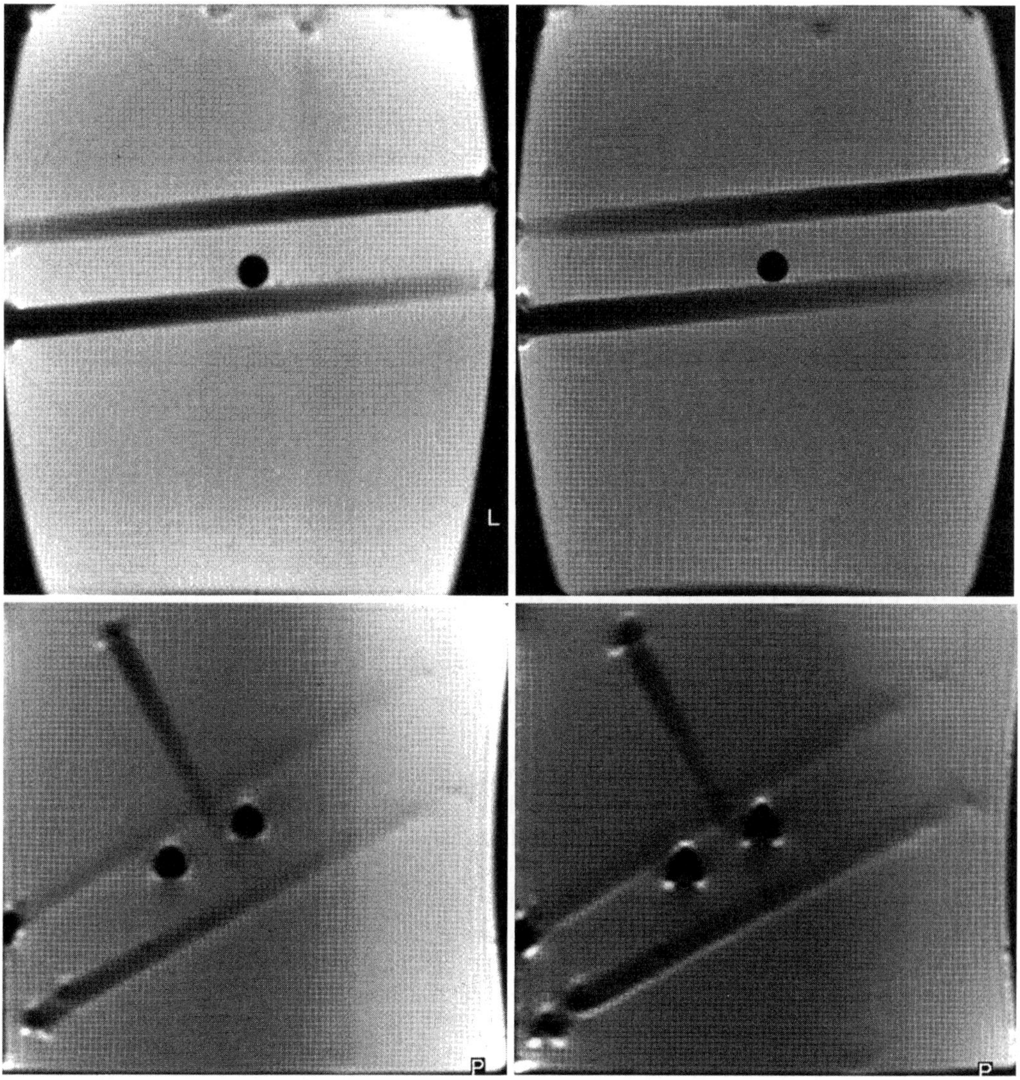


7 8

**Image Set II-8** (*Contd.*)

(6) Even for normal values of the read-out gradient (2.7 mT/m), at high field strengths, distortion and brightness modulation in the images will occur near small air pockets, like the tubes used in the phantom for this image set.

## II-9 Phase Distribution in Field-Echo Images as a Sign of a Main Field Inhomogeneity

(1) Ideally, at the centre of the echo, the spins all have phase zero. In Field-Echo methods, the phase depends in a sensitive way on the strength of the main field, see (2.45), and this condition will be reached only when this field is completely homogeneous. In real systems, the main field is inhomogeneous and the spin phase will differ considerably between voxels. This is the reason that for FE images of the type modulus are the only images that are useful for the study of tissue properties. The images of type phase do, however, reflect in an interesting fashion the inhomogeneity of the field. Coronal FE phase images of the head are compared in this image set.

Parameters: $B_0 = 1.5\,\mathrm{T}$; FOV $= 350\,\mathrm{mm}$; matrix $= 256 \times 256$; $d = 6\,\mathrm{mm}$; TR/TE/$\alpha$= 100/15/60; NSA = 4

| | | | | | | |
|---|---|---|---|---|---|---|
| image no. | 1 | 2 | 3 | 4 | 5 | 6 |
| volunteer | A | A | B | B | B | B |
| $z$ position (cm) | 0 | 6 | 0 | 0 | 0 | 0 |
| slice | 1 | 1 | 1 | 2 | 3 | 4 |

(2) Volunteer A was shifted along the $z$ axis over 6 cm between the scans. The slice number in volunteer B reflects a difference in the slice position in the AP direction. The slice in image 3 passes through the frontal nasal sinus; images 4, 5, and 6 are of slices shifted over 11, 22, and 33 mm relative to image 3.

(3) The usual display of a phase image is based on encoding the brightness per pixel proportional to the phase of the pixel value. The deviations in phase between the voxels are visible as a gradual change in image brightness. When the phase is zero, the pixel is mid grey. At $+180°$, the image is top white. At this moment a further increase in the spin phase results in a sharp step in brightness to top black, because the spin phase is recorded modulo 360° and has obtained a negative value. A threshold algorithm has replaced the uncertain brightness of the air pixels around the head and in the frontal nasal sinus by zero. A phase-correction algorithm has removed trends in the phase in the direction of the read-out gradient.

(4) With the precautions stated, the phase differences in the image reflect differences in the main field. One should be aware that a region of homogeneous brightness in the phase image does not correspond necessarily to a region of homogeneous main field. This is due to the phase-correction algorithm mentioned above. Per consequence, a region of homogeneous pixel brightness may also represent a region with a constant field slope. The existence of such a slope is, however, unlikely in a normal situation.

(5) At a field strength of 1.5 T and at an echo time of 15 ms, a phase deviation of 360° corresponds to a change in the main field of approximately

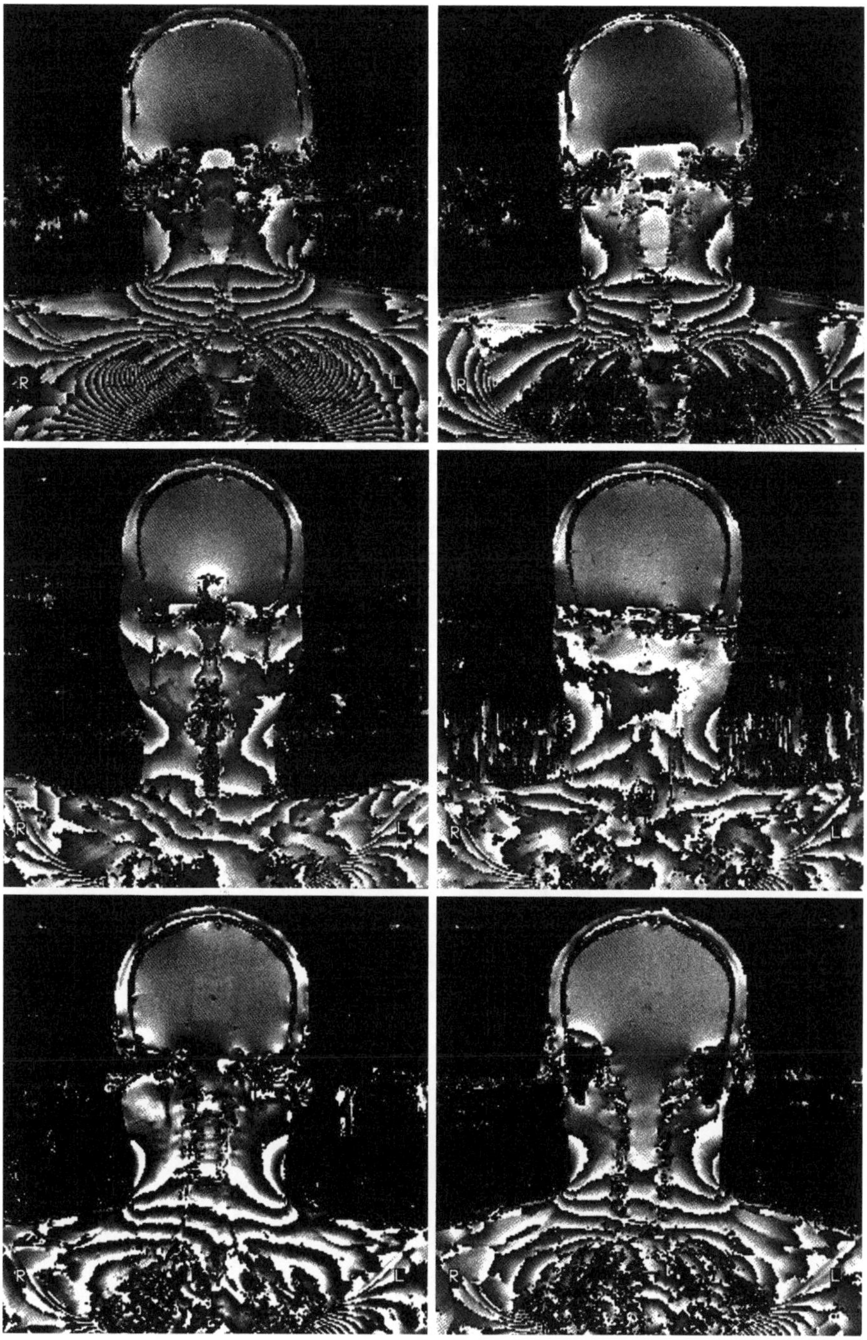


1 2 3 4 5 6

**Image Set II-9**

1 ppm (66 Hz). Hence, field deviations in ppm between each pair of points can be read from these images by counting the number of steps in the image brightness.

(6) Near the frontal nasal sinus, the main field pattern is disturbed by about 1 ppm in both volunteers. At the level of the spinal canal, this field pattern is much less disturbed (image 6). The field disturbance near the sinus appears not to be related to the shift in position of the volunteer with respect to the isocentre (compare images 1 and 2). However, moving from the head to the thoracical space corresponds to a field change of about 7 ppm in image 1 and of about 5 ppm in image 2. So this field change is related to the shift in the volunteer position.

(7) The most important contribution to the main field inhomogeneity in the region of the head is due to the difference between the susceptibility of the head and shoulders and that of the surrounding air. Its influence on the spin phase in a FE image is drastic. Practical consequences of this inhomogeneity are discussed with the help of image sets II-8 (distortion in SE), II-10 (intensity modulation in FE), and III-5 (distortion in EPI).

## II-10 Susceptibility Influencing FE Images of the Brain at Three Values of the Main Field Strength

(1) In Field-echo imaging, the signal strength depends on $T_2^*$. This parameter describes the signal decay due to dephasing of the spins in a locally inhomogeneous static field (Sect. 2.5). The definition of $T_2^*$ is not very strict. Its value depends on the distribution of the field over the voxel volume and the dephasing over this volume (intra-voxel dephasing). For large voxels, $T_2^*$ is small even when the gradient in the static field is small. $T_2^*$ can in such cases vary significantly from voxel to voxel. This is for instance the case when the static field is disturbed by local magnetic inhomogeneities such as the air pocket of the nasal sinus. The variation in $T_2^*$ in the neighbourhood of such a local magnetic inhomogeneity can be found in the image as a visible change of signal strength with distance to the locus. Of course this occurs more easily at large TE.
In this image set, FE images of a slice near the nasal cavities are shown for a volunteer, imaged at three field strengths. The FE methods were designed to be completely equal for each field strength.

Parameters: FOV = 230; matrix = 256 × 256; $d$ = 6 mm; TR/TE/$\alpha$ = 400/15/20. Other slices of the same scans are shown in image set VI-7.

| image no. | 1 | 2 | 3 |
|---|---|---|---|
| $B_0$ (T) | 0.5 | 1.0 | 1.5 |

(2) All images show some signal loss in the region above the nasal cavity. This is the region where the intra-voxel dephasing has lead to very short $T_2^*$. The signal loss is the strongest in the image made at the highest field strength. This is explained by the fact that the absolute value of the static field inhomogeneity caused by the nasal sinus is proportional to the field strength, so that the dephasing inside voxels in anatomically equivalent locations increases.

(3) In FE imaging, intra-voxel dephasing can give signal voids at places where the magnet field is disturbed. The dephasing is more disturbing at a high field strength and (not shown) at large echo times and for large voxels.

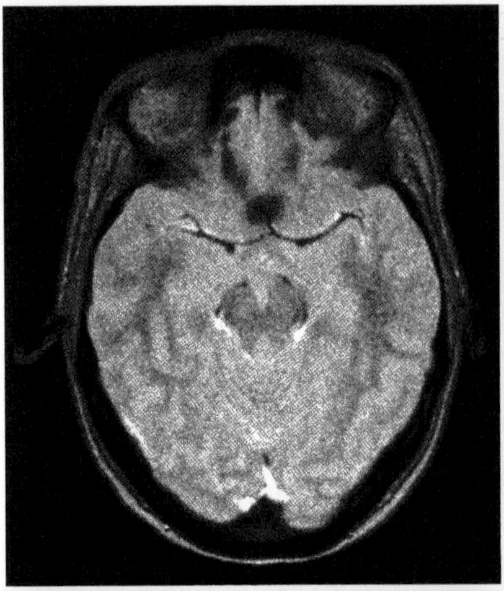

1

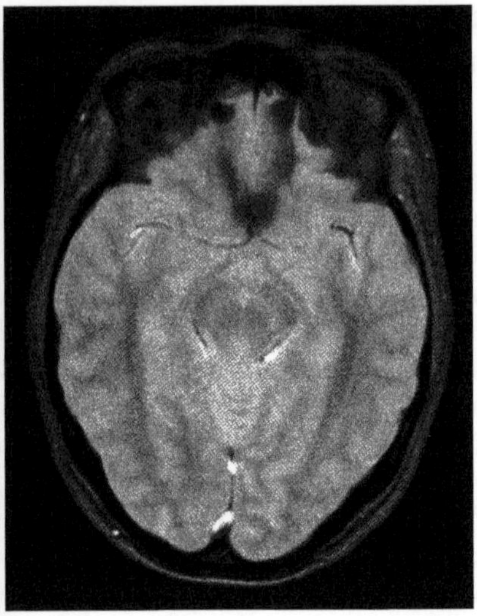

2

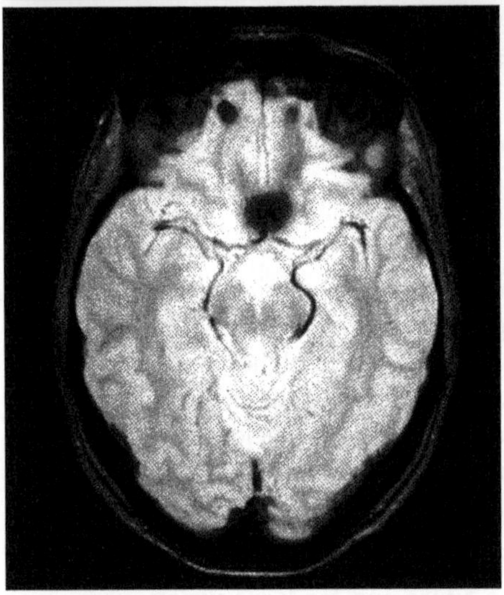

3

**Image Set II-10**

## II-11 Cross Talk in Multi-slice Spin-Echo and Field-Echo Imaging

(1) In multi-slice (MS) techniques, it has to be assumed that each slice is influenced to some extent by the excitation of neighbouring slices within the same repetition time. This assumption is based on the expected overlap of the slice profile, as determined by the spatial profiles of the RF pulses. The extent of overlap is not only a function of slice distance but also of the design of these pulses; see Sect. 2.3.2. Long RF pulses allow a more precise slice profile than do short pulses. In any realistic MR system, some compromise between RF pulse duration and slice overlap will have to be made. In Spin-Echo imaging, the problem regards both the excitation pulses and the refocusing pulses. This image set illustrates effects from slice overlap in some SE and FE scans. (Another cause of mutual influence between slices in multi-slice scans is magnetization transfer, illustrated in image set VI-4. The technical factors in the present set are chosen to avoid a strong influence of this last effect).

Parameters: $B_0$=1.5 T; FOV = 160 mm; matrix = 256 × 256; $d$ = 3 mm; TR/TE/$\alpha$ = 400/15/90; NSA = 2.

| image no. | 1 | 2 | 3 | 4 | 5 | 6 | 7 | 8 |
|---|---|---|---|---|---|---|---|---|
| method | SE | SE | SE | SE | SE | FE | FE | FE |
| no. slices | 9 | 9 | 9 | 1 | 1 | 9 | 9 | 1 |
| slice gap (mm) | 0 | 0.3 | 0.6 | na | na | 0 | 0.3 | na |
| scan mode | MS | MS | MS | MS | SS | MS | MS | MS |

na = not applicable; SS = single slice

The smallest slice gap allowed in our system is normally 10% of the slice thickness for SE imaging, as used for image 2, and zero for FE imaging, as used for image 6. The SE image with zero slice gap (image 1) is experimental. A SE image obtained in a single-slice scan mode is included in the set (image 5), because in this scan mode, contrary to the multi-slice mode, our system employs a refocusing pulse that is not slice selective. This image is included to inspect the influence of the slice-selective nature of the refocusing pulse on the quality of the image.

(2) Cross talk phenomena will depend on the time delay between the excitation of neighbouring slices. In our system, the default temporal order of addressing of the slices per TR is designed to approach a time delay that is equal for each neighbouring pair. In scans with an uneven number of slices, as the ones shown here, that equality is fully reached.

(3) The expected influence of slice cross talk in the first place is a reduction of signal strength. So, to inspect the cross talk, values of SNR for the subcutaneous fat observed in each of the images are compared:

| Image no. | 1 | 2 | 3 | 4 | 5 | 6 | 7 | 8 |
|---|---|---|---|---|---|---|---|---|
| SNR (fat) | 10 | 11 | 12 | 13 | 16 | 9 | 10 | 11 |

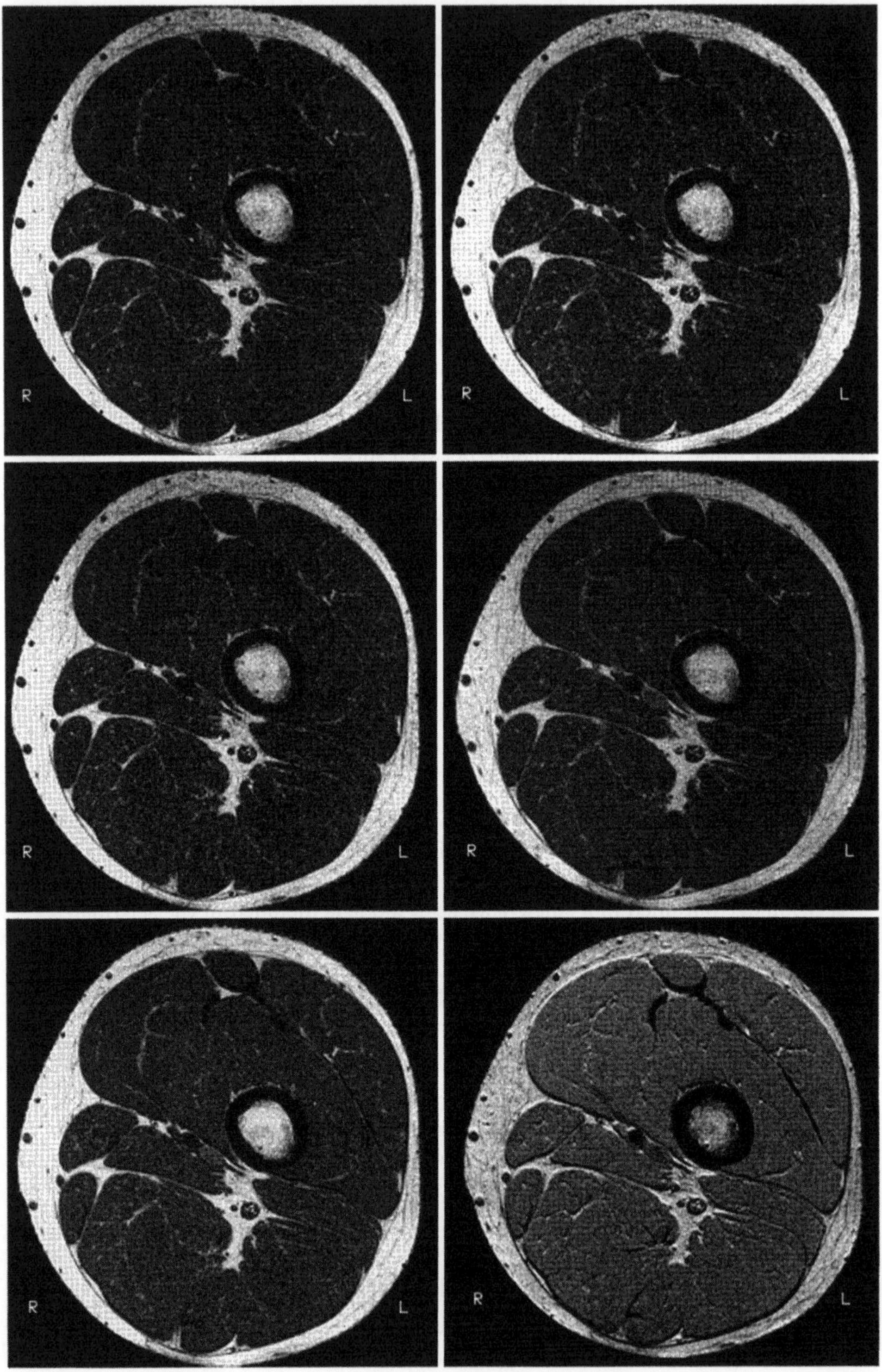


1 2 3 4 5 6

**Image Set II-11**

7 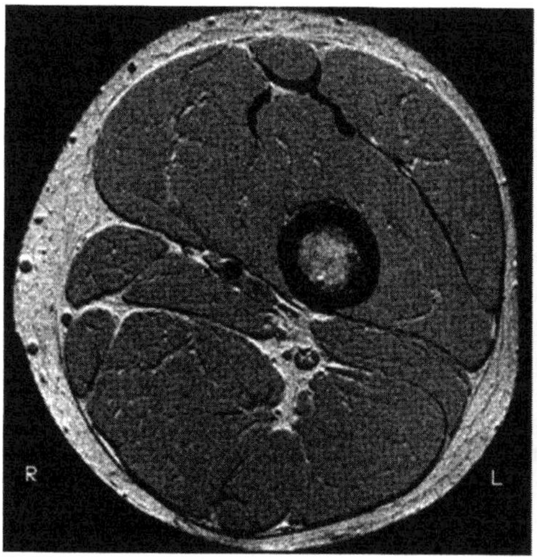


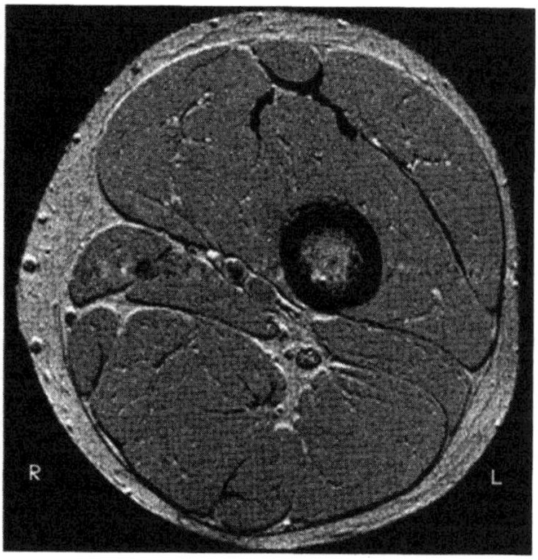 8


**Image Set II-11** (*Contd.*)

The values are seen to vary in the expected fashion. The SNR in the FE images is somewhat lower, which could be due to a short $T_2^*$ of lipids.

(4) A comparison of images 2 and 3 with image 1 shows that the multi-slice SE images have a relatively small amount of cross talk ($< 20\%$) as long as the slice gap is equal or larger than its minimum-allowed value; 10% of the slice thickness. Apparently the system maintains a sharp match between the slice thickness excited and the slice thickness refocussed. A comparison between images 3 and 4 shows that the use of a slice-selective refocussing pulse, compared to a non-selective ("hard") refocussing pulse, reduces the SNR by another 20%. The contrast of all SE images shown is not visually different.

(5) A comparison of images 6, 7, and 8 shows that cross talk in FE is present as well; its value is relatively small ($< 20\%$), even in the case of a zero slice gap. Apart from cross talk, performance differences between multi-slice and single-slice FE imaging are not expected, because the excitation pulse is slice selective in both cases. Such a performance difference exists between multi-slice and 3D gradient-echo imaging, as discussed in image set IV-2.

(6) The use of the multi-slice technique reduces the signal-to-noise ratio when compared with single-slice imaging. Part of this reduction is due to slice cross talk. In the images shown, SE and FE, slice cross talk reduces the SNR by less than 20%. Enlargement of the slice gap above the normally allowed value does not eliminate the effect. For SE, the use of a multi-slice technique reduces the SNR by a further 20% because the use of a slice-selective refocussing pulse is needed in this acquisition method.

(7) The cross talk phenomenon observed in this image set may differ with the relaxation properties of the tissue being imaged. These differences are not explored. If present, they would show up in a way that mimics normal partial volume effects.

## II-12 Cross Talk in Multi-slice Inversion Recovery Spin Echo

(1) The problem of cross talk between slices in multi-slice Inversion Recovery Spin-Echo sequences (see Sect. 2.7.1) is more complex than that in Spin-Echo sequences. (see image set II-11).
The causes are: (a) the temporal order of the slice addressing per TR in a multi-slice IR is more irregular than that in SE; (b) the inversion pulse in IR is a third slice-selective element in the sequence; next to the excitation and the refocussing pulses. This image set allows inspection of the consequences of cross talk in IR images.

Parameters: $B_0 = 1.5\,\mathrm{T}$; FOV = 160 mm; matrix = 256 × 256; $d = 4\,\mathrm{mm}$. TR/TI/TE = 1400/200/15; NSA = 1. The images are of modulus type.

| | | | | | | |
|---|---|---|---|---|---|---|
| image nos. | 1 | 2 | 3 | 4 | 5 | 6 |
| slice | 10 | 11 | 10 | 11 | | |
| no. of slices | 19 | 19 | 19 | 19 | 1 | 1 |
| slice gap (mm) | 0.8 | 0.8 | 5.6 | 5.6 | na | na |
| scan mode | MS | MS | MS | MS | MS | SS |

(na = not applicable; MS multi-slice)

(2) A single-slice image (image 6) is added because in our system the single-slice (SS) IR method, next to a non-selective refocussing pulse, is provided with a nonselective inversion pulse. Such a pulse is expected to have a relatively ideal inversion behavior. So, a comparison with this image allows inspection of the effectiveness of the slice-selective inversion pulse. From the first two scans, two neighbouring slices are shown per scan (slice 10 in images 1 and 3 and slice 11 in images 2 and 4). This is done because in our system for multi-slice IR, the default temporal order of slice addressing per TR gives a different temporal history for even and for odd numbered slices. The difference is a complicated function of the TI, TE, and TR and results from a shuffle algorithm that allows a large number of slices to be addressed per TR (in this image set 19 slices within a TR of 1400 ms).

(3) The images show a flow artifact that is generated by pulsatile arterial flow. In clinical images this artifact can easily be avoided by adding saturation pulses above and below the image volume.

(4) There are differences in contrast between the IR images. In image 5, from a multi-slice scan with one image, the muscles posterior to the tibial condyle are slightly darker than the surrounding fat. This difference is almost absent from images 1 to 4 obtained with multi-slice scans with 19 slices per TR. So, some cross talk from neighbouring slices can be noticed for the IR method. The differences in contrast between images 1, 2, 3, and 4 are relatively small, showing that the increase in the distance above the value normally allowed does not reduce the contrast difference easily.

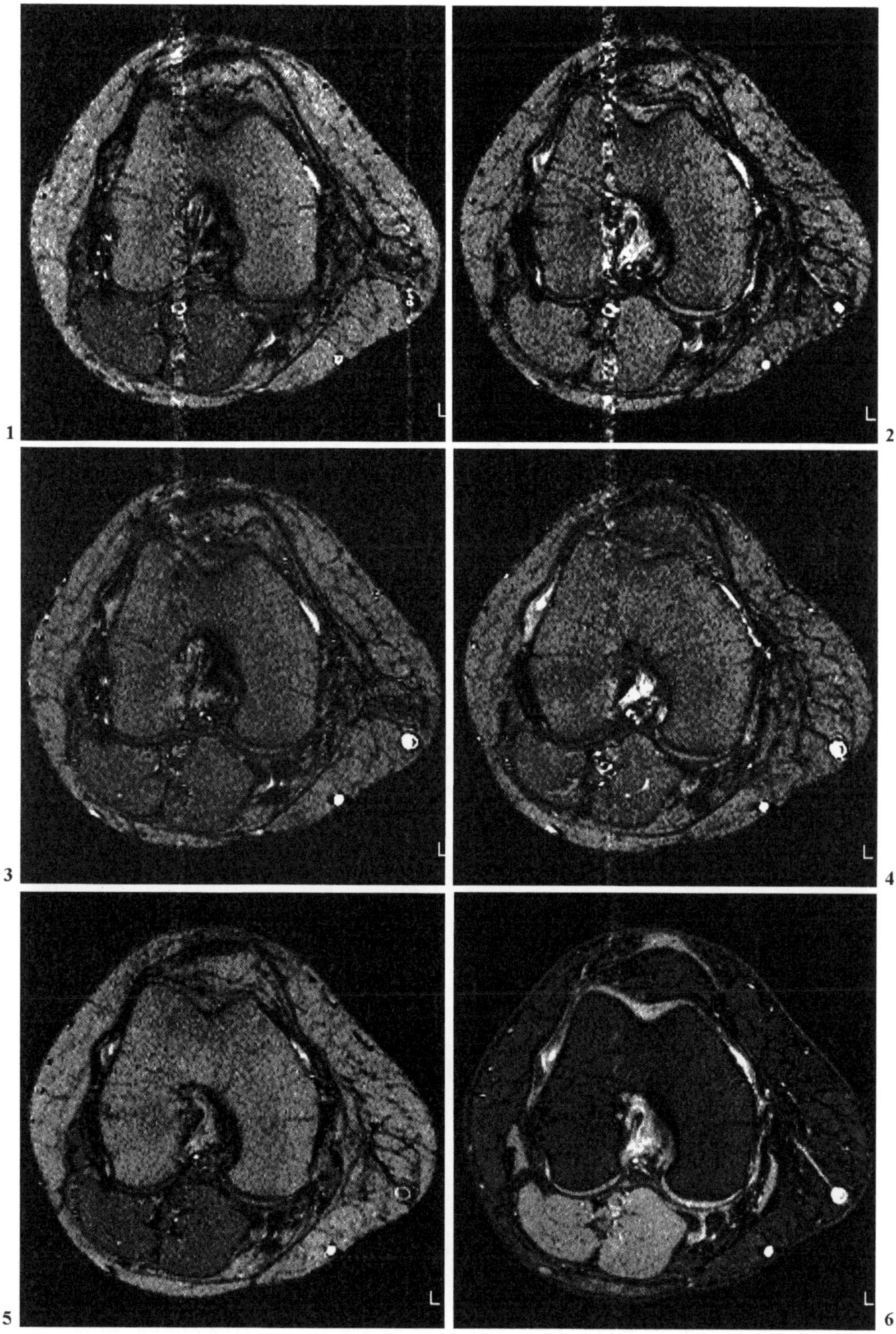


Image Set II-12

(5) The contrast of the single-slice image is noticeably different from that of the multi-slice images. Muscle is much brighter and areas of cartilage around the condyle now are much more conspicuously present. These differences show that the behaviour of the inversion pulse clearly depends on its design. The slice-selective inversion pulse creates smaller negative magnetization than the non-selective version of this pulse.

(6) Cross talk can also influence SNR in multi-slice IR imaging. A comparison of this aspect of the images by looking at the values of the SNR for marrow in each image is difficult to interpret, because the regrowing magnetization after inversion brings a large uncertainty in the signal value, that now depends strongly on the behavior of the inversion pulse. (See image set II-11 for remarks on SNR in multislice SE).

(7) The use of a multi-slice technique in IR imaging brings a noticeable change in contrast, when compared with single-slice imaging. The larger part of this change is due to the use of a slice-selective inversion pulse. A further but smaller change in contrast is the result of cross talk between neighbouring slices.

## II-13 Sensitivity Encoding (SENSE) in Receive Coil Arrays to Reduce the Number of Phase Encode Steps

Courtesy of KP Pruessmann, M Weiger, P Boesiger, Institute of Biomedical Engineering, ETH Zurich, Switzerland.

(1) Sensitivity encoding (SENSE) [21] utilizes information contained in the spatial distribution of the sensitivity of the individual coils of a receiver coil array (see Sect. 1.3.4.1) for signal localization. In standard Fourier imaging this allows the scan time to be reduced by increasing the distance of lines in $\vec{k}$ space. Then $\vec{k}$ space is basically undersampled, resulting in aliasing artifacts in conventional single-coil images. However, in voxel superposition, aliasing means each signal component is weighted according to the local coil sensitivities. Therefore, knowledge of the coil sensitivities allows pixels that are aliased up to a certain manifold to be separated by means of linear algebra. This can be shown by formal rewriting of (6.49), which gives the signal value of the object for a given coil $i$ as

$$S_{i,k} = \omega_0 \, \mathrm{d}V \int\limits_r m_{\mathrm{T}}(r) \; \beta_i(r) \; \exp(-\mathrm{j}k \cdot r) \; \mathrm{d}r, \tag{1}$$

the local coil sensitivity $\beta_i(r)$ is not considered as a modulation of the tissue contrast but as a hybrid encoding function together with the harmonic $\exp(-\mathrm{j}k \cdot r)$. By assembling the signal values of all coils in a signal vector $\bar{S}$ and the pixel values of the desired unaliased image in an image vector $\bar{V}$, hybrid reconstruction of $\bar{V}$ is described by the matrix equation

$$\bar{V} = \overset{\leftrightarrow}{B} \, \mathfrak{F} \, \bar{S} \tag{2}$$

$\mathfrak{F}$ is the discrete Fourier transform. The matrix $\overset{\leftrightarrow}{B}$ is calculated from the coil sensitivities and performs the inversion of sensitivity weighting and pixel superposition in the aliasing process. Thus the unaliased image is also intensity corrected.

For proper reconstruction accurate knowledge of the coil sensitivities is crucial. Therefore, sensitivity maps are created from in situ reference images by means of a particular polynomial fit procedure [21].

(2) The SNR in a SENSE reconstructed image is lower than the SNR in an image obtained with the same array with full Fourier encoding:

$$\mathrm{SNR}^{\mathrm{SENSE}} = \frac{\mathrm{SNR}^{\mathrm{fullFourier}}}{g(r)\sqrt{R}}, \tag{3}$$

where $R$ is the reduction factor. $R$ can have any value between 1.0 and the number of coil elements; it need not be an integer. The geometry factor $g(r)$ is a local quantity describing the ability of the coil array to separate pixels superimposed by aliasing. It depends on geometrical relations of coil sensitivities and on $R$. In practice, the maximum reduction is often unfavorable as the factor $g(r)$ usually grows rapidly when $R$ approaches the number of coils used.

(3) This set of images demonstrates SENSE reconstruction and its properties concerning noise enhancement. The elements of a flexible five-coil array were placed around a cylindrical phantom with a diameter of 200 mm. Data sets for various reduction factors are displayed per row.

| | | | | | |
|---|---|---|---|---|---|
| image row no. | 1 | 2 | 3 | 4 | 5 |
| reduction factor $R$ | 1.0 | 2.0 | 2.4 | 3.0 | 4.0 |

The degree of aliasing is visible in the conventional sum-of-squares reconstructions shown in column a. SENSE reconstructed images are shown in column b. Note the correct intensity distribution over the object. The noise maps (arbitrary scale) in column c were calculated according to (3). Note the correspondence between calculated noise enhancement and actual noise as visible in the images. Column d shows maps of the geometry factor $g(r)$ (absolute scale as displayed below the column).

(4) SENSE has the potential to speed up all conventional MR imaging techniques at the cost of loss in SNR. Therefore the most sensible applications are rapid and real-time acquisition, in particular in cardiac imaging. In some cases SENSE can even reduce artifacts and gain SNR, by shortening echo trains in single-shot sequences, for example, which may be significant for fMRI. Generally, when physiological and/or technical limits to the desired speed in $\vec{k}$ space are faced, sensitivity encoding is an alternative to enhancing gradient performance.

(5) In the related SMASH technique [22] image reconstruction is based on approximating harmonics of the field of view by linear combinations of the coil sensitivity functions. The assessment of the best-fitting combinations is a less laborious procedure than the sensitivity determination in SENSE. On the other hand, the need for proper approximation of harmonics imposes restrictions on the coil design and on imaging geometry, the slice orientation and the field of view being largely determined by array geometry and positioning.

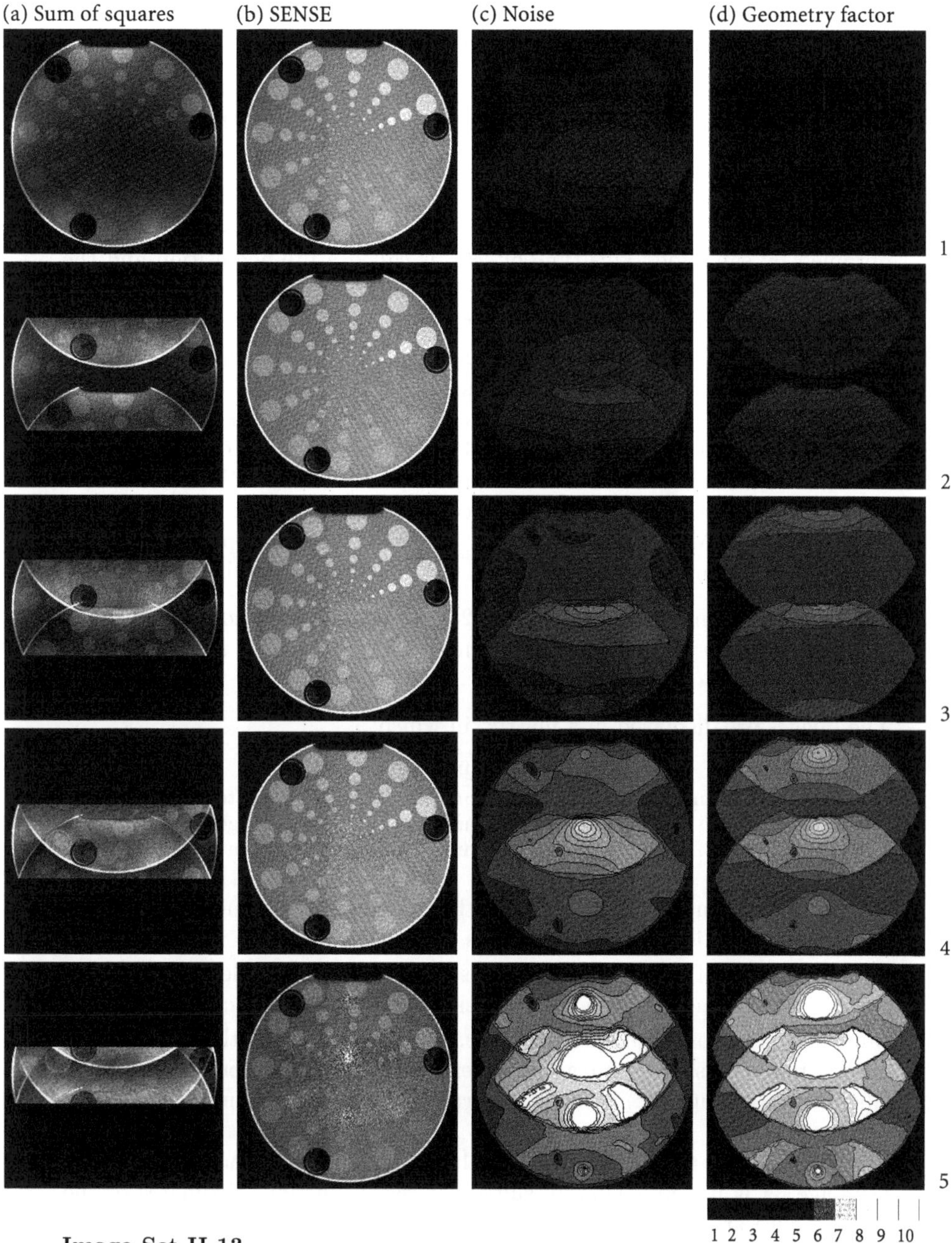


**Image Set II-13**

# II-14 Scan-Time Reduction Through Non-uniform Sampling

Courtesy of GJ Marseille, Technical University of Delft, The Netherlands

(1) In the choice of the parameters that define the scan, the selection of the reconstruction matrix together with the field of view determines the size of the image pixel. The nominal situation is that all necessary $\vec{k}$-space data are obtained by sampling. This type of scan is called a full acquisition and the properties of the obtained image governed by reconstruction of these data are described in Sect. 2.4.1. In the interest of saving scan time, $k_y$ values are frequently omitted, so that part of the $\vec{k}$ space is not sampled. The data set needed for reconstruction is then incomplete and has to be completed by estimation of the missing sample values. The resulting images differ from the full scan and offer a less precise presentation of the scanned object.

Apart from the well-known techniques of "half scan" (see Sect. 2.4.1.3 and 6.7.1) and "reduced scan", more advanced approaches in omission of samples are explored by Marseille in [23]. This image set compares performance aspects of a full scan and a reduced scan with one example of his approach.

Parameters: reconstruction matrix = 256 × 256, signal-to-noise ratio (original image) = 20. Transverse image through the brain of a normal healthy volunteer. Phase-encode direction: LR

The images are displayed in a matrix, in which:

| | |
|---|---|
| row 1: full scan | column a: sampled data |
| row 2: reduced scan | column b: $\vec{k}$ space as completed by estimation |
| row 3: non-uniform $k_y$ distribution | column c: reconstructed image (zoomed area) |
| | column d: difference with the full scan image. |

(2) The reduction in scan time for both the reduced scan in row 2 and the non-uniform $k_y$ distribution in row 3 is 30%, compared with the full scan.

The $\vec{k}$ space of the reduced scan as sampled is not suitable for immediate reconstruction because the sharp transients in these data (image 2a) would cause strong Gibbs ringing in the image. To prevent ringing, the $\vec{k}$-space data are filtered, as visible in image 2b, after which reconstruction is performed. The resulting image (2c) is slightly blurred as demonstrated in the difference image 2d.

Analogous to the reduced scan technique, the non-uniform distribution of image 3a is not immediately suited for reconstruction. Estimates of the missing data are shown in image 3b and the resulting reconstruction (image 3c) is much closer to the full scan image, as demonstrated in image 3d.

(3) Where the reduced scan technique is relatively straightforward, the alternative proposed in row 3 needs to be discussed further to understand its implications for practical MR scanning. This technique is based on elaborate

optimization strategies of which the value is defended in [23] and of which row 3 offers a representative example. Although a formal proof of the validity of these strategies was not given, it was shown that they are robust and in a test set of volunteer images lead to results that show minimal differences to the full scan, as long as the omitted fraction of $k_y$ values does not exceed 30%–40% (symmetric $k_y$ distributions).

The first step in this technique is the selection of the set of omitted $k_y$ values. The set shown in image 3a is symmetric around $k_y = 0$ and approaches the optimum in the sense that it provides that unique basis for subsequent data estimation that gives the smallest obtainable difference to the full scan image. This implies in theory that this set depends on the contents of the full scan image. In practice it is demonstrated in [23] that this set of omitted $k_y$ values is close to the optimum for a wide variety of MR images. The selected set is characterized by scarce omissions for low $k_y$ values, the omitted $k_y$ values are not periodically distributed along the $k_y$ axis and the set is mirror symmetric around $k_y = 0$.

The second step is the estimate of the values of the missing samples. The estimation operator is included after the first Fourier transformation (in the $k_x$ direction) and treats each column of data separately. With the estimation operator the final solution is reached by iterative steps that offer a stepwise maximization of the likelihood of the image. This likelihood has to be defined with prior knowledge of general properties of MR images rather than with knowledge of the full scan image that is approached by the estimation process, because the estimation operator has to perform when the full scan image is unknown. In [23] three important general properties of MR images are given: (i) the typical MR image possesses many flat regions bordered by sharp edges; (ii) the noise in the MR image is white (i.e. uncorrelated) and Gaussian with a known standard deviation; (iii) the noise-free $\vec{k}$ space data show complex conjugate symmetry around the origin. These properties are given in a precise mathematical formulation and are combined into a single operator that defines the role of the prior knowledge in the iteratively obtained estimates of the missing samples.
After a few iterations a satisfactory convergence to a final image estimate is demonstrated in [23]. Image 3c provides an example of such a result.

(4) Comparison of the reduced scan technique with the non-uniform sampling technique subjectively shows the attractiveness of the latter. The example given is for a symmetric distribution of omitted $k_y$ values. Asymmetric non-uniform distributions are discussed and demonstrated in [23] as well and compare favorable to the commonly used half scan technique in SNR and in the total scan time reduction that can be reached: up to 60%.

In analogy with the general prior knowledge described in [23], image-specific prior knowledge is available at all scans after the start of a dynamic series of MR images. The study of that prior knowledge and its use in optimiz-

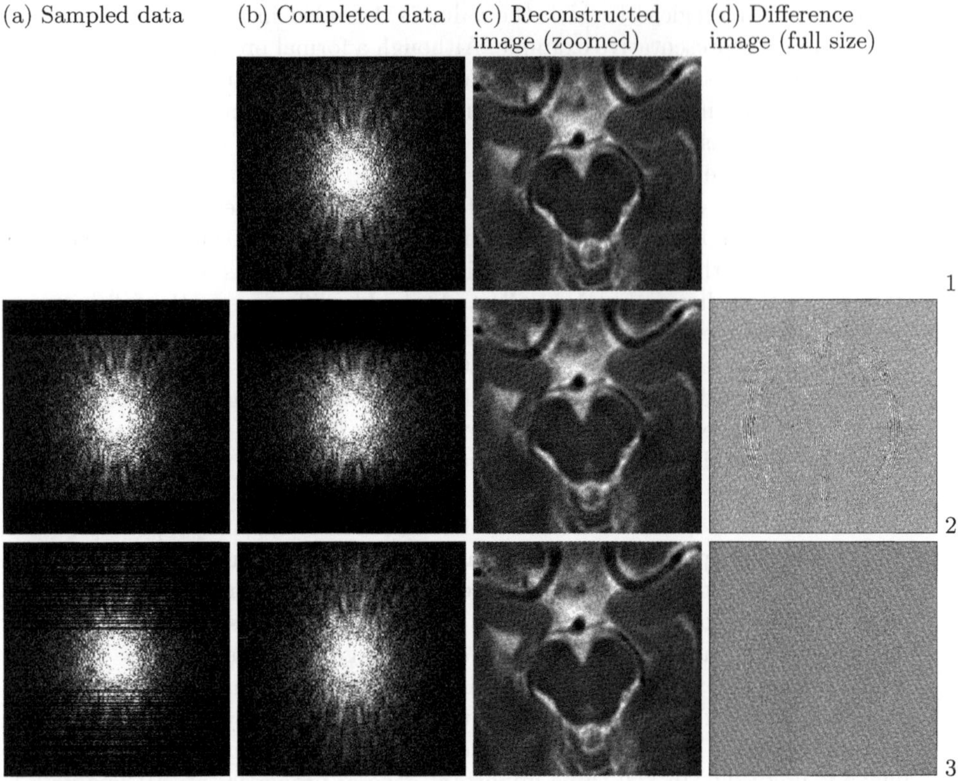


**Image Set II-14**

ing non-uniform sampling techniques is a related problem that is described in [11].

(5) The complexity of non-uniform sampling techniques such as described in [23] and [11] has up to now prevented its introduction in clinical use. However, their use appears to be completely feasible, at the cost of an increase in reconstruction effort only. For example, the reconstruction of image 3c took 27 times as long as that of image 2c. The balance between the gain in scan-time reduction and the increase in reconstruction time may in the foreseeable future tip over, driven by increased power of reconstruction hardware.

# 3. Imaging Methods with Advanced $\vec{k}$-Space Trajectories

## 3.1 Introduction

In the previous chapter we have shown that for MR imaging the transverse magnetization of Spin-Echo sequences is sampled on linear sets of the cartesian points in the $\vec{k}$ plane (see Fig. 2.6; for 2D methods we speak of the $\vec{k}$ plane, for 3D methods we use $\vec{k}$ space). An image is generated by "reconstruction", which is done by applying a fast Fourier transform to these sampled data ("raw data"). The temporal evolution of the $\vec{k}$ value after excitation, called the trajectory through the $\vec{k}$ plane, determines the time on which a certain cartesian point is reached. This trajectory is specified by the time integral of the gradient (see (2.26)). Other and/or faster trajectories through the $\vec{k}$ plane than those of conventional sequences are also possible. The choice of such trajectories determines the time at which the transverse magnetization is sampled and opens the possibility of finding faster imaging methods and/or shorter echo times. In this chapter we shall give a global survey of such imaging methods by looking at their trajectory through the $\vec{k}$ plane. This appears to be a clarifying way to introduce new sequences in those cases, in which all magnetization arises from one single previous excitation. When there is still magnetization present from earlier excitations (when $\mathrm{TR} < T_2$), the $\vec{k}$ plane description of this chapter is not adequate and must be complemented with a more complete description of the "steady state". This will be the subject of Chap. 4. The search for faster imaging methods is powered by the desired increase in the system throughput, by the possibility to "freeze" motion and thus avoiding motion artifacts, and by dynamic imaging, aimed at studying time-dependent processes (for example, the distribution of contrast agents through tissue as a function of time).

The first method to be discussed, the "Turbo Spin-Echo" method (TSE), is a fast imaging method for $T_2$ weighted scans. Originally [1], this sequence was called RARE (Rapid Acquisition of Repeated Echoes). The next method is "Echo Planar Imaging", (EPI) an ultra-fast scanning method, making in principle sub-second imaging possible [2]. A combination of EPI and TSE (called GRAdient Spin Echo or GRASE) is possible, it avoids a number of the problems encountered with EPI [3]. The last methods to be discussed are methods in which the trajectory through the $\vec{k}$ plane does not go through the cartesian points but follows radial paths [4] or spiraling paths [5], called "radial imaging" and "spiral imaging", respectively. The latter method relies on sinusoidal gradient pulse shapes, which generate the spiraling path through the $\vec{k}$ plane.

As has been shown in Chap. 2, (2.26), the position in the $\vec{k}$ plane at time $t$ (or the trajectory, when we consider $t$ as a variable) is given by the relations

$$k_x(t) = \gamma \int_{t_e}^{t} G_x(t')\mathrm{d}t', \quad \text{and} \quad k_y(t) = \gamma \int_{t_e}^{t} G_y(t')\mathrm{d}t', \tag{3.1}$$

where $t_e$ is the time of the last excitation. The velocity along the trajectory at a certain point in time is

$$\frac{\mathrm{d}k_x(t)}{\mathrm{d}t} = \gamma G_x(t), \quad \text{and} \quad \frac{\mathrm{d}k_y(t)}{\mathrm{d}t} = \gamma G_y(t), \tag{3.2a}$$

and its direction is given by

$$\frac{\mathrm{d}k_y(t)}{\mathrm{d}k_x(t)} = \frac{G_y(t)}{G_x(t)}. \tag{3.2b}$$

So with the time dependence of the gradient we can produce any trajectory through the $\vec{k}$ plane, as long as the gradient system is designed to generate the desired waveform. However, we must realize that for reconstruction with the familiar fast Fourier transform we need measurements in the rectangular (cartesian) grid, so for exotic paths through the $\vec{k}$ plane we must interpolate to obtain samples at the correct cartesian reference points (see Fig. 3.1). This interpolation is called "gridding". To our knowledge, except for projection reconstruction for radial scans, no other fast transformation of raw data on non-cartesian $\vec{k}$ plane coordinates into an image is possible.

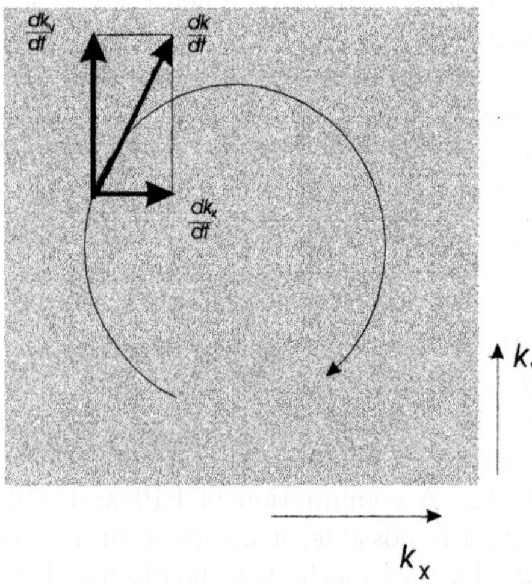


**Fig. 3.1.** Path through $\vec{k}$ plane

First, however, we shall discuss a number of scan methods for which the measured values are sampled directly on the points of a rectangular reference system. These scan methods are: Turbo Spin Echo (TSE) [1], Echo Planar Imaging (EPI) [2], Gradient and Spin Echo (GRASE) [3], and square spiral imaging [6]. Later on we shall also do some joyriding in the $\vec{k}$ plane, ending not in catastrophic events, but in potentially useful scan methods. Important examples are spiral imaging [5] and radial imaging [4]. For all scan methods to be discussed the essential problems (fundamental artifacts) will also be discussed.

## 3.2 Turbo Spin Echo

It is explained in Chap. 2 that the total scan time for the Spin-Echo method is long due to the waiting time for equilibrium between profiles (large TR). This could be improved if it were possible to acquire several profiles per excitation. This is indeed possible under certain circumstances. We use the fact that more than one echo can be formed from a single excitation. To understand this we can look again to Fig. 2.4 and realize that after the echo top is formed (see Fig. 2.4d), the spins dephase again when we leave the gradient unaltered. When the dephasing is equal to that just before the previous 180° pulse (but with opposite direction in the $x' y'$ plane) a second 180° pulse inverts the magnetization again, which will rephase in a newly applied gradient field, so as to cause a second echo. This process can be repeated until the $T_2$ relaxation has destroyed the signal. This multi-echo sequence is depicted in Fig. 3.2. It is frequently used for accurate $T_2$ measurements. Depending on the scan parameters, an "echo train" of 32 echos or more can be acquired with a single excitation.

We can use this multi-echo sequence for fast scanning by arranging that every echo is subjected to a different phase-encoding (preparation or evolution) gradient [1]. In this way we can acquire a number of echoes ("$\vec{k}$ profiles") following a single excitation and the total scan time is reduced. This method is the Turbo Spin Echo sequence (TSE or RARE). The number of profiles per excitation is the "turbo factor" (TF) and the total scan time for an $N^2$ image becomes $N$TR/TF, which is a factor of $F_{\mathrm{T}}$ smaller than the scan time of a conventional Spin Echo sequence. For example, when the scan time of a regular $256^2$ Spin Echo is about 256 s, with a turbo factor of 16 the scan time is reduced to 16 s. Note that, compared to a Spin-Echo sequence, in a multi-slice TSE sequence fewer slices can be addressed, because a larger part of the repetition time, TR, is needed for each slice.

The phase-encode gradient is added after each 180° refocussing pulse as shown in Fig. 3.3. We must, however, compensate for the phase-encode gradient before the next 180° pulse, so as to repeat the situation before the previous 180° pulse. So the net gradient surface between two 180° pulses in

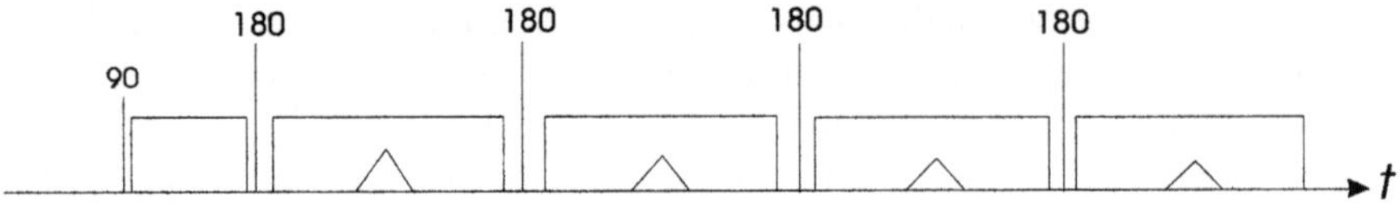


**Fig. 3.2.** Multi-echo sequence

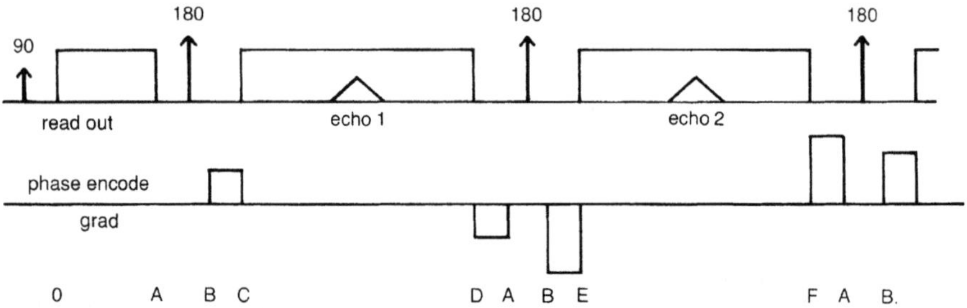


**Fig. 3.3.** Turbo Spin-Echo sequence, the points O, A, ..., F are shown in Fig. 3.4

the phase-encode direction is zero. Although the number of slices per scan is diminished it is a great achievement to have a $T_2$ weighted image in 16 s, which means the system has a higher throughput.

The Turbo Spin-Echo sequence is depicted in Fig. 3.3. We can now show how, in a Turbo Spin-Echo sequence, one moves through the $\vec{k}$ plane. This is shown in Fig. 3.4. At excitation $k = 0$ always, so we start in the origin. Due to the dephasing gradient the $\vec{k}$ vector moves to point A. The 180° degree

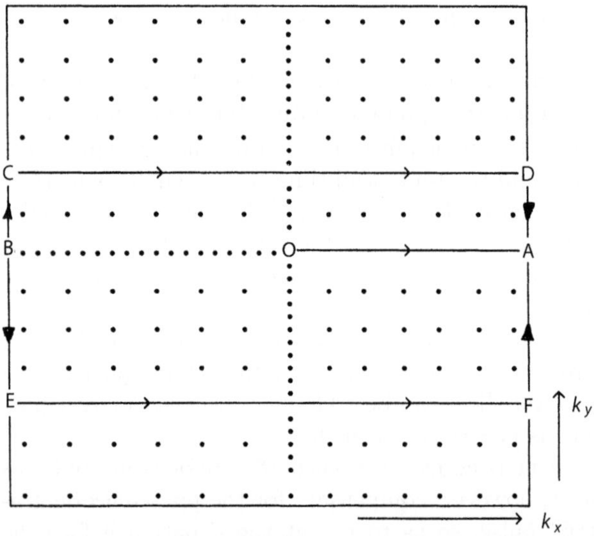


**Fig. 3.4.** Path through $\vec{k}$ space in a Turbo Spin-Echo sequence

pulse brings the $\vec{k}$ vector to point B. The first phase-encode gradient brings the $\vec{k}$ value to the point C and the read-out gradient causes the $\vec{k}$ vector to move from C to D, during which the total spin magnetization is sampled. The "rewinder" gradient moves the $\vec{k}$ vector back to point A, after which the next 180° pulse moves the $\vec{k}$ vector to point B. Then the phase-encode gradient brings the $\vec{k}$ vector to point E, after which the next profile can be measured. This can be repeated until $T_2$ decay has spoiled the signal. Assume that we can acquire 16 profiles per excitation. For a $256^2$ scan we then have to repeat the procedure only 16 times with different $k_y$ values to fill the $\vec{k}$ plane completely. The set of profiles collected per TR is named a "segment" and the acquisition of a complete set of profiles in several segments is called "segmentation". Image set III-1 illustrates the influence of the different scan parameters in a TSE image, especially echo spacing and segmentation.

The rewinder gradient is necessary to ensure the same phase encoding for all components of an echo. Until now we only considered the transverse magnetization, which is excited by the excitation pulse. In the ideal case the refocussing pulses do not add new transverse magnetization. However, it is practically impossible to make an exact 180° pulse, and therefore there will always be a deviation from the ideal flip angle. Therefore also FIDs are generated by each refocussing pulse and also stimulated echoes (see Sect. 4.2.3). So if the phase encoding would be cumulative over the intervals between the refocussing pulses, these new signals would have a wrong phase encoding. Therefore the phase encoding is uniquely defined in each interval and, after acquisition, compensated for by a gradient with the same strength but opposite sign, a "rewinder", within the same interval. With the theory of configurations as described in Chap. 8, it is possible to explain the underlying physics in more detail – see Sects. 8.2.4 and 8.4.1.

There is one application in which the values of $T_1$ and $T_2$ are sufficiently long to allow a single-shot TSE. This is in Magnetic Resonance Cholangiopancreatography (MRCP). $T_1$ and $T_2$ are around 4000 ms and 1000 ms respectively. This leaves sufficient time to acquire all $k$ profiles with a single excitation. The use of these single shot TSE methods or similar ones with long echo trains can cause a large power deposition, surpassing the limit that is imposed for reasons of safety of the patient. High power deposition will occur especially at high field strength or when the echo spacing is short. In such a case, it is possible to reduce the nominal flip angle of the refocusing pulses to a value below 180°, without an intolerable loss of SNR. This possibility is discussed in more detail in Sect 8.4.1 and illustrated in image set VIII-1.

### 3.2.1 Profile Order

We now have to think about the temporal profile order in each segment because the $T_2$ relaxation essentially influences the magnitude of the later echos. For example, if we take adjacent profiles per segment and within the segment take a linear profile order, then in the phase-encode direction we superimpose on the actual magnitudes of the profiles a periodic function due to $T_2$ decay in each segment. The effect of such a periodic function is best shown by looking at the image of a point object (described by a $\delta$ function) in the center of the field of view. As we have seen in the $\vec{k}$-space discussion, this object yields for all $k$ values equal signal amplitudes. However, due to our segmentation in the phase-encode direction a periodic signal is superimposed. As an illustration we take here a sinusoidal periodic function (see Fig. 3.5), and note that every periodic function can be written as a Fourier series of harmonics. We assumed the turbo factor, $F_T$, to be 16, so the periodicity in the phase-encode direction of the $\vec{k}$ plane is 16 profiles. In Fig. 3.5 we observe that because of this periodic overlay (filter) we obtain a ghost of the point object at 256:16 pixels away from the center. If more harmonics of the periodic function are taken into account we also get "ghosts" at 256:8, 256:4, and 256:2 pixels away from the object.

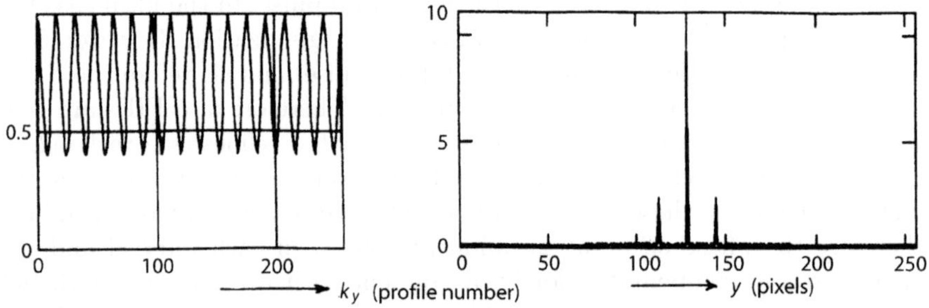


**Fig. 3.5.** (*Left*) superposition of a constant signal in $\vec{k}$ space and a periodic signal in the phase-encode direction. (*Right*) image of a point object with two ghosts

We can also follow another strategy in the profile order per segment [7, 8]. The profiles are numbered according to the order of the $k_y$ values. For example we can take the profile numbers 0, 16, −17, 32, −33, 48, −49, 64, −65, 80, −81, 96, −97, 112, −113, and 128 for the first segment and follow with similar segments, but now starting with another first profile number between −8 and +7 until all profiles are sampled. This strategy, called "centric" profile order, has the advantage that the lowest profiles are measured as soon as possible after the excitation and since the low profiles determine the contrast of an image (whereas the high profiles determine the resolution) the $T_2$ contrast is determined mainly by the first echos per excitation (short effective TE). As an example we assume that the $T_2$ relaxation reduces the signal of the 16th echo by a factor of about 2. The result is shown in Fig. 3.6: there is no "ghost", but there is some blurring and ringing due to the discontinuity at $k_y = 0$. It will be clear that with the centric profile order the $T_2$ decay is not smooth, as assumed in the figure, but appears as steps in the phase-encode direction. This can be considered as a superposition of a smooth decay (which causes blurring) and a periodic deviation from this smooth decay, which gives rise to ghosts as shown by (2.49).

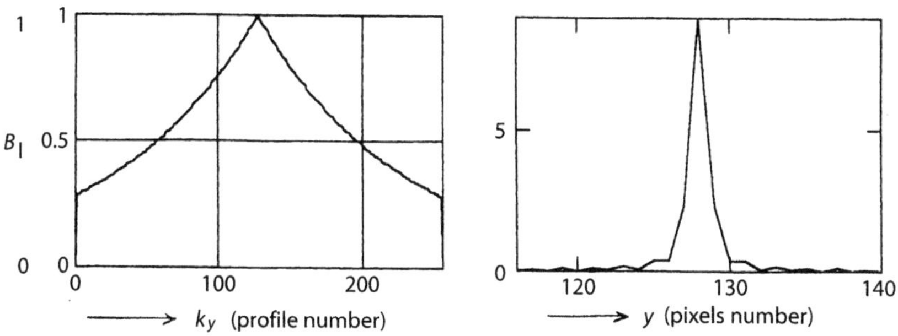


**Fig. 3.6.** (*Left*) segmented centric profile ordering, effect in $\vec{k}$ space. (*Right*) image of a point object; ringing and blurring can be observed

The "effective" echo time is defined as the time between excitation and the echo at which the lowest $\vec{k}$ profile is measured. The strategy in the profile order discussed here is called "centric" because we start with the lowest value of $|\vec{k}|$. There are of course more possible strategies of profile order. Typically TSE is used in combination with a long effective echo time. This allows a large turbo factor in combination with a "linear" profile order, running from $\vec{k}_{\text{max}}$ to $\vec{k}_{\text{min}}$, so that the effective echo time is halfway along the echo train. In this situation, multiple-slice acquisition can be used effectively, because the long TR needed for $T_2$-weighting can accommodate excitation of all slices. Image sets III-1 and III-2 are obtained in this way. $T_2$-weighted TSE can be combined with 3D acquisition and the efficiency of the scan can be maintained by multiple chunk acquisition [9]. An alternative and attractive method to maintain efficiency in 3D-TSE imaging by means of a reset pulse (see Sect. 2.7.2) is shown in image set III-3.

Examples of strongly $T_1$-weighted TSE with centric profile order are shown in image set III-4; the $T_1$-weighting is boosted by the use of inversion pulses that are interleaved in the multiple-slice acquisition.

### 3.2.2 Sources of Artifacts in TSE Images

The TSE sequence is an important modern method for obtaining fast $T_2$ weighted scans, but it is very sensitive to imperfections in the system. There are three possible causes of image degradation. In the first place it is practically impossible to make an exact 180° pulse and if we could make one there would always be transition regions at the edge of the slice, because of slice selection, where the flip angle is not equal to 180°. The effect of a sequence of a $90^\circ_x$ and a number of imperfect $180^\circ_x$ pulses is that the signal dies out much more rapidly than dictated by $T_2$. We can counteract this deteriorating effect by using a $90^\circ_x$ excitation pulse that causes rotation around the $x$ axis in the rotating reference frame followed by $n$ $180^\circ_y$ echo pulses (Carr–Purcell–Meiboom–Gill (CPMG) sequence). In Fig. 3.7 we show that when

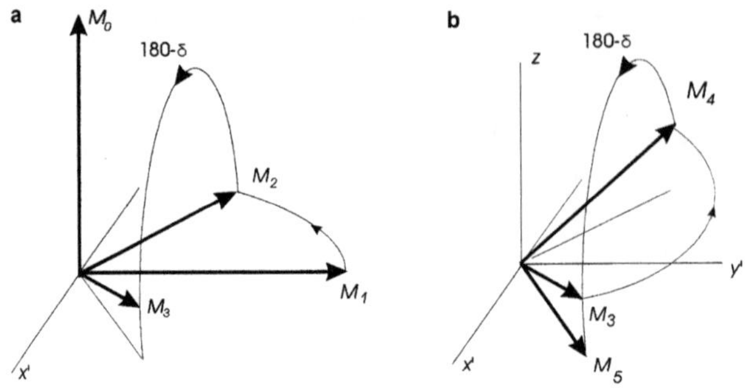


**Fig. 3.7.** Motion of the magnetization vector in the rotating frame of reference in case of a CPMG sequence. **a** $M_0 \rightarrow M_1$ by excitation pulse, $M_1 \rightarrow M_2$ by precession, $M_2 \rightarrow M_3$ by 180-$\delta$ pulse, $M_3$ lies above the $x' - y'$ plane. **b** $M_3 \rightarrow M_4$ by precession, $M_4 \rightarrow M_5$ by 180-$\delta$ pulse. The result is that $M_5$ lies again in the in $x'$–$y'$ plane

the excitation pulse results in a rotation of 90° around the $x'$ axis and the echo pulses in (180-$\delta$)° around the $y'$ axis, errors due to incorrect 180° pulses are automatically corrected. We leave it to the reader to show that when the refocussing pulses cause rotation around the same axis as the 90° pulse the signal level decreases faster than dictated by $T_2$. The use of the CPMG sequences is a sufficient precaution to avoid such fast decay and is used in the practical implication of TSE.

In the second place, severe artifacts may arise when the RF phase difference between the $90^\circ_x$ and $180^\circ_y$ pulses is not equal to 90°. Suppose the phase difference between the two pulses is $(90 - \alpha)^\circ$. The effect is depicted in Fig. 3.8. The $180^\circ_{y-a}$ pulses now cause rotation around the line $\lambda$ which is under an angle $\alpha$ with the $y'$ axis. The magnetization vector at the first echo now makes an angle $2\alpha$ with the $y'$ axis. After the second 180° pulse the echo coincides with the $y'$ axis and after the third echo the angle is $2\alpha$ again, etc. So we obtain a periodic phase change over the profiles that we measure, which results in a shifted ghost image as can be seen from (2.48), where the function $M(k_y)$ now multiplied with $\exp(j2\alpha)$ for all odd echo numbers. So, when we choose the centric profile order and a turbo factor of 16, all profiles between $-8$ and $+7$ are multiplied with $\exp(j2\alpha)$, the profiles $+8$ to $+24$ and $-9$ to $-25$ with 1, the next two adjacent groups of 16 profiles with $\exp(j2\alpha)$, etc. This means that the image will be convolved with the Fourier transform of this square-wave (block) function. To obtain an idea of what happens we assume $\alpha$ to be small and assume that the block function itself is written as a Fourier series with the fundamental wavelength $\Lambda$ in $\vec{k}$ space. This description allows the use of (2.48) and (2.49):

$$m(y) = \int_k M(k_y) \exp(\mathrm{j}k_y \cdot y) \exp\left(j2\alpha \sum \left(c_n \exp \mathrm{j}\frac{2\pi n}{\Lambda} k_y\right)\right) \mathrm{d}k_y. \quad (3.3)$$

Using the fact that $\alpha$ is small so that the exponential can be written as the first two terms of the series expansion, we find

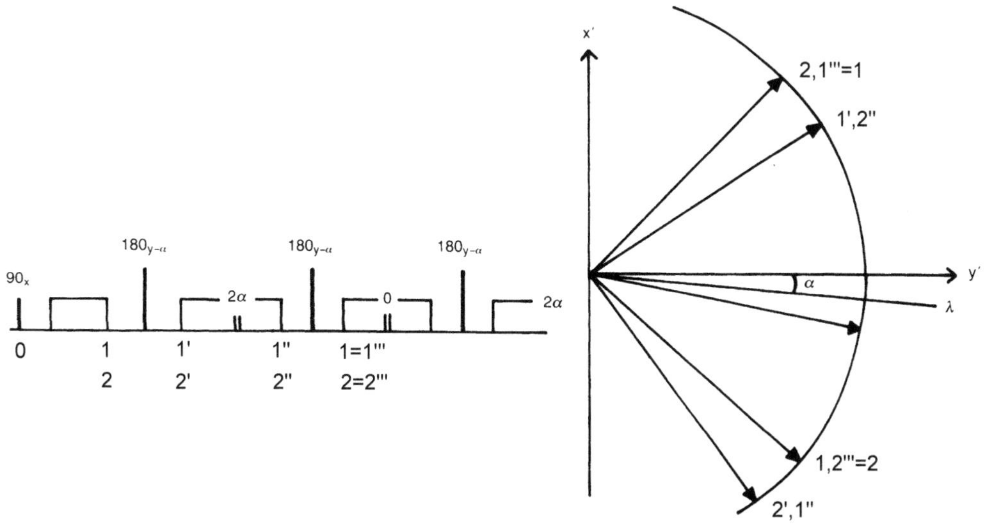


**Fig. 3.8.** Construction of the phase of the echo's for a TSE when the rotation axis of the refocussing pulses is about a line $\lambda$, which does not coincide with the $y'$ axis (phase error)

$$m(y)' = m(y) + 2j\alpha \sum C_n m\left(y - \frac{2\pi n}{\Lambda}\right). \tag{3.4}$$

The image is composed of the original image and a series of ghosts which are shifted over a distance $2\pi n/\Lambda$. The wavelength $\Lambda = 2(\vec{k}_{y\,\max})/F_\mathrm{T} = 2.2\pi N/(\mathrm{FOV}\,F_\mathrm{T})$, where $F_\mathrm{T}$ is the turbo factor and $N$ is the resolution. For a $256^2$ image, FOV $= 0.256\,\mathrm{m}$, $F_\mathrm{T} = 16$, we have a shift of 1.6 cm and higher orders at multiples of 1.6 cm. In modern systems accurate adjustment of the phase difference of the 90° and 180° RF pulses is possible and the ghosts can be adequately suppressed.

The third cause of artifacts covers the case when the gradient surfaces between the 180° pulses are unequal or are not equal to twice the gradient time integral between the excitation pulse and the first 180° pulse. This can be caused by both the selection gradient and the gradient in the read-out direction. The result is the superposition of an extra phase term in the $k_y$ profiles and artifacts in the image. In Fig. 3.9 it can be seen that when the gradient time integral before the first 180° pulse is less than half the gradient time integral between the 180° pulses, the spin echo following the first 180° refocussing pulse comes too early, the one after the second 180° pulse comes too late, etc. This means an extra alternating phase shift in the rotating frame of reference, which has the same consequence as the phase modulation due to incorrect RF phase of the $180^\circ_y$ pulses, giving rise to ghost images shifted over a certain distance (see (3.3) and (3.4)). The causes of the unwanted extra time-dependent gradient fields are eddy currents, which

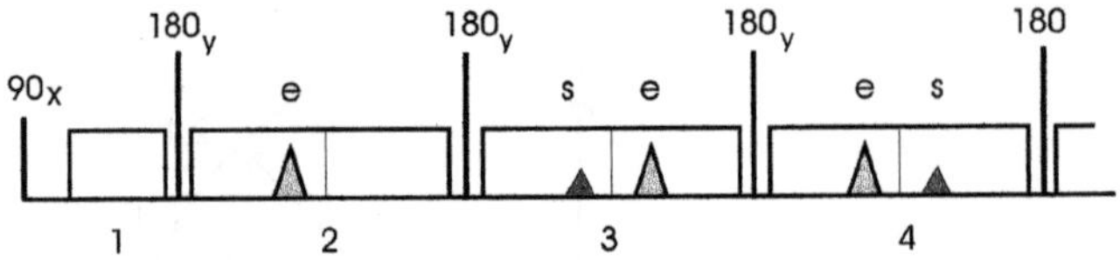


**Fig. 3.9.** Wrong echo timing when the gradient time integral in interval 1 is smaller than $^1/_2$ the gradient time integral in the later intervals. (e is spin echo, s is stimulated echo)

may reach some sort of equilibrium after a few 180° pulses, thus satisfying the requirement of constant gradient surfaces, but this equilibrium is more difficult to establish at the time of the $90^\circ_x$ and the first $180^\circ_y$ pulses. Eddy currents will also have higher-order components (spherical harmonics) so that the image distortion also becomes position dependent. So with eddy currents present, special measures are necessary to impose the required conditions. We shall come back to this point when we discuss applications.

The problem is further worsened by the "stimulated echo", which is caused by a sequence of three excitation pulses with flip angles not equal to 180°. This type of echo will be discussed in Sect. 4.2.3. Here we only remark that a stimulated echo in TSE is caused by imperfect 180° pulses and that the stimulated echo can be phase shifted with respect to the spin echo (see Fig. 3.9). Superposition of both echos gives rise to dark bands in the image due to local extinction.

## 3.3 Echo Planar Imaging

Also for field (gradient) echo sequences it is possible to measure more than one profile per excitation [2]. It is then necessary to reverse the read-out gradient at the end of the acquisition of a profile and to add a small phase-encode gradient surface as shown in Fig. 3.10. The trajectory through the $\vec{k}$ plane is then a meander and the small "blips" in the evolution gradient increase the $k_y$ value of the trajectory in steps, as shown in Fig. 3.11. The time since the last excitation increases along the meandering path. It will be clear that the total path through the $\vec{k}$ plane must be covered within the transverse relaxation time $T_2^*$ (comprising $T_2$ decay and off-resonance effects), otherwise the signal is destroyed before sufficient information is acquired. This means that the speed along the meandering trajectory must be sufficient to cover the whole $\vec{k}$ plane in about 100 ms or less. As we shall show later, the required speed and acceleration through the $\vec{k}$ plane determine the gradient system requirements.

The effective echo time TE* is the time between the excitation and the acquisition of the $k_y = 0$ profile. This profile is acquired rather late in the scan when a full matrix acquisition with a linear profile order is applied. However, other strategies are possible. If the total $\vec{k}$ plane cannot be traversed

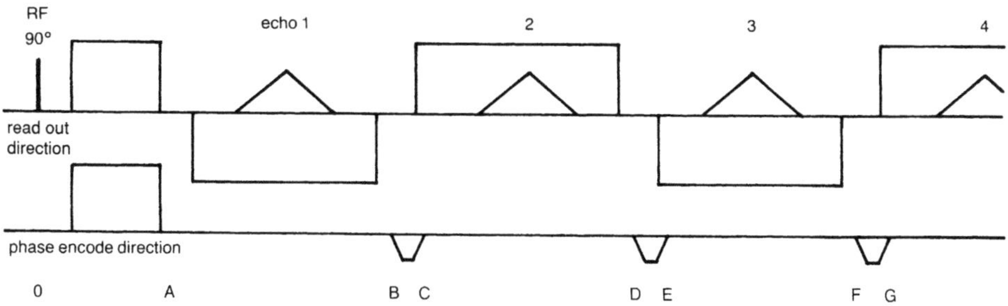


**Fig. 3.10a.** Sequence for blipped echo planar imaging

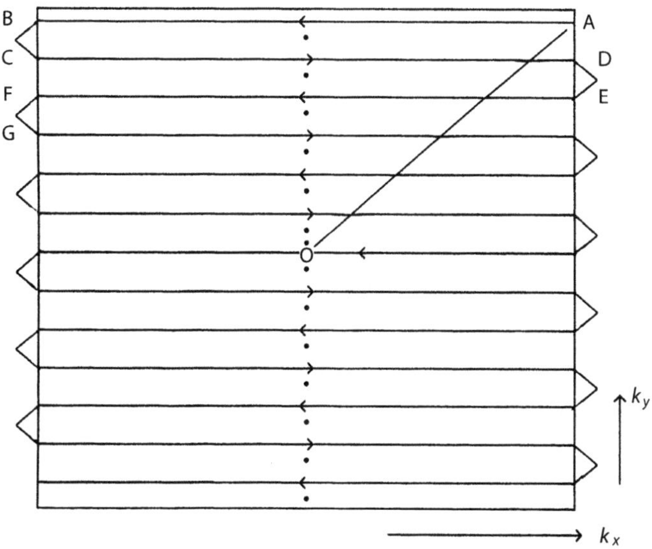


**Fig. 3.10b.** Trajectory through $\vec{k}$ space for a field-echo (EPI) sequence. Meaning of the letters is shown in Fig. 3.10a

within $T_2^*$, one can also apply segmentation (a smaller number of profiles per excitation), just as we have explained in connection with turbo spin echo. By using segmentation the gradient requirements are relaxed and segmented EPI scans can be performed on a standard whole-body system [10, 11]. Again the profile strategy determines the "filter" over the profiles in the phase-encode direction and thus determines the ghosts, blurring, or ringing.

Instead of the "blips" in the phase-encode gradient one can also use a small but constant phase-encode gradient. In this case we have oblique lines through the $\vec{k}$ plane. This is the first example in our treatment of a scan having its acquisition points off rectangular grid and so we have to interpolate. This process is discussed later.

Another way of looking at echo planar imaging is to say that we start with normal Spin-Echo or Field-Echo sequences but use a different measuring (readout) strategy. Instead of the constant read-out gradient with acquisition

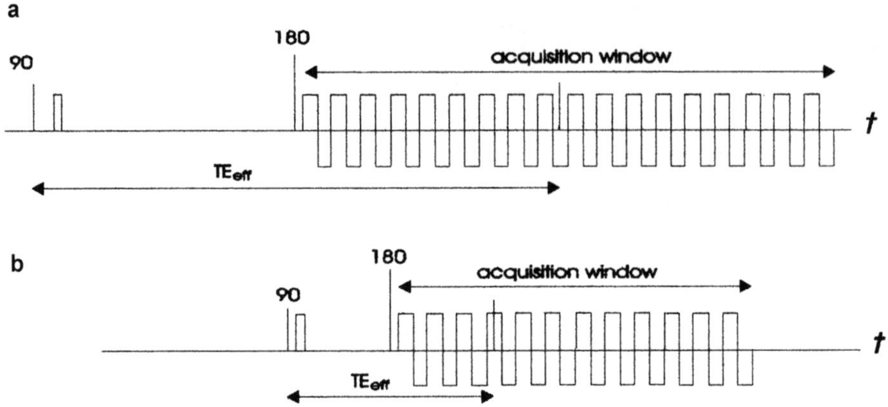


**Fig. 3.11.** $T_2$-weighted EPI (SE-EPI) with RF refocussing: **a** full matrix, **b** half matrix

along one single line in the $\vec{k}$ plane, we now have acquisition along several lines in the $\vec{k}$ plane, due to the meandering trajectory. Examples are given in Fig. 3.11, and in image set III-5. If the acquisition window (read-out period) cannot be long enough for complete $\vec{k}$ plane acquisition one can use several excitations to complete full $\vec{k}$ plane scanning and thus apply segmentation. Since the lowest $\vec{k}$ profiles are measured at the echo time TE and thus the influence of field inhomogeneity and susceptibility are compensated during their acquisition, these imaging methods are predominantly $T_2$ and not $T_2^*$ weighted.

### 3.3.1 Practical Example

In order to get an idea of the practical requirements imposed on the system by echo planar imaging we shall present a numerical example. We assume a system having a gradient strength of 20 mT/m and a rise time $R$ of 0.2 ms. So, the slope of the gradient-versus-time curve, the "slew rate", is $s = 100\,\mathrm{Tm}^{-1}\mathrm{s}^{-1}$. We furthermore assume that the acquisition window is restricted by $T_2^*$ to 60 ms. The field of view of the $128^2$ image is taken to be 0.256 m so the pixel size is 2 mm. A half-matrix is acquired for which we need 74 profiles. During the "blips" there is no acquisition so the acquisition time $t_{\mathrm{acq}}$ for a profile is equal to the time $T$ between the blips. The situation is drawn in Fig. 3.12. The total profile time is $T_{\mathrm{p}} = T + T_{\mathrm{b}}$, and is in our example equal to the acquisition window divided by the number of profiles (+1 to account for the excitation and the measurement compensation (dephasing) gradient just after the excitation pulse). So $T + T_{\mathrm{b}} = T_{\mathrm{p}} = 0.8\,\mathrm{ms}$. We now need to know $T_{\mathrm{b}}$, for which we can use (2.34), in which the factor 2 is omitted:

$$\gamma G_{y,\max} T\,\mathrm{FOV} = N, \tag{3.5}$$

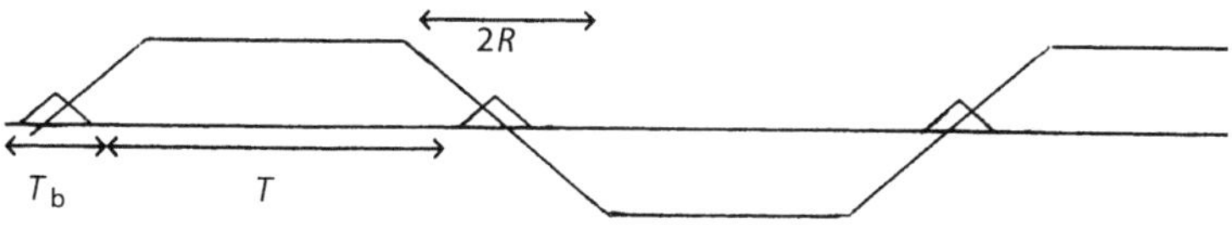


**Fig. 3.12.** Gradient waveform for echo planar imaging

where $G_{y,\max}T$ is the total surface of all 128 blips (of which only 74 are applied in this "half-matrix" acquisition). So the area A of a single blip is given by

$$A = (\gamma \mathrm{FOV})^{-1}, \tag{3.6}$$

which yields for our example: $92 \times 10^{-9}$ Ts/m. Because of the rise time, the blips will have a triangular form. The surface of such a triangular gradient form can be shown to be equal to $s\,T_b^2/4$. This results in $T_b = 60\,\mu$s and the maximum gradient field strength is 3 mT/m. This means that the acquisition time $T$ for a single profile is 0.74 ms and the sampling time $t_s = 5.8\,\mu$s.

As has been said, we sample during the time $T$ (including the ramps of the gradients). This means that the sample points are not equidistant in the $\vec{k}$ plane and interpolation to the cartesian grid must take place before reconstruction. When we define an average gradient value $\langle G_x \rangle$, the gradient surface needed to cross the $\vec{k}$ plane during acquisition is given by (2.38a):

$$\langle G_x \rangle T = N/(\gamma \mathrm{FOV}), \tag{3.7}$$

where $t_{\mathrm{acq}} = T$ and which yields $\langle G_x \rangle T = 12 \times 10^{-6}$ Ts/m (which is – of course – 128 times the surface of a single blip). The value of $\langle G_x \rangle$ is $12 \times 10^{-6}/0.74 \times 10^{-3} = 16.2$ mT/m. However, we need a higher maximum gradient field ($G_{x,\max}$) because of the finite rise time (the ramps). The area of read-out gradient is given by

$$\langle G_x \rangle T = G_{x,\max}(T - R + T_b - T_b^2/4R) \cong G_{x,\max}(T - R + T_b).$$

This means that the maximum gradient strength must be 20 mT/m. This is just possible with the system. It is clear that for EPI the system requirements are very severe. Therefore it is frequently necessary to apply segmentation.

### 3.3.2 Artifacts Due to $T_2^*$ Decay and Field Inhomogeneities

EPI suffers from $T_2^*$ decay during the acquisition and from magnetic field inhomogeneities. We shall discuss these effects in more detail and discover that these effects especially manifest themselves in the phase-encode direction. The calculation of the influence of these effects on the image is done in the same way as is discussed in Sect. 2.6 for the conventional acquisition methods. We start again with (2.42) and after introduction into (2.28) we find [12]

$$\begin{aligned} m'(x,y) &= \int_{k_x}\int_{k_x} M_{\mathrm{T}}(k_x,k_y)\exp(+\mathrm{j}(k_x x+k_y y))\exp(-\mathrm{t}/T_2^*) \\ &\quad \exp(-\mathrm{j}\gamma\delta Bt)\mathrm{d}k_x\mathrm{d}k_y. \end{aligned} \tag{3.8}$$

The factor $(2\pi)^2$ is assumed to be part of $m'(x,y)$. As in the discussion in (2.43), we can express $t$ in $k_x$ and $k_y$, since we know the path through the $\vec{k}$ plane as described by the parameter equations $k_x = k_x(t)$, $k_y = k_y(t)$. Therefore we can also express $t$ in $\mathfrak{T}_x(k_x)$ and $\mathfrak{T}_y(k_y)$, or in $\mathfrak{T}(\vec{k})$. Note that according to (2.43), for a constant measuring gradient, $t = k_x/\gamma G_x = \mathfrak{T}(k_x)$. For EPI we also have to consider the ramps of the measuring gradient but for ease of argument we shall neglect these ramps, since, as is explained in Sect. 2.6, the effect in the measuring direction is small anyhow. With the function $\mathfrak{T}(\vec{k})$ we can also describe more complicated dependencies, which we use later when we describe arbitrary paths through the $\vec{k}$ plane. Looking now at the Fourier transform we see that

$$m'(x,y) = m(x,y) \otimes \mathfrak{F}\left\{\exp\left(-\mathfrak{T}(\vec{k})/T_2^*\right)\right\} \otimes \mathfrak{F}\left\{\exp\left(-\mathrm{j}\gamma\delta B\mathfrak{T}(k)\right)\right\}, \tag{3.9}$$

where $\mathfrak{F}$ means Fourier transform and $\otimes$ means convolution. The first convolution describes artifacts due to $T_2^*$ decay and the second describes the artifacts due to resonance offset.

#### 3.3.2.1. Artifacts Due to $T_2^*$ Decay

We first concentrate on the effect of $T_2^*$ decay (so for the time being we take the inter-voxel variation $\delta B = 0$). $T_2^*$ relaxation describes the combined effect of the intrinsic $T_2$ relaxation and the intra-voxel dephasing due to inhomogeneity of the magnetic field, which is, for example, caused by susceptibility variations in the tissue. Suppose that, due to these susceptibility variations, the field-strength distribution $dB$ in a voxel is Lorentzian. Then the relation between an effective relaxation time $T_2'$ due to these fluctuations, and $dB$ is given by

$$\gamma dBT_2' = 1/2, \tag{3.10}$$

and the overall relaxation time is given by

$$\frac{1}{T_2^*} = \frac{1}{T_2} + \frac{1}{T_2'}. \tag{3.11}$$

Let us assume in our example that $T_2$ is 80 ms and that $T_2^*$ is 40 ms. In that case $T_2'$ is 80 ms, which corresponds to a $dB$ of 0.15 $\mu$T or 0.1 ppm at 1.5 T. In special cases, for example at the boundary of air and tissue, the $dB$ can be even 1 ppm, which can cause extinction of the signal in gradient echo sequences (see Sect. 2.5).

We now compare the effect of $T_2^*$ decay with the sampling point-spread function, caused by the finite measuring time in the measuring direction.

The point-spread function is shown in (2.35) to be $2\text{sinc}(\gamma\langle G_x\rangle x t_{\text{acq}})$, where $t_{\text{acq}} = T$ as defined in Fig. 3.12c. Here $\langle G_x\rangle$ is the average value of $G_x$ over the trapezoidal gradient waveform, which is an approximation. In our example of the previous section this value is 16.2 mT/m. The width at half maximum of the main lobe of the sinc function is given by, see (2.36), $\Delta x = 1.2\pi$ $(\gamma G_x \mathrm{T})^{-1}$. The Fourier transform of the exponential describing $T_2^*$ decay in (3.9) is

$$\mathfrak{F}^{-1}\left\{\exp\left(-k_x/\gamma\langle G_x\rangle T_2^*\right)\right\} = \frac{\gamma\langle G_x\rangle T_2^*}{1 + \mathrm{j}\gamma\langle G_x\rangle T_2^* x}. \tag{3.12}$$

It is interesting to note here that $T_2^*$ decay causes complex values of the signal, which makes the measurement of phase effects (for example due to flow; see Chap. 6) difficult. The width at half maximum of the transform (the Lorentz function) is given by

$$\Delta x = 2(\gamma\langle G_x\rangle T_2^*)^{-1}. \tag{3.13}$$

Since the signal $m(x, y)$ is convolved with the sinc and the Lorentz functions we must speak of the distance from a point $x$ and replace the $x$ by $\Delta x$. We can now write for the ratio of the width of the $T_2^*$ blurring and that of the sampling point spread function

$$\frac{\delta x_{T_2^*}}{\delta x_{\text{acq}}} \cong \frac{T}{2T_2^*}. \tag{3.14}$$

This is a small number since $T = t_{\text{acq}} \ll T_2^*$, so in the measuring direction the $T_2^*$ decay has an unimportant effect. As stated in Sect. 2.6, this is a general fact since the read-out gradient is large.

In the phase-encode direction, however, the effect of $T_2^*$ decay is much more important. Physically speaking this can be explained by the fact that the value of the average velocity through the $\vec{k}$ plane in the $y$ direction is much lower than that in the $x$ direction. Effects like $T_2^*$ decay (and also resonance offset, $\delta\omega = \mu\delta B$) therefore have more time to degenerate the measurements. The average velocity in the phase-encode direction of the $\vec{k}$ plane can be expressed in terms of the velocity in the phase-encode direction. We define the velocity $\nu_y$ by $k_y(t) = \nu_y t$, and the profile time by $T_{\mathrm{p}} = T + T_{\mathrm{b}}$ (see Fig. 3.12c). It follows that

$$\nu_y = \left(\frac{t_{\text{acq}}}{t_{\text{scan}}}\right)\nu_x = \left(\frac{T}{T_{\mathrm{p}}N}\right)\nu_x \cong \frac{\nu_x}{N},$$

where $N$ is the number of profiles taken (for example 128). The associated average gradient field, $\langle G_y\rangle$, which is proportional to the average velocity in the phase-encode direction of the $\vec{k}$ plane, see (3.2a), is found by taking $\langle G_y\rangle T_{\mathrm{p}}N = \langle G_x\rangle T$, which simplifies to $\langle G_y\rangle \cong \langle G_x\rangle/N$ when $T_{\mathrm{b}} \ll T = t_{\text{acq}}$.

Let us first look at the point-spread function. The total scan time is $t_{\text{scan}} = N T_{\mathrm{p}}$. The width $\Delta y$ of the point spread function is now given by

$$\gamma \langle G_y \rangle T_\mathrm{p}\, N\, \Delta y_\mathrm{acq} = 1.2\pi, \tag{3.15}$$

and since $\langle G_y \rangle \cong \langle G_x \rangle / N$, we have in the $y$ direction almost the same point-spread function as in the $x$ direction (as expected). The effect of the $T_2^*$ decay is formally the same as in the phase-encode direction, but now in a much lower pseudo gradient field $\langle G_y \rangle$ and a longer time. Equation (3.13) for the preparation direction now reads

$$\gamma \langle G_y \rangle \Delta y_{T2^*} T_2^* \cong 2. \tag{3.16}$$

For the ratio of the point spreading due to the finite sampling time and the spreading due to $T_2^*$ decay, we find for the preparation direction

$$\frac{\Delta y_{T2^*}}{\Delta y_\mathrm{acq}} \cong \frac{N\, T_\mathrm{p}}{2 T_2^*}, \tag{3.17}$$

which tells us that the width of the point-spread function, due to $T_2^*$ decay, is of the same order as the the width due to the finite acquisition time ($N\, T_\mathrm{p}$ is the total scan time), since $T_2^* = 40\,\mathrm{ms}$ and $N\, T_\mathrm{p} = 60\,\mathrm{ms}$.

Finally, we should mention that the effects described above are different when other strategies of scanning the $\vec{k}$ plane are followed. One can use segmentation, so as to effectively traverse the $\vec{k}$ plane faster. As shown in Fig. 3.12 it is also possible to use in the EPI method a spin-echo-like excitation (SE-EPI). The interesting point here is that at the time of the spin echo, $t = \mathrm{TE}$, the effect of the susceptibility and field inhomogeneity is zero and we have true $T_2$ decay. When the lowest $\vec{k}$ profiles are acquired around $t = \mathrm{TE}$ the image is predominantly $T_2$ weighted. The effect of $T_2^*$ must be described as $\exp(-|t - \mathrm{TE}|/T_2^*)$. This scan with $64^2$ resolution might be very interesting for dynamic imaging.

#### 3.3.2.2. Artifacts Due to Resonance Offset

We shall now consider resonance offset due either to the water–fat shift, susceptibility variations, or the main-field inhomogeneity. In the same argument we can also understand and calculate the influence of disturbances in the main field at the edge of the homogeneity sphere. We could start with (3.9), but in this case there is a more direct method. We start therefore with (2.27), which determines the Fourier transform in the ideal case. However, in a real case we also have, apart from the influence of the gradients, the influence of the $B_0$ field inhomogeneity and the susceptibility variations, $\delta B$, which can cause a position-independent offset, $\alpha$, and position-dependent terms, as described by [12]:

$$\delta B = \alpha + \beta_x x + \beta_y y. \tag{3.18}$$

The transformation in the read-out direction is now defined by ($k_y$ is constant and $k_x = \gamma\, G_x\, t$)

$$M'_{\mathrm{T}}(k_x) = \int_x m(x) \exp(-\mathrm{j}\gamma G_x(x + \alpha/G_x + \beta_x/G_x x)t)\mathrm{d}x, \tag{3.19}$$

which can be replaced by a regular Fourier transformation when we substitute $x' = x(1 + \beta_x/G_x) + \alpha/G_x$. We thus obtain

$$M'_{\mathrm{T}}(k_x) = \frac{1}{1 + \beta_x/G_x} \int_x m(x') \exp(-\mathrm{j}\gamma G_x x' t)\mathrm{d}x'. \tag{3.20}$$

From this equation we conclude that the signal strength is multiplied by the factor before the integral sign, that the position is shifted over a distance $x - x'$, and that the area of an object is magnified by $\mathrm{d}x'/\mathrm{d}x$; see also Sect. 2.6.

Let us take the example of offset due to the water–fat shift which is 3.3 ppm (so $\alpha$ is about 5 $\mu$T at 1.5 T). Then the fat image is shifted with respect to the water image over $\alpha/G_x = 5 \times 10^{-6}/20 \times 10^{-3}\,\mathrm{m} = 1/4\,\mathrm{mm}$ (a pixel is 2 mm in our example). Also the effect of the main field gradients is small (in Sect. 2.6 we stated that in the read-out direction these effects are generally small). However, in the phase-encode direction we can follow the same arguments but now the pseudo gradient field is active. We calculate the pseudo gradient field from $2k_{y,\mathrm{max}}/\gamma = G\,N\,T_{\mathrm{p}} = 12 \times 10^{-6}\,\mathrm{Ts/m}$. In the phase-encode direction the time in which this surface is reached is $128 \times 0.8 \times 10^{-3}\,\mathrm{s}$. So the pseudo field is 0.12 mT/m. The water–fat shift now becomes $5 \times 10^{-6}/0.12 \times 10^{-3} = 4.2\,\mathrm{cm}$. A gradient in the main field of 10 ppm/m (0.015 mT/m) causes a shift characterized by $x' = x(1 + 0.015/0.12) = 1.13x$ and the signal strength is decreased by 13%. Susceptibility-induced gradients can be of the same amplitude. We conclude that in the phase-encode direction, distortions of the EPI image are severe and limit the possibilities of EPI imaging. This fact is illustrated in image set III-7, where SE-EPI is compared with TSE and GRASE (gradient echo and spin echo; see Sect. 3.4).

#### 3.3.2.3. Artifacts Due to Gradient Field Properties and Errors

As can be seen in Fig. 3.11 the path through the $\vec{k}$ plane is meandering; the odd lines are scanned from right to left and the even lines from left to right. This means that we have to mirror the acquired samples of the odd lines in the reconstructor. A consequence is that the time delay between the acquisition of adjacent points in the phase-encode direction of the $\vec{k}$ plane varies, fortunately only within the time necessary for scanning two lines, which is relatively short in comparison with the total scan time. Therefore the errors due to varying $T_2^*$ decay and resonance offset due to the meandering path are small in comparison with errors that build up during the total scan, as described in the previous sections. One could, however, think of reconstructing twice, first all samples from lines traversed from left to right and later on the other samples.

An essential difficulty with EPI is that large gradients are applied and that instrumental and fundamental errors become more obvious than in the case of conventional sequences. With fundamental "errors" we mean that Maxwell's equations do not allow pure gradient fields, which have only components in the $z$ direction: these fields will always be accompanied by transverse fields as expressed by rot $\vec{B} = 0$. Both types of errors have the consequence that at time $t$ we do not know exactly in which point in the $\vec{k}$ plane we are. The actual phase at time $t$ is not only determined by the linear gradients $\vec{G}(t)$, but depend also on the eddy current fields $\vec{B}_{\mathrm{e}}(r,\, t)$ and the Maxwell field $\vec{B}_{\mathrm{M}}(r,\, t)$:

$$\begin{aligned} \varphi(\vec{r},t) &= \vec{r}\cdot\vec{k}(t) = \gamma\int \vec{r}\cdot\vec{G}_{\mathrm{act}}(\vec{r},t)\mathrm{d}t \\ &= \gamma\vec{r}\cdot\int \vec{G}(t)\mathrm{d}t + \gamma\int\left\{\vec{B}_{\mathrm{e}}(\vec{r},t)+\vec{B}_{\mathrm{M}}(\vec{r},t)\right\}\mathrm{d}t. \end{aligned} \tag{3.21}$$

We first discuss the instrumental errors. The actually excited field is shifted in time with respect to the intended field. This is caused by the fact that there is an unknown time delay between the demand input of the gradient amplifier and the actual output gradient field (propagation delay). The consequence is that the echos come later than expected. However, the meandering path through the $\vec{k}$ plane means the echos are shifted to the right for the even profiles and to the left for the odd profiles. This can be described as a $\vec{k}$-space shift alternating in the phase-encode direction. When the propagation delay is $\delta t$, the extra phase at the echo top is $\varphi_{\mathrm{p}} = \gamma \mathrm{G}\, x\, \delta t$, where $G$ is the read gradient. For 15 mT/m and a delay of 1 $\mu$s this causes a phase error of $\pm 0.4$ radian at 0.1 m from the isocenter. As described in Sect. 2.7 and (2.49) this alternating phase error causes a ghost, shifted over half a field of view in the phase encode direction. A constant propagation delay, $\delta t$, can be measured with MR methods (for example by measuring the position of the echo top), and correction by multiplying with $\exp\{-\mathrm{j}\gamma G\, x\, \delta t\}$ is possible. Such a linear phase correction is applied in practice.

The total pattern of the eddy current field is more complex than just a delay, because the spatial-harmonic content is unknown, so there is an unknown temporal and spatial dependence. The linear components of the eddy-current fields can be corrected for by applying a nonlinear phase correction, characterized by $\varphi_{\mathrm{el}}(x) = \gamma G_{\mathrm{el}}(t) x\, t$, where $G_{\mathrm{el}}(t)$ describes the temporal behavior of the linear components of the eddy-current gradient fields. However, the higher-order terms will cause unrepairable errors. The only defense here is to minimize the eddy currents by blocking their potential current paths. Note that roughly speaking for EPI only eddy currents with short time constants are of importance, long time constants, $\tau \gg 1$ ms, yield almost equal but reversed eddy currents as a result of the rise and decay of a short gradient pulse, so they add to almost zero. Eddy currents with short time constants, which have a non-zero effect, are not excited in the inner shield of the magnet, but in the RF shield between the RF transmit coil and the gradient coils or are due to redistribution of the current in the gradient wires.

We now come to the fundamental "error": the Maxwell field or concomitant field. The trend in imaging is to ever faster acquisition, and so to higher velocities along the trajectortes in the $\vec{k}$ plane, so that the total $\vec{k}$ plane is scanned as fast as possible. This means high measuring gradients, so high that the they bring us potentially into conflict with safety regulations [13] and signal-to-noise requirements. However, the higher the gradient the more the assumption that a gradient field has only a $z$ component is wrong.

A true gradient field, defined by $\vec{G}_r = \delta\vec{B}_z/\delta\vec{r}$, does not exist. Always such a gradient field will be accompanied by transverse field components as stated by Maxwell's equations [14]. For low gradients, used in conventional imaging, this transverse field component can be neglected. In EPI the measuring gradient can be very high and therefore be accompanied by a transverse field component that cannot be neglected. In order to estimate this transverse field, Maxwell's equation $\nabla \times \vec{B} = \varepsilon_0 \delta\vec{E}/\delta t$ is solved for the situation of a read-out gradient in the $x$ direction. The term describing the associated electric field can be disregarded for relatively low frequency fields used in MRI, so $\nabla \times \vec{B} = 0$. The $y$ component of this equation is [14]

$$\vec{\mathrm{j}} \cdot \left\{ \frac{\delta B_x}{\delta z} - \frac{\delta B_z}{\delta x} \right\} = 0, \quad \text{or} \quad \frac{\delta B_x}{\delta z} = G_x, \quad \text{or} \quad B_x \cong G_x\, z, \tag{3.22}$$

so the total magnetic field in the $z$ direction, which determines the precession frequency, is given by

$$B_z(x, 0, z) = \sqrt{(B_0 + G_x x)^2 + G_x^2 z^2} \cong B_0 \left\{ 1 + \frac{G_x^2 z^2}{2\, B_0^2} \right\} + G_x x, \tag{3.23}$$

which shows that an extra term depending on $z^2$ occurs, causing resonance offset, depending on position in the $z$ direction. In a transverse image this means that we have a resonance offset equal to $\gamma$ times the second term between the brackets. Since the direction in which the $k$ profiles are sampled alternates, this causes an alternating phase shift in the read out direction, which corrupts the image (ghosts) if no correction is applied. The correction is difficult, but the effect described so far is probably small because of the high $\vec{k}$-plane velocity in the measuring direction (high measuring-gradient field).

In the preparation direction the "effective" gradient field is much lower, so the effect is larger and can be described as a displacement:

$$\Delta y = \frac{G_x^2 z^2}{2\, B_0 \langle G_y \rangle} \cong \frac{G_x z^2\, N}{2\, B_0} \tag{3.23a}$$

where $\langle G_y \rangle = G_x/N$ ($N$ = matrix size). Still, since $z$ = constant for a transverse plane, this effect only shifts the entire image into the phase-encode direction.

In coronal or saggital images the effects are different. First we assume the read-out gradient to be in the transverse direction (perpendicular to the $z$ axis). In this case we find a distortion proportional to $z^2$ as given by (3.23a).

In the read out direction, with its high gradient, this is not too severe. For a coronal image we find a distortion that is quadratic in $z$. For a 1.5 T system with gradients of 20 mT/m at 0.1 m from the isocenter this displacement is $6.7 \times 10^{-2}$ mm, which is negligible. However, in the phase encode direction the effective gradient field $\langle G_z \rangle$ is about $N$ times smaller and the effect is therefore larger. The resulting distortion is described by an equation similar to (3.23a). On each side of the isocenter the distortion is in the same direction, which means that a circular object with its center in the isocenter deforms into an egg-shaped object. The distortion at 0.1 m from the isocenter is equal to 8.5 mm. Actually it is somewhat bigger since the time of the blips is not taken into account: $(80/74) \times 8.5\,\text{mm} = 9.2\,\text{mm}$.

When the read-out and phase-encode directions are interchanged, so now the read-out direction is the $z$ direction, of course a similar effect occurs. Since $\mathrm{d}B_z/\mathrm{d}z$ does not occur in $\nabla \times \vec{B} = 0$, we now must start with $\nabla \cdot \vec{B} = 0$, and write this equation in cylindrical coordinates. For a cylindrical situation this yields the $z$ gradient associated Maxwell field: $B_r = 1/2\, r\, G_z$. The dc offset due to this field is given by $\delta\omega = \gamma G_z\, r^2/8\, B_0$. The distortion due to this offset is calculated in the same way as above in (3.23) but is four times smaller for similar images. We must be aware of these effects when designing systems for ultra-fast imaging (apart from the consequences for the signal-to-noise ratio; see Chap. 6).

## 3.4 Combination of TSE and EPI: GRASE

By combining TSE and EPI it is possible to improve the scanning efficiency further [3, 15]. The idea is to take a TSE sequence in which a short EPI sequence is implemented between the 180° pulses. In this case more than one profile can be measured between two 180° pulses. Such an imaging method has the advantage over TSE that in total fewer 180° degree pulses are needed, so that the power dissipation in the patient is reduced. The sequence has the advantage over EPI that susceptibility variations and magnetic-field inhomogeneities are compensated for by the repeated refocussing pulses, so that mainly, only $T_2$ decay is relevant.

The duration of such a scan is determined by the product of the TSE turbo factor and the EPI factor, which is the number of gradient lobes between two refocussing pulses. The sequence is called GRadient And Spin Echo (GRASE) and is shown in Fig. 3.13.

The path through the $\vec{k}$ plane for a single segment of a GRASE sequence with a turbo factor of 4 and an EPI factor of 3 is depicted in Fig. 3.14. For a matrix size of $128^2$ and a repetition time TR = 2000 ms. the total scan time is about 22 s. The paths of the ten other segments are interlaced between the paths shown in Fig. 3.14.

The GRASE sequence can be considered as the "mother" of all sequences [16] (and should be named Grace), because if one takes all 180° pulses out

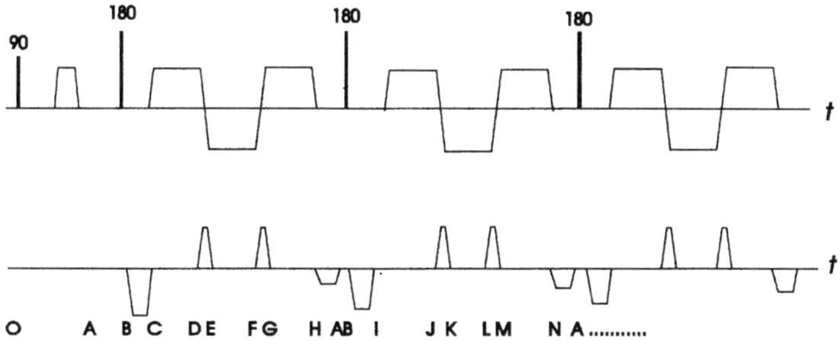


**Fig. 3.13.** GRASE sequence

one obtains EPI or when only one profile is measured one has FE or F(T)FE, when one takes out all pulses but one 180° pulse one obtains SE or SE EPI (Fig. 3.11) and when one increases the number of 180° pulses one has TSE or GRASE.

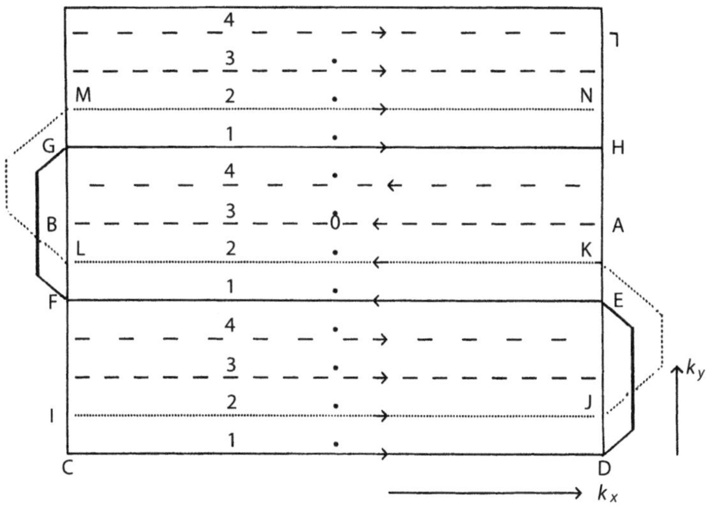


**Fig. 3.14.** $\vec{k}$-plane trajectory for GRASE

Artifacts in the GRASE sequence arise from the fact that trace 1 in Fig. 3.14 is acquired after the first refocussing pulse, trace 2 after the second refocussing pulse at a later time, when $T_2$ decay has taken place, and so on. In the phase-encode direction we therefore see a periodic function, somewhat like that shown in Fig. 3.5. This means that we can expect ghosts in the phase-encode direction. Also, the other artifacts discussed in connection with TSE and EPI will occur to some degree [17], as is also shown in image set III-5.

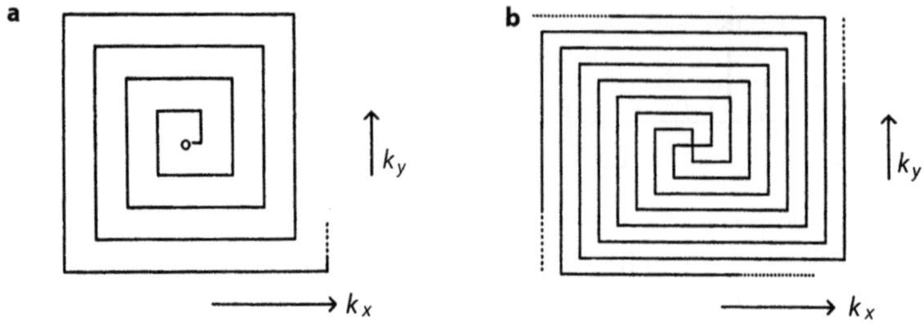


**Fig. 3.15.** **a** Square spiral trajectory through $\vec{k}$ plane. **b** Spiral trajectory consisting of four segments

## 3.5 Square Spiral Imaging

A special type of imaging is obtained, when the $\vec{k}$ plane is scanned along a square spiral (Fig. 3.15), which passes through all Cartesian points. This makes simulations of this type of spiral imaging relatively simple [6, 18]. The special feature is that the spiral starts at the point $k = 0$, so the magnetization at the low $k$ values is acquired directly after the excitation. Very roughly speaking, the magnetization measured for low $k$ values determines the contrast of the image and that measured at high $k$ values determine the resolution. When the magnetization at low values is measured, effects such as flow, susceptibility, $T_2^*$ decay, and so on, have not had time to develop serious errors, changing the contrast. On the other hand, the velocity in the $\vec{k}$ plane in the outward direction is relatively low, so blurring can occur and especially off-resonance effects (for example, those due to field inhomogeneities or the water–fat shift) cause point spreading and ringing. This imaging method therefore has new possibilities, but also has specific problems. We shall not go into detail here, since a similar discussion comes up in connection with spiral imaging where the trajectory is curved and does not pass through the Cartesian points.

## 3.6 Joyriding in $\vec{k}$ Space

The previous discussion suggests that there are many more possibilities to traverse the $\vec{k}$ plane. In Fig. 3.16 some examples are shown: a "rosette-like" trajectory, a spiral trajectory, and radial trajectories. These trajectories have in common that they generally do not pass through the Cartesian points.

Before we proceed, it is useful to realize that the gradient system must be able to generate the trajectories. In a very global way the requirements imposed on the gradients can be calculated from (2.26). The self-induction of the gradient coils is $L$ (for easy argument $L$ is assumed to be equal in the three directions). In Sect. 1.3.3 we introduced the coil sensitivity $C$, defined by the quotient of the gradient and the coil current, $C = G/I$. In order to follow a certain trajectory in the $\vec{k}$ plane, described by $k_i(t)$, the current ($I$) and voltage ($V$) have to satisfy the relations

$$I_i = C^{-1}G_i = \gamma^{-1}C^{-1}\frac{\mathrm{d}k_i(t)}{\mathrm{d}t}, \tag{3.24a}$$

and

$$V_i = L\frac{\mathrm{d}I_i}{\mathrm{d}t} = C^{-1}L\frac{\mathrm{d}G_i}{\mathrm{d}t} = \gamma^{-1}C^{-1}L\frac{\mathrm{d}^2k_i(t)}{\mathrm{d}t^2}, \tag{3.24b}$$

where the index $i$ stands for $x$, $y$, or $z$. This means that the velocity in the $\vec{k}$ plane, $\mathrm{d}\vec{k}(t)/\mathrm{d}t$, depends on the current of the gradient system and the sensitivity, $C$, of the coils. The acceleration, $\mathrm{d}\vec{k}^2(t)/\mathrm{d}t^2$, however, is restricted by the voltage of the gradient system, just as in the case with joyriding, where the available power is the limiting factor (and friction). It is therefore impossible to change the gradient momentarily. In EPI sequences, for example, the time it takes to reverse the gradient depends on the voltage of the gradient amplifier and during this reversal the distance to the next $\vec{k}$ profile must also be covered by means of a small preparation gradient (blip). This is characterized in Fig. 3.11 by the part of the trajectory outside the actual part of the $\vec{k}$ plane, which is the part of the $\vec{k}$ plane that is scanned. The part of the trajectory outside the $\vec{k}$ plane can be made smaller when a high voltage on

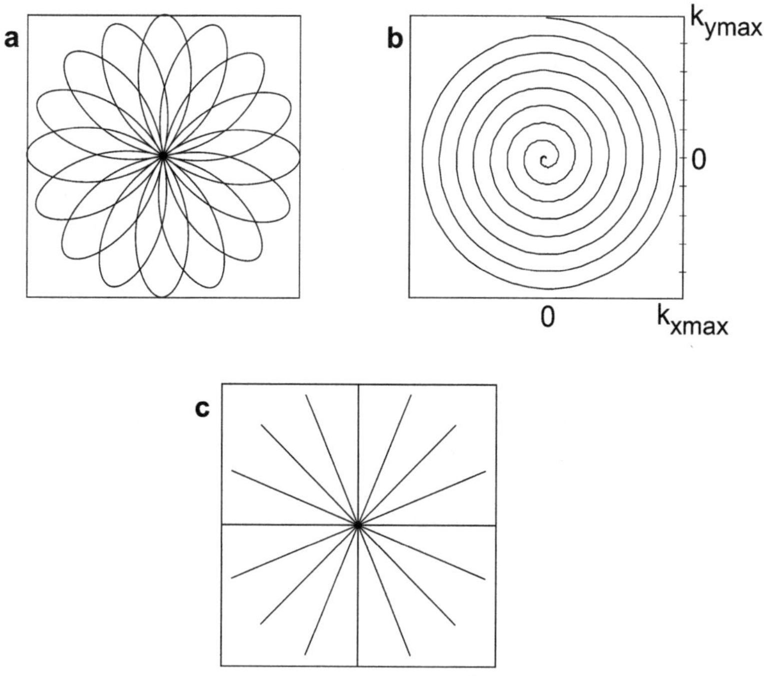


**Fig. 3.16.** Three trajectories through the $\vec{k}$ plane: **a** "rosette", **b** spiral, and **c** radial trajectories

the gradient coil is available. If it is decided to sample continuously (so also during the ramps of the necessary gradient fields) the deviations lie within the actual $\vec{k}$ plane. When one still wishes to use the fast Fourier transform, the sampled values of the magnetization must first be brought from the actual sampling points to the Cartesian grid by interpolation. This procedure is called "gridding" (see Sect. 3.6.1.2).

The trajectory through the $\vec{k}$ plane can be described in parameter form as $k_x = k_x(t)$ and $k_y = k_y(t)$ and differentiation of these relations yields the gradients (and gradient currents) needed in the $x$ and $y$ direction, respectively. Differentiating twice yields the gradient voltage needed. The maximum current and voltage depend on the specification of the gradient amplifier used.

### 3.6.1 Spiral Imaging

Spiral imaging, Fig. 3.16b, has a number of potential advantages over the fast imaging sequences discussed so far [5]. The most interesting one is the fact that the magnetization at the lowest $\vec{k}$ values, determining the contrast, is measured directly after excitation (see Fig. 3.17). This has the advantage that $T_2^*$ decay (so including susceptibilities) and flow (even turbulent flow) cannot develop large errors in the image of the larger structures (described by low $\vec{k}$ values) before essential information is acquired.

A further difference from other ultra-fast methods is that the path through the $\vec{k}$ plane is very smooth, no artifacts occur due to, for example, a meandering trajectory (Sect. 3.3.2.3). The exact position of the sampling points in the $\vec{k}$ plane, however, must be known in order to obtain a reliable reconstruction. This is a difficult requirement since it assumes that the exact total magnetic field is known at any time $t$, including the components of the gradient field (which may be delayed due to "gradient amplifier delay") and those due to eddy currents and main-field inhomogeneities. Special methods had to be developed to obtain the exact sampling points (for example, see [19]).

A third difference is that spiral imaging is less demanding for the gradient system since fast gradient-field changes as in EPI or TSE are not required. Traversing the complete $\vec{k}$ plane while acquiring sufficient sampling points with one spiral within the time $T_2^*$ nevertheless is very difficult because of the current and voltage limitations of the gradient amplifiers (or signal-to-noise ratio requirements). Therefore segmentation usually is applied by using a number of similar spiral arms, which are rotated with respect to each other.

We first look at the path through the $\vec{k}$ plane. For a spiral with constant increment per rotation, we have the general form

$$\vec{k}(t) = A\varphi(t)\exp(\mathrm{j}\varphi(t) + \mathrm{j}m\varphi_0), \tag{3.25}$$

where the relation $\varphi(t) = n2\pi$ determines the pitch of the spiral and $\varphi_0$ the angle under which the first spiral starts in the origin. The number of segments is given by $N_\mathrm{s} = 2\pi/\varphi_0$, which must be an integer and $0 < m < N_\mathrm{s}$.

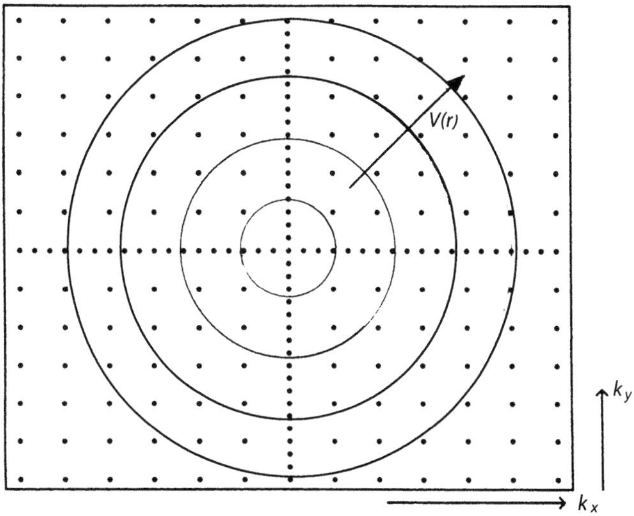


**Fig. 3.17.** Filling of the $\vec{k}$ plane with spiral scan

From (2.29) it can be deduced that for a conventional Cartesian acquisition sequence the size of the actual $\vec{k}$ plane to be acquired (defined as $2k_{\mathrm{max}} \times 2k_{\mathrm{max}}$, where the center of the actual $\vec{k}$ plane is taken at $k = 0$) is given by $2k_{\mathrm{max}} = \gamma G_x N t_{\mathrm{s}} = 2\pi N/\mathrm{FOV} = 2\pi/d_{\mathrm{v}}$ (see also (2.32)) where $d_{\mathrm{v}}$ is the dimension of a voxel. With spiral imaging, for the same FOV and matrix size, $N$, we acquire a circular region of the $\vec{k}$ plane with radius $\mathfrak{R} = k_{\mathrm{max}} = \pi/d_{\mathrm{v}}$. For FFT we need a square region, which can be obtained from the circular one by zero padding in the corners that are not scanned. The maximum radius is reached when $t = T_{\mathrm{m}}$, after $N_{\mathrm{r}}$ revolutions. The function $\varphi(t)$ is then $\varphi(T_{\mathrm{m}}) = 2\pi N_{\mathrm{r}}$. Introducing this in (3.25) yields, when $m = 0$,

$$A = \frac{\mathfrak{R}}{2\pi N_{\mathrm{r}}} = \frac{1}{2d_{\mathrm{v}} N_{\mathrm{r}}}. \tag{3.26}$$

Note that $\mathfrak{R}$ has the dimension of $\vec{k}$ ($= \mathrm{m}^{-1}$). Ideally the number of revolutions is $N_r \geq N/2$, to avoid aliasing. This, however, is still impossible on a commercial system with a gradient strength of 10 mT/m and a rise time of 1 ms. Therefore segmentation is applied by using interlaced spirals. The number of segments to be used is $N_{\mathrm{s}} \geq N/(2N_{\mathrm{r}})$.

The next problem is to determine which function $\varphi(t)$ will be used. In our discussion we restrict ourselves to two cases, namely $\varphi(t) = \Omega t$ and $\varphi(t) = \Omega' \sqrt{t}$. In practice a smooth transition between both functions during the acquisition period is made.

1. $\varphi(t) = \Omega t$. This is a spiral with a constant angular velocity. The linear velocity along the spiral in $\vec{k}$ space is $\mathrm{d}\vec{k}/\mathrm{d}t$. It can be seen that the velocity in the outer spiral arms becomes very large (meaning high gradient values,

which the gradient amplifier may not be able to generate). Another problem is that using very high gradients means that the receiver must have a high bandwidth (see (2.32) for high-read gradients). This also means that the signal-to-noise ratio is low, since the noise power is proportional to the square root of the bandwidth (see Chap. 6 on the signal-to-noise ratio).

**2.** $\varphi(t) = \Omega'\sqrt{t}$. For large values of $t$ this means a constant linear velocity spiral. However, in this case the problem is that $\mathrm{d}\vec{k}/\mathrm{d}t$ ($= \vec{G}$) has a pole at the origin, which means that in order to realize this spiral, the values of $\vec{G}$, and especially $\mathrm{dG}/\mathrm{d}t$, near $\vec{k} = 0$ must be impossibly high.

In practice mostly a combination of both spiral types will be used, the inner part of the spiral will be according to type 1 and the outer part will be like type 2. This is done by taking

$$\varphi(t) = \Omega t(1 + t/T)^{-1/2}, \tag{3.27}$$

where the parameter $T$ determines the transition from constant angular velocity, when $t \ll T$, to constant linear velocity, when $t \gg T$. One might think of an optimalisation procedure in which a value of $T$ is found such that both the velocity and the acceleration in $\vec{k}$ space remain just within the limits set by the gradient system (see (3.24)). Another optimization procedure would be to keep $|\vec{G}|$ within bounds set by the SNR requirements.

### 3.6.1.1. A Practical Example

If we choose $T$ to be small with respect to the total time available for acquiring one spiral arm (segment), $T \ll t_{\mathrm{acq}}$, most of this spiral arm will be a spiral with a constant linear velocity. On the other hand, if $T$ is larger than $t_{\mathrm{acq}}$, we have a spiral with (almost) a constant angular velocity. The advantage of this description of the spiral is that we now have an analytical form, which depends on two parameters $\Omega$ and $T$.

We assume the FOV to be 256 mm and the in-plane voxel size to be 2 mm, which means that we have a $128^2$ matrix. The time available for one single spiral arm is, say, 20 ms [5]. For each set of values for $\Omega$ and $T$ we can now see how many revolutions are made in 20 ms. For example, when $\Omega = 2\pi 1600$ rad/s and $T = 3$ ms $N_{\mathrm{r}}$ is 13. To cover the required $\vec{k}$ space we find from (3.26) that $A = 20$. Substituting this into (3.25) we find the maximum gradient field strength from $\vec{G}(t) = \gamma^{-1} \cdot \mathrm{d}\vec{k}/\mathrm{d}t$ to be 14 mT/m. From (3.24b) the voltage necessary to excite this spiral path through $\vec{k}$ space is calculated. As practical values we use $C = 3 \times 10^4$ A m/T and $L = 200\,\mu$H. The maximum voltage necessary is 350 V. The example described here is shown in Fig. 3.18.

The example given above may yield too-high requirements for the gradient system. In that case the requirements are relaxed when, for example, $\Omega$ is

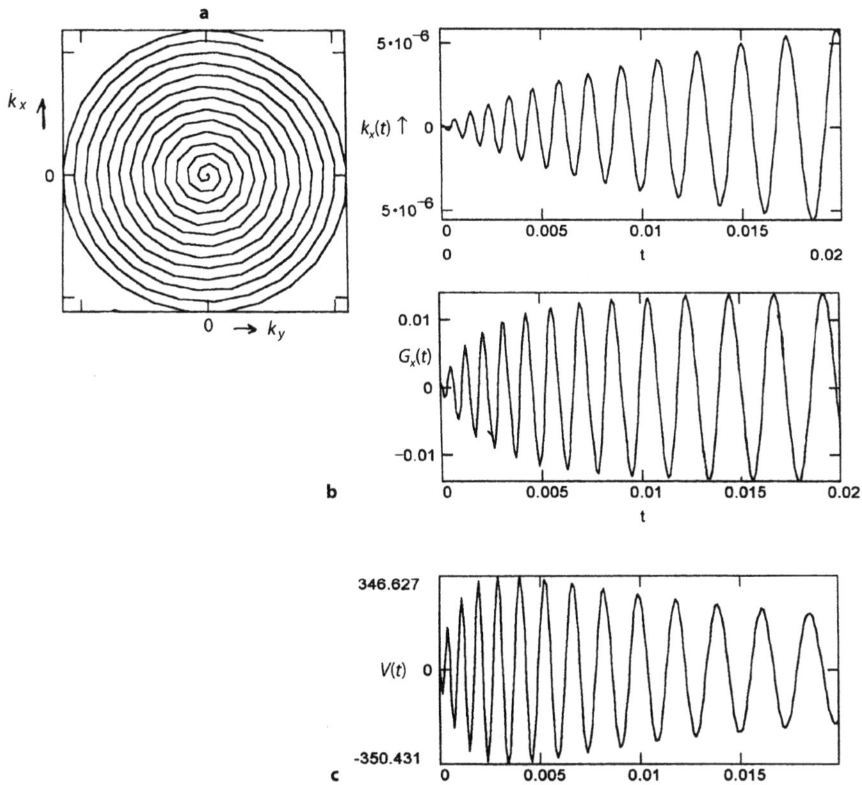


**Fig. 3.18.** Path through $\vec{k}$ space. **a** The actual path for $\Omega = 2\pi 1600\,\text{rad/s}$, $T = 3\,\text{ms}$, $A = 20\,\text{s/m}$. **b** the $x$ gradient necessary. **c** the voltage that the gradient must deliver to the gradient coil

lowered. The technique of spiral imaging is still in an experimental stage so our treatment is aimed at showing the design considerations from which practical situations can be derived.

In image set III-6, spiral imaging has been used to visualize the coronary arteries in a cardiac-triggered and respiration-gated (see Sect. 7.2) imaging method.

#### 3.6.1.2. Reconstruction by "Gridding"

For the first time we have encountered a scan method that does not sample on a Cartesian grid in the $\vec{k}$ plane. This makes a direct Fourier transform impossible, which is a pity because there is no other transform nearly as efficient. It is beyond the scope of this textbook to elaborate on reconstruction methods. However, we shall briefly describe a method to transform from the samples acquired at the points in the $\vec{k}$ plane reached by the spiral arms to an equivalent set of samples on the Cartesian grid, for which the efficient fast Fourier transform is possible. This method, called "gridding", is described in [21] and here we summarize this description.

We start again with (2.27) in which $t$ is replaced by the momentaneous $\vec{k}$ values:

$$M_{\mathrm{T}}(k_x, k_y) = \iint m(x, y) \exp -\mathrm{j}(k_x x + k_y y) \mathrm{d}x \, \mathrm{d}y. \tag{3.28}$$

In this equation $M_{\mathrm{T}}(k_x, k_y)$ is a continuous function. However only the sampling points on the spiral arms are acquired, given by $k_{xi}$ and $k_{yi}$, so the sampled function we actually measure can be described as

$$\begin{aligned} M_{\mathrm{TS}}(k_x, k_y) &= M_{\mathrm{T}}(k_x, k_y) \sum_{i=1}^{p} \delta(k_x - k_{xi}, k_y - k_{yi}) \\ &= M_{\mathrm{T}}(k_x, k_y) S(k_x, k_y), \end{aligned} \tag{3.29}$$

where $S$ is the parameter description of the path through the $\vec{k}$ plane. Now we have only a set of discrete points in the $\vec{k}$ plane in which the value of $M_{\mathrm{TS}}(k_x, k_y)$ is known and these points are certainly not the Cartesian points. Therefore the measured data is convolved with a function $C(k_x, k_y)$, which may be a Gausian function or a sinc function or just a finite window, in order to construct a continuous function:

$$M_{\mathrm{TSC}}(k_x, k_y) = M_{\mathrm{T}}(k_x, k_y) S(k_x, k_y) \otimes C(k_x, k_y). \tag{3.30}$$

Before sampling this function on Cartesian points, we must first correct for the inhomogeneous density of the original sampling points. It will be clear that in spiral imaging, this density is higher in the center of $\vec{k}$ space than in the outskirts, and the amplitude of $M_{\mathrm{TSC}}$ is proportional to this density of sampling points and not to the function we need for reconstruction. Therefore we replace $M_{\mathrm{TSC}}$ in (3.30) by a new function $M_{\mathrm{TSCW}}$ which is obtained by dividing $M_{\mathrm{TSC}}$ by the area density function

$$D(k_x, k_y) = S(k_x, k_y) \otimes C(k_x, k_y). \tag{3.31}$$

This yields

$$M_{\mathrm{TSCW}}(k_x, k_y) = \frac{M_{\mathrm{T}}(k_x, k_y) S(k_x, k_y) \otimes C(k_x, k_y)}{D(k_x, k_y)}, \tag{3.32}$$

which can be sampled on the Cartesian points by multiplying this function with a "comb function", describing the Cartesian points:

$$\mathrm{III}(k_x, k_y) = \sum_k \sum_l \delta(k_x - k_{xk}, k_y - k_{yl}), \tag{3.33}$$

where $k_{xk}$ and $k_{yl}$ are the Cartesian points in the $\vec{k}$ plane. The result is

$$M_{\mathrm{TSCWS}}(k_x, k_y) = \mathrm{III}(k_x, k_y) M_{\mathrm{TSCW}}(k_x, k_y), \tag{3.34}$$

where $M_{\mathrm{TSCWS}}$ is the sampled function that we need for reconstruction by Fourier transform. It will be clear that this function depends on the choice of the convolution function $C(k_x, k_y)$ and this is the subject of [20] and other papers.

### 3.6.1.3. Artifacts in Spiral Imaging

In spiral imaging the measured region extends in a way similar to the wave a stone makes when thrown into the water (see Fig. 3.17). $T_2^*$ decay gradually reduces the measured values in a radial direction. In the tangential direction there is also decay, but since the velocity in this direction is much higher its effect is lower, just as discussed earlier with respect to, for example, EPI, where the artifacts in the phase-encode direction are much more important than those in the read-out direction. Therefore we neglect the $T_2^*$ and off-resonance effects in the tangential direction as a first approximation. Then we can write the imaging equation, (2.28), in vector form:

$$m(\vec{r}) = \frac{1}{2\pi} \int_{\vec{k}} M(\vec{k}) \exp(\mathrm{j}\vec{k} \cdot \vec{r}) \exp(-t/T_2) \exp(\mathrm{j}\delta\omega\, t) \mathrm{d}\vec{k}, \tag{3.35}$$

where the second exponential describes the $T_2$ decay and the third describes the resonance offset (or chemical shift). Using the time dependence of $|\vec{k}|$ as described previously in Sect. 2.6 with $\varphi(t) = \Omega t$ and $\varphi(t) = \Omega\sqrt{t}$, we can eliminate the time from the integrand in (3.24):

$$m(\vec{r}) = \frac{1}{2\pi} \int_{\vec{k}} M(\vec{k}) \exp(\mathrm{j}\vec{k} \cdot \vec{r}) \exp\left(-\frac{|\vec{k}|}{\Omega T_2^* A}\right) \exp\left(-\mathrm{j}\delta\omega \frac{|\vec{k}|}{\Omega A}\right) \mathrm{d}\vec{k}, \tag{3.36}$$

when $\varphi(t) = \Omega t$, and for $\varphi(t) = \Omega' \sqrt{t}$ we have

$$m(\vec{r}) = \frac{1}{2\pi} \int_{\vec{k}} M(\vec{k}) \exp(\mathrm{j}\vec{k} \cdot \vec{r}) \exp\left(-\frac{|\vec{k}|^2}{(\Omega' A)^2 T_2^*}\right) \exp\left(-\mathrm{j}\delta\omega \frac{|\vec{k}|^2}{(\Omega' A)^2}\right) \mathrm{d}\vec{k}. \tag{3.37}$$

So the image function $m(\vec{r})$ is convolved with the Fourier transforms of the two exponentials. The exponential with a real exponent, describing $T_2^*$ decay, is easily recognized as a two-dimensional Lorentz function for a spiral with a constant angular velocity, and a two-dimensional Gaussian function for a spiral with a constant linear velocity, respectively. Since they are independent of the tangential position they (almost) describe isotropic blurring [18].

The function with an imaginary exponent describes radially undulating functions in the $\vec{k}$ plane, which cannot be Fourier transformed analytically. For example, when we have a small object in the center of the object plane, and also an off-resonance situation, instead of measuring constant values

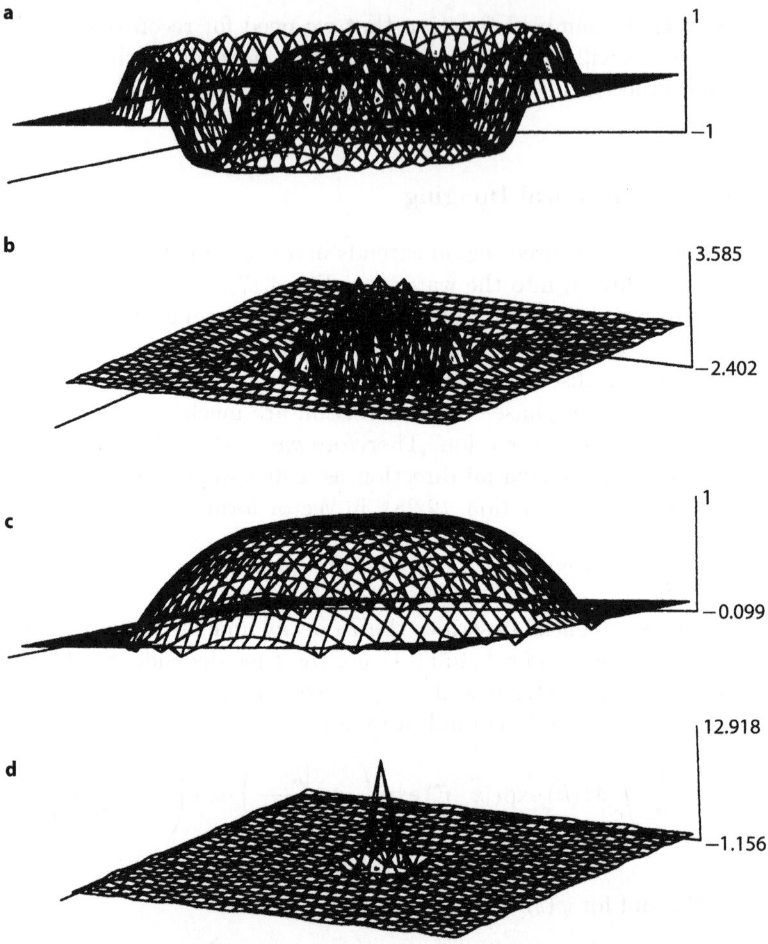


**Fig. 3.19.** **a** Magnetization as position in the $\vec{k}$ plane measured with large resonance offset. **b** Image amplitude in object space (presented inside out, so the corners represent the center). **c** Same as a but with small offset. **d** Image after 2D Fourier transform

for the magnetization as a function of position in the $\vec{k}$ plane for $|\vec{r}| < k_{\max}$ (outside this circle we apply zero padding), we measure a magnetization function like the one shown in Fig. 3.19a. We used MathCad to show this function and to perform the 2D Fourier transform for an example with a $32 \times 32$ matrix-imaging situation. The point object is imaged as a wider point (due to the finite dimensions of the actual $\vec{k}$ space) surrounded by a number of rings (Fig. 3.19b). An intuitive requirement for the resonance offset could be that the first zero point of the ondulating function is, at or beyond the edge of the $\vec{k}$ plane (at $\mathfrak{R}$). This situation is shown in Fig. 3.19c,d. With these ideas it is simple to find design requirements for systems with spiral imaging. For example the resonance offset must, for the situation of Fig. 3.19c, be smaller than found from $\delta\omega/\Omega = 1/4N_r$). In Fig. 3.19d the "point-spread function" is much sharper (higher) and more restricted in area than in Fig. 3.19b.

In [21] a method is proposed to correct for resonance offset, which is generally a function of position in the object plane, by reconstructing the measured raw data with several different resonance frequencies. In the images so obtained the sharp areas are selected so as to combine these "sub-images" into a complete sharp picture.

### 3.6.2 "Rosette" Trajectory

A "rosette-like" trajectory, Fig. 3.16a, through the $\vec{k}$ plane can be described by assuming a radial trajectory (a straight line through the origin), along which the sampling point moves harmonically in time as sin $\Omega t$; the line rotates slowly around the origin at the same time. This situation is described in [22]. It is possible to reconstruct the measured values for the magnetization (raw data) at the sampling points on such a trajectory into an image. However, the reconstruction process requires much computing power (see Sect. 3.6.4). A possibility to solve this problem is to translate the measured values into values on the Cartesian points. This is described in Sect. 3.6.1.2.

### 3.6.3 Radial Imaging

The path through the $\vec{k}$ plane is described by (see Fig. 3.16c)

$$k_x(t) = \frac{k_m t}{T_0} \sin\alpha, \quad \text{and} \quad k_y(t) = \frac{k_m t}{T_0} \cos\alpha, \tag{3.38}$$

where $0 < t < T_0$ and $\alpha$ is incremented after each profile in the region $0 < \alpha < 2\pi$. With this type of scan the sampled data on each line through the origin can be Fourier transformed and the obtained images (which are only position-dependent in the direction of the sampled line) superimposed (back-projected) to obtain the projection-reconstruction image. The other method used to reconstruct a radial scan is gridding followed by Fourier transformation.

Artifacts can be described in the same way as in spiral imaging and the assumption that only the radial velocity counts is true in case of radial imaging when each radial profile is acquired after excitation from the equilibrium situation (long TR). However, there are many different radial scanning methods possible, showing different artifacts. One can choose TR to be smaller than $T_1$, so that $T_1$ weighting is introduced and the total scan time is shortened. Even TR $\cong T_2$ is possible so that a dynamic equilibrium is built up (see next chapter). A special possibility is to rotate the radial line in the $\vec{k}$ plane while advancing the slice selection in the selection direction. At each moment one can make an image of the last acquired complete $\vec{k}$ plane. This is helical imaging [23]. This can also be done with a single slice and the time as the parameter for dynamic imaging.

Radial imaging can also be combined with TSE, as is described in [24], where the artifacts due to this special method of scanning the $\vec{k}$ plane are also discussed. Using the methods to visualize the acquisition of raw data in the $\vec{k}$ plane, as dependent on the trajectory followed and as used earlier for spiral scanning, one can also obtain a feel for the artifacts in radial TSE.

### 3.6.4 Some Remarks on the Reconstruction of Exotic Scans

The 2D Fourier transform is the ideal reconstruction method for those scans that have acquisitions in a rectangular grid in the $\vec{k}$ plane. For the more exotic paths through the $\vec{k}$ plane a more general method, known as weighted correlation, is proposed.

In general we have to solve (2.23) for an arbitrary path and its gradient $\vec{G}(t)$:

$$M_{\mathrm{T}}(t) = \int_{\mathrm{slice}} m(\vec{r}) \exp(-\mathrm{j}\vec{r} \cdot \vec{k}(t))\mathrm{d}\vec{r}, \quad \vec{k}(t) = \gamma \int \vec{G}(t)\mathrm{d}t. \tag{3.39}$$

We measure $M_{\mathrm{T}}(t)$ but we wish to know $m(\vec{r})$. To find $m(\vec{r})$ a new function $m^*(\vec{r}')$ is defined by

$$m^*(\vec{r}') = \int_{\mathrm{t}} M_{\mathrm{T}}(t) \exp(+\mathrm{j}\vec{r'} \cdot \vec{k}(t)) w[\vec{k}(t)]\mathrm{d}t, \tag{3.40}$$

where $w[\vec{k}(t)] = w(t)$ is a weighting function that must be chosen for the situation under consideration. For each point $\vec{r}$ this integral can be evaluated.

When the velocity through the $\vec{k}$ plane is time dependent, one may assume that the signal must be weighted with $w(t) = |\mathrm{d}\vec{k}(t)/\mathrm{d}t|$. In a projection reconstruction scan we can choose $w(t) = |\vec{k}(t)|$, since the density of measured points is high near the origin. Introducing (3.39) into (3.40) we find

$$m^*(\vec{r}') = \int_t \int_{\vec{r}} m(\vec{r}) \exp[-\mathrm{j}\vec{k}(t) \cdot (\vec{r} - \vec{r}')] w[\vec{k}(t)]\mathrm{d}\vec{r}\,\mathrm{d}t, \tag{3.41}$$

which can be seen as a two-dimensional convolution of $m(\vec{r})$ with a point spread function $h(\vec{r})$, when we change the order of integration:

$$\begin{aligned} m^*(\vec{r}') &= m(\vec{r}) \otimes \otimes h(\vec{r}), \\ h(\vec{r} - \vec{r}') &= \int_t \exp(-\mathrm{j}\vec{k}(t) \cdot (\vec{r} - \vec{r}')) w[\vec{k}(t)]\mathrm{d}t, \end{aligned} \tag{3.42}$$

where $\otimes\otimes$ means a two-dimensional convolution.

The function $h(\vec{r})$ is the point-spread function, as can be shown easily in a one-dimensional case, assuming that one line in the $\vec{k}$ plane is traversed with constant velocity. If we take $k(t) = k_{\mathrm{m}}(t - T/2)/(T/2)$ and $w(t) = \text{constant}$, we find

$$h(x - x') = \frac{2\sin[k_\mathrm{m}(x - x')]}{k_\mathrm{m}(x - x')}$$

which we knew already from the conventional Fourier transform (see (2.35)). However, one can also traverse this line with a sinusoidal gradient. If we now take $w(t)$ = constant we find a bessel function divided by its argument as a point-spread function instead of a sinc. Such a point-spread function (PSF) has a larger half-maximum width than does a sinc. Now the reconstruction can be improved by taking $w(t) = |\mathrm{d}k(t)/\mathrm{d}t|$. We then again obtain a sinc function, so this weighting function gives a better solution.

The price we pay for these exotic reconstruction methods is that for each point $\vec{r}'$ the integral in (3.40) must be evaluated, which takes a lot of computing power. The other option is gridding, which requires less computer power.

## 3.7 Two-Dimensional Excitation Pulses

Following our discussion on spiraling paths through the $\vec{k}$ plane we come back to our promise in Sect. 2.3.3 to discuss a method to design two-dimensional excitation pulses (giving a cylindrical excited region). We waited until now, since we shall use the ideas explained in the previous section. The treatment we follow is taken from [25]. For excitation pulses we also have a "$\vec{k}$ plane" (see Sect. 2.3.2). However, the definition of $\vec{k}$ is slightly different, since here we have to take the integral over the gradient from the actual time $t$ until the *end* of the pulse (see (2.15) and (2.17)).

We assume again that we have a small flip angle. This restriction is made for educational purposes, a change to a spinor description (see [24]) can relieve this restriction and also allow for arbitrary starting conditions and for a large angle between the $z$ axis and the actual magnetization, as long as the effect of the RF field is small compared with that of the magnetic field. Our example with a small flip angle is a special case of this more-general approximation. We start with (2.15) with $t = 0$ at the beginning of the pulse:

$$M_\mathrm{T}(\vec{r}) = \mathrm{j}\gamma M_0 \int_0^T B_1(t)\exp\left(-\mathrm{j}\gamma\vec{r}\cdot\int_\mathrm{t}^T \vec{G}(t')\mathrm{d}t'\right)\mathrm{d}t \tag{3.43}$$

and introduce, as in the imaging case, a $\vec{k}$ vector defined by

$$\vec{k}(t) = -\gamma\int_t^T \vec{G}(t')\mathrm{d}t', \tag{3.44}$$

where the time in this case runs *from* $t$ until the end of the excitation pulse. The magnetization can now be written as

$$M_\mathrm{T}(\vec{r}) = \mathrm{j}\gamma M_0 \int_0^T B_1(t)\exp(\mathrm{j}\vec{r}\cdot\vec{k}(t))\mathrm{d}t, \tag{3.45}$$

where $\vec{k}(t)$ is the parametric description of the path through the $\vec{k}$ plane. We want to see something similar to a Fourier transform, containing the term $\exp(\mathrm{j}\vec{k}\cdot\vec{r})$, which can be realized by writing (3.45) as

$$M_\mathrm{T}(\vec{r}) = \mathrm{j}\gamma M_0 \int_0^T B_1(t) \int_{\vec{k}} {}^3\delta(\vec{k}(t)-\vec{k}) \exp(\mathrm{j}\vec{r}\cdot\vec{k})\mathrm{d}\vec{k}\,\mathrm{d}t, \tag{3.46}$$

where ${}^3\delta$ is the symbol for the three-dimensional $\delta$ function. By changing the order of integration we obtain

$$M_\mathrm{T}(\vec{r}) = \mathrm{j}\gamma M_0 \int_{\vec{k}} \left\{ \int_0^T B_1(t)^3\delta(\vec{k}(t)-\vec{k})\mathrm{d}t \right\} \exp(\mathrm{j}\vec{r}\cdot\vec{k})\mathrm{d}\vec{k}. \tag{3.47}$$

We now denote the function between brackets by $f(\vec{k})$, so the result is

$$M_\mathrm{T}(\vec{r}) = \mathrm{j}\gamma M_0 \int_{\vec{k}} f(\vec{k}) \exp(\mathrm{j}\vec{r}\cdot\vec{k})\mathrm{d}\vec{k}. \tag{3.48}$$

Discussing the function $f(\vec{k})$ in more detail,

$$f(\vec{k}) = \int_0^T \frac{B_1'(\vec{k}(t))}{|\gamma \vec{G}'(\vec{k}(t))|}\, {}^3\delta(\vec{k}(t)-\vec{k}) \left| \frac{\mathrm{d}\vec{k}(t)}{\mathrm{d}t} \right| \mathrm{d}t, \tag{3.49}$$

and using the rule $f(\vec{k}(t))\,{}^3\delta(\vec{k}(t)-\vec{k}) = f(\vec{k})\,{}^3\delta(\vec{k}-\vec{k}(t))$, we can write this expression as

$$f(\vec{k}) = \frac{B_1(\vec{k})}{|\gamma G(\vec{k})|} \int {}^3\delta(\vec{k}(t)-\vec{k}) \left| \frac{\mathrm{d}\vec{k}(t)}{\mathrm{d}t} \right| \mathrm{d}t = W(\vec{k})S(\vec{k}). \tag{3.50}$$

The first factor can be understood as an RF weighting function, $W(k)$, and the integral as a function that describes the path through the $\vec{k}$ space, $S(\vec{k})$. The integral $S(k)$ is unity when the path covers the $\vec{k}$ space in a sufficiently dense way (otherwise "aliasing" occurs, we then see more excited regions). So the result in (3.48) is equal to

$$M_\mathrm{T}(\vec{r}) = \int_{\vec{k}} W(\vec{k})S(\vec{k}) \exp(\mathrm{j}\vec{k}\cdot\vec{r})\mathrm{d}\vec{k}. \tag{3.51}$$

We shall now apply this equation to a two-dimensional pulse, which is realized using a round spiral type of path trough $\vec{k}$ space. We have seen in the treatment of a slice-selective pulse in Sect. 2.3.2, that for a refocused pulse the weighting with RF power along the path in the $\vec{k}$ plane must be symmetric around the origin and the path must end in $\vec{k} = 0$.

For the $\vec{k}$-space trajectory in the form of a round spiral, starting at the outside of the "$\vec{k}$" space so that we can end in the origin to make a refocused pulse, we can write (see Fig. 3.20a)

$$k_x(t) = a(1-t/T)\sin\frac{2\pi nt}{T}, \quad k_y(t) = a(1-t/T)\cos\frac{2\pi nt}{T}. \tag{3.52}$$

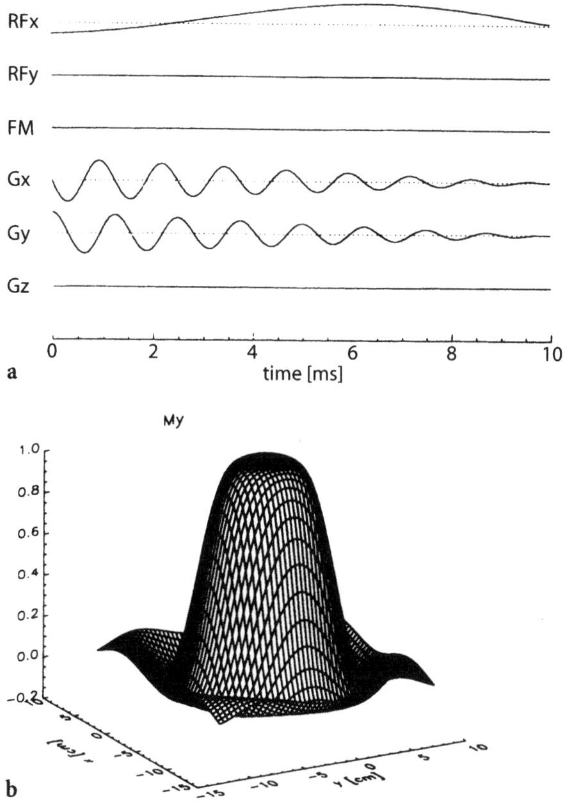


**Fig. 3.20.** **a** RF and gradient waveform of a 2D RF excitation pulse. **b** The $y$ component of the magnetization after application of the pulse sequence of Fig. 3.20a

The gradients can now easily be found by differentiation. We now must find the desired RF weighting function for a specified profile, of the transverse magnetization. From the theory of Fourier transforms we know that a Gaussian function transforms into another Gaussian Function. So when we require a Gaussian profile, the function $f(\vec{k})$ in (3.48) must be a Gaussian function:

$$f(\vec{k}) = W(\vec{k})S(\vec{k}) = \alpha \exp\left(-\beta|k|^2/a^2\right). \tag{3.53}$$

If the spiral is sufficiently densely wound (to avoid aliasing) the function $S$ is unity, so

$$B_1(t) = W(\vec{k}(t))|\gamma G(t)|, \tag{3.54}$$

which yields for the RF field strength

$$B_1(t) = \gamma\alpha\frac{a}{T}\exp\left(-\beta^2(1-t/T)^2\right)\left\{(2\pi n(1-t/T))^2+1\right\}^{1/2}. \tag{3.55}$$

This waveform is shown in [25] and results in transverse magnetization in the $y$ direction only, which means that the excitation pulse is refocused. Although the theory is formally correct only for small flip angles it can be used for flip angles up to 90°. Examples are shown in Fig. 3.20, the result of a numerical simulation (Courtesy H. Slotboom, private communication), and in image set III-8.

A similar theory can also be formulated for combined spatial spectral pulses, which can be used for simultaneous slice-selective excitation and fat suppression, see [26] and image set IV-4.

# Image Sets
# Chapter 3

## III-1 Influence of Echo Spacing on the Contrast in TSE Imaging

(1) In Turbo Spin-Echo (TSE) imaging, the echo spacing can be adjusted while TE, the effective echo time, is kept constant (see Sect. 3.2).
This allows the inspection of the subtle influence of the choice of the echo spacing. In this image set TSE scans are compared with a decreasing echo spacing, accompanied with an increasing value of the turbo factor, so that the effective echo time remains constant.

Parameters: $B_0 = 1.5\,\mathrm{T}$; FOV $= 200\,\mathrm{mm}$; matrix $= 256 \times 256$; $d = 4\,\mathrm{mm}$; TR/TE $= 2000/120$. In each of the scans, the bandwidth per pixel was the same.

| image no. | 1 | 2 | 3 | 4 | 5 | 6 |
|---|---|---|---|---|---|---|
| method | SE | TSE | TSE | TSE | TSE | TSE |
| echo spacing (ms) | na | 60 | 30 | 15 | 13 | 10 |
| turbo factor (FT) | na | 3 | 7 | 15 | 20 | 27 |
| NSA | 1 | 4 | 4 | 8 | 12 | 16 |
| scan time | 8′ 38″ | 11′ 26″ | 4′ 54″ | 4′ 54″ | 4′ 30″ | 4′ 22″ |

na = not applicable

In all TSE images a linear profile order is used and the set of profiles collected had nearly symmetrically distributed $k_y$ values. The scan time is kept roughly constant by adjustment of NSA (except image 1 and image 2).

(2) For the images shown, the contrast between grey matter and frontal white matter is not different between SE and TSE and is only marginally influenced by the echo spacing. For these tissues, the effective echo time is a useful indicator for the contrast of TSE images. Although fat is much brighter in the TSE images (see below), the use of the acronym TE for the effective echo time appears to be allowable; it labels the TSE images in a way that builds on the experience of reading SE images.

(3) The signal-to-noise ratio of each of the images was measured for some tissue types. This parameter was divided by $\sqrt{\mathrm{NSA}}$. Complete physical equivalence of the methods would, per tissue, lead to a constant value of this last ratio (Chap. 6).

| image no. | 1 | 2 | 3 | 4 | 5 | 6 | |
|---|---|---|---|---|---|---|---|
| CSF | 44 | 38 | 37 | 33 | 32 | 33 | $\frac{\mathrm{SNR}}{\sqrt{\mathrm{NSA}}}$ |
| grey matter | 21 | 21 | 17 | 16 | 16 | 16 | |
| frontal white matter | 13 | 14 | 11 | 9 | 10 | 10 | |
| internal capsule | 8 | 11 | 10 | 8 | 9 | 9 | |
| subcutaneous fat | 15 | 18 | 20 | 31 | 31 | 35 | |

For CSF, grey matter and white matter, the SNR decreases when changing from SE to TSE. Moreover, the decrease is larger for short echo spacings.

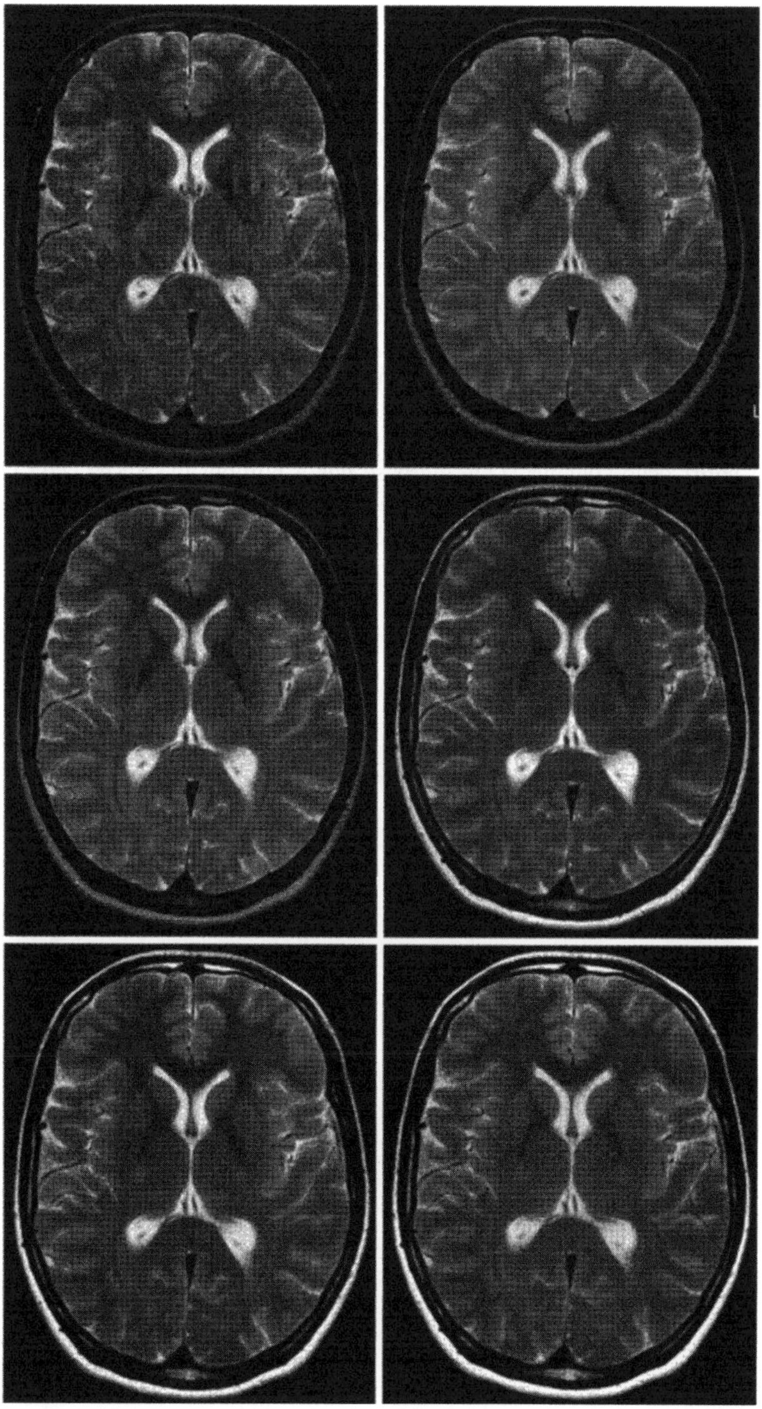


**Image Set III-1**

An explanation for this effect is the occurrence of magnetization transfer (Sect. 6.3.3 and image set VI -4).

(4) In SE and in TSE with a large echo spacing (echo spacing = 60 ms; image 2), the globus pallidus is clearly darker than the frontal white matter. This signal difference disappears at shorter echo spacings. The effect is visible also by comparing the SNR of this tissue with that of other white matter. The cause of this phenomenon is related to microscopic field inhomogeneities in iron containing nuclei in the brain and signal loss due to diffusion of spins in those regions [27]. Large echo spacings give an increased sensitivity to such signal loss.

(5) The brightness of lipids increases with a decreasing echo spacing. This effect is visually apparent from the images and numerically from comparison of the SNR. The phenomenon is ascribed to the apparent increase of $T_2$ of lipids at short echo spacing. This increase has a number of causes [28]. The dominant cause may be the loss of phase coherence at long echo spacings by diffusion of the spins in the microscopically inhomogeneous magnetic field in the fat. Another cause is the elimination of J coupling of the CH2 spins at short echo spacings.

# III-2 Blurring and Ghosts from $T_2$ Decay During the TSE Shot

(1) When long echo trains are used in TSE, a significant $T_2$ decay of the signal during the progress of the echo train will occur. The late echos will be much smaller than the early ones. This will have a negative effect on the quality of the image. The extent of the effect will depend on the length of the echo train and the choice of the profile order (see Sect. 3.2.1). This image set compares sagittal multi-slice TSE images of the head with different echo-train durations.

Parameters: $B_0 = 1.5\,\text{T}$; FOV $= 230\,\text{mm}$; matrix $N_x \times N_y = 256 \times N_y$; $d = 4\,\text{mm}$; TR/TE $= 4000/80$. (In TSE imaging, TE indicates the effective echo time).

| | | | | |
|---|---|---|---|---|
| image no. | 1,5 | 2,6 | 3,7 | 4,8 |
| turbo factor | 8 | 16 | 26 | 36 |
| $N_y$ | 256 | 256 | 234 | 256 |
| echo train duration (ms) | 96 | 192 | 312 | 432 |
| NSA | 4 | 6 | 12 | 16 |

Images 1–4 are of the mid-sagittal slice in the volunteer's head. Images 5–8 are obtained in the same scans as images 1–4, but describe another slice. NSA was adapted to keep the scan time at approximately 5 minutes.

(2) In all scans the echo spacing was 12 ms and the profile order was linear (see Sect. 3.2.1). When with these conditions the echo train becomes longer, the step in $k_y$ per echo is smaller and for a symmetric set of profiles TE would become larger. To keep TE at its constant value (80 ms), the scans were based on sets of profiles with a different degree of asymmetry. In the scan for image 1, the echo-train duration (96 ms) was only slightly larger than TE; here almost all profiles with positive $k_y$ values were left out. To obtain image 2, an equal number of negative and positive $k_y$ values were used. For image 3, profiles with large negative $k_y$ values were left out, so that the echo-train duration (312 ms) was longer than twice the effective echo time. The scan of image 4 was obtained in a similar way, but with a still more extreme lengthening of the echo-train duration (432 ms).

(3) Pulsatile flow of the venous sinus was the cause of ghosting in the mid-sagittal images 1–4. The periodicity of the artifacts depends on the turbo factor. A ghost that depends on the turbo factor is expected in all TSE images, because the $T_2$ weighting of the profiles has a periodic dependence on $k_y$, as discussed in Sect. 3.2.1. The level of this ghost usually is negligible as can be seen in images 5–8.

(4) In all images the brain and cerebellum are sharply delineated from the bright CSF. The area of the nose and mouth, however, becomes unsharp

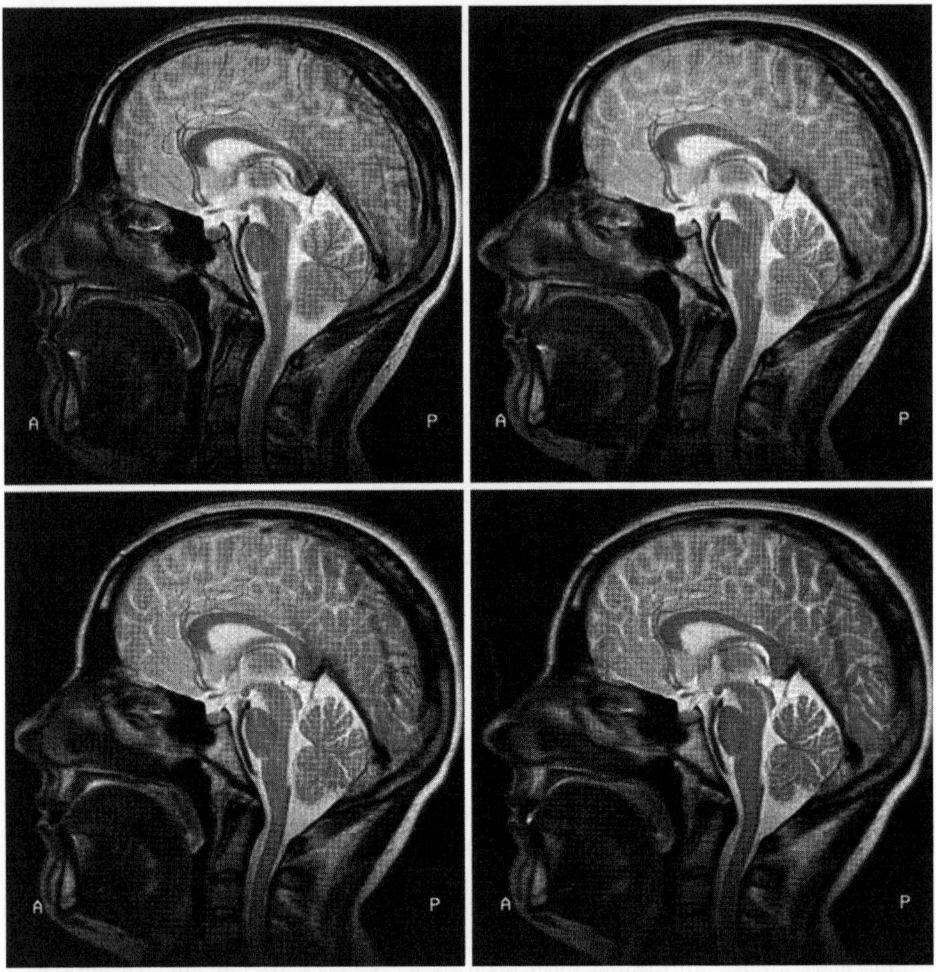


1 2

3 4

**Image Set III-2**

with an increasing turbo factor. The sharp delineation of brain tissue can be explained by the long $T_2$ of CSF; its estimated value is 4 seconds. So, even in the scan of image 4, the signal from CSF is not attenuated by $T_2$ decay. Tissue outside the brain, such as the tongue, has a much lower $T_2$. For such tissue, signal attenuation during long echo trains, as used for example for images 3 and 4, is considerable, especially at the late part of the train when the high $k_y$ values are collected. This attenuation is equivalent to a reduction of the spatial resolution.

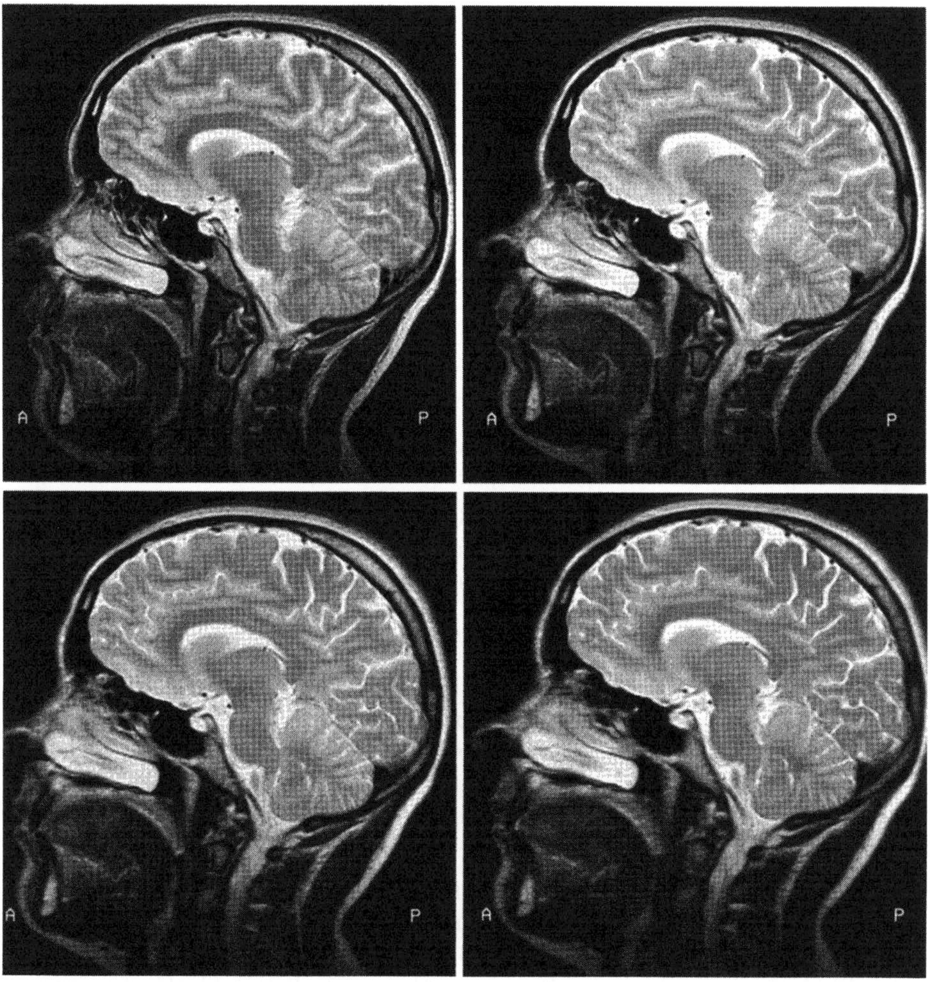


5 6 7 8

**Image Set III-2** (*Contd.*)

(5) The image set demonstrates that TSE with large effective echo times in combination with very long and asymmetrical echo trains can be used for imaging of the brain. However, the technique depends on the presence of CSF for sharp delineation of tissues; outside the brain a strong blurring of the image occurs.

## III-3 3D TSE-DRIVE, TSE with Reset Pulses

(1) An RF pulse that is given after signal acquisition to convert the remaining transversal magnetization into positive longitudinal magnetization is called a reset pulse (see Sect. 2.7.2). Imaging methods in which these pulses are used are called DRIVE (DRIVen Equilibrium). The use of a reset pulse is required to reach strongly $T_2$-weighted images at relatively short TR. In $T_2$-weighted multi-slice TSE, the need for short TR's is not always present. Long TR's can be used without loss of efficiency, for instance when the number of slices is large and all slices are excited within one TR. In $T_2$-weighted 3D TSE, the reset pulse is an important tool to shorten scan time, as demonstrated in this image set. All images are of the same region of the cervical spine of a normal volunteer. The window and level are kept constant in all three images.

Parameters: $B_0 = 1.5\,\mathrm{T}$; FOV $= 230\,\mathrm{mm}$; matrix $= 256 \times 189$; $d = 3\,\mathrm{mm}$; TE $= 120\,\mathrm{ms}$; turbo factor $= 27$; echo spacing $= 8.8\,\mathrm{ms}$; NSA $= 1$.

| | | | |
|---|---|---|---|
| image no. | 1 | 2 | 3 |
| reset pulse | no | no | yes |
| TR (ms) | 3250 | 750 | 750 |
| scan time | 11′56″ | 2′37″ | 2′37″ |

The echo train lasts 240 ms; the reset pulse requires one additional echo and is followed by a spoiler gradient. The repetition time chosen in images 2 and 3 is not the shortest one possible; a period of 500 ms is allowed for the longitudinal magnetization to regrow. As a result the images contain not only a signal in areas of long $T_2$ (reset signal), but also in areas of short $T_2$ and short $T_1$ (re-growth signal). This should result in a resemblance of the contrast in images 1 and 3.

(2) When examination of the nerve roots is required, strong $T_2$-weighting for a bright spinal fluid signal is needed. Traditionally this is reached by using a long TR, as shown in image 1. Image 3 shows that a similar contrast can be reached by a reset pulse in combination with a much shorter TR. Comparison of images 1 and 3 shows a roughly equal signal in most parts of the image, but the spinal fluid is clearly brighter in image 3, corresponding to a 15% difference in signal strength. Together with the slightly lower spinal cord signal, this results in a fluid-to-cord contrast in image 3 that has increased by almost 40% compared to that in image 1.

(3) The 3D TSE images do not show artefacts from pulsatile flow of the spinal fluid. These artefacts can easily arise in multi-slice TSE, as shown in image set VII-8, and can seriously degrade the visualization of the nerve roots.

(4) The example shows that the 3D TSE-DRIVE method combines the properties needed to give clinically useful images for nerve root inspection: freedom of flow artefacts, good contrast between nerve and spinal fluid, and attractively low scan time. In image set V-3 another modern technique for cervical spine imaging is shown.

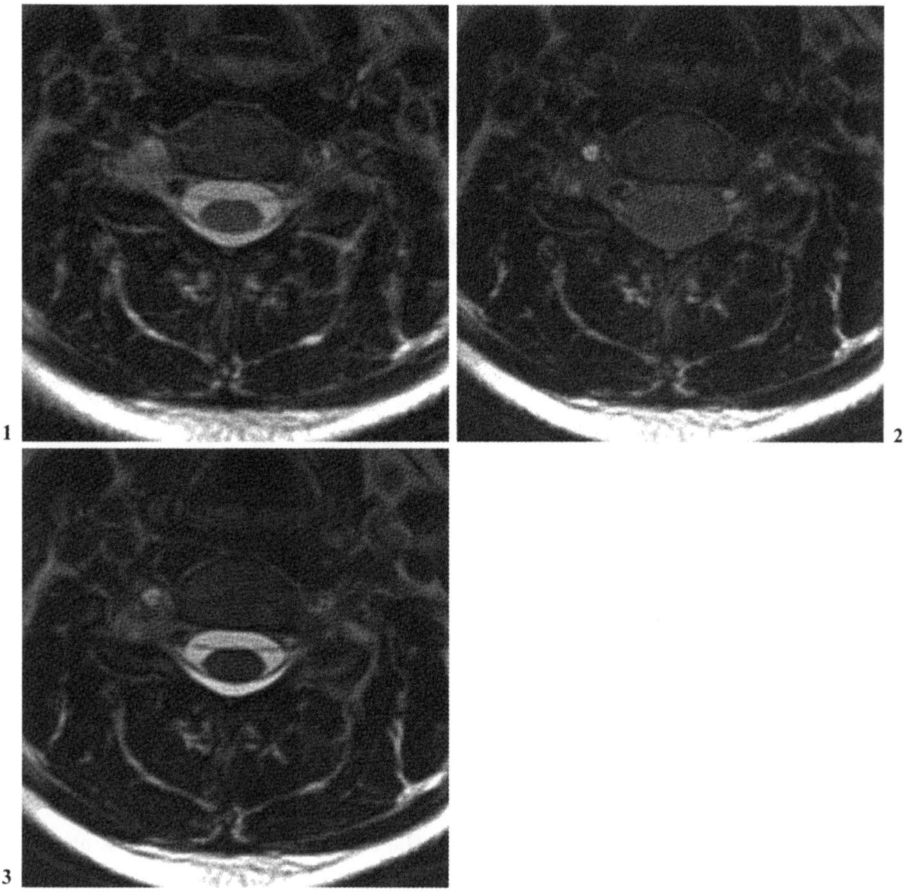

Image Set III-3

## III-4 Contrast in IR-TSE

(1) Multi-slice Turbo Spin Echo (TSE) can in our system be combined with inversion pulses (IR-TSE). Per TR the sequence contains for a number of slices both the slice-selective inversion pulses and the excitation pulses. The images compare the result of some combinations of TR and TI.

Parameters: $B_0 = 1.5\,\text{T}$; FOV $= 200\,\text{mm}$; matrix $= 256 \times 256$; $d = 4\,\text{mm}$; TE $= 20\,\text{ms}$; NSA $= 2$.

| image no. | 1 | 2 | 3 | 4 | 5 | 6 |
|---|---|---|---|---|---|---|
| TR (ms) | 1400 | | 1400 | | 2000 | |
| TI (ms) | 100 | | 280 | | 280 | |
| imagetype | R | M | R | M | R | M |
| scan time | 4′ 06″ | | 4′ 06″ | | 5′ 50″ | |

The profile order is centric (see Sect. 3.2.1) and the turbo factor is 3. Both real (images 1, 3, 5) and modulus images (images 2, 4, 6) are given. The images are obtained in a multi-slice mode with eight slices per TR. NSA = 2.

(2) When based on a SE technique, the type of imaging shown in this image set is known as STIR (short tau inversion recovery; see Sect. 2.7.1). The use of the faster turbo technique allows the realisation of an IR-TSE scan with contrast that is similar to STIR, while reducing the scan time to clinically more acceptable values, even when NSA = 2 and even for a value of TR that is longer than usual.

(3) The use of a long TR contributes to the signal. First we will look at the modulus images (images 2, 4, and 6). At TI $= 100\,\text{ms}$ and TR $= 1400\,\text{ms}$ (image 2), the signal level of CSF is close to that of white matter. Its relative brightness has increased in images 4 and 6 (TI $= 280\,\text{ms}$). At this TI the contrast between grey and white matter has also increased. In image 6, from a scan with the longest TR, in addition both grey and white matter are brighter than in image 4. The use of the combination of a long TR (2000 ms) and a long TI (280 ms) allows a clear display of grey matter, white matter, and CSF in one image.

(4) The real-type images (images 1, 3, and 5) show that for all scans the brain tissue magnetization is negative (the surrounding air is brighter than the tissue). In these images a long $T_1$ (e.g., that of grey matter) results in a large negative magnetization and a dark signal. Darkness of the signal in a real image, however, is not equivalent to a weak signal, as can be seen in the corresponding modulus images (images 2, 4, and 6).

The inversion pulse in the IR-TSE images will give $T_1$ weighting, but the unusual aspect of the negative magnetization is that increase in $T_1$ will give rise to an increased signal. In many tissues, such as grey and white matter and

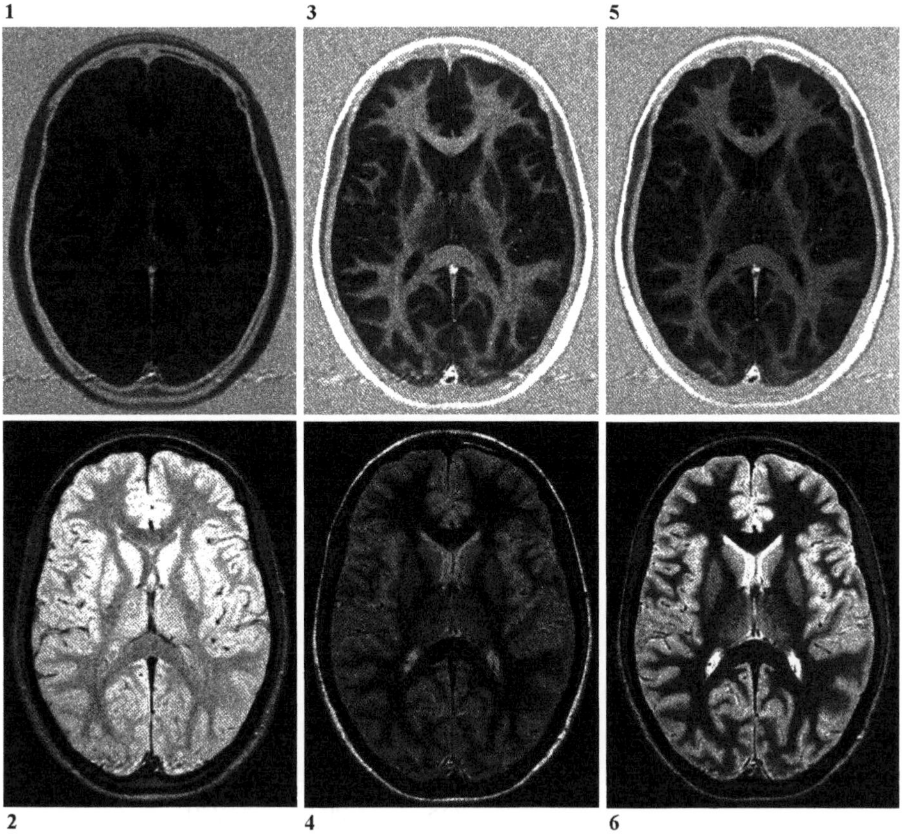


**Image Set III-4**

CSF, the tissue with the highest $T_1$ also has the highest $T_2$ and the highest proton density. As a result, in Real type IR-TSE images, these tissues are shown in strong contrast. Note that with Modulus images a bright aspect of CSF is obtained with an echo time of only 20 ms.

(4) The use IR-TSE imaging, based on a combination of Turbo Spin Echo and an inversion pulse, appears to offer an interesting contrast in a clinically acceptable scan time. Between many tissue pairs the contrast is strong and is best characterised as a combined $T_1$, $T_2$, and proton-density weighting.

## III-5 Comparison of Contrast in FE-EPI, SE-EPI, and GRASE

(1) Acquisition methods based on repeated gradient echos are introduced in Sects. 3.3 (EPI) and 3.4 (GRASE). The extreme version of such an acquisition method is the classical single-shot EPI method, with a rather atypical use.

Much closer to normal imaging tasks are methods with a lower number of gradient echos per excitation (low EPI factor). In this image set, some of these methods are compared.

Parameters: $B_0 = 0.5\,\mathrm{T}$; FOV = 230 mm; matrix = $256 \times N_y$; $d = 6\,\mathrm{mm}$; NSA = 1. All methods are set up in a multi slice mode with 22 slices per TR.

| image no. | 1 | 2 | 3 |
|---|---|---|---|
| method | SE-EPI | FE-EPI | GRASE |
| turbo factor | – | – | 10 |
| RF echo spacing (ms) | – | – | 18 |
| EPI factor | 15 | 7 | 3 |
| gradient echo spacing (ms) | 2.5 | 2.5 | 2.5 |
| TE (ms) | 100 | 27 | 100 |
| TR (ms) | 2002 | 1360 | 2231 |
| flip angle (degree) | 90 | 15 | 90 |
| WFS (pixels) | 5 | 5 | 2.7 |
| fat suppression | no | no | no |
| scan time (s) | 96 | 33 | 88 |
| $N_y$ | 248 | 203 | 240 |

(2) The water–fat shift (WFS) given for the SE-EPI and the FE-EPI image is the value for the phase-encode direction. In these acquisition methods, as explained in Sect. 3.3.2.2, this is the direction with the largest value of the WFS. The values of TR are the shortest possible, given the number of slices per TR.

(3) In each of the scans the image contrast between white and grey matter and CSF is $T_2$ weighted and roughly comparable. In the case of image 2, with its much shorter effective echo time, this contrast is possible because of the use of a small flip angle (see image set II-3). The misregistration of water and fat in the EPI scans (images 1 and 2) is greater than that in the GRASE scan technique (image 3), but in all cases it appears to be clinically acceptable.

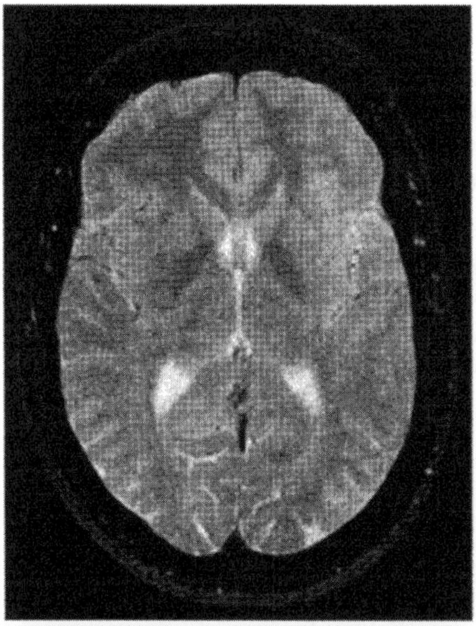
1

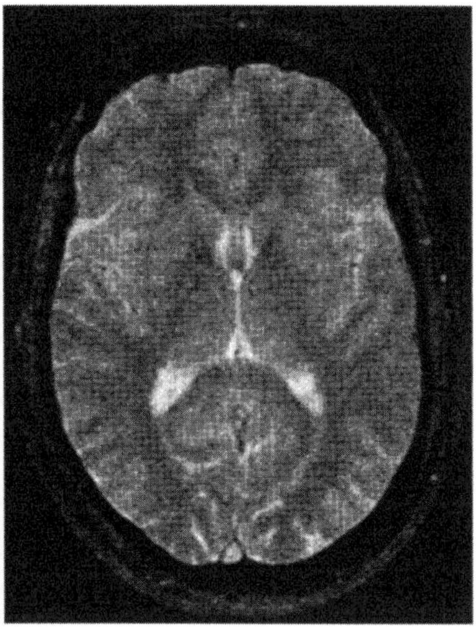
2

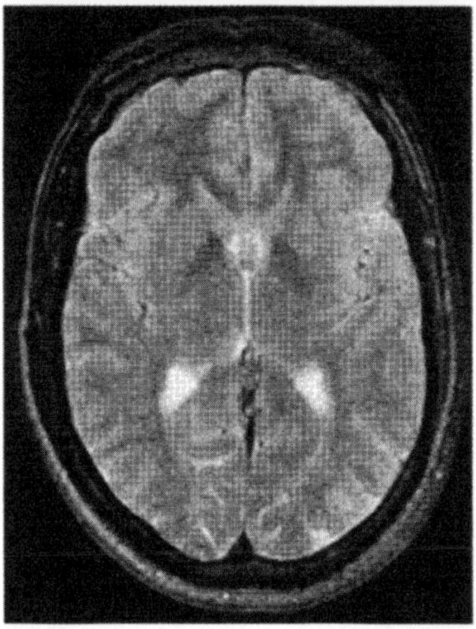
3

**Image Set III-5**

## III-6 Spiral Imaging

(1) One of the most important uses of a read-out strategy based on a number of spirally wound trajectories in $\vec{k}$ space is spiral FID imaging (see for example Fig. 3.18 or for a segmented spiral scan Fig. 3.15). In these scans, the echo time can be very short, because the trajectories start at the lowest $\vec{k}$ values. As a consequence, the sequence is insensitive to flow artifacts. The large number of samples on each spiral trajectory means that the total scan time can be short, even when only one trajectory per excitation is read. In this image set, some spiral FID images of the heart are shown. The images are of an experimental system (courtesy of Dr Holz of Philips Research Laboratory in Hamburg).

Parameters: $B_0 = 0.5$ T; FOV = 400 mm; matrix = 256 × 256; $d = 7$ mm; TE = 4 ms; NSA = 1.

The images are obtained in diastole with a cardiac-triggered TR of 1 beat. The flip angle is 90°. A spectrally selective fat-suppression pulse was added before each excitation pulse. Each image is the result of 64 trajectories read in 64 TR (64 seconds). Using a respiration sensor, data collection was restricted to periods that had a constant respiration phase. The trajectories are distributed evenly over the angular directions and describe a spiral path in $\vec{k}$ space with constant angular velocity. Each trajectory took 9.8 ms to acquire 1536 samples. The receive coil had six elements connected to six separate receiver channels; see Sect. 1.3.4.1. The bandwidth was 350 Hz. Gridding (Sect. 3.6.1.2) and zero padding is used to adapt the data to Fourier transformation.

(2) The experimental nature of the scan implied that a full flexibility in choice of scan parameters was not available. The spiral scan allows a more efficient use of the acquisition time than can be reached in normal FE imaging.

(3) By using fat suppression, the contrast between the coronary vessels and the pericardial fat is increased. Image 1 shows a part of the left coronary

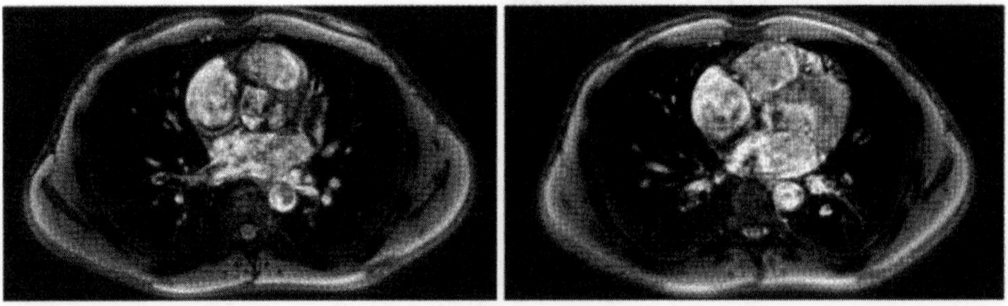

1 2

**Image Set III-6**

artery. Although the volunteer is breathing normally during the scan, the artefact level is low enough to visualize this artery (see also image set VII-13).

(4) The images show, based on a cardiac-triggered, respiration-gated spiral FID acquisition method, that combines high resolution, short scan time, and low artifact level.

# III-7 Distortion Near the Nasal Sinuses at Three Field Strengths

(1) Brain acquisition methods have to be resistant to the main field inhomogeneity from air pockets such as the nasal sinuses. Field-echo imaging is vulnerable in this respect, as illustrated in image set II-10, because this method makes use of gradient echos. Accordingly, the use of RF echos in brain imaging is preferred. In Chap. 3 various methods have been introduced that are based on the use of gradient echos as well as RF echos. To inspect the extent in which such methods are vulnerable to main field inhomogeneities, two of them will be compared in this image set. These are: Spin-Echo Planar imaging (SE-EPI, Sect. 3.3), and Gradient and Spin Echo (GRASE, Sect. 3.4). Comparison with Turbo Spin Echo (TSE, Sect. 3.2) is added for purpose of reference.
Each method is used to scan in three MR systems, running at 0.5, 1.0, and 1.5 T. Each method is of completely equal design at the different field strengths. Transverse images were made of the brain of the same volunteer in all systems. The images shown represent tissue just above the nasal sinuses. Other scans of the same slice are shown in image set II-10. Other slices of the same scans are shown as image set VI-7.

Parameters (all scans): FOV = 230 mm; $d$ = 6 mm; matrix $256 \times N_y$; TR/TE = 2700/80; NSA = 2. All scans were in multi-slice mode with 17 slices. Per TR, all slices were addressed.

| image no. | 1 | 2 | 3 | 4 | 5 | 6 | 7 | 8 | 9 |
|---|---|---|---|---|---|---|---|---|---|
| Field strength(T) | 0.5 | 0.5 | 0.5 | 1.0 | 1.0 | 1.0 | 1.5 | 1.5 | 1.5 |
| method | TSE | SE-EPI | GRASE | TSE | SE-EPI | GRASE | TSE | SE-EPI | GRASE |
| turbo factor | 9 | na | 6 | 9 | na | 6 | 9 | na | 6 |
| EPI factor | na | 13 | 3 | na | 13 | 3 | na | 13 | 3 |
| $N_y$ | 252 | 247 | 252 | 252 | 247 | 252 | 252 | 247 | 252 |
| WFS (pixels) | 0.55 | 3.4 | 1.0 | 1.1 | 6.8 | 2.0 | 1.7 | 10.2 | 3.0 |
| BW (Hz) | 130 | 250 | 220 | 130 | 250 | 220 | 130 | 250 | 220 |
| scan time | 2:33 | 1:48 | 1:21 | 2:33 | 1:48 | 1:21 | 2:33 | 1:48 | 1:21 |

na = not applicable

The turbo factor is the number of RF echos per excitation; the EPI factor (EF) is the number of gradient echos between each pair of refocussing pulses. WFS is the water–fat shift in the image; BW is the bandwidth per pixel.

(2) The values of the turbo factor in TSE and GRASE and of the EPI factor in SE-EPI and GRASE are arbitrary. For the combination of factors used for GRASE, TR could not be reduced further, given the strategy of 17 slices per TR. The GRASE and SE-EPI scans were made at the maximum available BW. The value of BW used per method was the same for each field strength. The value of this parameter used for TSE was arbitrary, but represents a sensible approach for clinically useful imaging.

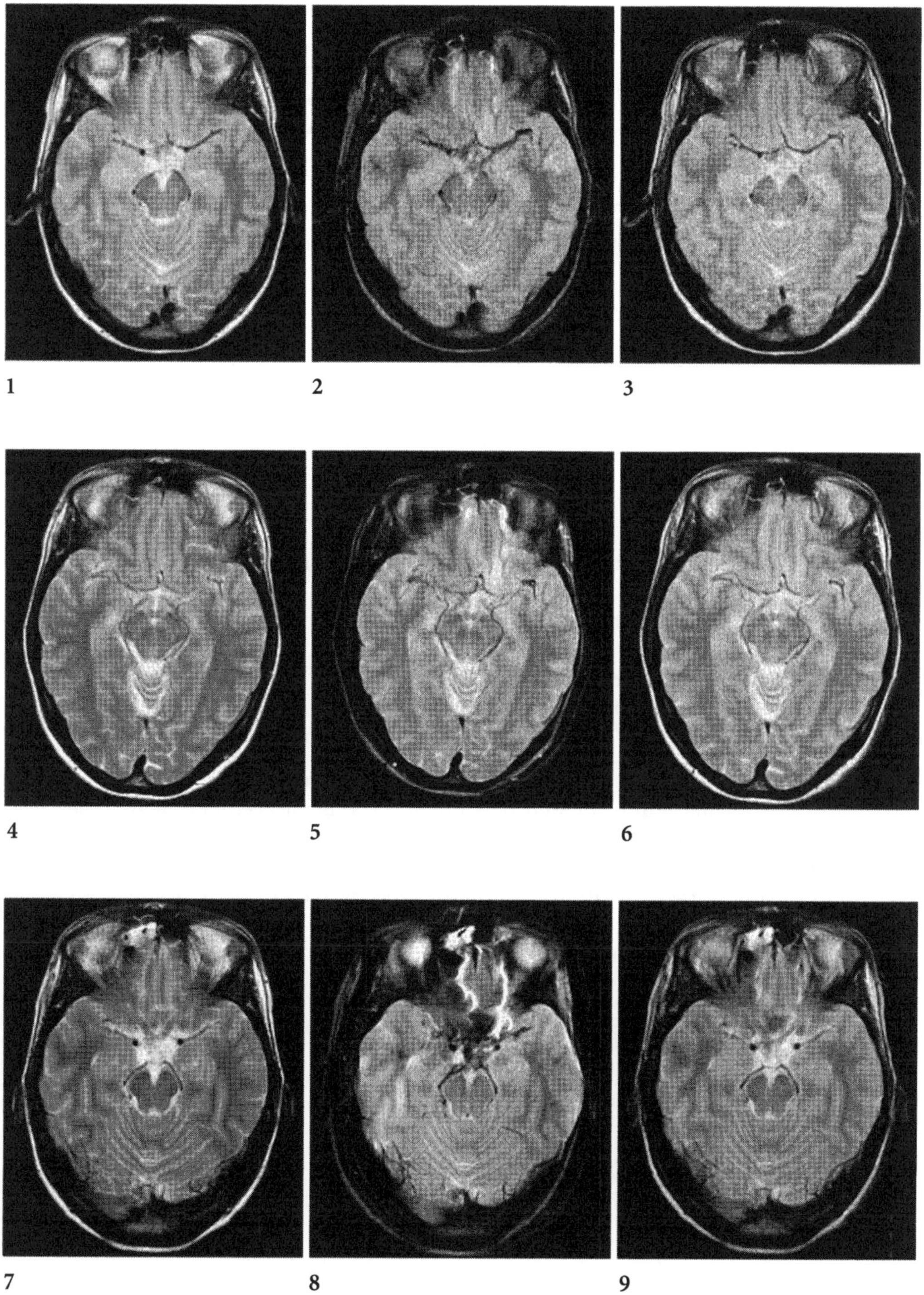


Image Set III-7

(3) In our system, control of BW is offered to the user through the parameter of the water–fat shift in pixels, WFS. For methods like SE, FE, and TSE, WFS relates to a shift in the read-out gradient direction. For such methods the relation between WFS and BW can be derived from (3.18) by substituting for $\alpha$ the main field $B_0$ and the relative shift $\delta$ of water and fat: $\alpha = \delta_{\text{wf}} B_0$, giving:

$$\text{WFS} = (\delta_{\text{wf}}\, B_0/G_x)(N_x/\text{FOV}).$$

Now using (2.32) and (2.38) to substitute $G_x$, $N_x$, and FOV, the result, valid for example for a TSE scan, is

$$\text{WFS} = \gamma\!\!\!{}^{-}\,\delta_{\text{wf}}\, B_0/\text{BW}.$$

For SE-EPI and GRASE, the WFS relates to a shift in the phase-encode gradient direction and is proportional to the pseudo gradient $G_y$, (see Sect. 3.3.2.2) so that for the same bandwidth, field strength, and matrix size an approximate relation exists that can be used to estimate the water–fat shift in these cases:

$$\text{WFS(SE-EPI and GRASE)} = \text{EF WFS(TSE)} = \text{EF}\, \gamma\!\!\!{}^{-}\,\delta_{\text{wf}}\, B_0/\text{BW}.$$

(4) The deviating susceptibility of the nasal sinus gives a field disturbance that is proportional to the main field. So WFS can be used to estimate from the scan parameters the relative degree of distortion that will occur in the images in this image set. Clearly, the SE-EPI images should have the largest distortion, especially those at high field. The distortion in the GRASE image at 1.5 T is still smaller than that in the SE-EPI image at 0.5 T. The distortion in the TSE images (in the read-out direction) is the smallest.

(5) Comparison of the images confirms this assessment. The distortion in the SE-EPI image at 1.5 T and possibly at 1.0 T is unacceptably high. The distortion in the SE-EPI image at 0.5 T is acceptable. The GRASE images and the TSE images have negligible distortion at all field strengths.
In conclusion, the SE-EPI acquisition method, when used with a large EPI factor and especially at high field strengths, is less suitable for imaging of the brain near the nasal sinuses.

## III-8 Practical Aspects of 2D Selective Excitation Pulses

(Courtesy of J Smink, Philips Medical Systems, Best, The Netherlands)

(1) The theory given in Sect. 3.7 is used to construct 2D selective RF pulses for a pencil beam excitation. Such pencil beams can be used to generate navigator echoes through the diaphragm (see Sect. 7.2.2 and Image Set VII-13). The purpose of these navigators is to detect the position of the diaphragm and with that to correct the position of the slice during acquisition of structures in the heart or lungs (prospective navigation). Some of the engineering aspects of pencil beam excitation will be discussed.

(2) Ideally the pencil beam has a constant flip angle $\alpha$ within a radius $\rho$ and zero outside. Using the small-flip-angle approximation, this ideal shape is reached by using an RF weighting function $W(k)$, see (3.50),

$$W_k = \frac{J_1(k\rho)}{k\rho},$$

with $J_1$ a first-order Bessel function. The ideal $\vec{k}$ space weighting cannot be reached in practice. The first compromise is to limit the $\vec{k}$ space coverage to within this first zero crossing, ($k_{\text{max}} = 3.83/\rho$). This will limit the RF pulse time at the cost of some degradation of the pencil beam profile. A useful approach for the $\vec{k}$ space trajectory is a round inward spiral as described in (3.52), modified as suggested in (3.27) to make efficient use of the maximum available gradient values and slew rate. The number of windings $N_{\text{w}}$ of the spiral that is used gives the second compromise. Too few will cause problems with aliasing of the excitation profile in the image space, the aliases being cylinders coaxial to the pencil beam. Analogous to (2.34), the number of windings determines the radius $\rho_{\text{a}}$ of the aliasing cylinders as

$$\rho_{\text{a}} = \frac{N_{\text{w}}\pi}{k_{\text{max}}} = \frac{N_{\text{w}}\rho\pi}{3.83}.$$

(3) In this image set images of a water-filled phantom with a diameter of 380 mm after spiral excitation with a 2D selective radial RF pulse are compared to images of similar excitations in the thorax of a volunteer. Between the images, the pencil diameter and number of spiral arms is varied.

Parameters: $B_0 = 1.5\,\text{T}$; $G_{\text{max}} = 23\,\text{mT/m}$; slew rate $= 100\,\text{T/ms}$; $B_{1,\text{max}} = 26\,\mu\text{T}$. For imaging gradient echoes are read. Orientation of the pencil beam is FH. The spiral RF pulses used have a duration that is the smallest obtainable.

| image no. | 1 | 2 | 3 | 4 | 5 | 6 | 7 | 8 | 9 |
|---|---|---|---|---|---|---|---|---|---|
| image orientation | tra | tra | tra | cor | cor | cor | cor | cor | cor |
| pencil beam radius (mm) | 12.5 | 12.5 | 12.5 | 10 | 10 | 10 | 15 | 15 | 15 |
| number of spiral windings | 4 | 7 | 10 | 6 | 12 | 16 | 6 | 12 | 16 |
| RF pulse duration (ms) | 2.6 | 4.1 | 5.6 | 4.1 | 7.7 | 10.3 | 3.3 | 6.2 | 8.1 |
| flip angle (degrees) | 10 | 10 | 10 | 25 | 25 | 25 | 25 | 25 | 25 |

(4) Comparison of the phantom images (images 1– 3) shows the cross-section of the pencil beam in the isocentre. Also visible are the position and intensity of the aliases with a typical structure associated with the spiral excitation. An increase in the number of spirals increases the radius of the first alias.

(5) Images 4 to 9 show the longitudinal aspect of the pencil beam at its typical position across the diaphragm. It is located to the right of the isocentre; arrows indicate the pencil beam axis. It should be noted that, to obtain the excitation at such a non-isocentric position, the RF frequency has to be modulated synchronously with the local field variation from the spiraling gradients at that position. In the extreme caudal and cranial regions of the images, the RF frequency modulation is no longer adequate because of inhomogeneity of the main field so that in these regions there is loss of localization of the excitation, causing blurring of the excited region. This is visible especially for the larger pencil radius and for the longest RF pulses used, such as in images 8 and 9 at the lower extreme side of the pencil.

(6) In images 4 to 9, the contour of the thorax of the volunteer is visible as a low-intensity background. The brightness of these contours is due to aliases of the pencil beam as well as to regions of tissue where the gradient and main field inhomogeneity add up to give spins in the Larmor frequency. This situation is similar to that in Image Set II-7, but the pattern is less sharp because the excitation is not restricted to a slice. For a small number of windings and a small pencil radius, intense aliasing regions prohibit the use of the pencil as navigator echo (images 4, 5 and 7).

(7) The images illustrate that a 2D-selective RF pulse obtained by spiral excitation can be used for generation of navigator echoes that define the position of the diaphragm. For best performance the choice of RF pulse parameters is critical. Short RF pulse durations and a sufficient number of windings have to be combined and images 6, 8 and 9 show examples of pencil beams that can be used for reliable determination of the diaphragm position.

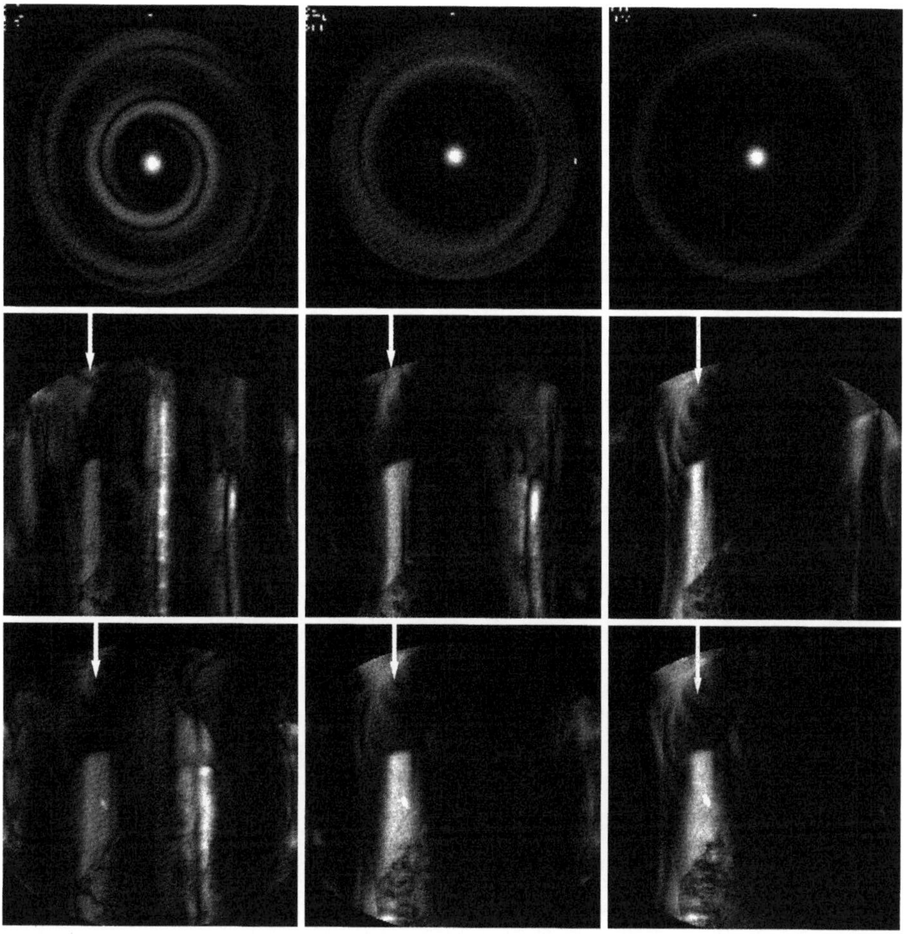

1,2 3

4,5 6

7,8 9

**Image Set III-8**

# 4. Steady-State Gradient-Echo Imaging

## 4.1 Introduction

Conventional MRI, based on methods like Spin Echo (SE) or Field (Gradient) Echo, has the important drawback that the scan time is long to very long compared with time constants of cardiac, peristaltic, and respiratory motion in the patient, giving rise to artifacts in the images. This long scan time is caused by the time it takes to allow the spins after acquisition to relax back to the equilibrium situation before the next excitation pulse is applied (Chap. 2). In many cases the problems of patient motion can be circumvented by applying ECG triggering and/or respiratory motion detection, but in cases of abdominal imaging, heavy arrythmia of the heart function, fast dynamic processes, or very ill and traumatized patients, very fast scanning is necessary.

One solution to this problem is to acquire more profiles (TSE or EPI) or covering more sampling points in the $\vec{k}$ plane after a single excitation (spiral imaging) (see Chap. 3). In this chapter we shall deal with another solution: the total scan time is made short by taking the repetition time between the excitation pulses, TR $\ll T_1$ and even TR $\leq T_2$. In the first case the longitudinal magnetization does not reach its equilibrium value before each excitation, in the latter case also the transverse magnetization does not relax back to zero. The magnetization before an excitation pulse is in this case caused by several previous excitation pulses and has both longitudinal and transverse components. However, after a number of excitations a "dynamic equilibrium" builds up, which means that before each excitation the same magnetization situation exists. Although this idea can be applied to both SE and FE, it is most frequently applied to FE sequences. The family of imaging methods based on this dynamic equilibrium situation is called Fast-Field-Echo Imaging (FFE) and is the subject of this chapter.

The theoretical description of the dynamic equilibrium was already known in the literature on nuclear magnetic resonance long before the FFE sequences for MR imaging were developed [1, 2]. The recent literature on FFE imaging methods is abundant, see for example [3–6], and the names of these methods are confusing [7, 1.10]. In this chapter the theory of the FFE imaging methods will be given and the mutual relations between these FFE methods, with their surprising acronyms such as FISP, FLASH, FAST, GRASS, PSIF, turbo-FLASH, FADE, SSFP, etc [1.10] will be clarified. A more systematic classification will be proposed.

In order to describe the fast-gradient-echo sequences we start with a model [8] in which the object to be imaged is subjected to a periodic train of excitation pulses with fixed distance (the repetition time TR) and with arbitrary flip angle $\alpha$, which is usually smaller than 90°. The gradient–time integral has the same value in each interval between the excitation pulses. In our model we describe the effect of the excitation pulses by a simple instantaneous rotation around an axis perpendicular to the static magnetic field; the length of the excitation pulse is neglected. This is a surprising assumption, since in some sequences available nowadays, this pulse takes up to 20% of the repetition time between excitations. However, we are not in the first place interested in the details of the excitation pulses but in the overall effects of the periodic excitation.

A spin system subjected to a periodic series of excitation pulses reaches a dynamic equilibrium situation ("steady state") after a time of the order of the spin lattice relaxation time $T_1$. In this situation the magnetization just after the $n$th $\alpha$ pulse is equal to the magnetization just after pulse $(n+1)$ or $(n-1)$. Note that the magnetization has components both in the longitudinal direction, parallel to the static magnetic field $B_0$, and in the transverse direction, perpendicular to $B_0$. In this chapter such a steady state will be assumed. It is also possible to start scanning before the steady state is reached. This will be the subject of the next chapter.

The signal induced by the transverse magnetization just after the $n$th pulse, comes in the first place from the fresh longitudinal magnetization, existing just before the $n$th $\alpha$ pulse, which is partly rotated into the transverse plane. We call this signal the "free induction decay" (fid). However, there is also a transverse component of the magnetization *before* the $n$th pulse. This is not so surprising, since (for example) the spin echo of the $(n-2)$th pulse, refocused by the $(n-1)$th pulse (which occurs even when the pulses have flip angles not equal to 90° and 180°) coincides with the $n$th pulse, as does the stimulated echo of the $(n-3)$th, $(n-2)$th, and $(n-1)$th pulses. This transverse magnetization before the next pulse, called the ECHO, apparently consists of a number of echoes, caused by the preceding pulses. This ECHO is rotated partly out of the transverse plane by the nth excitation pulse and therefore a transverse magnetization component persists after that pulse and is added to the fid. We call the resulting sum signal the FID. The essential difficulty in fast-field-echo imaging is that the transverse magnetization vector of the fid just after the $n$th pulse is independent of the position in the object plane, whereas the magnetization due to the ECHO is both in length and in direction dependent on the local gradients experienced by the spins before that pulse, and is therefore position dependent. So the magnetization forming the FID also becomes position dependent.

Both the FID and the ECHO (see Fig. 4.1a) can be turned into a gradient echo in the usual way by dephasing the spins by a gradient in the read-out direction and by subsequently rephasing them by reversing this gradient (see Fig. 2.14). From the FID signal a normal gradient echo is formed (remember, we use capitals for the original FID and ECHO to distinguish them from a fid following a single pulse and from gradient echoes); see Fig. 4.1b. In the ECHO case the read-out gradient can be shaped so that the time seems to run "reversed", which is understood by realizing that the echo and its subsequent ECHO, due to appear at the next $\alpha$ pulse, have zero phase difference and are related like a FID and its echo when $\int G(t)\mathrm{d}t$, integrated between TE and TR, is zero (see Fig. 4.1c). Both the echo of the FID and the "echo" of the ECHO are used for imaging purposes. It will be shown later that due to the position dependence of the magnetization in the FID and ECHO in the image, bands with varying contrast are formed. The different FFE imaging methods are distinguished by the method in which these bands are suppressed or made invisible.

In the next section we shall first try to build up more understanding of what happens in FFE sequences and what are the problems, without using mathematics. This will also familiarize us with the different echos possible. In Sect. 4.3 a formulation of the mathematical model of fast-gradient echos is presented. In Sect. 4.4 this model is applied to practical FFE sequences in which a steady state of the spins has build up and in which the varying contrast bands are suppressed. In Sect. 4.5 the different FFE imaging methods are discussed and a new nomenclature is proposed.

## 4.2 On FIDs and ECHOs

We first have to form an idea of what happens when a spin system in an homogeneous object is subjected to a long train of excitation pulses. To achieve this we shall first consider the effect of a small number of pulses ($T_1$ relaxation will be neglected for the time being). Let us start with a single pulse. If the magnetic moment of the spins is not parallel to the main magnetic field $B_0$, the vector of the magnetic moment will precess around $B_0$, which is always taken to be parallel to the $z$ axis of the laboratory frame of reference. In the rotating frame of reference $(x', y', z)$, rotating with respect to the laboratory frame of reference around the $z$ axis with the Larmor frequency $\omega = \gamma B_0$, there is no precession. An RF excitation pulse generates a $B_1$ field, for example along the $-x'$ axis (see Sect. 1.1.3). As a result the magnetization rotates around the $-x'$ axis over an angle $\alpha$ so as to form two components, $M_y$ and $M_z$ (see Fig. 4.2). Only the transverse component induces a signal in the antenna of the receiver.

The induced signal decays because the alignment along the $+y'$ axis is lost due to spin–spin relaxation ($T_2$ relaxation) and also due to the inhomogeneity of the main magnetic field, $\delta B(\vec{r})$, which causes the phase of the spins at

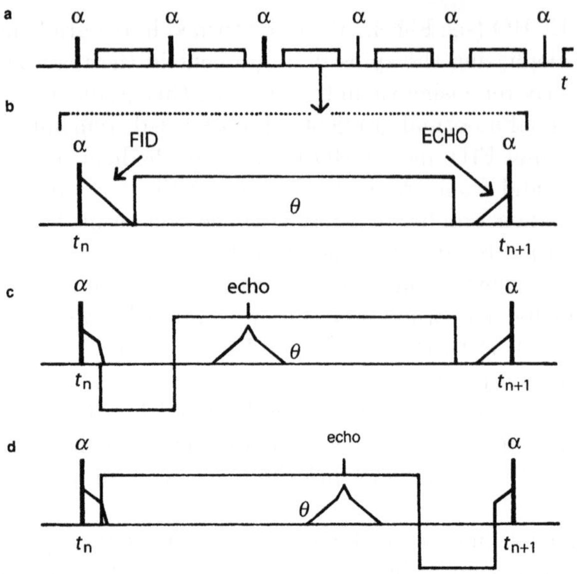


**Fig. 4.1.** **a** FFE sequence; **b** FID and ECHO; **c** echo generated by the FID; **d** echo "generated" by the ECHO

different locations to spread out in the $(x', y')$ plane of the rotating frame of reference. The combination of both effects is called $T_2^*$ relaxation (see Sect. 1.1.5).

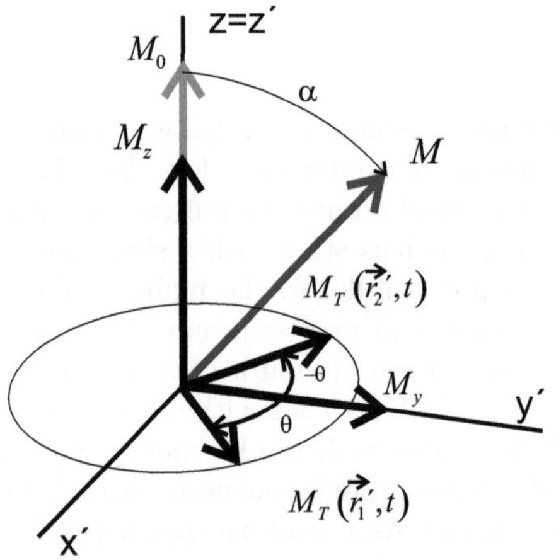


**Fig. 4.2.** The original magnetization $M_0$ is brought out of its equilibrium situation in the $z$ direction to form a transverse component $M_y$ and a decreased longitudinal component $M_z$. After a time $t$ the transverse magnetization moved away from the $y'$-axis with an angular velocity depending on the local magnetic field. When we have a local field higher than $B_0$ the magnetization moves into the $(+x', +y')$ quadrant; see $M_T(r_1', t)$. Deviations from the $B_0$ field are due to gradient fields as well as field inhomogeneity

### 4.2.1 Spin Echo

To prepare for the discussion of multi-pulse sequences, we first study the effect of two excitation pulses. Therefore we first consider the well-known Spin-Echo sequence [9] (see Sect. 2.4). After the short 90° excitation pulse, which brings the magnetization along the $+y'$ axis, we switch on a gradient so that the local magnetic field is dependent on the position in a known way (we disregard for the moment the $B_0$ inhomogeneity). At places where the local magnetic field is higher than $B_0$ the transverse magnetization moves into the $(+x', +y')$ quadrant with a precession velocity depending on the local value of the gradient field (see $M_{\mathrm{T}}(\vec{r}_1, t)$ in Fig. 4.2, where the magnetization is shown for two positions, one with a positive and one with a negative gradient field on each side of the isocentre). After a certain time $t$ the precession angle

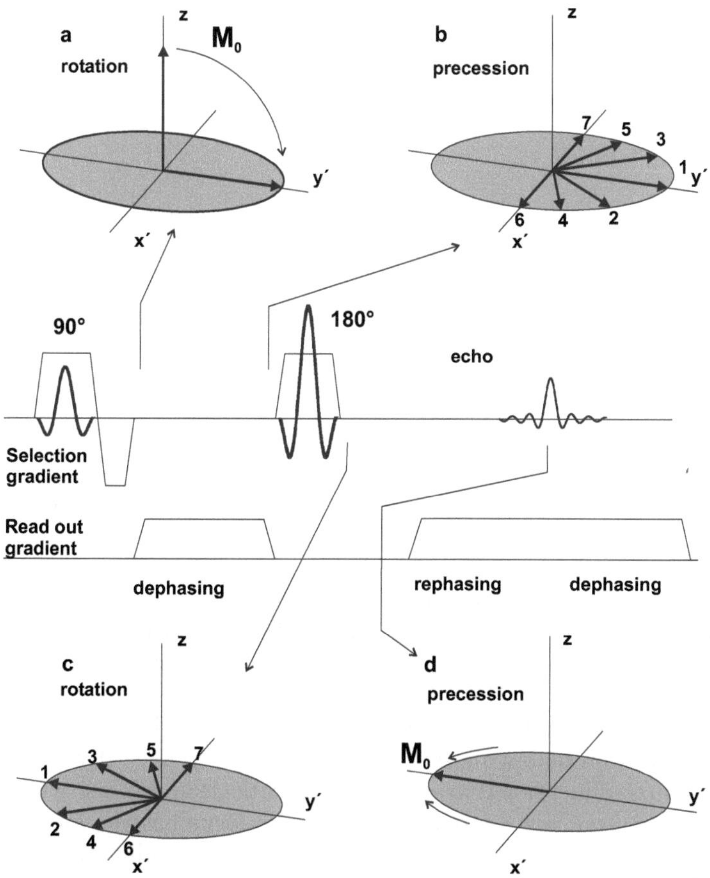


**Fig. 4.3.** Spin-Echo sequence. **a** Rotation. **b** Precession. **c** Rotation. **d** Precession.

for a certain position is $\theta$ degrees. For negative values of the gradient field the transverse magnetization moves backwards into the $(-x', +y')$ quadrant (see $M_T(\vec{r}_2, t)$ in Fig. 4.2). For ease of argument we assume in the following that the maximum precession is restricted to 180°. This restriction is easily removed and does not change the conclusions.

We now consider the effect of the 180° pulse causing a rotation of 180° around (for example) the $-x'$ axis. The magnetization vectors in Fig. 4.2 now move into the $(+x', -y')$ and the $(-x', -y')$ quadrants, respectively. This is shown in Figs. 4.3b and 4.3c where the vectors 1 to 5 change position with respect to 6 and 7. We then again switch on the same gradient as present between the 90° and 180° pulses; the transverse magnetization vectors proceed their precession and will meet each other on the $-y'$ axis. For all positions in the object plane the time at which this happens is the time at which the gradient–time integral $\int G(t)\mathrm{d}t)$ after the 180° is equal to that before this refocussing pulse, because the magnetization vectors then have moved through the same angle. So all spin vectors add to form a maximum signal. We have repeated here the description of the Spin-Echo sequence from Chap. 2 to prepare the description of other echo types.

### 4.2.2 "Eight-Ball" Echo

Now suppose that we replace the 180° pulse by a 90° pulse around the $-x'$ axis. Then the transverse magnetization vectors rotate around the $-x'$ axis into the $(+x', -z)$ and the $(-x', -z)$ quadrants, respectively, for those magnetization vectors that precessed less than 90°, and to the $(-x', +z)$ and $(+x', +z)$ quadrants for those that precessed between 90° and 180° (see Fig. 4.4b). For further discussion we only consider the transverse components of the magnetization (which coincide with, respectively, the $-x'$ and $+x'$ axes: Fig. 4.4c).

After the second pulse the $x'$ components resume their precession and form an echo when the same phase angle as before the second pulse is covered. Some effort will show that now at the moment of highest ordering, given that the signal is maximum (the echo), the points of the transverse magnetization vectors lie on a circle through the origin with the $x$ axis as tangent (Fig. 4.4d). The curve that connects these points is called the "locus". For the eight-ball echo the sum vector of all the transverse magnetization vectors at the echo $t = \mathrm{TE}$ is smaller than that for the Spin Echo (50%).

What is essential here is that for each position in the object plane the transverse magnetization makes another angle with the $-y'$ axis. If we look at the complete three-dimensional locus of the magnetization vectors (which is easily constructed from Fig. 4.4b) we find that the locus has the form of a figure eight situated on a sphere around the origin with radius $M_0$. We therefore call this echo an "eight-ball" echo [9].

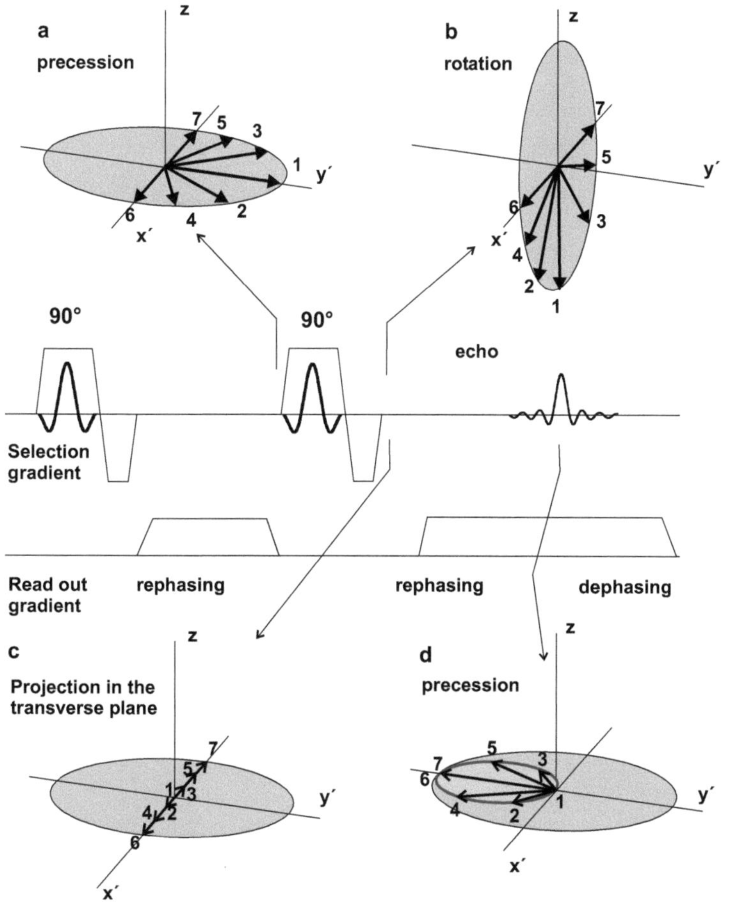


**Fig. 4.4.** "Eight-ball" Echo. **a** Precession. **b** Rotation. **c** Projection in transverse plane. **d** Locus of transverse magnetization vector at the time of the echo

### 4.2.3 Stimulated Echo

We now have to introduce a new echo, which occurs after a third $90^\circ_{-x}$, pulse (the subscript denotes the axis of rotation). To understand this echo we have to look at the magnetization along the $z$ axis after the second 90° pulse, see Fig. 4.4b. This longitudinal magnetization differs from "fresh" longitudinal magnetization, because it carries the history built up before the second 90° pulse. The third 90° pulse rotates this magnetization again towards the $y'$ axis (see Fig. 4.5c). The position of the magnetization along this axis now depends on the precession the spins experienced before the first 90° pulse. Those that precessed less than 90° will be located along the $-y'$ axis and those with more than 90° along the $+y'$ axis. After the third pulse they precess away from this axis in the same way as before the first pulse and

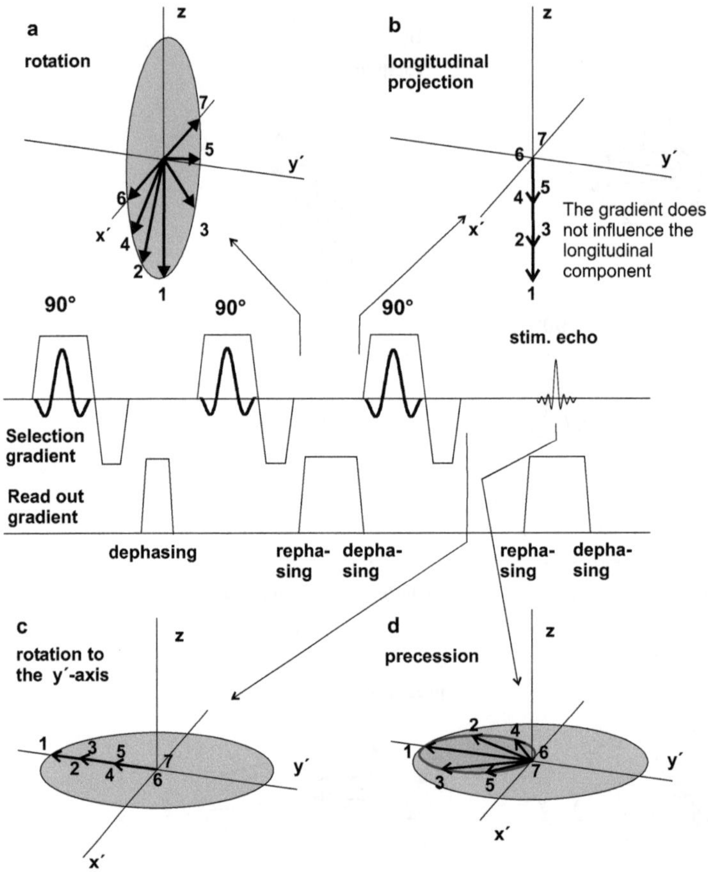


**Fig. 4.5.** Stimulated Echo. **a** Rotation. **b** Longitudinal projection. **c** Rotation to $-y'$ axis. **d** Precession

at the moment of maximum ordering, when they have experienced the same gradient surface ($\int G(t)\mathrm{d}t$) as before the first pulse, all magnetization vectors are in the $(-x', -y')$ and $(+x', -y')$ quadrants and again on a circle through the origin and with the $x'$ axis as a tangent. We call the echo after three pulses the stimulated echo [9].

In this case, for every position in the object plane the transverse magnetization also has a different direction and magnitude. This is very important for the understanding of an essential difficulty in fast-gradient-echo imaging. Note that in the stimulated echo locus the position of the subvoxels has rotated with respect to that in the locu of the eightball echo, so that now the strongest contribution is from subvoxel 1, situated in the centre of the voxel.

### 4.2.4 RF Phase

So far we have assumed the axis of rotation of the RF pulses always to be along the $-x'$ axis. It is interesting to see what happens when we take another direction. In Fig. 4.6 we look at the locus of an eight-ball echo when the phase of the second 90° pulse is advanced by 45°. The result is that the locus of the magnetization is rotated over 90° in the $(x', y')$ plane. Generally it can be stated that an RF phase difference $\phi$ rotates the locus of the eight ball around the origin of the $(x', y')$ plane over an angle of $2\phi$.

In Figs. 4.4 and 4.5 we see that the loci of the eight-ball echo and the stimulated echo are both located in the negative $y'$ half plane. When we now introduce a phase difference of 180° between the pulses we obtain a $90^\circ_{-x'}$, $90^\circ_{x'}$, and a $90^\circ_{-x'}$, $90^\circ_{x'}$, $90^\circ_{-x'}$ sequence. In Figs. 4.4b, 4.5a the vectors 1, 2 and

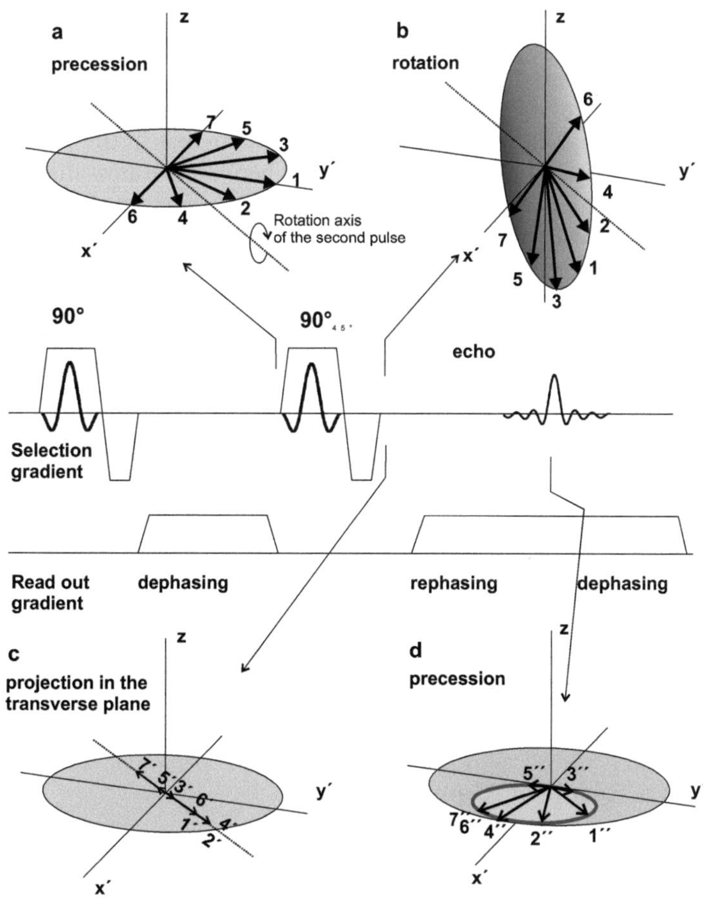


**Fig. 4.6.** Phase advance at 45°. After a first excitation and precession, the transverse magnetization vectors are denoted by $1, \ldots, 7$. After a second excitation the vectors become $1', \ldots, 7'$, after the second precession period we have $1'', \ldots, 7''$

3 are now "up" instead of "down". The result is that the eight-ball echo remains unchanged (because in Fig. 4.6 the locus rotates with the double phase change around the $z$ axis, which is 360°), but that the stimulated echo moves to the positive $y'$ half plane. In this case the stimulated echo of three pulses will interact destructively with the eight-ball echo of the last two pulses, see also Sect. 8.2.4.2.

It can be concluded from this discussion that the RF phase has a marked influence on what happens with the locus of the echos in the $(x', y')$ plane. Actually this can be used to influence the interaction between different echos in such a way that they extinguish each other. This effect, called "RF spoiling", will be discussed in more detail in Sect. 4.2.6, where more than three excitation pulses are taken into account.

### 4.2.5 Response to RF Pulses with $\alpha < 90°$

So far we have discussed only the response to a limited number of 90° and 180° pulses. In most fast-gradient-echo imaging experiments, however, the periodic sequence of excitation pulses have smaller flip angles, which results in a rotation through $\alpha$ radians, where $\alpha < \pi/2$, around an axis in the $(x', y')$ plane, determined by the RF phase of the pulse. Therefore we now discuss the effect of a limited number of pulses with flip angles a smaller than 90°. Such an excitation pulse divides the original magnetization vector into a longitudinal and a transverse component (see Fig. 4.2).

To illustrate the effect of two RF pulses we consider the resulting eight-ball echo. Looking at Fig. 4.4 the rotation of the transverse magnetization between **a** and **b** is not 90° but less, and the projection on the $(x', y')$ plane of the magnetization vectors therefore describes an ellipse (see Fig. 4.7). The results of the precession during the measuring gradient is now a locus in the form of an ellipse around the origin (see Fig. 4.7b). We leave it to the reader to show that the locus of the stimulated echo still goes through the origin, but that the circle has a smaller size than in Fig. 4.5d, when the flip angle is smaller than 90°.

As we have seen the locus of the magnetization vectors form a closed curve (circle, ellipse, etc.) at the echo, $t = \mathrm{TE}$. A shift between two points on the locus corresponds to a shift in position in the object plane in the direction of an increasing gradient field – time integral, $\int \vec{r} \cdot \vec{G}(t)\, \mathrm{d}t$, between two excitations (see Fig. 4.6).

### 4.2.6 Echoes as a Result of Many Excitations

We now reach the point where we can understand what happens when a large number of equidistant excitation pulses with $\alpha < 90°$ are applied to the spins in an object. Let us first consider the longitudinal magnetization. When the excitation pulses have flip angles of 90°, the response of the longitudinal

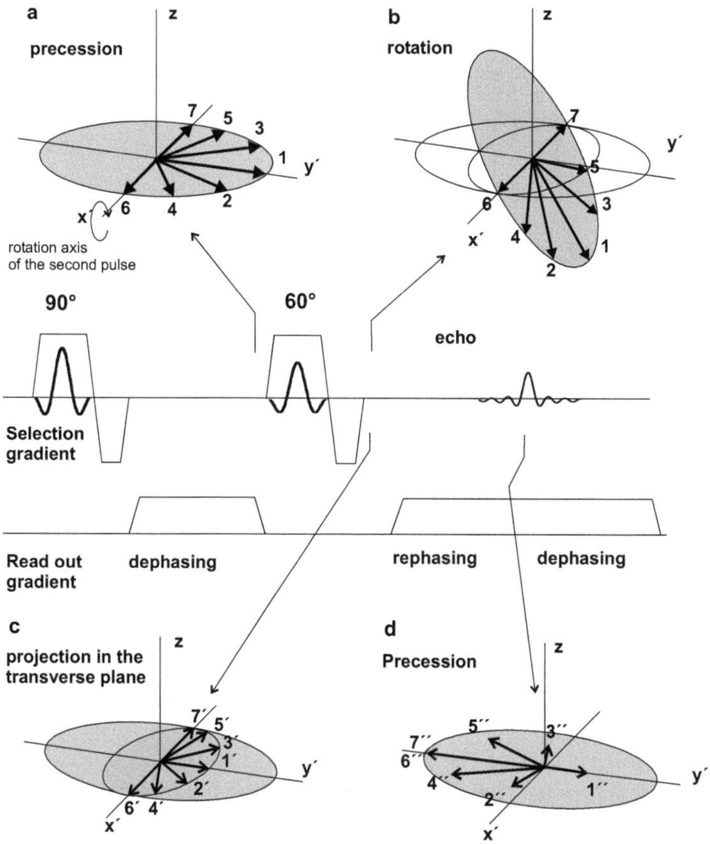


**Fig. 4.7. a** When $\alpha = 60°$, the locus of the transverse magnetization becomes an ellipse. **b** The second precession period results in an ellipse through the points $(1'' \dots 7'')$ enclosing the origin of the $(x', y')$ plane

magnetization is as sketched in Fig. 4.8a for two tissues with different $T_1$. The response is clearly $T_1$ weighted. When $\alpha$ is smaller than 90° the system will seek the dynamic equilibrium such that the longitudinal magnetization that is lost due to an excitation pulse can just be regained by $T_1$ relaxation in the interval between two excitation pulses, TR (see Fig. 4.8b). For the moment we neglected the influence of the transverse component, which partly rotates into the longitudinal direction due to an excitation pulse. This influence is easily included in the mathematical model, presented later in this chapter. For now it is essential that the fid due to the excitation pulses is always $T_1$ weighted.

In Fig. 4.9 the response to a number of excitation pulses in a continuous gradient field in the measuring direction is sketched. Here a simple example is chosen in which the response is easily constructed; however, extension to a continuous sequence of excitation pulses is easily understood. It will be clear

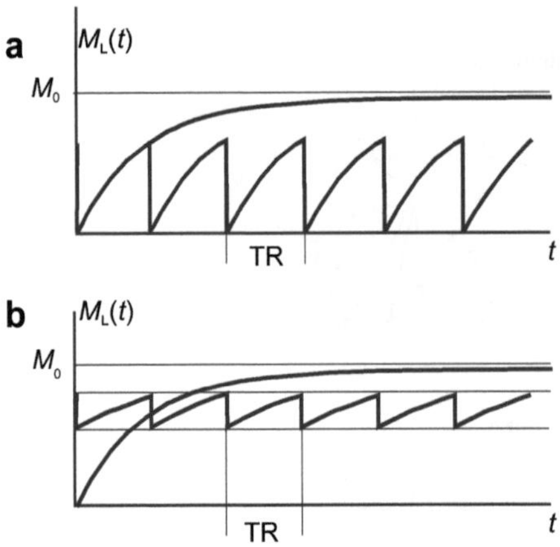


**Fig. 4.8.** Longitudinal Magnetization as function of time during a train of excitations, with **a** $\alpha = 90°$, or **b** with $\alpha < 90°$

that when the time TR between the pulses is constant, always there will be an echo coinciding with an excitation pulse. This echo is a superposition of many components, which result from a combination of earlier excitation pulses and is named ECHO, in correspondence with the definition in Sect. 4.1. The magnetization of the ECHO carries the history of its different components and is therefore position dependent (has a locus).

The magnetization vectors of the ECHO are also rotated by the excitation pulse and after the pulse the pertaining transverse magnetization adds vectorially to that of the fid (which is formed from the longitudinal magnetization and has no phase dispersion directly after the pulse), resulting in the FID. Now the FID is also position dependent since the signal due to the ECHO sometimes adds to the fid and sometimes subtracts from the fid. In FFE imaging this FID is transformed into a gradient echo, which now also has a position-dependent magnetization, described by its locus at the echo time.

More specifically, as will be shown later, the signal of the FID is a periodic function of the phase angle $\theta(\vec{r}) = \gamma \int_0^{\mathrm{TR}} \vec{r} \cdot \vec{G}(t)\mathrm{d}t$ and therefore of the position in the object plane. This periodicity of the FID describes a changing

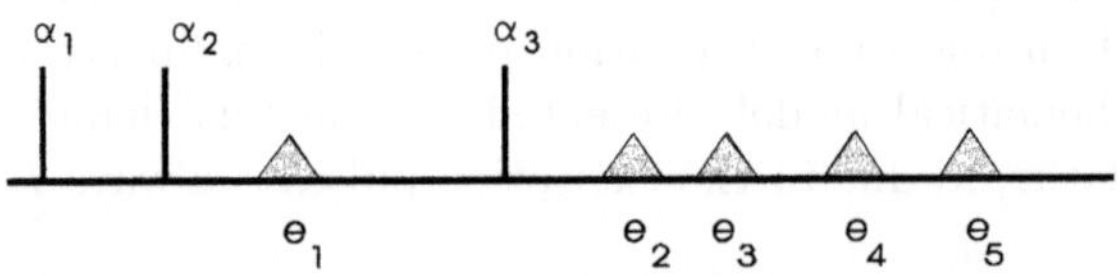


**Fig. 4.9.** Three excitation pulses yield four echos. $e_1$ is the eight-ball echo of $\alpha_1$ and $\alpha_2$; $e_2$ is the stimulated echo of $\alpha_1$, $\alpha_2$, and $\alpha_3$; $e_3$ is the refocussed (by $\alpha_3$) echo $e_1$; $e_4$ is eight-ball echo of $\alpha_2$ and $\alpha_3$ and $e_5$ is the eight-ball echo of $\alpha_1$ and $\alpha_3$

$T_1$ and $T_2$ weighting, because the fid is mainly $T_1$ weighted and the ECHO is also heavily $T_2$ weighted. For positions with an equal phase of fid and ECHO the respective magnetizations add in the FID, for positions with an opposite phase they subtract. This manifests itself in the image of an FFE scan as bands of varying contrast in a direction perpendicular to the direction of the increasing gradient-versus-time surface (increasing $\theta$).

The different FFE methods to be discussed in this chapter are distinguished by the different methods implemented to avoid these bands of varying contrast. One of these methods relies on varying the RF phase of the excitation pulses as already suggested in Sect. 4.2.4. In the next section this method will be described and since it depends on the interaction of the different components from which the ECHO is built up, a new method of describing these components must be introduced.

### 4.2.7 ECHO Components and RF Phase Cycling

Using a different approach we can gain further insight into the reaction of a spin system to a large number of excitations [10, 11]. The idea is that a vector in the $x', y'$-plane can always be decomposed in two components in that plane at arbitrary angles $+\eta$ and $-\eta$ with the $y'$-axis. Applied to the case of a small subvoxel in which the magnetization phase before a certain excitation pulse has increased by precession to an angle $\theta$, this means that after that pulse the transverse component of its magnetization can be decomposed into two components with phase angles $\theta$ and $-\theta$, as is illustrated in Fig. 4.10 [12]. This decomposition has the interesting result that the component with negative $\theta$ has rephased after the next evolution period and the one with positive $\theta$ has continued its dephasing as if no excitation pulse had been given. Figure 4.11 makes use of this concept and shows for the magnetization of a single subvoxel the branched evolution of its components over a number of repetition times.

In this discussion and also for the mathematical treatment later on it is useful to introduce the concept of isochromats. An isochromat is the line that connects all points $\vec{r}$ in the object plane in which the spins have equal phase evolution $\theta(\vec{r})$ during a segment TR of the FFE sequence. The shape of the isochromats depends on the gradients used in the imaging sequence considered. For example, when the effective gradient time integral in the individual segments is in the $x$ direction (the other gradients' directions show zero net surface) the lines $x$ = constant are isochromats. At this point it should be noted that also inhomogeneity of the main magnetic field has an influence on the precise form of the isochromats, defined by

$$\theta_c(\vec{r}) = \int_0^{\mathrm{TR}} \left[\delta\vec{B}(\vec{r}) + \vec{r}\cdot\vec{G}(t)\right] dt = \text{constant}$$

for a steady-state situation. Depending on the scan parameters, there may be a large spread of phase within one voxel and the total signal from one voxel

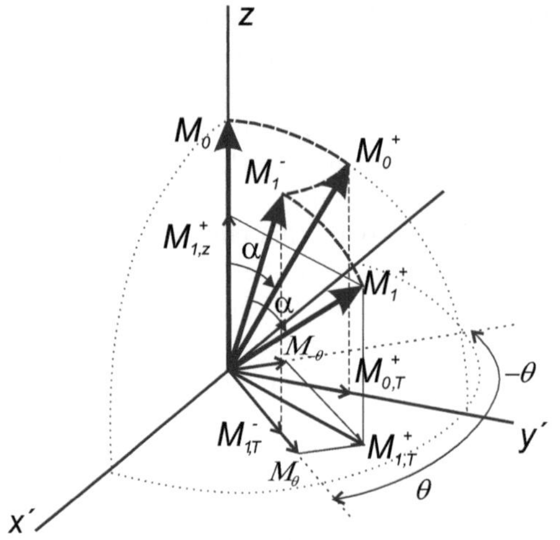


**Fig. 4.10.** Decomposition of the magnetization vector in dephasing and rephasing components. The initial magnetization $M_0$ is longitudinal only. At the zeroth RF pulse it rotates around the $x'$-axis over an angle $\alpha$, leading to $M_0^+$ with a transverse component $M_{0,T}^+$. In the interval to the next RF-pulse, the phase of this component increases over an angle $\theta$, leading to $M_{1,T}^-$. Then again the excitation rotates over $\alpha$, leading to $M_1^+$ with a longitudinal component $M_{1,z}^+$ and a transverse component $M_{1,T}^+$. $M_{1,T}^+$ can be decomposed in two components $M_\theta$ and $M_{-\theta}$ at angles $+\theta$ and $-\theta$ to the $z$'-axis, of which in the next precession period $M_\theta$ dephases and $M_{-\theta}$ rephases

is therefore the signal averaged over the isochromats contained by the voxel considered.

It is clear that Fig. 4.11 holds for each isochromat and that the $n$th ECHO consists of the sum over all isochromats of the components that then have zero phase. That is easily understood by realizing that at $t_n$ every isochromat in the $m$th component of the ECHO has a different phase, $(\theta_{c,m}(t_n)$, so that the result will average out, except for those components for which all isochromats end up at $\theta_{c,m}(t_n) = 0$.

One can see the connection with the discussion on "eight-ball echo" and "stimulated echo" by realizing that the trajectories 1 and 2 in Fig. 4.11 characterize those echos at the $n$th pulse. The segments in which $\theta_c > 0$ for a component (for isochromats experiencing a positive gradient field) are counted by $j$, during these segments the magnetization is dephased. The segments during which $\theta_c < 0$, are counted by $i$: then the magnetization vectors are rephased. For an "eight-ball" echo (trajectory 1 in Fig. 4.11) we have $j = 1$ (segment $n-2$) and $i = 1$ (segment $n-1$). For a stimulated echo (trajectory 2 in Fig. 4.11) we have $j = 1$ (segment $n-3$) and $i = 1$ (segment $n-1$). For trajectory 3 we have $j = 2$ (segments $n-5$ and $n-4$) and $i = 2$ (segments $n-2$ and $n-1$), since the time during which the magnetization was longitudinal (segment $n-3$) does not count for precession. So for components

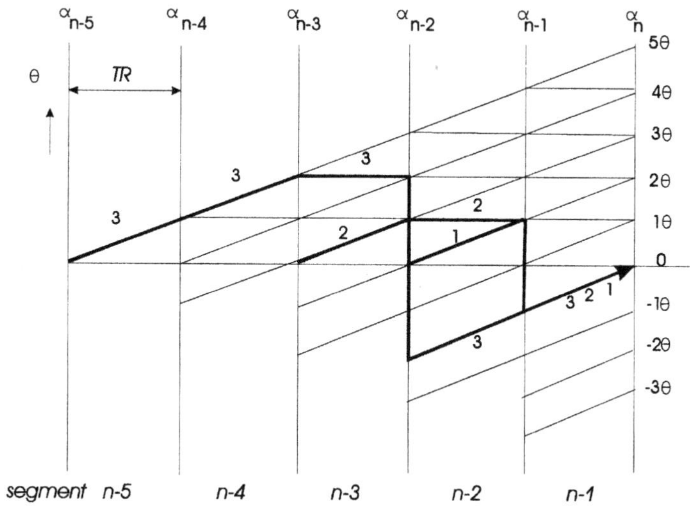


**Fig. 4.11.** Evolution of the different components during segments. Every RF pulse splits the magnetization of an incoming component in three new components, one with the same phase, one with opposite phase and one (longitudinal) with equal but constant phase

that contribute to the echo we can conclude that the number of $i$ segments is equal to the number of $j$ segments. There are many components with different trajectories as can be seen in Fig. 4.11, resulting from earlier excitations. The components that have more or less $i$ segments than $j$ segments average out in a pixel because their isochromats have different phases distributed over the full $2\pi$ range.

We now come to the technique of phase cycling. This name is generally used but is misleading: the phase of the excitation pulses is advanced according to a certain recipe from pulse to pulse. We know from the previous discussion that for the components contributing to the echo we have $\sum_I \theta_{c,I} + \sum_J \theta_{c,J} = 0$ at $\alpha_n$. Here $I$ is actual number of rephasing segments and $J$ is the number of dephasing segments. By advancing the phase of every next RF pulse by $\phi(n)$ we now increase the total phase angle of the $n$th interval. For those components which contribute to the echo at the $n$th pulse, the resulting phases are

$$\sum_I \phi(I) - \sum_J \phi(J)$$

where many paths with different $I$ and $J$ combinations are possible.

In Sect. 4.2.5 we have shown that by using phase cycling we can rotate the components of the transverse magnetization around the $z$ axis (Fig. 4.6). This fact can be used to arrange the loci of the components that contribute to the ECHO around the center of the rotating frame of reference, ($x' = 0$, $y' = 0$), in such a way that they interact destructively, so that all transverse magnetization before an $\alpha$ pulse is destroyed (to a large extend). This, however, is only useful if, notwithstanding the phase shift of the RF pulses, a stationary situation is still built up. This means that the phase of the ECHO (built up by those components for which the total phase increment due to precession is equal to zero) is independent of $n$. It is straightforward to show that this condition can be satisfied with

$$\phi(n) = n\phi_\mathrm{a} + \phi_\mathrm{b},$$

where $\phi(n)$ is the phase shift of the $n$th excitation pulse and $\phi_\mathrm{a}$ and $\phi_\mathrm{b}$ are constants.

A constant phase jump, $\phi_\mathrm{b}$, is trivial since it only rotates all the resulting transverse magnetization, which does not help in destroying it. But a linearly increasing phase jump is very interesting, since it rotates the loci of the different components contributing to an ECHO, in a different way in the transverse $(x', y')$ plane (see Fig. 4.6) so that they can – for certain values of $\phi_\mathrm{a}$ – interact destructively. This type of phase cycling is known as RF-spoiling. The actual value of $\phi_\mathrm{a}$ useful for spoiling must be found by simulation, but from the insight we have now we may conclude that when $\mathrm{TR} \cong T_2$ only specific values can be found (in the literature values of 117° to 150° are mentioned [13, 14]), but that when $\mathrm{TR} \ll T_2$, when many components interact, smaller values of $\phi_\mathrm{a}$ can be applied (see also Sect. 8.4.2).

### 4.2.8 Suppressing the Spatial Variation of the Signal

Let us now discuss the contrast. The fact that we have a periodic train of $\alpha$ pulses that bring the magnetization vector out of its longitudinal equilibrium position and between which the longitudinal magnetization is allowed to relax (partly) to the equilibrium value, due to spin–lattice relaxation, means that the magnitude of the fid after an excitation pulse is dependent on $T_1$ (and of course on $\rho$, the spin density); see Fig. 4.8.

As we said earlier, the FID just after an excitation pulse is the local sum of the fid (which always has the same phase for all spins just after formation) and the part of the ECHO that remains after an excitation (built up of many components in which the different isochromates have different phase). Since the components of the ECHO originated from earlier fids, they are originally $T_1$ weighted. However, the time elapsed since their origination also causes $T_2$ weighting. This time can span several TR periods depending on the component and the ECHO therefore is built up of components with different $T_2$ weighting. Since we have to take the vector sum of the fid and ECHO to evaluate the FID as a function of position in the object plane, for certain positions the magnetization of the ECHO is parallel to the fid and must be added to the fid, whereas for other positions it must be subtracted from the fid. Therefore in the image a periodic structure of bands of different $T_1$ and $T_2$ weighting occurs (banding artifact). These contrast bands diminish the diagnostic value of the image.

One solution for making the periodic variations in $T_1$ and $T_2$ weighting invisible is to make the wavelength (width) of the contrast bands smaller than one voxel. In that case the total signal from a voxel is an average signal having mixed $T_1$ and $T_2$ weighting. The small wavelength is accomplished when within each voxel the isochromates of all components of the ECHO have a large phase dispersion. We can cause such a large dispersion by making the time integral of the gradients between two excitation pulses very large, see Sect. 4.5.3.1.1. The dispersion within a pixel is proportional to the time integral of the gradient and to the voxel dimension.

For the ECHO the same reasoning holds, so in the ECHO we also find similar artifacts. We may expect, however, that the $T_2$ weighting is more pronounced than in the FID, since the primary fid is not present.

In the following section we shall see that the different methods used to get rid of the bands of varying contrast give rise to the different fast scanning sequences.

### 4.2.9 Conclusions of the Qualitative Description

So far we have seen that the problem with fast-gradient echo imaging is due to the transverse magnetization (sometimes called "transverse coherences") present just before the excitation pulses. We have also seen that these coherencies give a banding structure across the image with contrast bands of varying $T_1$ and $T_2$ weighting, making the images undiagnostic. The different fast-gradient echo imaging methods can be classified by the way this banding structure is avoided.

#### 4.2.9.1. N-FFE and $T_2$-FFE

The uncompensated gradient surface between two excitations is large enough to make the difference in the phase angle $\theta$ for the different isochromates of all components forming the ECHO within one voxel more than $2\pi$. One can say that the "wavelength" of the band structure is smaller than the voxel dimensions, so the band structure cannot be seen (is averaged out). There are two possibilities: either the FID or the ECHO is measured (see Fig. 4.1).

When the FID is measured we call the scan a fast field echo or N-FFE [15, 16] (we add the prefix N for normal, because we already used the name FFE for the family of fast-gradient echo imaging methods). The contrast is mainly $T_1$ contrast, mixed with "$T_2$" contrast due to the different components of the ECHO. The quotation marks denote that this contrast depends on the number of TR periods that a component exists so $T_2$ weighting differs per component and the sum of the components (the ECHO) has a mixed $T_2$ weighting.

When the ECHO is measured we still have some $T_1$ contrast, since the transverse magnetization is always born from a fid, but the $T_2$ contrast is more dominant due to the extra relaxation periods TR and the absence of a primary fid. This scan is named $T_2$-FFE, or $T_2$-enhanced Fast-Field Echo [17, 18].

#### 4.2.9.2. $T_1$-FFE

If phase cycling is applied to spoil the ECHO, the contrast is mainly $T_1$ contrast, with only a little $T_2$ contrast due to the finite time between the excitation pulse and the echo. This sequence is $T_1$-FFE ($T_1$-enhanced Fast-Field Echo).

#### 4.2.9.3. R-FFE

It is also possible to make a zero net gradient time integral between two excitations. Then all components of the transverse magnetization add. This is also an interesting sequence because it can be compensated for in-plane flow. However, if the effect of the gradients is zero, the inhomogeneity of the main magnetic field becomes predominant and causes a banding structure over the image, but now the bands are curved as determined by the precise form of the main magnetic field. The sequence we call R-FFE (rephased fast-field echo) and one of its realisations is called balanced-FFE or "true" FISP (free induction by stationary precession) [19]. In Sect. 4.5.4.1 we shall explain how the sequence can still be applied.

So far we have used only qualitative arguments to explain the different phenomena that occur with fast-field imaging. In the remainder of this chapter the same phenomena will be described quantitatively.

## 4.3 Mathematical Model

In this section we shall develop the mathematical model for stationary precession. The reader can skip this mathematics and resume reading at the end result, (4.10), which shows in mathematical form the results that we described more intuitively in Sect. 4.2.

In this section we consider gradient-echo sequences that have reached a stationary situation. Such a sequence is represented by a series of excitation pulses of arbitrary flip angle $\alpha$ separated by time intervals characterized by a constant net-gradient surface (see Fig. 4.12). This means that the variable phase encoding (preparation) gradients necessary for imaging must always be compensated for with a gradient in the opposite direction so that the gradient time integral in the phase-encoding direction is zero (this extra gradient is called the "rewinder"). As we said earlier the description of the pulses is restricted to a simple rotation around the direction of the $B_1$ field in the

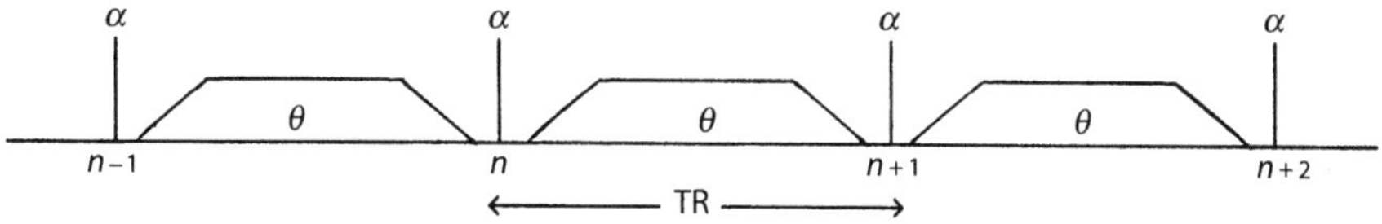


**Fig. 4.12.** The equivalent of a FFE sequence

rotating frame of reference and the duration of the pulse, and its selection gradient (inclusive its refocussing part) is neglected. The model to be developed follows the work published in [1, 2, 5, 6, 19]; we prefer the notation of [6].

At the start of the $\alpha$ pulses – of course – no stationary situation exists. There will first be a transient state during a time of the order of $T_1$ after which the spin system settles to a dynamic equilibrium (steady state). In this chapter we shall not discuss the transient state.

The net (uncompensated) gradient can be in any direction and we define the phase angle $\theta$ as the angle through which the spins at a certain position precess with respect to the "rotating frame of reference" during an interval TR, due to the fact that they sensed a total magnetic field not equal to the static magnetic field $B_0$. The rotation velocity of this frame of reference is given by $\omega = \gamma B_0$, where $\gamma$ is the gyromagnetic constant. The extra magnetic field can be caused by gradient fields, $\vec{G} \cdot \vec{r}$, and by an inhomogeneity of the static magnetic field, $\delta B_0(x, y, z)$.

$$\begin{aligned}\theta(x, y, z, \mathrm{TR}) &= \int_{t_n}^{t_n+1} \gamma[G_x(t)x + G_y(t)y + G_z(t)z]\mathrm{d}t \\ &\quad + \gamma\, \delta B_0(x, y, z)\mathrm{TR}.\end{aligned} \tag{4.1}$$

It is interesting to note that $\theta(x, y, z)$ increases in the direction:

$$\nabla\theta = \gamma \int_{t_n}^{t_n+1} \vec{G}(t)\mathrm{d}t + \gamma\nabla\delta B_0(x, y, z). \tag{4.2}$$

In the following we shall disregard the second part of this equation, which is due to $\delta B_0$, in cases where the first part is dominant. When all gradients are compensated for ($\int G(t)\mathrm{d}t = 0$), however, the field inhomogeneity is important and generates artifacts (a problem with "true FISP" to be discussed later). The integral does not include the RF pulses. In a more precise theory one would have to include what happens during the RF pulses. This, however, would make the theory more complicated and would not add essential new information. The RF pulses are therefore thought of as rotations occurring instantaneously.

### 4.3.1 Rotation and Precession Matrix

We wish to describe a rotation of a degrees $\alpha$ around the $-x'$ axis with a simple rotation in the rotating frame of reference (see Sect. 2.3.1):

$$\vec{\vec{\mathbb{R}}}_{\alpha,-x'} = \begin{pmatrix} 1 & 0 & 0 \\ 0 & \cos\alpha & \sin\alpha \\ 0 & -\sin\alpha & \cos\alpha \end{pmatrix} . \tag{4.3}$$

The precession around the $z$ axis due to the gradient field in one TR interval (we still disregard the main field inhomogeneity for the time being) can be written similarly:

$$\vec{\vec{\mathbb{P}}}_{\theta,+z} = \begin{pmatrix} \cos\theta & \sin\theta & 0 \\ -\sin\theta & \cos\theta & 0 \\ 0 & 0 & 1 \end{pmatrix} , \tag{4.4}$$

where we must realize that $\theta$, the angle between the transverse magnetization and the $y'$ axis, is position dependent as shown in Fig. 4.2.

In Sect. 4.2 we have shown that the RF phase can have an important effect. If we advance the phase by $\phi$ radians for every next excitation this can be described with a rotation of the reference system such that the next pulse is again along the $-x'$ axis. We assume that the phase of the receiver is also advanced so that we do not have to rotate the reference system back to describe the measurements. In this case the phase advance can be described by (4.4) with $\theta$ replaced by $\phi$. The result of both effects, precession over $\theta = \theta(x, y, z)$ radians and phase advance over $\phi$ radians (equal for all isochromates), is given by

$$\vec{\vec{\mathbb{P}}}_{\theta,+z}\,\vec{\vec{\mathbb{P}}}_{\phi,+z} = \begin{pmatrix} \cos(\theta+\phi) & \sin(\theta+\phi) & 0 \\ -\sin(\theta+\phi) & \cos(\theta+\phi) & 0 \\ 0 & 0 & 1 \end{pmatrix} . \tag{4.5}$$

The next processes that have to be included in the equations are the spin–spin relaxation, $T_2$, and the spin–lattice relaxation, $T_1$.

### 4.3.2 Relaxation Matrix

Spin–spin relaxation causes the transverse magnetization to relax back to zero, according to:

$$M_{\mathrm{T}}(t) = M_{x'}(t) + \mathrm{j}\,M_{y'}(t) = M_{\mathrm{T}}(0) \exp\left(-\frac{t}{T_2}\right) . \tag{4.6}$$

The longitudinal relaxation is somewhat more complicated, since in this case the magnetization relaxes back to the equilibrium magnetization, $M_0$:

$$M_z(t) - M_0 = (M_z(0) - M_0) \exp\left(-\frac{t}{T_1}\right) . \tag{4.7}$$

We can combine these two equations in a matrix equation for the relaxation between two excitation pulses ($t = \mathrm{TR}$):

$$\begin{pmatrix} M_{x'}(t_{n+1}^-) \\ M_{y'}(t_{n+1}^-) \\ M_z(t_{n+1}^-) \end{pmatrix} = \begin{pmatrix} E_2 & 0 & 0 \\ 0 & E_2 & 0 \\ 0 & 0 & E_1 \end{pmatrix} \begin{pmatrix} M_{x'}(t_n^+) \\ M_{y'}(t_n^+) \\ M_z(t_n^+) \end{pmatrix} + (1-E_1) \begin{pmatrix} 0 \\ 0 \\ M_0 \end{pmatrix}, \tag{4.8}$$

where $E_1 = \exp(-\mathrm{TR}/T_1)$, $E_2 = \exp(-\mathrm{TR}/T_2)$, and $t_{n+1}^-$ is the time just before $(n+1)$th pulse and $t_n^+$ the time just after the $n$th pulse.

With (4.4) or (4.5), and (4.8) the collective phenomena as the result of precession, phase shift, and relaxation are completely described. Since precession and relaxation occur simultaneously we describe the processes by a single equation:

$$\begin{pmatrix} M_{x'}(t_{n+1}^-) \\ M_{y'}(t_{n+1}^-) \\ M_z(t_{n+1}^-) \end{pmatrix} = \begin{pmatrix} E_2\cos(\theta+\phi) & E_2\sin(\theta+\phi) & 0 \\ -E_2\sin(\theta+\phi) & E_2\cos(\theta+\phi) & 0 \\ 0 & 0 & E_1 \end{pmatrix} \begin{pmatrix} M_{x'}(t_n^+) \\ M_{y'}(t_n^+) \\ M_z(t_n^+) \end{pmatrix} + (1-E_1)\vec{M}_0, \tag{4.9}$$

or $$\vec{M}(t_{n+1}^-) = \overset{\Rightarrow}{\mathbb{Q}}(E_1, E_2, \theta, \phi)\vec{M}(t_n^+) + (1-E_1)\vec{M}_0. \tag{4.9a}$$

Note that $\vec{M}_0$ is the equilibrium magnetization in the direction of the main magnetic field. This equation is equivalent to (2.9) and describes what happens between two RF pulses.

## 4.4 Steady State

The magnetization just after the $(n+1)$th pulse can now easily be calculated by multiplying $\vec{M}(t_{n+1}^-)$ with the rotation matrix $\overset{\Rightarrow}{\mathbb{R}}_\alpha$. Now "steady state" means that the magnetization just after the $(n+1)$th pulse is equal to the magnetization just after the $n$th pulse. By substituting

$$\overset{\Rightarrow}{M}(t_{n+1}^+) = \overset{\Rightarrow}{\mathbb{R}}_\alpha\, M(t_{n+1}^-) = \overset{\Rightarrow}{M}(t_n^+), \tag{4.9b}$$

we express the magnetization just after the $n$th pulse in the equilibrium magnetization $M_0$, using $\overset{\Rightarrow}{\mathbb{U}}$ as the unit matrix, by

$$\left(\overset{\Rightarrow}{\mathbb{U}} - \overset{\Rightarrow}{\mathbb{R}}_\alpha\, \overset{\Rightarrow}{\mathbb{Q}}\right) \vec{M}(t_n^+) = (1-E_1)\, \overset{\Rightarrow}{\mathbb{R}}_\alpha \vec{M}_0. \tag{4.10a}$$

This is a simple set of three coupled linear equations for the components of the magnetization, which can be solved with Cramer's rule. Since the signal measured is induced by the transverse components we shall solve this set of equations for the transverse magnetization just after an $\alpha$ pulse, $M_{\mathrm{T}}(x, y, t_n^+) = M_{x'} + \mathrm{j}M_{y'}$ [6]:

$$\begin{aligned} M_{\mathrm{T}}(x,y,t_n^+) &= M_0(x,y)\frac{(1-E_1)\sin\alpha(1-E_2\exp(-\mathrm{j}\theta))}{C\cos\theta+D} \\ &= M_0(x,y)F^+[\alpha,\theta(x,y),E_1,E_2] \end{aligned} \tag{4.10b}$$

where $M_0(x,y)$ is the equilibrium spin density distribution in the slice studied,

$$C = E_2(E_1-1)(1+\cos\alpha), \tag{4.10c}$$

and

$$D = (1-E_1\cos\alpha)-(E_1-\cos\alpha)E_2^2. \tag{4.10d}$$

Equation (4.10b) describes the FID as defined in Sect. 4.2. Since $\theta$ and $\phi$ are additive (see (4.5)) we did not mention $\phi$ separately in (4.10). We find that the measured magnetization is proportional to the equilibrium magnetization in the slice to be imaged. However, due to the fact that equilibrium is not present between the excitation pulses, we find an additional factor $F^+$, which describes the effects discussed in Sect. 4.2, viz. the bands of varying $T_1$ and $T_2$ weighting. Its numerator consists of a constant factor $(1-E_1)\sin\alpha$, dependent on $T_1$ weighting only, and the factor $(1-E_2\exp(-\mathrm{j}\theta)$, describing the varying $T_2$ weighting. In the complex plane this latter factor describes a vector with length $l$ to which a circle with radius $E_2$ is added (see Fig. 4.13). Note that $E_2 < 1$. Figure 4.13 reminds us of the situation depicted in Figs. 4.4d and 4.5d, when we realize that the loci of the echoes of previous pulses are added to a fid of the latest pulse to form a FID. Since the denominator of (4.10b) is also a function of $\theta$, and $C$ is smaller than $D$ (because it is proportional to $E_2$), the circle will be somewhat deformed. Its precise form is not important for further discussion.

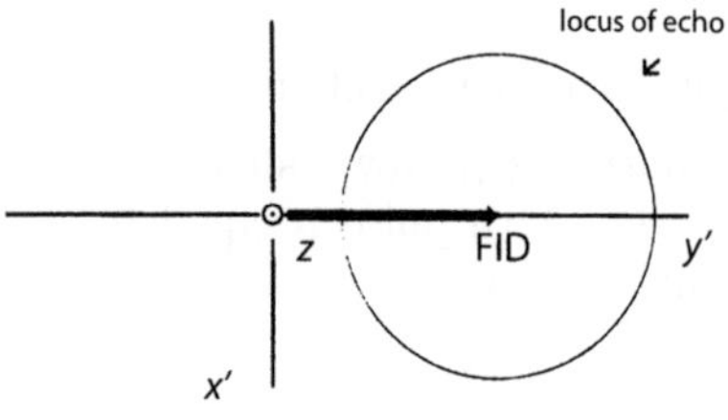


**Fig. 4.13.** Locus of the FID (= fid + ECHO) in the transverse $(x',y')$ plane

The equations derived so far are correct for any slice in the object. For ease of argument we call the $x$ direction the read-out (frequency-encoding or measuring) direction, the $y$ direction the phase-encoding (preparation) direction, and the $z$ direction the (slice-)selection direction. For oblique slices a simple rotation of the laboratory system of reference is required.

During the time interval between the $\alpha$ pulses, the transverse magnetization develops in the slice studied as

$$M_{\mathrm{T}}(x,y,t) = M_{\mathrm{T}}(x,y,t_n^+)\exp\left(-\frac{t}{T_2}\right)\exp(\mathrm{j}\varphi(t)), \tag{4.11}$$

where $\varphi(\vec{r},t) = \gamma\vec{r}\cdot\int_{t_n}^{t}\vec{G}(t)\mathrm{d}t$ and $t_n < t < t_{n+1}$, so for $t = t_{n+1}$, $\varphi(t) = \theta$.

A gradient echo is generated by dephasing the FID with a gradient and subsequently refocussing the FID by reversing this gradient (see Fig. 4.1b, FFE method [3]). These gradients determine $G(t)$ in (4.11). Furthermore for the different profiles there is also the phase encode gradient as part of $G(t)$. If the gradient echo is formed on $t$ = TE the signal measured is (see also (4.10b))

$$\begin{aligned} S(t') &= \exp\left(-\frac{\mathrm{TE}}{T_2}\right)\iint\limits_{x,y} M_0(x,y)F^{+}(\alpha,\theta) \\ &\quad \exp(\mathrm{j}\gamma(G_x t' x + G_{eN}T_y\, y))\mathrm{d}x\,\mathrm{d}y, \end{aligned} \tag{4.12}$$

where $t' = t - t_n$, $T_y$ is the duration of the phase-encoding gradient, and $G_{eN}$ is its value for the $N$th profile. (4.12) would be the normal imaging equation [20], see (2.26) and (2.27), if the term $F^{+}(\alpha,\theta)$ were constant so that the transverse magnetization just after the $n$th excitation pulse is equal or proportional to $M_0(x,y)$, the equilibrium magnetization in the static magnetic field. As we see in (4.12), however, we have an additional factor $F^{+}(\alpha,\theta)$, which is a complicated, but periodic, function of $\theta$ $(x,y,z)$, and (4.2) shows that $\theta$ increases in the direction of $\nabla\theta = \int_0^{\mathrm{TR}}\vec{G}\,\mathrm{d}t$ if the inhomogeneity of the static magnetic field is for the moment neglected. Note that $\theta$ is not a function of $t'$.

Equations (4.10b) and (4.12) show therefore, that when we perform a fast scan with stationary precession, the object $M_0(x,y)$ is overlaid by the imaging method with a periodic function $F^{+}(\theta)$. The direction of the periodicity is in the direction of the uncompensated gradient surface as shown in Fig. 4.14, and

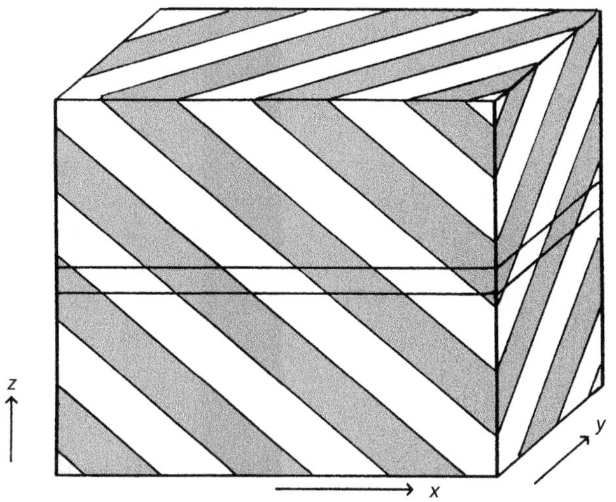


**Fig. 4.14.** Object space traversed by bands of varying $T_1$ and $T_2$ weighting perpendicular to the direction of the gradient

shows itself as bands of varying $T_1$ and/or $T_2$, weighting in the object space. In conventional imaging, such as spin-echo or gradient-echo imaging there is also $T_1$ and $T_2$ weighting, but in this case the weighting is homogeneous over the image.

Therefore we only obtain an image of $M_0(x,y)$ when using fast-gradient-echo methods if we find a way to avoid or average out the periodic behavior of the second factor in (4.10b). If not, we obtain the overlay of a band structure over the image in the direction perpendicular to the vector

$$\left(\vec{i}\int_0^{\mathrm{TR}} G_x(t)\mathrm{d}t,\ \vec{j}\int_0^{\mathrm{TR}} G_y(t)\mathrm{d}t,\ \vec{k}\int_0^{\mathrm{TR}} G_z(t)\mathrm{d}t\right),$$

with a periodicity given by

$$\gamma\vec{r}\int_0^{\mathrm{TR}} \vec{G}(t')\mathrm{d}t' = 2\pi. \tag{4.13}$$

All scan methods described hereafter are aimed at making the influence of $F^+(\theta)$ harmless by either making it constant or by finding situations in which the wavelength of the periodicity of $F(\theta)$ is smaller than the dimensions of a pixel, so that the periodicity is averaged out.

At this point we shall give some practical examples of $F^+(\theta)$ based on the following assumptions: TR = 40 ms, $\alpha = 1/4$, $1/2$, and 1 radians, $T_1 =$ 400 ms, $T_2 =$ 60 ms (vertebral marrow), and $T_1 =$ 2400 ms, $T_2 =$ 160 ms (CSF). The absolute values of $F^+(\theta)$ and its locus in the complex plane are shown in Fig. 4.15.

The occurrence of the band structure is due to the build up of transverse magnetization ("coherence") over a number of TR segments, depending on which component in the ECHO is considered. It is therefore very difficult, if at all possible, to estimate the consequences of effects such as flow and motion in FFE sequences in which the ECHO is not suppressed.

A special case of "phase cycling" can be studied when we add $\pi$ radians to each TR interval by advancing the phase of every next pulse with $\pi$ radians. As we have shown, we can in (4.10b) replace $\theta$ by $\theta + \pi$ which changes the sign of the exponential in the numerator and of the cosine term in the denominator. However, although it changes the contrast of the scan, it does not open essential new possibilities, as is also clear from the discussion in Sect. 4.2.8.

We now consider the situation just before an excitation pulse:

$$\begin{aligned} M_\mathrm{T}(x,y,t_n^-) &= M_0(x,y)\frac{(1-E_1)\sin\alpha(\exp(\mathrm{j}\theta(x,y)-E_2)E_2}{C\cos\theta + D} \\ &= M_0(x,y)F^-\left[\alpha,\theta(x,y),E_1,E_2\right]. \end{aligned} \tag{4.14}$$

In order to form a gradient echo at time $t' =$ TE out of this ECHO at $t' = 0$, we must assume that the net gradient surface between TE and the

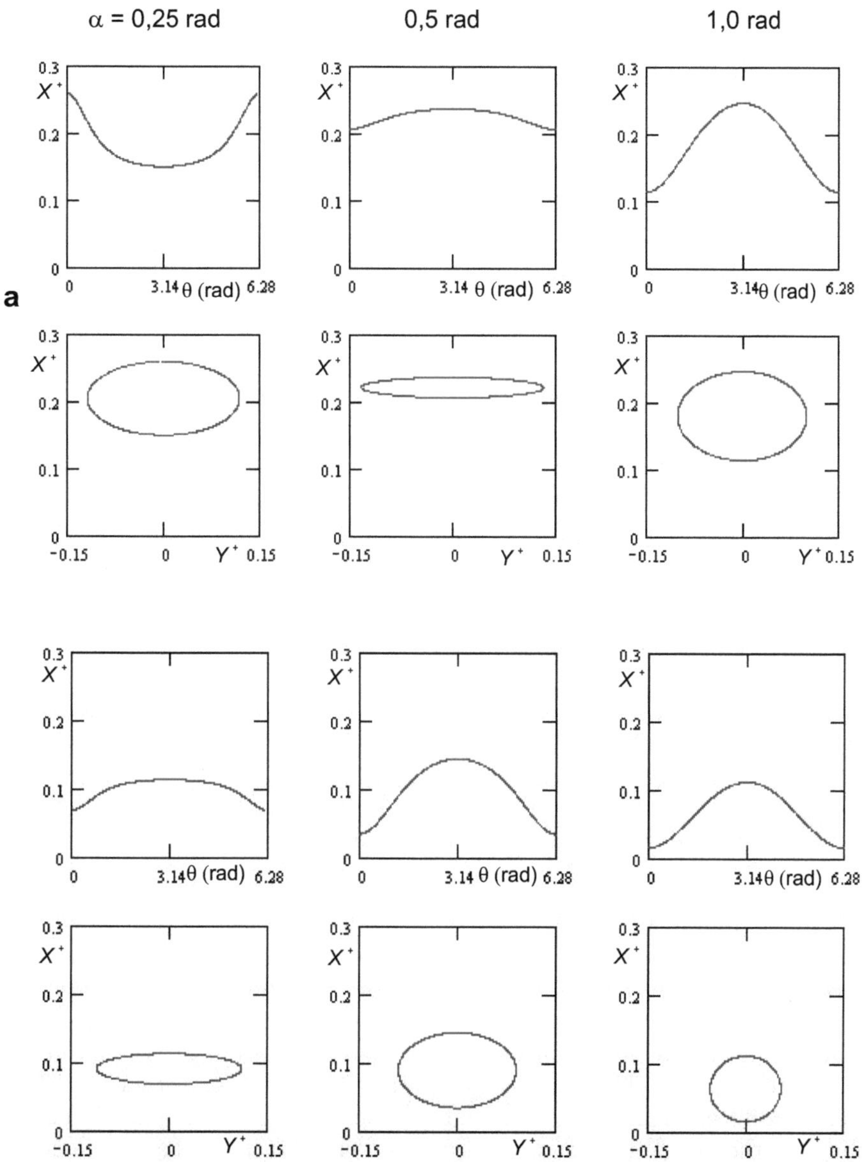


**Fig. 4.15.** Behaviour of the function $F^+$ (4.10b), shown as $X^+$ versus $\theta$, and as $X^+(\theta)$ versus $Y^+(\theta)$ with $X^+$ and $Y^+$ the Real and Imaginary part of $F^+$ respectively. Results are given for three sequences with TR $= 0.04$ s and with three values of the flip angle $\alpha$: 0.25 (left), 0.5 (middle) and 1 rad (right), and for two tissues with relaxation times: **a** $T_1 = 0.4$ s and $T_2 = 0.06$ s and **b** $T_1 = 2.4$ s and $T_2 = 0.16$ s

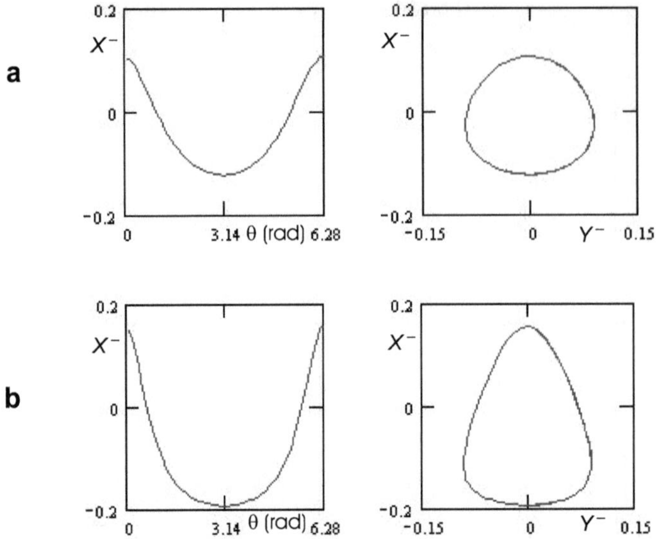


**Fig. 4.16.** The Real part $X^-$ of $F^-$ and the locus of $F^-$ in the complex plane as function of $\theta$. Results are given for a sequence with TR $= 0.04$ s and a flip angle of 0.5 rad, and for two tissues with relaxation times: **a** $T_1 = 0.4$ s and $T_2 = 0.06$ s and **b** $T_1 = 2.4$ s and $T_2 = 0.16$ s

next a pulse is zero, while containing a single gradient reversal. In that case the measured signal is

$$S(t') = \exp\left(\frac{\mathrm{TR} - \mathrm{TE}}{T_2}\right) \iint_{x,y} M_{\mathrm{T}}(x, y, \mathrm{TR}) \exp\left[\mathrm{j}\gamma(G_x(t')t'x + G_{e,N}\, T_y\, y)\right] \mathrm{d}x\, \mathrm{d}y, \qquad (4.15)$$

where $G_{e,N}$ is the preparation gradient of the $n$th profile (note that here TE is measured from the previous excitation pulse, this also explains the positive exponential for $T_2$ relaxation, which was overestimated in (4.14). Inspection of (4.15) and its conditions reveals that it is as if the gradient echo at TE is formed from the ECHO, which comes later, with the time reversed. 2D FFT then transforms the measured data into the desired function $M_{\mathrm{T}}(x, y)$, whose function (the image) unfortunately again is overlaid by a band structure, due to $F^-(\theta)$.

Equation (4.14) shows that in the $\theta$-dependent factor of the numerator, the constant part is now equal to $E_2$, and that the radius of the circle is larger than the constant part. The circle (again deformed due to the denominator) now encloses the origin of the $x', -y'$ plane. Figure 4.16 shows the values of $F^-(\theta)$ as a function of $\theta$ for some of the parameters also used for Fig. 4.15.

We shall now consider the different situations in which the periodic terms in (4.10) and (4.14) become unimportant or averaged out so that $M_{\mathrm{T}}(x, y, t' =$

0) and $M_T(x, y, t' = TR)$ become really proportional to the distribution of the spins, weighted only in a spatially homogeneous way by $T_1$ and $T_2$ relaxation.

There are a number of scan methods by which this can be accomplished. In our nomenclature we use the name "Field Echo" (FE) for gradient-recalled echo methods, as opposed to methods where the echo is formed by a refocusing pulse, Spin Echo (SE). We arrange the sequences of the FE family by decreasing values of TR:

1. $TR \gg T_1$. The magnetisation does not depend on the previous RF pulses.
2. $T_1 > TR > T_2$. The fid after each RF pulse does not depend on the transversal magnetisation generated by any previous RF-pulse.
3. $TR \cong T_2$. The FID is no longer equal to the fid and is clearly influenced by the ECHO from previous RF pulses. To cope with that influence, a variety of solutions exist and these will be treated in this chapter. The steady state arises only after a transient period with a duration of the order of $T_1$. Usually, this time is covered as a "run-in period" in which the sequence is run, but data acquisition is suppressed.
4. $TR \ll T_2$. These TR values lead to very short scan times, for which a run-in period is usually considered not acceptable. Data acquisition will therefore start during the transient and may even be completed before the steady state has arrived. These sequences are called Transient Field Echo (TFE) and will be treated in Chap. 5.

This division of sequences based on their repetition times of course is not sharp. For instance, a run-in period can still be acceptable for TR's in region 4, when a 3D acquisition scheme is used that corresponds to a long scan time, and on the other hand, for TR's in region 2, a run-in period may be left out when the artefact level can be kept sufficiently low.

## 4.5 Steady-State Gradient-Echo Methods (FE and FFE)

The pulse sequences that belong to this family of scans are shown in Fig. 4.17. In this table the "conventional" fast-gradient-echo methods, which are acquired in the steady state are shown. (In practice one uses dummy excitations to wait for the steady state.)

In Fig. 4.17 the different members of the FFE family of sequences are shown. The names of the scan methods as used in this book are given on the appropriate place in the table of Fig. 4.17 and some of the names used in the literature are added between brackets. A more general nomenclature is given in the Appendix.

### 4.5.1 Sequences with Very Long TR

Here it does not make sense to sample before the next excitation, so reference is made to (4.10). Furthermore in this case both $E_1$ and $E_2$ are very small,

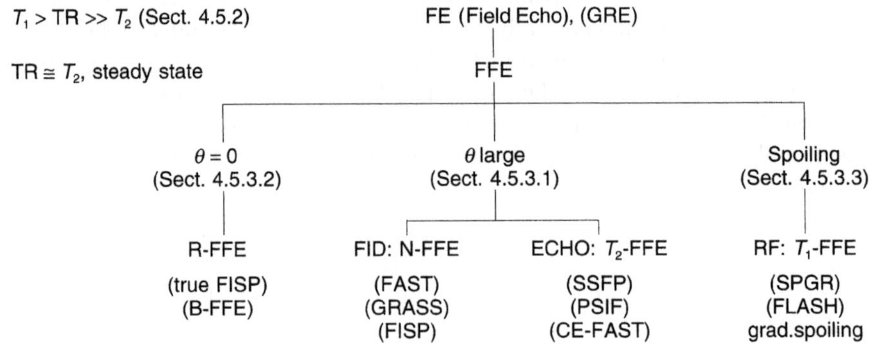


**Fig. 4.17.** Steady-state free precession methods

which means that $C = 0$ and $D = 1$, and the amplitude of the contrast bands is to zero. Since for gradient echos TE must be kept small, to avoid intra-voxel dephasing, there is little $T_2$ weighting. $T_1$ weighting is weak due to the long TR. This sequence is of limited pretical interest, because usually in this case spin echo can be applied, with its advantage of compensation of field inhomogeneities. An example of its use can be found in the imaging of the cervical spine, where the long $T_1$ of CSF yields interesting contrast (see image set II-3).

### 4.5.2 Sequences with $T_1 > \mathrm{TR} > T_2$

In this case the sampling is after the excitation, so again (4.10) applies. Here $E_1$ is approximately unity and $E_2$ is much smaller than unity. Then $C = 0$ and $D = 1 - E_1 \cos\alpha$. The full equation for the transverse magnetization after an a pulse simplifies to

$$M_{\mathrm{T}}(x,y) = M_0(x,y)\frac{\sin\alpha(1-E_1)\exp\left(-\frac{\mathrm{TE}}{T_2}\right)}{1-E_1\cos\alpha}, \tag{4.16}$$

and the periodicity in $\theta$ has disappeared. This is a useful sequence, giving a $T_1$-weighted image in a reasonable scan time if TE is chosen to be sufficiently short. When we choose TR = 200 ms, in a "half-matrix" $256^2$ scan (≅ 140 profiles) we have a scan in 28 s (see image set II-3). Note that there is no ECHO when $E_2 = 0$. This remark is equivalent to saying that all transverse magnetization has disappeared before the next excitation pulse.

### 4.5.3 Sequences with Small TR (TR≅$T_2$)

In this case neither $E_1$ nor $E_2$ are zero and the contrast bands, caused by $F(\theta)$, will be present in the images. Actually this is due to the fact that transverse magnetization builds up over several TR periods. As we explained earlier the transverse magnetization is built up from the eight-ball echo of the previous two excitation pulses and from stimulated echos formed by the previous three $\alpha$ pulses and – of course – from the more complicated components influenced by more than the three previous $\alpha$ pulses. Together all components determine the magnitude of the FIDs and ECHOs. However, there are several methods used to avoid or neutralize the bands due to $F(\theta)$. These methods are the subject of the next paragraphs.

### 4.5.3.1. Large Net-Gradient Surface

We have here two possibilities, either we are interested in the FID (and measure its refocused echo) or we are interested in the ECHO and sample its "reversed" echo as explained before. In the first case, N-FFE (FAST, GRASS, SPGR), we have a $T_1$-weighted fid and a $T_1$- and $T_2$-weighted ECHO forming together the FID. In the latter case, $T_2$-FFE (CE-FAST, SSFP, PSIF), we have only an ECHO and thus a relatively heavier $T_2$ weighting (see (4.14)).

In both cases we apply a large uncompensated-for gradient surface $\int G(t)\mathrm{d}t$ between the excitation pulses in order to make the periodicity of the contrast banding structure smaller than a voxel so that the signal from this voxel is an average over all possible isochromates, $0 < \theta < 2\pi$. We shall explain the FID measurement first.

#### *4.5.3.1.1. Measurement of the FID: N-FFE (FAST, GRASS)*

Suppose we have a scan in which only the gradient in the acquisition direction has a large uncompensated-for component. In that case (see Fig. 4.18) we have

$$\theta = \gamma x \int G_x(t)\mathrm{d}t = \gamma G_x t_{\mathrm{uncomp}} x. \tag{4.17}$$

From this equation it can be seen that the periodicity of $F^+(\theta)$ is described by its wavelength $\lambda$:

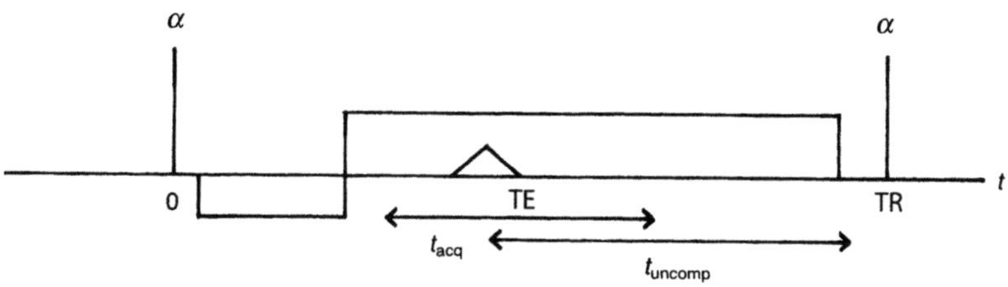


**Fig. 4.18.** Gradient echo sequence with long read-out gradient

$$\lambda = \frac{2\pi}{\gamma G_x t_{\text{uncomp}}}. \tag{4.18}$$

On the other hand the voxel dimension $\mathrm{d}x$ is given by

$$\mathrm{d}x = \frac{2\pi}{\gamma G_x t_{\text{acq}}},$$

so that

$$\frac{\lambda}{\mathrm{d}x} = \frac{t_{\text{acq}}}{t_{\text{uncomp}}}. \tag{4.19}$$

This means that the periodicity of $F^+(\theta)$ is averaged out within one voxel, when the uncompensated-for gradient surface is larger than the gradient surface during acquisition. To find the contrast we have to integrate the transverse magnetization over all possible angles $\theta$. This averaging can be performed analytically by integrating (4.10) over all values of $\theta$ from $-\pi$ to $+\pi$:

$$\langle M_{\mathrm{T}}^{+}(x,y,t_n^{+})\rangle = \frac{1}{2\pi}\int_{-\pi}^{+\pi} \frac{M_0(x,y)\sin\alpha(1-E_1)(1-E_2\mathrm{e}^{\mathrm{j}\theta})}{C\cos\theta + D}\mathrm{d}\theta, \tag{4.20}$$

which we can solve after realizing that the $\sin\theta$ term in the exponential of the numerator is anti-symmetric, so only the cosine term counts. Therefore the integral can be rewritten as

$$\langle M_{\mathrm{T}}^{+}(x,y,t_n^{+})\rangle = \frac{\sin\alpha}{2\pi}(1-E_1)M_0(x,y)\int_{-\pi}^{+\pi} \frac{(1-E_2\cos\theta)}{C\cos\theta + D}\mathrm{d}\theta, \tag{4.21}$$

in which the integral can be solved by introducing, as a new variable, $x = \operatorname{tg}\theta/2$. The result is

$$\langle M_{\mathrm{T}}^{+}(x,y)\rangle = M_0(x,y)\frac{(1-E_1)\sin\alpha}{C}\left(\frac{C + D\,E_2}{(D^2 - C^2)^{1/2}} - E_2\right). \tag{4.22}$$

This equation can be used to calculate the contrast, which is now homogeneous over the image. To do this we calculate the transverse magnetization as a function of $\alpha$ for two values of both relaxation times characteristic of grey and white matter. The results are shown in Fig. 4.19 (contrast increases with TR). Examples of N-FFE scans are given in image set IV-1. Since the echo contains the contribution of the ECHO, and is formed as a result of many components formed by several earlier excitation pulses, the sequence is very sensitive to motion.

#### *4.5.3.1.2. Measurement of the ECHO: $T_2$-FFE (CF-FAST, SSFP)*

For the measurement of the ECHO before the $\alpha$ pulse the contrast can be calculated in the same way, but now based on (4.14). Since $M_{\mathrm{T}}^{-}(x,y)$ is proportional to $E_2$ the scans based on the acquisition of the ECHO are more $T_2$ weighted. The result of averaging over all $\theta$s in one pixel is

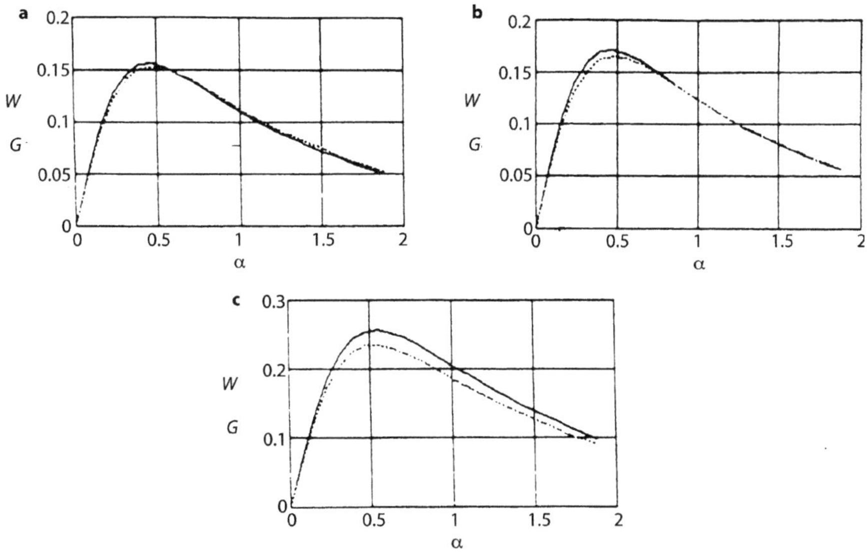


**Fig. 4.19.** Relative signal strength of grey matter G ($T_1 = 920\,\text{ms}$, $T_2 = 105\,\text{ms}$, $\rho = 0.94$) and white matter W ($T_1 = 760\,\text{ms}$, $T_2 = 75\,\text{ms}$, $\rho = 1$) for an N-FFE sequence, as a function of $\alpha$ for TR, which is **a** 30, **b** 40, and **c** 100 ms

$$\langle M_{\mathrm{T}}^{-}(x,y)\rangle = M_0(x,y)(1-E_1)\frac{E_2}{C}\sin\alpha\left(1-\frac{D+C\,E_2}{(D^2-C^2)^{1/2}}\right). \tag{4.23}$$

Since the ECHO is the only source of the signal, the sequence is strongly sensitive for motion, even more so than the N-FFE sequence.

In Fig. 4.20 we present an example of the contrast as a function of the flip angle with TR as parameter.

### *4.5.3.1.3. Combined Measurement of FID and ECHO: FADE*

It is also possible to excite both signals, the FID and the ECHO. This yields two images with different contrast. In this sequence, the methods of Sects. 4.5.3.1.1 and 4.5.3.1.2 are combined and the extra gradient, which takes care of sufficiently large $\theta$, is inserted between both echos. Surprisingly, the method has only been realized at low field strengths (0.08 T), but a clear contrast difference between the two echos is visible. In the experiments reported in [21] the acquisition gradient has a large uncompensated for part so that the bands of different contrast are averaged out over a pixel. From the images it is clear that the $T_2$ weighting in the ECHO is more pronounced than in the FID as predicted by theory.

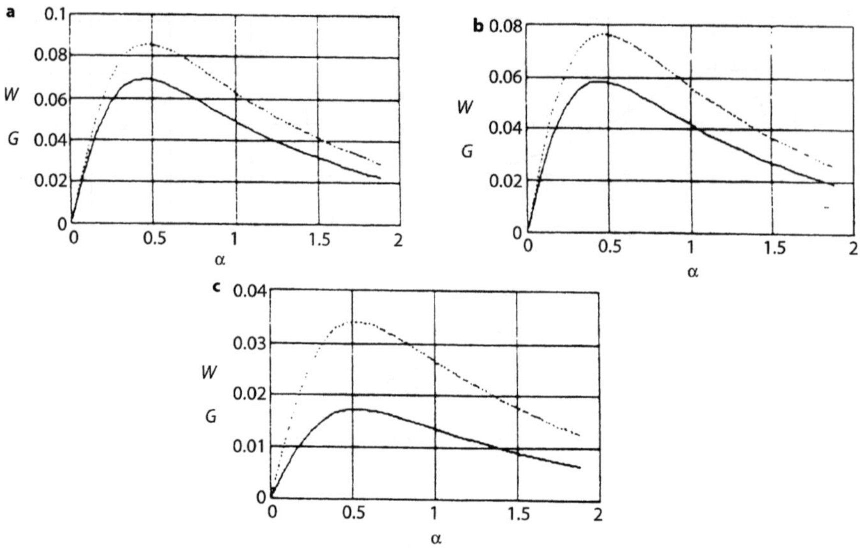


**Fig. 4.20.** Relative signal strength of grey ($T_1$ = 920 ms, $T_2$ = 105 ms, $\rho = 0.94$) and white ($T_1$ = 760 ms, $T_2$ = 75 ms, $\rho = 1$) matter for a $T_2$-FFE sequence as a function of $\alpha$; TR is **a** 30, **b** 40, and **c** 100 ms

### 4.5.3.2. Rephased FFE

When all gradients are completely compensated for between two RF pulses, the value of $\theta(x, y)$ is zero. This means that in the function $F^+(E_1, E_2, \alpha, \theta, \phi)$ there is no dependence of the magnetization on position and the contrast bands therefore do not occur. Of course the dependence on $M_0(x, y)$, the proton density distribution, remains. We added the dependence on $\phi$, to allow for a phase advance between the RF pulses. From (4.9) it is clear that in (4.10). the exponent $\theta$ should be replaced by $(\theta + \phi)$. The transverse magnetization just after an RF pulse then follows from (4.10a):

$$M_T(x, y, t_n^+) = M_0(x, y) \sin\alpha \frac{(1 - E_1)(1 - E_2 \exp(-j\phi))}{C \cos\phi + D} . \tag{4.24}$$

This group of sequences is named Rephased FFE (R-FFE).

A special member of this group is obtained when the RF pulses are alternated, so when $\phi = \pi$. Although its image contrast resembles that of N-FFE, its signal-to-noise ratio is about twice as high in practical circumstances. It is alternatively named "B-FFE" (balanced FFE) or "true FISP" (Fast Imaging with Steady Precession). The prefix "true" is used because the name FISP is sometimes also used for N-FFE. This use of the name "FISP" has a historical reason, it illustrates the problem with B-FFE: when $\theta$ is zero, the term $\gamma\delta B_0$TR in (4.1) becomes important, giving rise to bands in the direction perpendicular to $\nabla\delta B_0$ (see (4.2)). In view of that problem B-FFE was long considered an impractical sequence. The similarity of the image contrast between N-FFE (FISP), where the gradient surface is not compensated (balanced), and B-FFE (true-FISP) for tissue with not too long $T_2$ explains the historical choice.

Intuitively one can say that when $\gamma\delta B_0\mathrm{TR}$ is larger than about $\pi$ the resulting image is compromised by bands due to the magnetic field inhomogeneity. That means that when $\delta = 2\,\mathrm{ppm}$ and $B_0 = 1\,\mathrm{T}$, the repetition time TR must be smaller than about 6 ms [22]. To obtain this short repetition time very high gradients and short rise times are required, which were not available until recently. Furthermore, it was realized long ago that a scan with $\theta = 0$ will be flow insensitive, so that a stationary situation can build up even for moving spins (see Sect. 7.3.2). Next to its high signal-to-noise ratio this property makes B-FFE very desirable and explains the recent interest in this sequence with very short TR, see also Sect. 4.5.4.1.

We shall discuss the flow insensitivity of R-FFE sequences first and shall later in this section we show some possibilities for circumventing the problems due to the inhomogeneous magnetic field at long TR. In Sect. 5.3.2 we turn back to the R-FFE sequence, then named R-TFE sequence, for systems having sufficiently strong gradient systems to allow for the short TR, necessary to apply R-TFE in clinical applications.

It will now be shown that in the case of in-plane flow no stationary situation can be reached unless $\theta = 0$ [20]. The spins, moving like $x + \nu t$, accumulate a phase $\Delta\phi(n)$ in the interval between the $n$th and $(n+1)$th excitation equal to

$$\begin{aligned}\Delta\phi(n) &= \int_{n\mathrm{TR}}^{(n+1)\mathrm{TR}} (x+\nu t)\gamma G(t)\mathrm{d}t = \gamma x \int_0^{\mathrm{TR}} G(t')\mathrm{d}t' \\ &\quad + \gamma\nu \int_0^{\mathrm{TR}} t' G(t')\mathrm{d}t' + \gamma\,\nu\, n\,\mathrm{TR} \int_0^{\mathrm{TR}} G(t')\mathrm{d}t', \end{aligned} \tag{4.25}$$

where use is made of the substitution $t' = t - n\mathrm{TR}$. We conclude that the expression for $\Delta\phi$ is only independent of $n$ if the last term is zero, that is, when $\int_0^{\mathrm{TR}} G(t)\mathrm{d}t = 0$. We call this latter integral the zero-order moment, $m_0$. Because of the high flow sensitivity even small values of $\theta$ are not allowed, so $m_0$ must really be zero. Unfortunately this means that only in the R-FFE situation, where $\delta B_0$ inhomogeneity is the dominant effect, can flow insensitivity be obtained. Note that for accelerating matter the phase shift per TR period changes due to the second term of (4.25), which is proportional to the velocity.

There are some (complicated) methods used to avoid the band structure generated by $\delta B_0$, always at the cost of taking several scans and combining the results. The simplest method is to take several ($K$) scans with an extra RF phase shift in the $k$th scan ($0 < k < K$) equal to $\phi(k) = 2\pi k/K$. Adding the results yields an average over all phases, which reduces the effect of the magnetic field inhomogeneity.

There is a less intuitive method to set up in-plane flow-insensitive scans, which do not show the artifacts due to the magnetic field inhomogeneity. For better understanding let us first look at a sequence in which a small uncompensated-for moment $m_0$ in the $x$ direction, gives rise to the phase angle $\theta(x)$ for a certain isochromat. Since the function $F(\theta)$ is periodic in $\theta$ with period $2\pi$, it can be written as a Fourier series in $\theta$:

$$M(t^+) = M_0 F(\theta) = M_0 \sum_n A_n \mathrm{e}^{\mathrm{j}n\theta}, \tag{4.26}$$

in which the $A_n$ are independent of $\theta$. Together with (4.11), (4.26) yields, for the time evolution of the signal,

$$M(t) = M_0 \sum_n A_n \exp\left[\mathrm{j}(n\theta(x) + \varphi(x,t))\right] \exp\left(-\frac{t}{T_2}\right), \tag{4.27}$$

where $\varphi(t)$ is defined after (4.11). Now an echo arises when $\theta(x) + \varphi(x,t) = 0$ (for all isochromates). Since $\theta$ is assumed to be small, with increasing time we see the different echoes $A_2$, $A_1$, $A_0$, $A_{-1}$, $A_{-2}$, etc, as is shown in Fig. 4.21.

It is interesting to note that the different components can only be seen when $m_0$ is small. When $m_0$ is zero, all components coincide. When $m_0$ is large (the situation described in Sects. 4.5.3.1.1 and 4.5.3.1.2), we can acquire only $A_0$ and $A_{-1}$ describing FFE and $T_2$-FFE, respectively. The inverse Fourier transforms for these two terms are just (4.20) for $A_0$ and (4.23) for $A_{-1}$. The higher-order components are shifted too far apart to be seen within a single TR interval.

We now come to a measuring method used to avoid the bands of varying contrast due to the inhomogeneous main magnetic field in the flow-insensitive situation, $m_0 = 0$. This means that the gradient surface in a TR period is equal to zero. However, $\theta$ is not zero in this case since the term describing the main field inhomogeneity is now dominant (see (4.1)). We take $N$ scans in which the phase of the RF pulses is incremented by $\psi_k$ in every TR period,

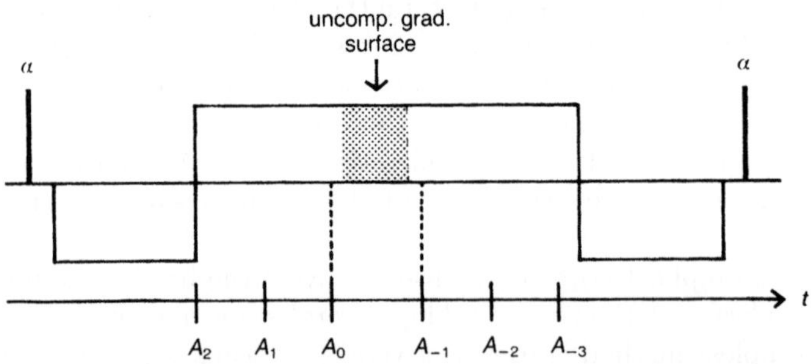


**Fig. 4.21.** Times of occurrence of the different components. When the uncompensated-for gradient surface is zero, all components coincide

with $\psi_k = 2\pi k/N = \psi k$. As shown in (4.5) such a phase shift may be added to the phase shift due to precession. In this case (4.27) becomes

$$M_k(t) = M_0 \sum_{n=-\infty}^{\infty} A_n \exp\left[j(n\theta + \psi_k + \varphi(t))\right] \exp\left(-\frac{t}{T_2}\right), \tag{4.28}$$

where the amplitudes $A_n$ are independent of $k$ and $\theta$. In this way we obtain $N$ different raw data sets from which the desired $\delta B_0$ independent data set can be found. We therefore write (4.28) in a simpler way

$$M_k(t) = M_0 \sum_{n=-\infty}^{\infty} B_n(t) \exp(jn\psi_k). \tag{4.29}$$

Equation (4.29) consists of $N$ equations each with $n$ terms. The $m$th coefficient $B_m(t)$ is found by multiplying the $k$th equation with $\exp(-m\psi_m)$, so that the exponents in the $m$th terms disappear for all $N$ equations, and by subsequently adding all equations obtained in this way. Only the $m$th terms add constructively, all others average out due to their phases. We thus find

$$B_m(t) = M_0 \frac{1}{N} \sum_{m=1}^{N} M_k(t) \exp(-j\psi_k m). \tag{4.30}$$

Substituting (4.29) into this equation and reversing the order of the summations we find

$$B_m(t) = M_0 \frac{1}{N} \sum_{n=-\infty}^{\infty} B_n(t) \sum_{m=1}^{N} \exp(j\psi(n-m)). \tag{4.31}$$

Looking now at the exponent, we see that only when $\psi(n-m)$ is an integer times $2\pi$ do the terms under the second sum add constructively. This is the case when $(n-m)/N$ is an integer, $I$, and the terms under the second sum add to $N$. So (4.30) simplifies to

$$B'_m(t) = M_0 \sum B_n(t), \quad \text{with } n = NI + m. \tag{4.32}$$

Since higher-order components decrease in amplitude it can be seen that when $N$ is large enough the only term that is important in (4.32) is the term with $n = m$. It appears that $N = 6$ is mostly sufficient.

From (4.32) it can be seen that the terms of the Fourier series of $F^+(\theta)$ can be isolated: $B_0$ is the FID and $B_{-1}$ is the ECHO. Examples of scans are shown in [13].

#### 4.5.3.3. FID Measurement with Spoiling of $M_T^-$: $T_1$-FFE (FLASH)

The transverse magnetization before the $\alpha$ pulse, $M_T^-$, is said to be "spoiled" when the different components of magnetization excited by previous pulses interact destructively. In this case we obtain images which are primarily $T_1$ weighted, since transverse magnetization, carrying $T_2$ information, is suppressed.

In the equations for the transverse magnetization this spoiling is taken into account by assuming $E_2$ to be zero, expressing the loss of coherence of the transverse magnetization caused by the spoiling. In this situation the transverse magnetization is given by (4.16),

$$M_{\mathrm{T}}(x,y) = M_0(x,y)\frac{\sin\alpha(1-E_1)\exp\left(-\frac{\mathrm{TE}}{T_2}\right)}{1-E_1\cos\alpha}, \tag{4.16}$$

from which it is seen that the $\theta$ dependence and $T_2$ weighting have disappeared. The $T_2$ weighting reappears due to the delay between the FID and its echo as the last term in the numerator of (4.16). The optimum flip angle, for maximum signal, can be found by differentiating (4.16) with respect to $\alpha$. We then find for the optimum value of $\cos\alpha$:

$$\begin{aligned} \cos\alpha &= E_1, \quad \text{and} \\ M_{\mathrm{T}}(x,y) &= M_0(x,y)\frac{\sin\alpha}{1+\cos\alpha}\exp\left(-\frac{\mathrm{TE}}{T_2}\right). \end{aligned} \tag{4.33}$$

The optimum flip angle is called the “Ernst angle”.

The parameters with which spoiling can be made are as follows:

- RF spoiling, as discussed in Sect. 4.2.6 [12]. In this section we have shown that effective spoiling is achieved when the RF phase $\phi$ is advanced by

  $$\phi(n) = n\phi_{\mathrm{a}} \tag{4.34}$$

  in the $n$th interval. Simulation has shown that the value of $\phi_{\mathrm{a}}$ is preferably between 117° and 150°. This is the most successful $T_1$-FFE method [3, 15, 16].
- Gradient spoiling by, for example, not refocussing the phase encode gradient. If we take a linear profile order, the phase evolution of each isochroinat during a TR interval increases linearly as required by (4.34). A disadvantage of this method is that in the isocenter almost no gradient field is excited so that in the center transverse coherence can indeed build up giving rise to a band around the center line in the measuring direction in which $T_2$ weighting occurs due to transverse coherence [14]. Actually there are also less strongly $T_2$-weighted bands at positions where stationary precession can build up over two or three TR periods (see Fig. 4.2 in [14]).
- The introduction of time jitter on TR (jiTR), random RF phase difference, or $B_0$ field fluctuations. It has been shown, however, that random fluctuations do have much less efficient spoiling properties [23].

The contrast is shown in Fig. 4.22 and examples of a $T_1$-FFE study is shown in image set IV-2. The “spoiled” $T_1$-FFE scan is the most popular scan method of the Fast Field Echo's and not particularly flow-sensitive.

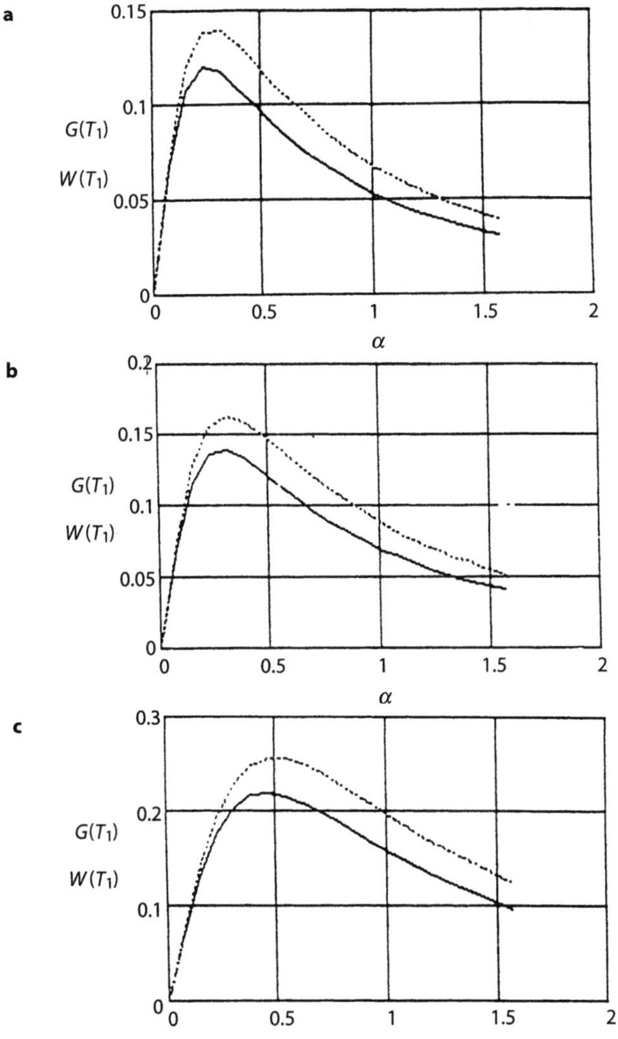


**Fig. 4.22.** Relative signal strength in a $T_1$-FFE sequence for grey matter ($\rho = 0.941$. $T_1 = 920\,\text{ms}$) and white matter ($\rho = 1$. $T_1 = 760\,\text{ms}$) for **a** TR = 30 ms, **b** TR = 40 ms, and **c** TR = 100 ms

### 4.5.4 FFE with Short TR in Steady State

In this section we shall describe gradient-echo-imaging methods in which the repetition time, TR, is reduced to far below $T_2$, say 10 ms or less. As long as we consider imaging under steady-state conditions the mathematical theory of the previous sections is applicable, but the results can be simplified. In this case we also disregard the finite length of the excitation pulses, although this approximation becomes increasingly questionable. $T_1$-FFE and N-FFE with short TR are frequently used for angiographic scans based on the inflow method [24] and for scans in the abdominal region, where we have respiratory and peristaltic motion [8].

In the contrast equations for FFE imaging (4.22) for N-FFE, (4.23) for $T_2$-FFE, (4.24) for R-FFE, (4.16) for $T_1$-FFE) certain approximations are possible due to the very short TR:

$$E_1 = 1 - \frac{\mathrm{TR}}{T_1}, \quad E_2 = 1 - \frac{\mathrm{TR}}{T_2}, \tag{4.35}$$

where we must realize that $T_1$ is in general 5–15 times larger than $T_2$ (except for fat), and larger than 400 ms (remember TR is 10 ms or smaller). There are two possible situations, $\alpha \gg \alpha_{\mathrm{E}}$ or $\alpha \ll \alpha_{\mathrm{E}}$, where $\alpha_{\mathrm{E}}$ is the Ernst angle, (4.33), for which approximate equations can be deduced.

### 4.5.4.1. N-FFE, $T_2$-FFE, and R-FFE with TR $\ll T_2$

We start with N-FFE, having contrast described by (4.22), and the definitions of $C$ and $D$ under (4.10b). With the approximations mentioned above and $\alpha$ assumed to be larger than $\alpha_{\mathrm{E}}$, the "Ernst" angle, $(1 - \cos\alpha) \gg (1 - E_1)$, we can write for $C$ and $D$:

$$C = E_2 \frac{\mathrm{TR}}{T_1}(1 - \cos\alpha), \quad D = (1 - \cos\alpha)\frac{2\mathrm{TR}}{T_2} . \tag{4.36}$$

In the approximation for $D$ we have used the fact that $E_1$, is almost equal to unity and that $\cos\alpha$ differs appreciably from unity. Equation (4.22) now becomes ($\alpha \gg \alpha_{\mathrm{e}}$)

$$\langle M_{\mathrm{T}}^{+}(x,y)\rangle = M_0(x,y)\frac{\sin\alpha}{1 - \cos\alpha}\frac{T_2}{2T_1} \tag{4.37}$$

The factor $\exp(-\mathrm{TE}/T_2)$, describing the $T_2$ decay between an excitation and the echo, must always be added, but because of the small TE (<TR) this factor is close to unity. For an N-FFE image with $\alpha \gg \alpha_{\mathrm{E}}$. the contrast is dependent on the ratio of the relaxation constants, which is a well-known fact. This may be a useful contrast, but since it is different from the usual contrast in conventional spin echo-echo and field-echo methods, in practice its use has been limited.

For $\alpha \ll \alpha_{\mathrm{E}}$ we find for (4.22)

$$\langle M_{\mathrm{T}}^{+}(x,y)\rangle = M_0(x,y)\frac{\alpha}{1 + E_2^2} = M(x,y)\frac{\alpha}{2} , \tag{4.38}$$

which shows proportionality with $\alpha$ and a negligible influence of relaxation on the contrast. The approximations made here are shown in Fig. 4.23. It must be remembered that for small $\alpha$ the signal is very small and this method is barely used in general. For inflow techniques in angiography, however, it is very important.

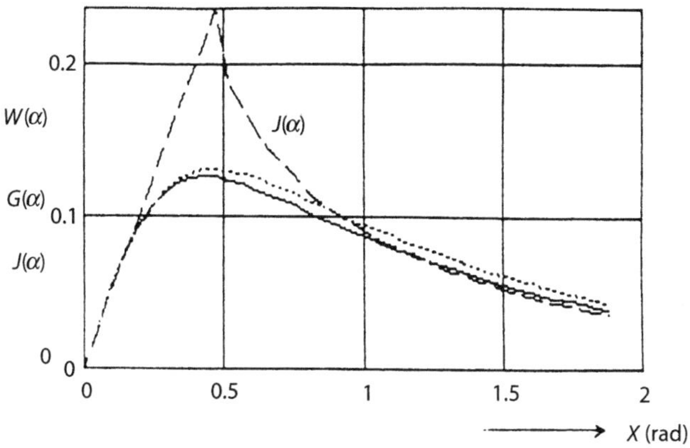


**Fig. 4.23.** Relative signal strengths of grey matter and white matter in an N-FFE sequence as a function of the flip angle $\alpha$. TR = 10 ms, $B = 1.5$ T. For white matter the approximations of (4.37) and (4.38) also are shown, $J(\alpha)$. The approximations are reasonable for $\alpha < 0.25$ and $\alpha > 0.7$ radians

The contrast in a $T_2$-FFE scan with short TR can be calculated in a similar way and the only difference with the N-FFE equation, (4.37) for $\alpha \gg \alpha_{\mathrm{E}}$ is an additional factor $E_2$, describing the heavier $T_2$ contrast. However, since TR is small, $E_2$ is close to unity. Pure $T_2$ weighting cannot be expected, because an ECHO is always "born" from a fid, which already has $T_1$ weighting, giving rise to the $T_1/T_2$ contrast mentioned. For very small flip angles, $\alpha \ll \alpha_{\mathrm{E}}$, we again have only density contrast (see (4.38)).

It will be clear that for the special case of R-FFE, named B-FFE, with alternating RF pulses ($\phi = \pi$), (4.24) leads to:

$$M_{\mathrm{T}}(x, y, t_n^+) = M_0(x, y) \sin \alpha \frac{(1 - E_1)(1 + E_2)}{D - C} \,. \tag{4.24a}$$

With $E_1$ and $E_2$ close to one, we have $(1 - E_1) = \mathrm{TR}/T_1$ and $(1 - E_2^2) = 2\mathrm{TR}/T_1$, $D = (1 - \cos\alpha)(1 - E_2^2) = (1 - \cos\alpha)2\mathrm{TR}/T_2$ and $C = (1 + \cos\alpha)\mathrm{TR}/T_1$. Large flip angles can be applied, making the signal high, and since $\cos(\alpha) \ll 1$, this leads to a $T_2/T_1$ contrast when $T_1 > T_2$. Especially in the case of liquids this latter value is high, which makes B-FFE with its inherent flow compensation an interesting candidate for imaging of body regions with liquid-filled spaces, especially when the liquid is not stationary. An example is shown in image set V-3.

There is not much difference left in the contrast of the other unspoiled FFE imaging sequences with short TR, namely N-FFE and $T_2$-FFE and R-FFE; for the flip angles $\alpha \gg \alpha_E$, they have equal mixed contrast, determined by the quotient of $T_1$ and $T_2$. For very small flip angles, relaxation has almost no influence on the contrast of the image, only proton density contrast remains.

We conclude from this discussion that, apart from the exciting new possibilities of B-FFE, the regular steady-state N-FFE, $T_2$-FFE and R-FFE imaging methods with very short TR do not allow much control of the contrast.

A further penalty in the use of steady-state gradient echo methods at short TR is the relatively long time needed for the development of the steady state. In the next chapter, gradient echo imaging sequences acquired during the transient state between an initial state and the steady state will be described.

#### 4.5.4.2. $T_1$-FFE with TR $\ll T_2$

In a $T_1$-FFE scan, where the transverse coherence is destroyed (spoiled) by phase cycling, the situation is different because in principle we have nearly pure $T_1$ contrast. We now have to work with (4.16). With $E_1$ again close to unity, we now find for $\alpha \gg \alpha_E$:

$$\langle M_T^+(x,y)\rangle = M_0(x,y)\frac{\sin\alpha}{1-\cos\alpha}\frac{TR}{T_1}, \tag{4.39}$$

which describes pure $T_1$ contrast for all flip angles (again the term $\exp(-TE/T_2)$ is omitted). This is a very useful sequence, which takes only a short time (0.5 to 1.5 s/slice) and yields the well-known $T_1$ weighting. This fast scan can also be used in 3D studies since the scan time remains reasonable. For instance a 3D $T_1$-FFE study with TR $= 10$ ms and a $256^3$ matrix takes $10 \times 256 \times 160$ ms $= 6.8$ min (half matrix scan).

It would be nice if we had a similar "pure" $T_2$-weighted FFE method, but based on the above discussion we can conclude that a pure $T_2$-weighted gradient echo in a steady-state situation does not exist. For a fast $T_2$ weighted image we must use turbo-spin echos (as described in Sect. 3.2, [3.1,8]) or GRASE (as described in Sect. 3.4, [3.3]).

We finally look at the approximation for the contrast for very small flip angles $(1-\cos\alpha) \ll (1-E_1)$. Equation (4.16) now reduces to

$$M_T(x,y) = M_0(x,y)\alpha, \tag{4.40}$$

as can also be seen in the graph in Fig. 4.24. This $T_1$-FFE sequence with small $\alpha$ is an ideal sequence for use with magnetization preparation, so that, when the excitation pulses start, the required information is already present in the magnetization of the spin system. This scan method will be discussed in the next chapter.

### 4.5.5 Slice Profile

In all FFE sequences the slice profile is not as simple as in the conventional imaging methods such as SE and FE. One reason is that the time allowed for the excitation pulses is short, so that only the main lobe and one side lobe in front of it are usually applied. Therefore, the RF profile is not nearly a

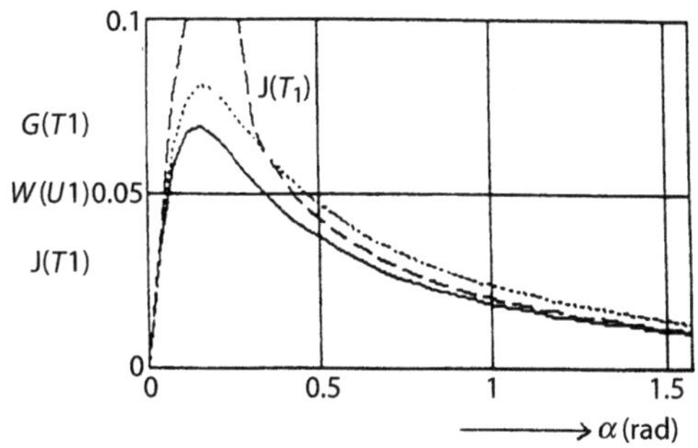


**Fig. 4.24.** Relative signal strength of grey matter and white matter (see Fig. 4.18) for $T_1$-FFE, with TR $= 10\,\text{ms}$, and $B_0 = 1.5\,\text{T}$, as a function of $\alpha$. $J(T_1)$ shows the approximations of (4.39) and (4.40)

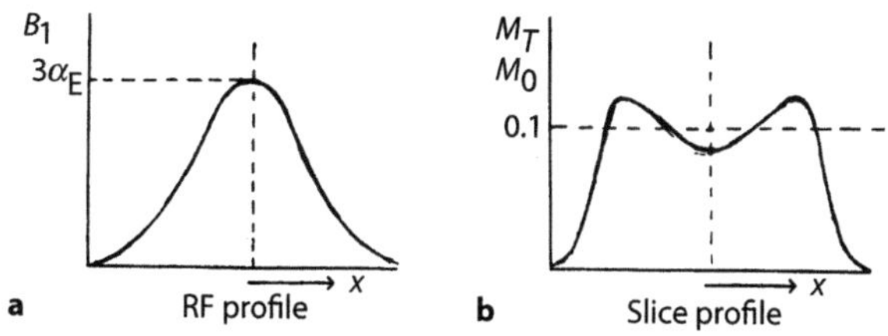


**Fig. 4.25. a** RF profile. **b** Slice profile

square profile as shown in Fig. 2.2, but has a form like a Gauss function (see Fig. 4.25).

This RF profile, however, leads in FFE sequences to a more complicated slice profile. We shall illustrate this for a $T_1$-FFE sequence, using Fig. 4.22. When the flip angle (maximum of the RF profile) is larger than the Ernst angle, the middle of the slice will obtain less magnetization than its flanks because the flip angle in the middle of the slice is on the decreasing flank of the magnetization curve of Fig. 4.22. The average contrast is easily calculated by sub-dividing the slice into a number of sub-slices and obtaining in each sub-slice the magnetization in the steady state for the relevant tissue from Fig. 4.22 and summing up the values found over the slice.

For all FFE sequences, when the RF profiles of the excitation pulses are known, the actual measured contrast can be calculated on the basis of the contrast equations (4.16), (4.22), and (4.34) [5].

## 4.6 Survey of FFE methods

FFE methods have found their way into clinical practice. The short TR presents a logistical advantage for applications that require a very short scan time. In this respect, the FFE scan method is especially competitive for $T_1$-weighted scans, and in such cases the $T_1$-FFE-method is preferred. An example is the single-slice dynamic study of bolus passage. Another important example is sequential single-slice inflow angiography where the need for short scan times follows from the large number of slices that have to be acquired. $T_1$-weighting in this case is needed to suppress the static tissue signal (see Sect. 7.5.2.1 and image set VII-12).

Almost always the clinical application requires more than one slice. The combination of the 3D acquisition method with short TR-FFE methods makes efficient use of the time, so that the scan time need not be shorter than that of multi-slice FE scans (with much longer TR). An additional advantage is that the partitions that can be obtained with 3D acquisition are easily thinner that the slice width obtained in multi slice. In 2D acquisitions the thinnest slice is limited by the length of the RF pulse and the maximum available selection gradient (2.19) and at a typical RF pulse length of 1 ms, the thinnest slice is of the order of 1.5 mm ($G_z = 20\,\mathrm{mT/m}$). This has lead to extensive use of $T_1$-weighted 3D-FFE scans. Examples of their application are given in image sets VI-3 and VII-12. In the 3D scan the excitation profile is flat over the image area and the resulting difference in contrast with 2D scans is shown in image set IV-2.

$T_2$-weighted scans, or rather $T_2/T_1$-weighted scans can be obtained in 3D N-FFE as shown in image set IV-3. The recent increasing role of B-FFE is associated with the use of short TR, and an example is given in image set V-3.

# Image Sets
# Chapter 4

## IV-1 Comparison of Two Fast-Field-Echo (FFE) Methods for Imaging of the Brain

(1) In Fast Field-Echo (FFE) scans, TR is short compared to $T_2$. Hence it can be expected (see Chap. 4) that the signal observed after each excitation (FID) reflects a steady state that has a contribution from longitudinal magnetization, converted to transverse magnetization by this excitation (fid) and a contribution from a number of earlier excitations (ECHO), as discussed in Sect. 4.2.6. The contribution of the earlier excitations can be made ineffective by a spoiling technique. With this spoiling technique, the method is called

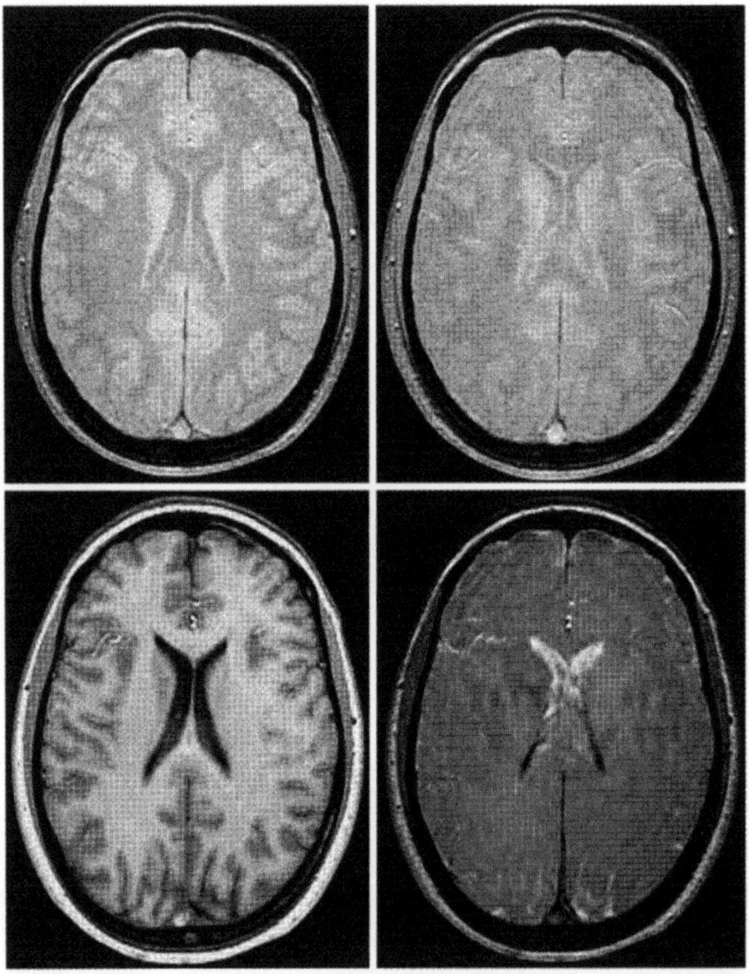

1 2
3 4

**Image Set IV-1**

$T_1$-enhanced FFE or $T_1$-FFE (Sect. 4.5.3.3); without spoiling, the method is called normal FFE or N-FFE (Sect. 4.5.3.1.1). The image contrast will depend on the spoiling. In this image set, this influence of spoiling on the image contrast is shown for 3D FFE images with a range of flip angles.

Parameters: $B_0 = 1.5\,\mathrm{T}$; FOV $= 230\,\mathrm{mm}$; matrix $= 256 \times 256$; $d = 3\,\mathrm{mm}$; TR/TE $= 15/5.2$; NSA $= 4$

| | | | | | | | | |
|---|---|---|---|---|---|---|---|---|
| image no. | 1 | 2 | 3 | 4 | 5 | 6 | 7 | 8 |
| flip angle (degrees) | 4 | 4 | 20 | 20 | 40 | 40 | 90 | 90 |
| RF spoiling | yes | no | yes | no | yes | no | yes | no |

The spoiling is obtained by programmed alteration of the RF phase of the excitation pulses (RF spoiling; see Sect. 4.2.7). Where the use of spoiling is

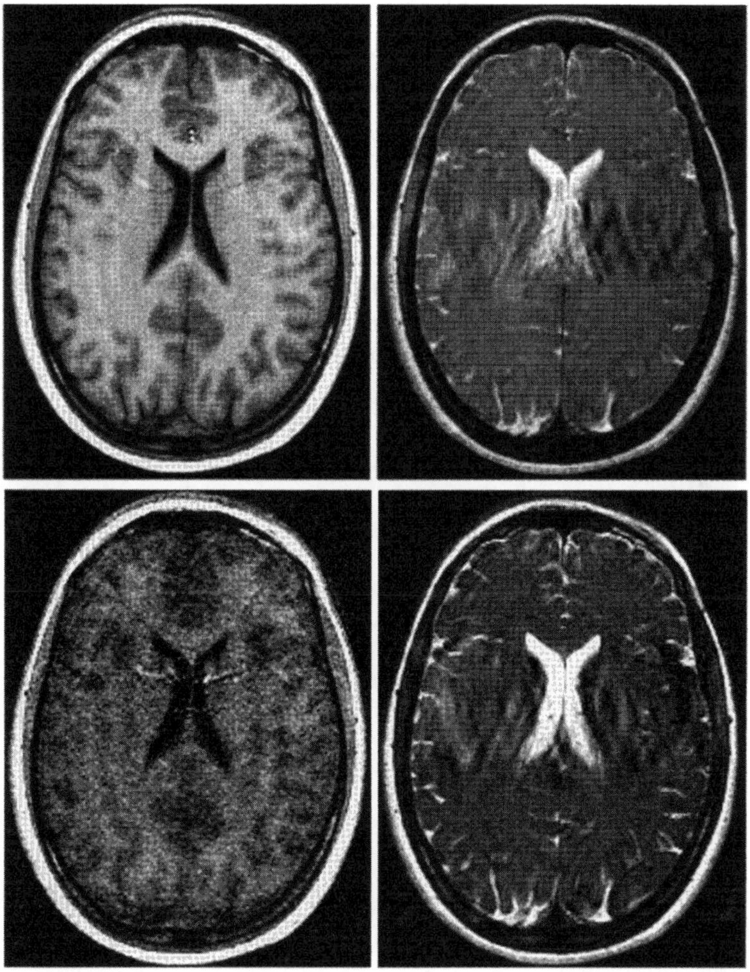

5 6

7 8

**Image Set IV-1.** (*Contd.*)

indicated, the acquisition method is $T_1$-FFE. The method without the use of RF spoiling is the normal FFE (N-FFE).

(2) At small flip angles, such as in images 1 and 2, the influence of spoiling is hardly noticeable. During these scans the longitudinal magnetization remained close to its equilibrium value, which made the signal independent of $T_1$, so that the contrast in both images is proton-density weighted (see (4.38) and (4.40) and Figs. 4.23 and 4.24). This causes grey matter to be somewhat brighter than white matter. In all other N-FFE images, the contrast between grey and white is lost, indicating a transition to a contrast dominated by the ratio $T_2/T_1$, with little dependence on flip angle (see Sect. 4.5, Fig. 4.23). The $T_1$-FFE images with larger flip angles have a $T_1$-weighted contrast between grey and white matter (see Sect. 4.5, Fig. 4.24). In these images the signal level decreases strongly with increasing flip angle, resulting in images that are more noisy than the N-FFE images made with the same flip angle.

(3) In N-FFE the intensity of CSF increases strongly with the flip angle due to its long $T_2$. Strong ghosting of the ventricles occurs in some of these images and reflects the sensitivity of this scan for motion. In $T_1$-FFE, the ventricles are dark for all flip angles; which indicates $T_1$ weighting and hence the success of the RF spoiling provision in this scan type.

(4) $T_1$-FFE appears to give better results than N-FFE for brain imaging. The scan can be used to obtain $T_1$ contrast at a low-motion-artefact level. The signal-to-noise ratio is optimal when the flip angle is not too large. The image set used to demonstrate these effects is made with 3D-FFE methods. Comparison with 2D-FFE is given in image set IV-2.

# IV-2 Difference in Contrast for 2D and 3D $T_1$-Enhanced Fast-Field-Echo Imaging in the Brain

(1) The need for short echo times in FFE imaging arises from the interest in short repetition times (short scan times), the degradation of signal by $T_2^*$ decay and by dephasing from flow (see image set VII-2). In the design of the slice-selective excitation pulses for FFE in our system, this need is reflected. The excitation pulse is short, and as a consequence its spatial profile and the resulting slice profile are not ideal. At the edges of the slice, the flip angle decreases gradually to zero.

In our system, the adjustment of the flip angle is automatic. The algorithm that is used for that purpose will approach a flip angle that equals the user-defined value in the centre of the slice. So it can be expected that for 3D acquisition the flip angle used will be equal to that value, but for 2D acquisition the effective flip angle, averaged over the entire slice profile, will be lower than the value defined by the user. This can be expected to influence both the contrast and the SNR of these images. A comparison between 2D $T_1$-FFE and 3D $T_1$-FFE is shown in this image set.

Parameters: $B_0$ = 1.5 T; FOV = 230 mm; matrix = 256 × 256; $d$ = 3 mm; TR/TE = 15/5.2.

| | | | | | | |
|---|---|---|---|---|---|---|
| image no. | 1 | 2 | 3 | 4 | 5 | 6 |
| scan mode | 3D | 2D | 3D | 2D | 3D | 2D |
| flip angle (degrees) | 15 | 15 | 30 | 30 | 60 | 60 |
| no. slices | 32 | 1 | 32 | 1 | 32 | 1 |
| NSA | 1 | 32 | 1 | 32 | 1 | 32 |

NSA was adjusted to give an equal number of excitations per image in the 2D and the 3D scans, so that nominally the signal-to-noise ratio (SNR) should be equal for a given flip angle for each of these scan types.

(2) Comparison of the contrast shows relatively minor changes. In the small-flip-angle images (1 and 2), CSF is somewhat brighter in the 2D image (image 2). In the large-flip-angle cases (images 3, 4 and 5, 6), the grey–white matter contrast is somewhat milder in the 2D images. The overall effect can be ascribed to an effective flip angle in 2D that is somewhat lower than the flip angle used in 3D. Inflow enhancement is of course more pronounced in the 2D images.

(3) *Comparison of the SNR.* At small flip angles, where the SNR will be proportional to the flip angle, the low value of the effective flip angle in 2D FFE will be the cause of a relatively low SNR for this technique.

At large flip angles, in 2D $T_1$-FFE a considerable contribution to the signal originates from the tails of the spatial profile of the excitation pulse (see Fig. 4.22). These low flip-angle tails extend in a region that is thicker than the slice. As a result, in those circumstances the signal-to-noise ratio of 2D $T_1$-FFE images will be relatively large.

(4) In the image set, NSA is so adjusted that apart from the effect of the spatial profile of the excitation pulse the expected value of the SNR is equal for images 1 and 2, 3 and 4, 5 and 6. The ratio of the experimental values of the SNR between these pairs of images is the following:

| | | | |
|---|---|---|---|
| image no. | 1:2 | 3:4 | 5:6 |
| flip angle (in degrees) | 15 | 30 | 60 |
| Ratio of SNR values (white matter) | 1.3 | 1.1 | 1.0 |

The expected trend in the SNR is visible. At 15° the best signal-to-noise is found in the 3D image; at 60° this difference has vanished.

(5) Clear differences in the behaviour of 2D $T_1$-FFE and 3D $T_1$-FFE result from the difference in spatial profile of the excitation pulse. The 3D technique leads to a better definition of the flip angle over the slices.

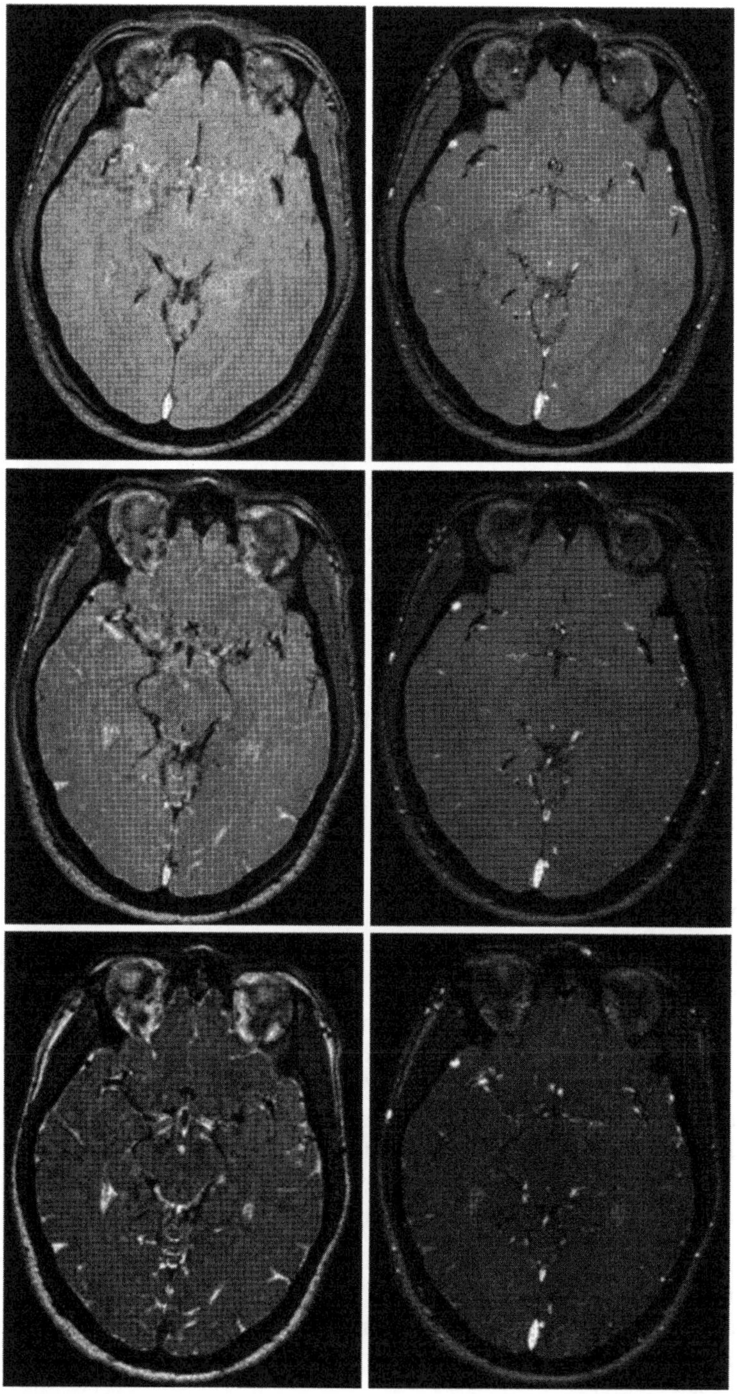

1 2 3 4 5 6

**Image Set IV-2**

# IV-3 In-Phase and Opposed Phase of Water and Fat in Gradient-Echo Imaging

(1) In Gradient-Echo imaging, spins from water and from lipids are not necessarily in phase. The relative phase evolution of these two classes of spins depends on their chemical shift and on the echo time used. Although spins in lipids can have different chemical surroundings, their frequency difference with spins in water is always about 3.4 ppm. At 1.5 T for instance, this corresponds to 217 Hz.
In gradient-echo imaging, this different phase evolution of lipids and water is recorded as a phase difference between water and fat in the image. Modulus images show per pixel a value that equals the vector sum of the contribution to the signals of water and fat. The extreme situations are the in-phase image, where this vector sum is a normal addition; and the opposed phase image, where the water and lipids have signals of opposed sign and the pixel value is the difference between the signals. With an increase of echo time, periodically the in-phase and the opposed-phase situation will occur. At 1.5 T, the first echo time for the opposed phase is $0.5 \times 1000/217 = 2.3\,\text{ms}$. The first in-phase echo time is at a TE of 4.6 ms.

In this image set in-phase and opposed-phase images of the kidney are compared. All images are 3D Normal Fast-Field-Echo (N-FFE) images; the scans were obtained under breath hold.

Parameters: $B_0 = 1.5\,\text{T}$; FOV $= 300\,\text{mm}$; matrix $= 256 \times 256$; $d = 10\,\text{mm}$; TR/$\alpha = 29/30$.

| image no. | 1 | 2 | 3 | 4 |
|---|---|---|---|---|
| TE (ms) | 4.6 | 6.9 | 9.2 | 11.5 |

(2) In images 1 and 3, water and fat are in-phase, whereas in images 2 and 4 these tissues are in opposed phase. The opposed-phase images are characterized by pixels in which the signal from fat cancels that of water. These pixels are found at the borders of lipids-free and fatty tissue and accentuate these borders with a black line; ("ink line"). In images 1 and 3, the transition from lipid-free to fatty tissue is much less obvious and is due only to the misregistration of water and lipids. Lipid tissue is never free of water and the signal from the subcutaneous fat in images 2 and 4 (opposed phase) is less bright than that in images 1 and 3 (in phase).

(3) In the choice of the echo time in gradient-echo imaging, attention should be paid to its influence on the relative phase of water and fat. Opposed phase imaging will show "ink lines" at tissue borders, when one of the tissues contains fat.

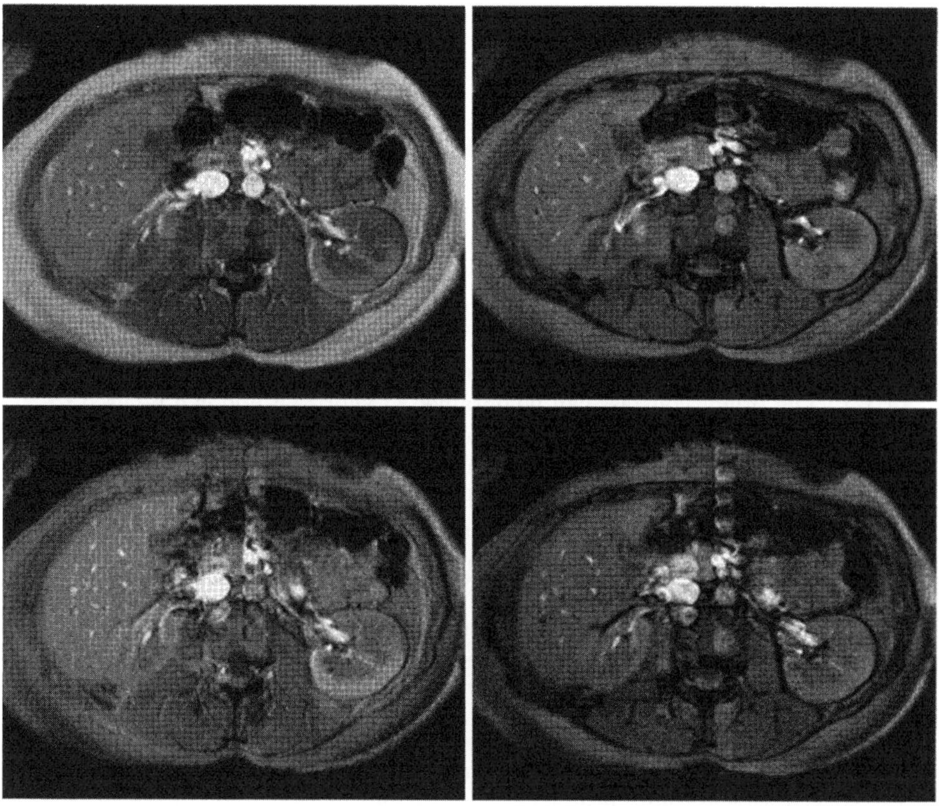

1 2 3 4

**Image Set IV-3**

## IV-4 Cartilage Delineation in N-FFE Combined with Water-Selective Excitation

(1) The imaging of degenerative cartilage, for instance in the knee, requires clear demarcation of this tissue against the bony structures as well as against fat, joint fluid and menisci. Strong $T_1$ weighting, resulting in low joint fluid signal can be used, but does not give information on the water content of the joint. Strong $T_2$-weighting by selection of a long TE cannot be used in view of the short $T_2$ of cartilage. Non-spoiled Gradient Echo techniques such as N-FFE (FISP, GRASS) or $T_2$-FFE (PSIF, SSFP) in principle allow the generation of bright contrast for tissues in which the ratio $T_2/T_1$ is large [see (4.37)]. The bright contrast can be obtained at short TE and can be used to provide the combined effect of bright cartilage and bright joint fluid. However, when these types of sequence are combined with a fat saturation prepulse, the signal of joint fluid is reduced as well. This occurs because the large gradient spoiler blocks accompanying the spectral selective fat saturation pulse create a high diffusion sensitivity of the sequence so that the effective $T_2$ of the freely diffusing joint fluid is reduced [25]. Direct water excitation does not require this usage of gradient spoiler blocks and therefore presents an interesting alternative, especially when other large-area gradient waveforms are avoided as much as possible.

(2) In this image set, 3D N-FFE images of the knee of a healthy volunteer (age 59) are compared. Excitation was obtained with a slab- and spectral-selective 1-3-3-1 water excitation pulse.

Parameters: $B_0 = 1.5\,\mathrm{T}$, TR/TE = 20/6.6 ms, FOV = 160 mm, matrix = 256 × 256, 36 partitions of 3 mm.

| image no. | 1 | 2 | 3 | 4 | 5 | 6 |
|---|---|---|---|---|---|---|
| Flip angle (°) | 15 | 30 | 45 | 60 | 75 | 90 |

(3) In all images the signal from fat and bone marrow is nearly completely absent, demonstrating the effectiveness of the selective water excitation pulse. The images show an increase in the signal of tissue fluid with increasing flip angle, as expected. Images 1 and 2 are $T_1$ weigthed and the contrast between cartilage and joint fluid is not sufficient for diagnostic purposes. In images 5 and 6, the contrast is $T_2/T_1$ weighted and, while the joint fluid is very bright, the signal of the cartilage is becoming too low to be useful. Images 3 and 4 show both a good signal from the cartilage and a strong contrast between cartilage and joint fluid. Numerical comparison of SNR of both tissues (mean signal divided by background noise) shows that the optimum situation exists at a flip angle of 45°; both the signal from cartilage and the signal difference between cartilage and joint fluid are high.

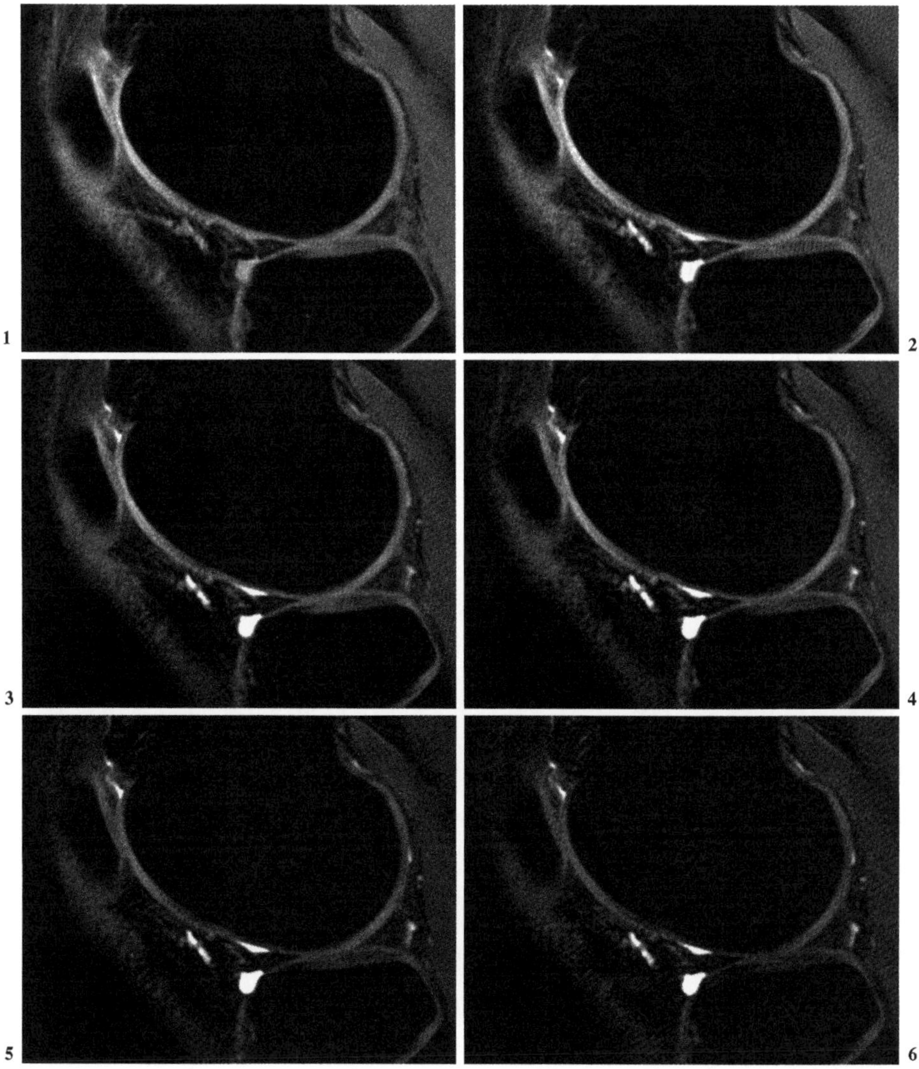


**Image Set IV-4**

| image no. | 1 | 2 | 3 | 4 | 5 | 6 |
|---|---|---|---|---|---|---|
| SNR cartilage | 108 | 106 | 80 | 63 | 47 | 37 |
| SNR joint fluid | 97 | 139 | 164 | 159 | 152 | 131 |

The ratio of the signals of joint fluid and cartilage approaches a value of 3.5 at large flip angles. This is the region where the signals of both tissues are dominated by $T_2/T_1$. Assuming for cartilage a short $T_2$ of 20 ms and a $T_1$ of 600 ms, $T_2/T_1$ for that tissue equals 0.03, which implies that for joint fluid $T_2/T_1 = 3.5 \times 0.03 = 0.1$. Given a value $T_1$ of 4000 ms, it follows that the effective $T_2$ of joint fluid is about 400 ms. This demonstrates that the influence of diffusion on the signal indeed is small.

(4) A lack of strong contrast between free water and all other tissue in the images shown will occur in situations where diffusion sensitivity of the sequence is too high, for instance when SPIR is used for fat suppression. In the sequence used for images 1 to 6, the diffusion sensitivity was low because of the absence of large-area gradient waveforms such as the spoiler blocks accompanying SPIR and REST pulses. Although limitation of diffusion sensitivity in Gradient-Echo images has previously not obtained much attention, the images indicate that such limitation is a clinically important performance aspect of the sequence.

# 5. Transient-State Gradient-Echo Imaging

## 5.1 Introduction

The steady-state existing in SE, FE, and FFE imaging methods is the consequence of the strict periodicity of the sequence. Quite commonly, in these steady-state methods the first repetitions of the sequence are used not for the acquisition of data, but only to run in the steady state.

For gradient-echo sequences with short TR, say 10 ms, a $128^2$ image is acquired in 1.28 s or even less when a half matrix method is used. This acquisition time is of the order of $T_1$. For such scans a steady state is not always reached during acquisition, for a variety of reasons.

**a.** The addition of a preparation pulse, such as an inversion pulse (compare Sect. 2.7), leads to the need for a "silent" (no excitation pulses) preparation delay after which the gradient-echo sequence starts immediately.
**b.** When the data acquisition is concentrated in a limited number of TRs in a certain cardiac phase in cardiac-triggered sequences (segmentation), the spin system is in a transient state during acquisition [1].
**c.** In dynamic imaging, wait times can arise when the dynamic scans are triggered to a non-periodic event such as a respiratory phase or manual trigger pulses [2].

The short TRs for the data acquisition comes in groups, "shots" immediately following a "silent" period. During these shots the spin system goes from its (initial) equilibrium state to a dynamic equilibrium (steady state). Acquisition during this transient is called "transient-state imaging".

So new imaging methods are introduced, in which the acquisition occurs in shots of gradient echos excited by RF pulses with a rapid rate during the transient from the initial magnetization state to the steady state (dynamic equilibrium). This initial state can be the equilibrium state with magnetization $M_0$, but it can also be a state "prepared" by certain preparation pulses [3]. Interestingly enough the duration of the transient from the initial state to the steady state is in its turn heavily influenced by the flip angle and the repetition rate of the excitation pulses of the sequence [4, 5].

The sequences with acquisition during the transient from the initial state to the steady state are called N-TFE, $T_1$-TFE, $T_2$-TFE, and R-TFE, where the T in TFE stands for "transient". The prefixes indicate the provisions during the shots, analogous to what can be used in N-FFE, $T_1$-FFE, $T_2$-FFE, or R-FFE. The group name for all transient-state gradient-echo sequences is Transient Field Echo (TFE). To avoid artifacts, the approach to the steady state has to be smooth (which already causes blurring), but, as will be shown, without proper precaution the approach tends to be oscillatory, which leads to ghosts.

For $T_1$-TFE imaging sequences we have seen that the flip angle for maximum signal is the Ernst angle, which is dependent on the tissue considered. For the other types of FFE sequences the optimum flip angle lies near 0.5 rad (see Figs. 4.19 and 4.20). These values do not hold for a scan taken during the transient. In order to make the measured signal as big as possible one might think of using large flip angles, especially when the shots are short. However, then the longitudinal magnetization is rapidly eroded leading to small fids. So in the transient case there also appears to be an optimum flip angle. Values of twice or three times the "Ernst" angle are suggested [7]. For still-larger flip angles much of the longitudinal magnetization is depleted so rapidly that the signal is concentrated in the first profiles and the approach to equilibrium is fast and oscillatory [4, 5]. Instead of using a large flip angle an increasing flip angle is frequently applied so as to increase the signal of the later profiles [6, 7]. This yields a situation in which the measured signal is more constant over the profiles and oscillatory behavior is suppressed.

Since the profiles with the lowest $k$ values determine the gross appearance of the image (and so the contrast) and the magnetization is in a transient state during a large part of the shot, it now becomes very important to know at what time during the shot these low-order profiles are acquired. Therefore the order of the profile acquisition is very important. For instance, when the profile order is centric and when the approach to dynamic equilibrium (steady state) is smooth, blurring in the phase-encoding direction will result. When the approach to steady state is oscillatory, ghost artifacts occur. These artifacts are due to the fact that in the phase-encoding direction the profiles are "weighted" with the momentaneous longitudinal magnetization at the time of the excitations, just as is explained in connection with TSE (see Sect. 3.2.1).

When only part of the profiles is acquired in one shot, for example in a cardiac-triggered scan where there is only limited time in one heart beat, the profile order is arranged per shot. Analogous to similar strategies mentioned in Chap. 3 (TSE or EPI), this is called segmentation. It is clear that segmentation, the profile order, and the wait time between the segments determine the contrast [9, 10].

Finally some remarks have to be made on the methods used to "prepare" the magnetization with some sort of preparation pulse or a sequence of pulses so as to bring the spins into a predefined state before the TFE shot starts [3]. Preparation pulses are usually applied to enforce an initial state, depending on $T_1$ [8] or $T_2$ weighting [9]. However, there are many more possibilities;

even preparation pulses defining an initial state dependent on diffusion are suggested [10, 11]. As an example we mention an inversion pulse at a time TI before the start of the TFE shot, giving a pronounced $T_1$ contrast – see image sets V-1 and V-2.

It is impossible to develop a closed analytical model for the transient state, as was the case for the steady state. For more insight one must turn to numerical simulation, for which the mathematical model of Sects. 4.3 and 4.4 can still be used [4], but for which the assumption of a steady state (4.9b), is not allowed. The full Bloch equation must be solved for each TR interval, with the magnetization at the end of the previous interval as the initial condition. Some global properties, however, can be derived analytically using a simplified model, which adds some feeling for the dependence of the signal on certain parameters.

In Sect. 5.2 we shall discuss the influence on the approach to steady state of scan parameters such as the flip angle and TR. In Sect. 5.3 the available preparation pulses are discussed and in Sect. 5.4 the importance of the profile order for the contrast will be studied. Finally in Sect. 5.5 a survey of all TFE scan methods will be given.

## 5.2 Signal Level During Transient State

In the previous chapter we described imaging methods with acquisition during the steady state. As has been stated earlier, the steady state builds up during a time (determined by $T_1$, $T_2$, TR, and the excitation flip angle, $\alpha$), which can be longer than the shot duration of a short-TR TFE imaging sequence. The acquisition then occurs during the transient from the initial state to the stationary situation.

At first the approach of the longitudinal magnetization to the steady state will be described. To obtain a closed analytical solution, in Sect. 5.2.1, we shall make the crude assumption that the transverse magnetization is "spoiled" before each excitation. Indeed, we may argue that we apply phase cycling, which was shown in Sect. 4.2.7 to give complete spoiling. However, the theory on phase cycling as described in that section requires a steady state for the individual components adding to the signal, which is not true during the transient. So the assumption is probably wrong for short TR ($\cong T_2$). We can only learn some general trends from its use and must be careful with the conclusions. Fortunately the more general case, taking proper account of transverse magnetization from previous excitations, can be described by the theory of Sect. 4.3 [4, 5]. However, in this case, the evaluation of the equations cannot be done in closed form and therefore depends on numerical solutions, which makes it difficult to find the dependence on individual parameters. This will be described in Sect. 5.2.2.

### 5.2.1 Approach to Steady State by Assuming RF Spoiling

As a result of an excitation pulse (0th pulse), the initial magnetization, $M_\mathrm{S}$, is changed to $M_\mathrm{S}C$, where $C$ stands for $\cos\alpha$. The relaxation between the 0th pulse and the first one is described by (4.7):

$$M_{z1}^{-} = M_0 + (M_\mathrm{S}C - M_0)E_1, \tag{5.1}$$

where $E_1 = \exp(-\mathrm{TR}/T_1)$. The second pulse rotates this magnetization by $\cos\alpha$ so that the longitudinal magnetization becomes

$$M_{z1}^{+} = M_{z1}^{-}C. \tag{5.2}$$

Subsequent relaxation in the second TR period can be found by again applying (5.1) with $M_\mathrm{S}C$ replaced by $M_{z1}^{+}$. This yields

$$M_{z2}^{-} = M_\mathrm{S}C^2E_1^2 - M_0E_1^2C + M_0E_1C - M_0E_1 + M_0. \tag{5.3}$$

We leave it to the reader to show that for the longitudinal magnetization before the $n^\mathrm{th}$ pulse we have

$$M_{zn}^{-} = M_\mathrm{S}C^nE^n + M_0(1 - E_1)\frac{1 - C^nE_1^n}{1 - CE_1}. \tag{5.4}$$

Since both $C$ and $E$ are smaller than unity, $C^nE^n$ approaches zero and the limiting value for the longitudinal magnetization is

$$M_{zss}^{-} = M_0\frac{(1 - E_1)}{1 - E_1\cos\alpha}, \tag{5.5}$$

which, after multiplying both sides with $\sin\alpha$ and $\exp(-\mathrm{TR}/T_2)$, is identical to (4.16) as should be expected. Equation (5.4) can now be rewritten as

$$M_{zn} = M_{zss} + (M_\mathrm{S} - M_{zss})(\cos\alpha)^nE_1^n. \tag{5.4a}$$

The steady-state value of $M_z$, as described by (5.5), does not depend on the initial condition, $M_\mathrm{S}$, as expected. The approach to the steady state is similar to the unperturbed relaxation process (note the equivalence with (4.7)), but with a shorter time constant. As was said earlier, the lowest $k$ profiles (determining the contrast) should be measured at a point in time during the transient, which yields the desired contrast. This can be done by playing with the time of their acquisition during the transient (see Sect. 5.4).

From (5.4a) we also learn that when $\cos\alpha = 0$ ($\alpha = 90°$), the equilibrium situation is reached immediately after the first excitation. When $\cos a = 1$, ($\alpha = 0$), we have relaxation that is not disturbed by excitations, that is undisturbed $T_1$ relaxation. This gives us a chance to compare the apparent relaxation in the presence of excitations ($C \neq 1$), with unperturbed relaxation ($C = 1$). The apparent relaxation time during excitations is found from $E_{1\mathrm{app}} = \cos\alpha E_1$:

$$\frac{\mathrm{TR}}{T_{1\mathrm{app}}} = \frac{\mathrm{TR}}{T_1} - \ln(\cos\alpha), \tag{5.6}$$

which results in $T_{1\mathrm{app}} = 0$ when $\alpha = 90°$, as should be expected.

For small excitation angles the apparent relaxation time is

$$T_{1\mathrm{app}} = \frac{T_1 \mathrm{TR}}{\mathrm{TR} + T_1 \frac{\alpha^2}{2}}, \tag{5.7}$$

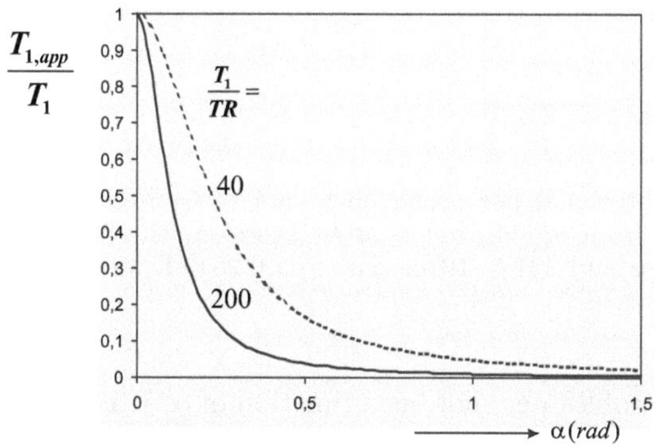


**Fig. 5.1.** Apparent relaxation time during excitations as a function of the flip angle when efficient RF spoiling is assumed. The curves show the relative reduction of the relaxation time when $T_1 = 2000\,\mathrm{ms}$ and $T_1 = 400\,\mathrm{ms}$, respectively, and $\mathrm{TR} = 10\,\mathrm{ms}$

approaching $T_1$ for small $\alpha$. In Fig. 5.1 we show the "apparent $T_1$", which describes the approach to equilibrium under the circumstances described above. The effect of accelerating the relaxation by excitation pulses is overestimated by the model discussed so far, because for short TR the longitudinal magnetization can be enhanced by the transverse magnetization due to previous pulses, which is partly rotated into the $+z$ direction by the excitation pulse.

The flip angle for maximum signal strength of a scan during the transient is, of course, not equal to the Ernst angle, which describes its value only under stationary conditions. An estimate for this optimum flip angle has been proposed in [6], where – as is done in this section – effective spoiling is assumed. Also the spin-lattice relaxation ($T_1$) has been neglected; only the maximum overall signal is considered. Under these circumstances the flip angle for maximum signal is found to be larger than the Ernst angle by a factor that is depending on the number of profiles measured during the scan [12].

When spin-lattice relaxation is not neglected and an initial situation is caused by a preparation pulse (5.4) can be used to evaluate the average value of the magnetization, as a function of the flip angle, as the sum

$$S_{\mathrm{av}} = \frac{\sin(\alpha)}{N-1} \sum_{0}^{N-1} M_{zn}^{-},$$

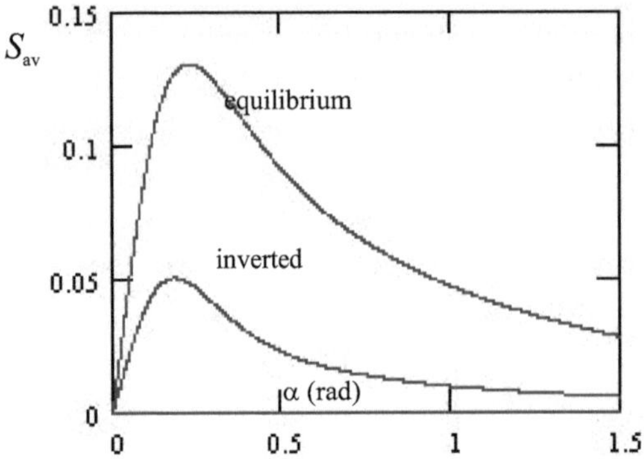


**Fig. 5.2.** Average strength of the signal per profile in a shot of 64 excitations, during a transient starting either from equilibrium or after inversion, as a function of the flip angle $\alpha$ for $T_1 = 941\,\text{ms}$ and TR $= 10\,\text{ms}$. For $\alpha \cong 0.25\,\text{rad.}$ we find an optimum value for both cases

where $N-1$ is the number of profiles per shot, as a function of $\alpha$. It is easily seen that this sum depends on the initial condition, on the flip angle, and on the spin-lattice relaxation time. Examples are shown in Fig. 5.2.

As was stated before, for small flip angles the signal obtained grows with the flip angle. However at large angles the available longitudinal magnetization is rapidly depleted, so that the signal decreases rapidly especially when the number of profiles per shot is large. This can be counteracted, so as to give more constant signal strength per profile, by taking flip angles that increase during the scan, starting at, for example, 6°–10° and gradually (for instance proportionally to the square root of the number of RF pulses passed) increase to 40°–60° [6, 7]. This precaution means the oscillations in the approach to steady state are flattened out and the profile signal is less dependent on the profile number.

### 5.2.2 Approach to Steady State Without RF Spoiling

A more realistic description of the transient, without the assumption that the transverse magnetization is spoiled, can be based on (4.3) and (4.9a):

$$\vec{M}(t_{n+1}^{+}) = \vec{\vec{\mathbb{R}}}_{-x',\alpha}\,\vec{\vec{\mathbb{Q}}}(E_1, E_2, \theta, \phi) \cdot \vec{M}(t_n^{-}) + (1-E_1)\,\vec{\vec{\mathbb{R}}}_{-x',\alpha}\,M_0. \quad (5.8)$$

This equation is the formal Bloch equation for a single TR interval and can be used for the study of the transient situation in gradient-echo methods. The condition for steady state (4.9b), is not used here. For every TR interval (5.8) has to be solved and the result is the initial condition for the next interval. An analytical solution is impossible and (5.8) can only be solved for specified parameter combinations using numerical methods [4, 5].

In Fig. 5.3 the result of such a numerical evaluation is shown for an N-TFE sequence. For this evaluation we used "MathCad". The uncompensated-for

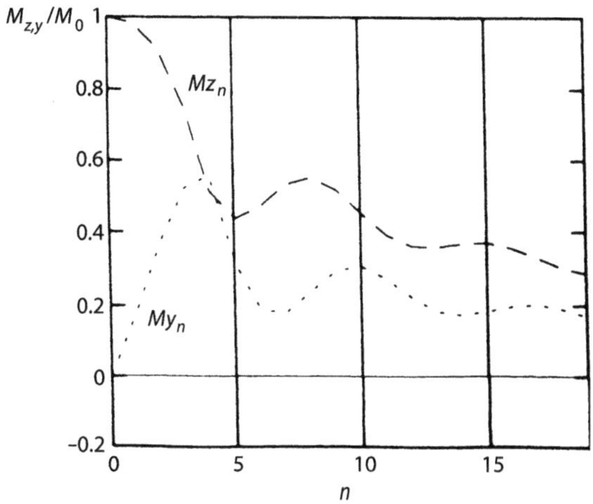


**Fig. 5.3.** Longitudinal $M_{zn}$, and transverse, $M_{yn}$, magnetization versus profile number, $n$, for an N-TFE sequence with TR = 10 ms, $T_1$ = 800 ms, $T_2$ = 50 ms. The flip angle is about 60° after the fourth pulse (see text)

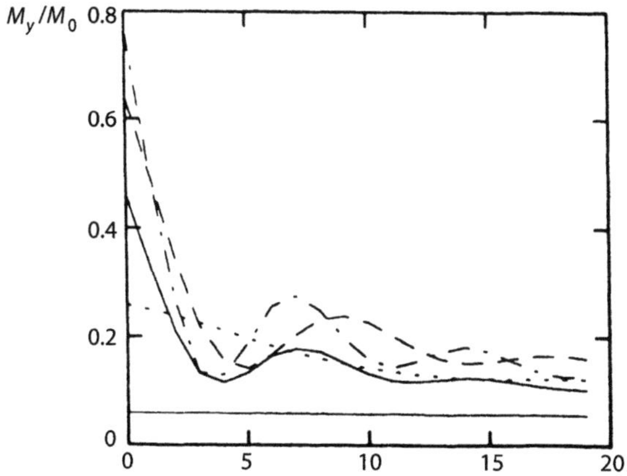


**Fig. 5.4.** Transverse magnetization $M_y$ as a function of the excitation number. The magnetization of several sub-slices is shown separately (*broken lines*), as well as the value $M_y$ for the complete slice. The RF profile is given in the text

gradient-time integral per TR period is taken to be so large that in each voxel all values of $\theta$ $(= \int_0^{\mathrm{TR}} G(t)\mathrm{d}t)$ between 0 and $2\pi$ occur. We therefore must solve the Bloch equations for many values of $\theta$ between 0 and $2\pi$ separately and subsequently add the magnetization vectors of the different contributions. The results are normalized to $M_0$, the equilibrium situation.

The flip angles are assumed to increase for the first three excitations, $\alpha_1 = \pi/18$, $\alpha_2 = \pi/8$, and $\alpha_n = 5\pi/18$ ($\cong 60°$) for subsequent excitations. This is done to damp the initial violent oscillations, as is also frequently done in practice. It is shown in Fig. 5.3 that for the large flip angles assumed in our example and notwithstanding our precaution of increasing initial flip angles, the approach to equilibrium is oscillatory. This is due to the conversion of transverse magnetization before an excitation pulse into new longitudinal magnetization that is adding to or subtracting from the longitudinal magnetization remaining after erosion by that pulse. When a number of profiles is acquired during these oscillations, this gives rise to artifacts in the phase encoding direction since the strength of these profiles is proportional to the momentary value of the transient curve (which means a convolution with the image; see Sects. 2.6 and 3.3.2). This will be further described in the next section.

The equilibrium situation reached after somewhat more than 20 profiles at $M_y$ is about $0.2\,M_0$. So the apparent relaxation time is about 100 ms, well below the true relaxation time of 800 ms, but much longer than shown in Fig. 5.1, obtained under the assumption of spoiling.

The numerical evaluation method can also be applied to estimate the effect of the RF profile on the slice profile, as was done for a steady state in Sect. 4.5.5. The flip angle is now taken to be constant for all excitations and

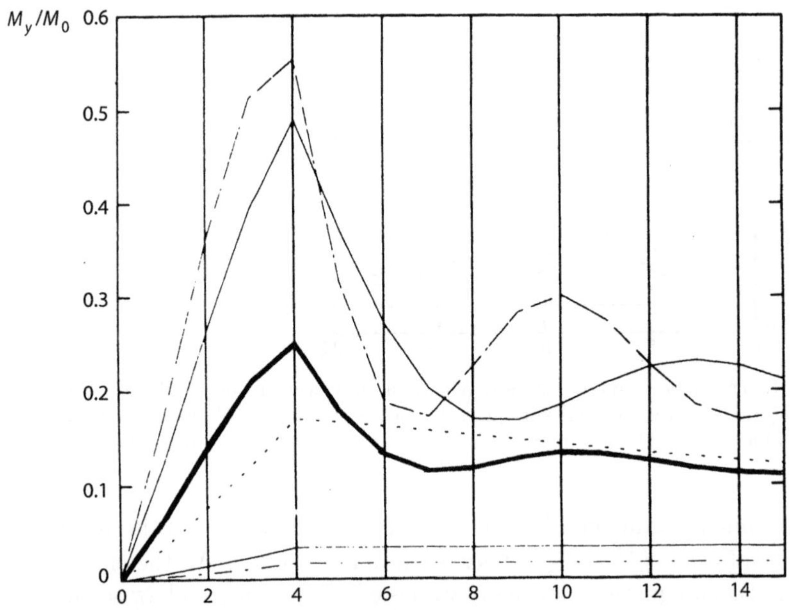


**Fig. 5.5.** Approach to steady state for an increasing flip angle with the RF slice profile taken into account. The slice profile is $a_0 = a_9 = 0.02$, $a_1 = a_8 = 0.04$, $a_2 = a_7 = 0.02$, $a_3 = a_6 = 0.7$, $a_4 = a_5 = 0.98$. Thick line shows $M_y$

the slice is sub-divided into sub-slices, each with its own flip angle corresponding to the RF profile (the short time allowed for the RF pulse in these fast imaging sequences always results in a shallow RF profile, see Sect. 2.3.2). The number of sub-slices used in the example of Fig. 5.4 is 7, with excitation flip angles equal to $\alpha = a_n\, 5\pi/18$ where $a_0 = 0$, $a_1 = 0.07$, $a_2 = 0.3$, $a_3 = 0.8$, $a_4 = 0.95$, and $a_5 = a_6 = 1$. The other parameters are equal to those of the previous examples (Fig. 5.3).

Figure 5.5 shows an example in which both the slice profile and the increasing flip angles are taken into account. The flip angle is given by $\alpha = a_s \cdot e_n \cdot \pi/180$, where $\alpha_s$ describes the slice profile, as shown in the figure, and $e_n$ the scale factor per excitation: $e_0 = 0$, $e_1 = 10$, $e_2 = 22$, $e_3 = 35$ and $e_n = 50$ degrees for $n > 3$. The oscillations are violent in some sub-slices, the average value over the slice shows a more quiet approach to equilibrium and after six excitations (60 ms) the average value is already close to its equilibrium value.

It should be noted that with the theory of Configurations (Chap. 8) the numerical calcualtions described here can be simplified.

## 5.3 Magnetization Preparation

In Sect. 4.5.5 it was mentioned that for small flip angles in the steady state the contrast is mainly density weighted. We assume that this is also the case for small flip-angle scans during the transient. Likewise for higher flip angles ($\alpha > \alpha_E$) we obtain $T_2/T_1$ contrast for TFE, $T_2$-TFE, and B-TFE, and $T_1$ contrast for $T_1$-TFE.

In order to obtain a useful contrast for small-flip-angle TFE scans, which depends on one of the relaxation processes, it is possible to generate an initial state of the spin system that contains the desired contrast at the start of the shot [3]. This is accomplished by applying one or several RF pulses and gradient wave forms before the actual shot is started. Imaging methods in which these preparation pulses are used are called "magnetization prepared". Sequences with larger flip angles have their own contrast weighting ($T_1/T_2$), but still the contrast enforced by the preparation pulses can be dominant.

In [3] it is proposed that a number of preparation pulses are applied, after which the magnetization at all sample points in the $k$ plane is acquired. However, the scans described in [3] are taken under conditions that are not usual in a commercial whole-body system: a small-bore system with a 30 mT/m gradient system having a rise time of 0.2 ms, which makes very fast (< 100 ms) imaging possible. Under the conditions imposed by a regular commercial whole-body system (12 mT/m, 0.6 ms) this very fast imaging is impossible, so a total scan in a time of the order of $T_2^*$ cannot be made. Yet the idea of using a preparation pulse followed by a TFE scan is used with success in commercial scanners by applying segmentation, where only part of the profiles is scanned after a preparation pulse.

The gross appearance of the image is determined by the type of pre-pulse, the profile order in the shot, the delay between the pre-pulse [15] and the central profile and the shot interval. Image set V-1 and V-2 demonstrate some of these influences.

Remarks on the profile order are given in Sect. 5.4. Various types of pre-pulse have already been listed in Sect. 2.7.2. Some pre-pulses that are special to TFE sequences are discussed in the next two subsections.

### 5.3.1 Pre-pulse to Avoid the Transient State in $T_1$-TFE

As stated earlier, acquisition during the transient state gives rise to artifacts. For $T_1$-TFE (spoiled gradient echo sequence) it appears to be possible to bring the magnetization directly into the steady-state situation by applying a pre-pulse [9]. The resulting images will be less disturbed by blurring and ghosts. This transient suppression is very important in applications like Contrast-Enhanced MR Angiography (see Sect. 7.5.2.2), where the acquisition must start directly upon arrival of the contrast agent in the region of interest or must be interrupted periodically, as is the case in imaging of the coronary arteries, when acquisition only occurs in a 100 ms interval during diastole (see Sect. 7.2.1).

The steady state of the transverse magnetization of a $T_1$-TFE sequence is described by (4.16). So the longitudinal magnetization before each RF pulse in the steady state is:

$$M_{\mathrm{SS}} = M_0 \frac{1 - E_1}{1 - E_1 \cos\alpha} . \tag{5.9}$$

where $E_1 = \exp(-\mathrm{TR}/T_1)$.

When an excitation pulse is applied to a spin system with longitudinal magnetization $M_s$, the longitudinal magnetization after this pulse is given by (5.1). We now assume the spin system to be in equilibrium, so $M_s = M_0$, and that a pulse with flip angle $\beta$ is applied $T_{\mathrm{rec}}$ s before the start of a $T_1$-FFE sequence. In that case (5.1) shows that the longitudinal magnetization at the start of the $T_1$-FFE sequence is:

$$M_{\mathrm{L}} = M_0 \left[ 1 + (\cos\beta - 1) \cdot \exp\left( -\frac{T_{\mathrm{rec}}}{T_1} \right) \right] . \tag{5.10}$$

Now it is possible to equate the longitudinal magnetization at the start of the $T_1$-TFE sequence $M_L$ to $M_{\mathrm{SS}}$. This yields:

$$T_{\mathrm{rec}} = T_1 \cdot \ln \frac{(E_1 - \cos\alpha)(1 - \cos\beta)}{1 - \cos\alpha} . \tag{5.11}$$

It appears that $T_{\mathrm{rec}}$ is a function of $T_1$. However, a numerical inspection of (5.11) shows that when $\beta = 90°$ $T_{\mathrm{rec}}$ is nearly constant for a wide range of $T_1$ values [15]. The value found is nearly 30 ms. The result of applying this pre-pulse is a marked reduction of blurring and ghosting artifacts, which improves the application of $T_1$-TFE.

### 5.3.2 Balanced-TFE Sequences

B-TFE sequences (excitation angle $\alpha$ and repetition time TR) frequently have an oscillatory approach to steady state, giving rise to ghosting, when the acquisition already starts during the transient. The approach can be made faster and smooth by applying a pre-pulse with flip angle $\alpha/2$ followed by a TR/2 interval, during which a gradient waveform equal to the gradient waveform in the second half of a TR interval of the B-TFE sequence is applied. This $\alpha/2$, TR/2 pre-pulse was first proposed first in [16]. The idea is that the pre-pulse rotates the equilibrium magnetization, $M_0$, away from the $z$'-axis in the, say, $z$'–$y$'-plane, over a flip angle $\alpha/2$. The transverse magnetization due to the pre-pulse will be rephased just before the first RF pulse of the B-TFE sequence. The first $-\alpha$ pulse in the B-TFE sequence rotates the magnetization vector in the $z'$–$y'$ plane to the direction $-\alpha/2$ away from the $z'$-axis, the next $+\alpha$ pulse to $+\alpha/2$, and so on. The transverse magnetization is therefore not converted to longitudinal magnetization by the alternating RF pulses, but remains constant and is only changed to the opposite direction. Consequently the approach to the steady state of the longitudinal magnetization is continuous, as explained in Sect. 5.2.1, and not oscillatory, as calculated in Sect. 5.2.2, where transverse magnetization is not spoiled and is partly converted in longitudinal magnetization. Now during the smooth approach to the steady state, the tissues with low $T_2/T_1$ values still give a high signal. So with centric (low–high) profile order, when the low profiles are measured first, these tissues are clearly visible, as is shown in image set V-3. Due to the smooth approach to steady state, ghosting artifacts are suppressed but blurring may still be present.

The $\alpha/2$, TR/2 pre-pulse can be combined with other pre-pulses that influence the longitudinal magnetization only. As an example we will discuss the insertion of a spectral-selective fat suppression pulse (SPIR) in a segmented B-TFE sequence with minimal time distance between the shots. It has recently been shown, that it is also quite possible to interrupt a 3D B-TFE sequence, for inserting a SPIR pulse, followed by a spoiler gradient, without compromising the steady state of the water spins [17]. This is accomplished by ending each train of TR intervals, separated by RF pulses with flip angle $\alpha$, with an interval of length TR/2 followed by an RF pulse with flip angle $\alpha/2$. After the SPIR block, the sequence is restarted with an $\alpha/2$ pulse and an interval TR/2, after which the next B-TFE block follows. The idea is that during the fat saturation the coherent transverse magnetization is safely stored along the longitudinal axes. After the SPIR block, this magnetization is brought back in the transverse plane, by the $\alpha/2$ pulse. The SPIR pulse has to be repeated frequently, for instance every 160 ms [17], where the combination of strong $T_2/T_1$-weighting of the steady-state magnetization of water spins and the suppressed fat signal is demonstrated in abdominal images of the bowel and spinal cord.

## 5.4 Profile Order

As has been stated before the gross appearance of the image, specifically the contrast, is determined by the lowest $k$ profiles and therefore by the period of time during the transient process in which the lowest $k$ profiles are acquired. Therefore the contrast obtained in TFE scans can be influenced by the profile order. The most frequently used profile orders are as follows.

"*Linear*" or "sequential", from profile $k = (-127, \ldots, -1, 0, +1, \ldots, +128)$, in which the central (low-$k$) profiles are measured half way along the scan, so late in time during the transient, when the influence of the initial condition has almost died out (the more so since the transient is accelerated because of the excitations of the scan).

"*Centric*" or low–high, such as $k = (0, +1, -1, +2, -2, \ldots, +128, -128)$, for which the lowest $k$ profiles are measured immediately, so the the initial condition is well-reflected in the image contrast. Some dummy excitations are mostly advisable before starting the actual acquisition, because during the first excitations the relaxation effect is not very well behaved, as is shown in Fig. 5.5.

For completeness we also mention the "*cyclic*" profile order, $k = (0, +1, +2, \ldots, +128, -128, -127, \ldots, -2, -1)$, although it is not frequently used since the lowest-order profiles are acquired at different times, which causes a step at $k = 0$.

If the apparent relaxation (during excitations) is too fast in comparison with the shot length, so that almost no signal depending on the initial magnetization is left during the acquisition of the last profiles, one can use shorter shots. This means that per shot a smaller number of profiles are acquired. The essence of the profile order is maintained within each shot. For example eight profiles per shot are measured and repeated sixteen times so as to acquire all 128 profiles. This can be done for all the profile orders mentioned above. As an example we mention a segment in the centric order $k = (0, +8, -8, +16, -16, +24, -24, \ldots$. In the other shots $k$ starts with $(+1, -1, -9, +9, \ldots)$, $(+2, -2, +10, -10, \ldots)$ etc. This segmentation is necessary anyhow when not much time is present for acquisition, for example in cardiac scans, where at every trigger pulse one segment is measured, and for 3D scans.

The gradual approach to the steady state during the shot changes the sensitivity of the measurement during the scan. This can be described as a "filter" over the profiles in the phase-encoding direction [13, 14]. After the Fourier transform the image is convolved with the Fourier transform of this filter, which manifests itself as blurring in the image when the transient is gradual and even as ghosts when the transient is oscillatory, as has already been explained in connection with TSE in Sect. 3.2.

We wish to get some feeling for the effect of this filter by assuming a specified form of the object in the phase-encode direction under the influence of different filters due to relaxation perturbed by the excitations. The object chosen is a $\delta$-function-like object in the center of the field of view. Although this object is artificial, it is known that it yields for all $k$-values equal magnetization values (see Fig. 2.11), and that the Fourier transform of the measurements yields the "point spread" function (Sect. 2.4.1.2). For a stationary situation the weighting of each profile is equal and we find for the point-spread function the well-known sinc function as the image of the $\delta$-function object.

In an actual scan during a transient the (perturbed) relaxation profile gives a varying weighting to the profiles. When we now perform the Fourier transform to find the image of the $\delta$-function object we observe that the point-spread function is disturbed, as illustrated in Fig. 5.6. In Fig. 5.6a we show the measured magnetization in the phase-encoding direction of the $k$ plane and the point-spread function when the measurements are taken during a steady state. The image is a simple sinc function, which is shown as a simple peak, due to the limited discrete number of points in the $k$ plane used in the simulation process.

In Fig. 5.6b a situation is simulated, where the transient is smooth (no oscillations) and the profile order is linear. We see that the point-spread function widens and that the maximum is lower (sensitivity), which means that blurring occurs. We also see ringing.

In Fig. 5.6c again a smooth transient is assumed (small flip angles), but the profile order is centric. We observe an important widening of the point-spread function and some ringing due to the discontinuity at $k = 0$.

Finally in Fig. 5.6d a centric scan and an oscillatory approach to the steady state is assumed. We observe that the point-spread function also widens and shows ringing as in the cases depicted in Figs. 5.6b and 5.6c, but that now a ghost is also visible at 10 pixels away from the object. In all cases the sensitivity is reduced during the transient.

## 5.5 Survey of Transient Gradient Echo Methods

When all Gradient Echo methods, magnetization preparation pre-pulses and profile orders are considered, an unwieldy number of scan methods are possible. In the first place there are in principle four types of TFE sequences, N-TFE, $T_1$-TFE, $T_2$-TFE and R-TFE. Then there are many different types of magnetization preparation, as listed in Sect. 2.7.2, and another possibility is, of course, no preparation pulse. Finally, we have defined three different profile orders. Not all methods are useful, as can be concluded from the discussions in earlier sections. In our experience, $T_1$-TFE is the workhorse amongst the various types of TFE sequences. It is useful for magnetization-prepared $T_1$-weighted studies as well as in combination with cardiac triggering and CE-Angio, see Sect. 7.5.2.2. Noteworthy is the special type of pre-pulse that is possible for $T_1$-FFE and that results in a near perfect immediate transfer to the steady state, as discussed in Sect. 5.3.1. The application of N-TFE and

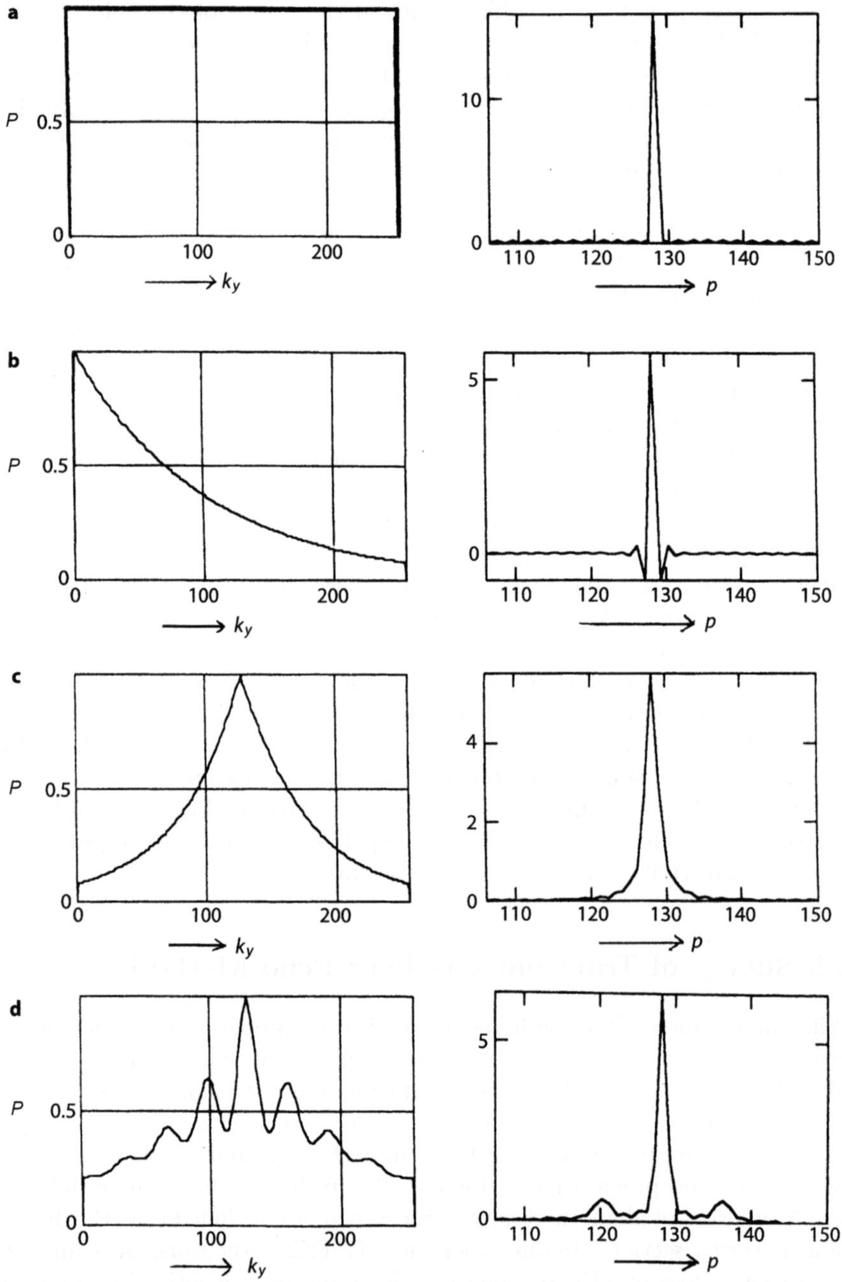


**Fig. 5.6.** Relative profile intensity $P$ vs profile number, $k$ (*left*) and point-spread function vs pixel number, $p$ (*right*). **a** Steady state. **b** Smooth relaxation and sequential profile order. **c** Smooth relaxation and centric profile order. **d** Oscillatory relaxation and centric profile order

$T_2$-TFE is less widespread, although it is used for special clinical purposes, such as cardiac-triggered coronary angiography; see, for instance, image set VII-13.

The utilization of B-TFE sequences is expanding. Apart from the practical availability of systems that allow short TR, the recently proposed $\alpha/2$, TR/2 pre-pulse is important, as discussed in Sect. 5.3.2. This sequence combines the high SNR, the strong $T_2/T_1$-weighted contrast for fluids, and the absent flow sensitivity of B-FFE with increased control of soft tissue contrast weighting, shown, for instance, in image set V-3.

# Image Sets
# Chapter 5

## V-I Inversion Pulses in Transient Field Echo Imaging: Different Shot Lengths

(1) In Transient Field Echo (TFE) methods, preparation pulses can be added to influence the contrast. A group of excitation pulses, separated typically by a short TR, follows after a preparation delay. This group is usually called a "shot". The profiles covered in the shot have a user selectable order (see Sect. 5.5 for a description of some of these orders). When the complete image is not collected in one shot, the sequence of preparation pulse, delay, and shot is repeated until all $k_y$ values are covered. The images of this set show the influence of the variation in the length of the shots in $T_1$-TFE imaging.

Parameters: $B_0 = 1.5\,\mathrm{T}$; FOV $= 340\,\mathrm{mm}$; matrix $= 256 \times 256$; $d = 10\,\mathrm{mm}$; TR/TE/$\alpha = 10.4/5.1/30$; shot-repetition time $= 800\,\mathrm{ms}$; profile order is linear.

| image no. | 1 | 2 | 3 | 4 | 5 | 6 |
|---|---|---|---|---|---|---|
| preparation pulse: | | none | | | inversion | |
| inversion delay (ms) | – | – | – | 400 | 400 | 400 |
| shot length (TR) | 8 | 16 | 32 | 8 | 16 | 32 |
| NSA | 1 | 2 | 4 | 1 | 2 | 4 |

(2) Each image was scanned in 26 sec and was obtained during breath hold. In the images with the largest shot length, images 3 and 6, this could be combined with NSA = 4. In part of the images a preparation pulse was added. The pulse used was a non-selective ("hard") inversion pulse. The preparation delay, defined as the time between the preparation pulse and the echo that is encoded with $k_y = 0$, is 400 ms (images 4, 5, and 6). Within the shot, the excitation pulses were dephased for RF spoiling ($T_1$-TFE). The images shown are modulus images.

(3) When no preparation pulse is given, the images show a weak contrast. Vessels show bright and emit a pulsation ghost.

(4) When a preparation pulse with a delay of 400 ms is added to the image sequence (images 4, 5, 6), the image contrast is $T_1$-weighted. Liver and spleen have a different signal level. The pancreas is clearly visible. Also, the magnetization of the upstream blood, that after inversion with the non-selective inversion pulses has flown into the slice at excitation will be nulled, resulting in a dark aspect of the vessels without pulsation artifacts.

(4) The image contrast varies only slightly with the shot length. For images 1, 2, and 3, obtained without preparation pulse, the contrast improves with shot length. For images 4, 5, and 6, obtained with an inversion preparation pulse at 400 ms, the contrast for the largest shot length (images 5 and 6) is less pronounced than that in image 4; for instance the spleen is somewhat less dark; the pancreas and liver are more equal in brightness. The images with the longest shot length have the highest NSA and because of that the highest signal-to-noise ratio.

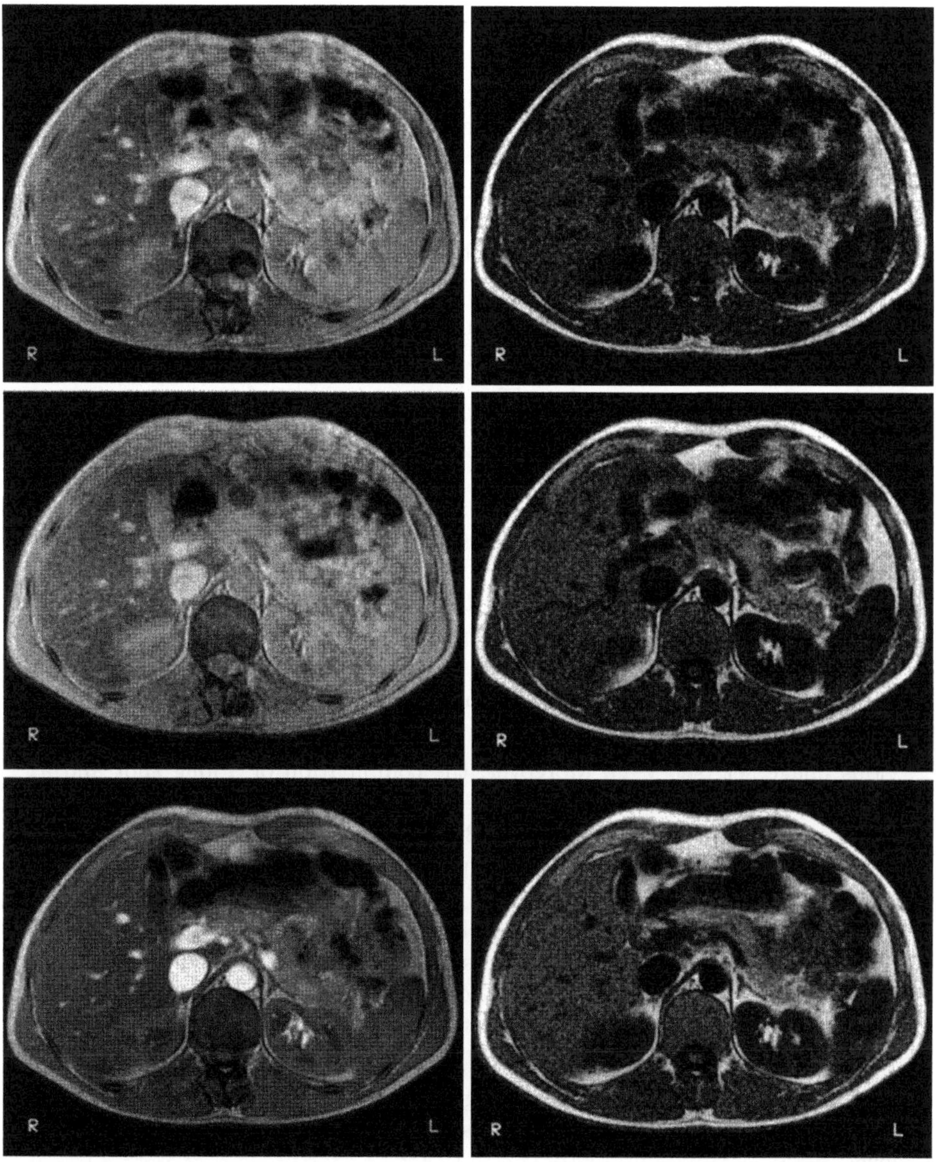


1 4

2 5

3 6

**Image Set V-I**

(5) In conclusion, the addition of an inversion pulse is desirable for the purpose of contrast. For a scan time compatible with breath holding, a long shot length appears to offer the best compromise between signal-to-noise and contrast.

The optimal timing of TFE scans is further inspected with help of image set V-2.

## V-2 Inversion Pulses in Transient Field Echo Imaging: Different Delay Times

(1) In the previous image set (V-1), the use of a Transient Field Echo method (TFE) with a long "shot" was demonstrated to be advantageous to the image quality. In addition, the use of an inversion pulse with a delay of 400 ms was shown to give a useful contrast. A closer inspection of the combination of inversion pulse and preparation delay is presented in this set of $T_1$-TFE images.

Parameters: $B_0 = 0.5\,\text{T}$; FOV $= 430\,\text{mm}$; matrix $= 128 \times 256$; $d = 10\,\text{mm}$; TR/TE/$\alpha = 9.5/4.3/25$; profile order is linear; NSA $= 6$

| image no. | 1 | 2 | 3 | 4 | 5 | 6 |
|---|---|---|---|---|---|---|
| shot length (TR) | 43 | 43 | 64 | 64 | 64 | 64 |
| inversion delay | 300 | 500 | 400 | 500 | 900 | 1300 |
| shot repetition time | 520 | 720 | 730 | 830 | 1230 | 1630 |

The inversion delay is defined as the time from the inversion pulse to the excitation of the profile with $ky = 0$.

(2) The scan time varies from 8.4 to 13 seconds; all compatible with breath holding, which was used in the scan. In all cases the shot repetition time had the shortest adjustable value, so that the inversion pulses followed immediately (25 ms) after the end of each shot.

(3) The shot lengths used are larger still than the values used in the previous image set (set V-1). Moreover, the combination of a long shot and a short shot-repetition time reduces the magnetization that is available at the moment of the next inversion pulse. Both effects reduce the $T_1$ weighting of the contrast. This is visible when comparing image 1 of the present set with image 6 of image set V-1.

There is a further decrease of the $T_1$ weighting of the contrast with an increase of the preparation delay above 400 ms (for instance visible as a decrease in signal difference between the liver and spleen in images 4, 5, and 6). Nevertheless, all images in the set have a stronger contrast than TFE images without preparation pulse (see images 1, 2, 3 from image set V-1).

(4) In conclusion, breath-hold imaging of the abdomen with TFE at 0.5T is possible with acceptable contrast and signal-to-noise ratio, and a low artifact level.

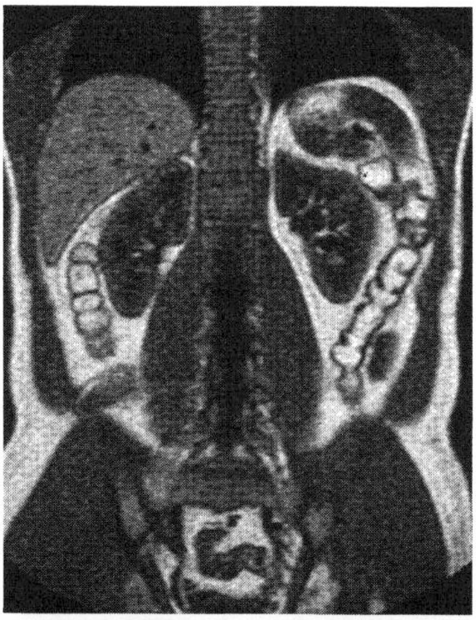
1

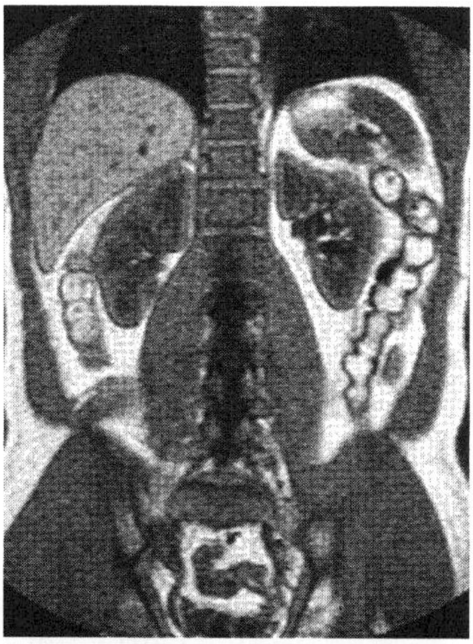
2

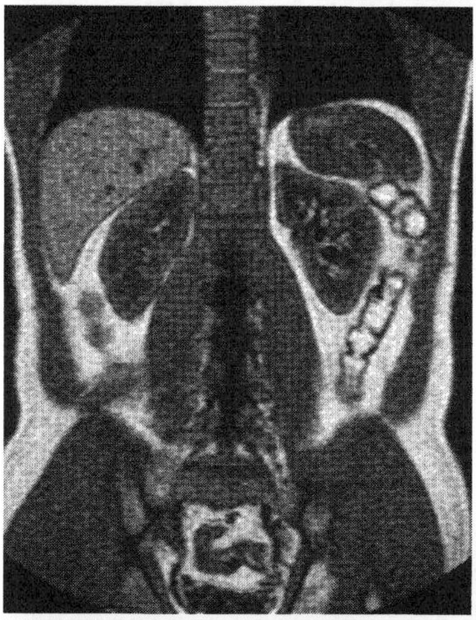
3

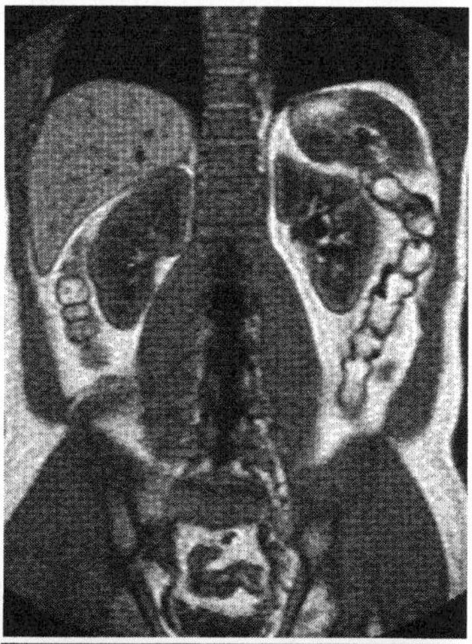
4

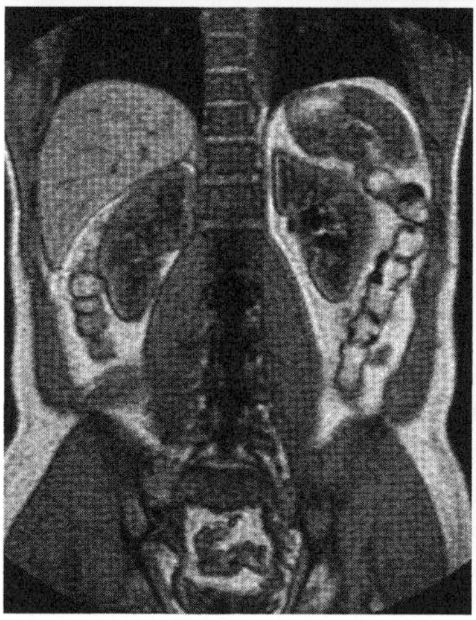
5

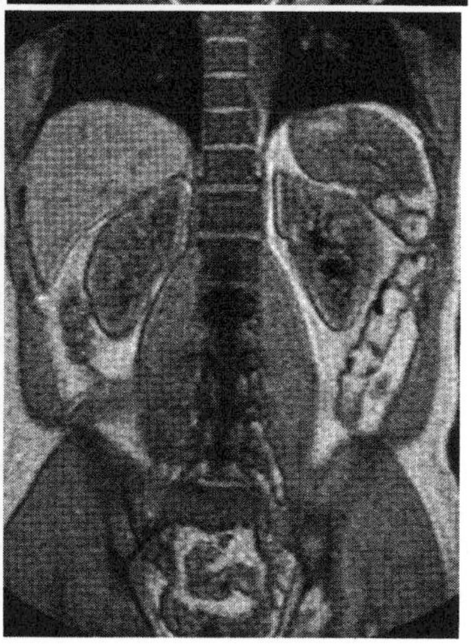
6

**Image Set V-2**

# V-3 Balanced FFE and Balanced TFE

(1) Recently interest in the use of balanced FFE (B-FFE) has grown. The absence of flow sensitivity and the $T_2/T_1$ contrast that is offered by this sequence at large flip angles (Sect. 4.5.4.1) are notable features. The interest in the sequence grew when TR values of less than 10 ms became available. These TR values are within reach in modern MR systems with strong gradients. For tissue with a sufficiently large value of $T_2/T_1$, the signal-to-noise ratio of this sequence is high so that short TR values of only a few ms can still be used at high resolution. The same range of TR values also is needed, because otherwise B-FFE, a rephased sequence, would become overly vulnerable to banding caused by $B_0$ inhomogeneity (Sect. 4.5.3.2).

Typical for the use of B-FFE is the imaging of body regions with liquid-filled spaces, especially when the liquid is not stationary. One example of such a body region is the cervical spine in which the CSF flow hampers the use of transverse multi-slice TSE, as illustrated in image set VII-8.

Although attractive, the $T_2/T_1$ contrast, a consequence of the steady state in the B-FFE sequence, does not give much signal in most soft tissues because of their low $T_2/T_1$ values. This leads to the interest in the use of a transient state version of the B-FFE sequence, balanced TFE (B-TFE), as a means of generating additional signal from soft tissue. The basis for this approach, as introduced in Sect. 5.3.2, is an $\alpha/2$, TR/2 prepulse that avoids the need for a long run-in period while allowing imaging free of ringing artefacts.

In this image set, 3D B-FFE of the cervical spine is compared to 3D B-TFE with an $\alpha/2$, TR/2 prepulse in a healthy volunteer.

Parameters: $B_0 = 1.5\,\mathrm{T}$; FOV $= 225\,\mathrm{mm}$; matrix $= 352 \times 256$; $d = 3\,\mathrm{mm}$; number of partitions $= 32$; flip angle $= 45°$

| image no. | 1 | 2 |
|---|---|---|
| scan method | 3D B-FFE | 3D B-TFE |
| TR | 7.8 | 5.4 |
| TE | 3.9 | 2.7 |
| shot length | n.a. | 256 |
| shot interval | n.a. | 4000 ms |
| scan time | 2′33″ | 4′17″ |

n.a. means not applicable. The shot interval is that between the start times of the shots.

(2) Both images show the SCF in the spinal canal as the brightest signal in the image, characteristic for the $T_2/T_1$ contrast. In image 1 (B-FFE), all muscular tissue has a low signal value, making it difficult to distinguish the borders. In image 2 (B-TFE) a much higher muscle signal is reached, and its contrast with the bony structures and other anatomical details is easier to read. The contrast between CSF and nerve tissue has not suffered. Both images show a pair of anterior nerve roots leaving the spinal canal.

The increased soft tissue signal in image 2 is caused by the combination of the distribution of the scan into shots and the use of a low-high profile order in each shot. To arrange the central profiles early in the shots, the $k_y$–$k_z$-plane is distributed into a number of concentric zones that equals the number of excitations per shot (256) and that each cover the same number of $k_y$–$k_z$ combinations (32). This creates the table of $k_y$–$k_z$ values to be used per shot. The shot interval (4000 ms) is much longer than the shot duration ($276 \times 5.4 = 1490$ ms), so that the longitudinal magnetization at the start of the shot is large for all tissue. After the $\alpha/2$, TR/2 pulse, the longitudinal magnetization remains larger than in steady state for a number of TR's. This number suffices to generate a soft tissue signal that has a $T_2$-weighted contrast and a signal strength that is much higher than that in the (steady-state) B-FFE scan. Start up cycles are used to select the level of the soft tissue signal at a signal strength that is higher than that in the (steady-state) B-FFE scan, but that is sufficiently different from the liquor signal to maintain a clear visualization of the spinal canal.

(3) In image 1, a dark band is visible inside the anterior surface of the neck. This is caused by the banding phenomenon and indicates a shift of the main field of a half-cycle per TR, i.e. about 100 Hz. In image 2, the TFE sequence is less visibly influenced because most of the signal is due to the transient state.

(4) The images show a high-resolution high-signal-to-noise representation of the nerve roots of the spinal cord obtained in an acceptable image time. Flow artefacts are absent and, especially in image 2, soft tissue signal is sufficient to visualize clearly the anatomical structure. Next to the 3D TSE with reset pulses (image set III-3), 3D B-TFE is a modern and adequate approach for clinical spinal cord imaging.

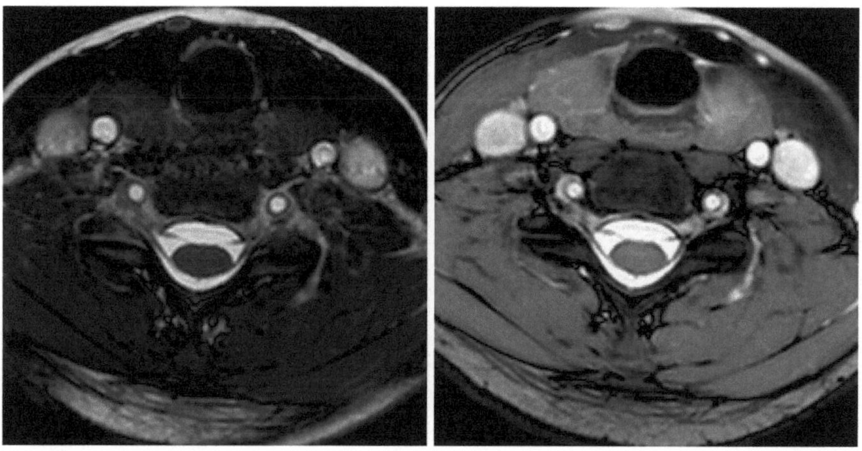

1 2

**Image Set V-3**

Image Set V.3

# 6. Contrast and Signal-to-Noise Ratio

## 6.1 Introduction

In an MR image, information is contained in the variation across the image of one single parameter: the grey level, which is a proportional to the signal level at that position. This parameter, called contrast, is the result of three properties of the imaged tissue: the proton density, $\rho$; the spin–lattice relaxation time, $T_1$; and the spin–spin relaxation time, $T_2$. It is therefore necessary to understand how the contrast depends, for a certain scan sequences, on the properties of the tissue in order to recognize the diagnostic information in the image. This is the topic of Sect. 6.2. In Sect. 6.3 a survey is given of the basic physics of relaxation. This survey is not meant to be complete (there is excellent literature available [1, 4.2]) but it is also needed in order to introduce two techniques which influence the values of the relaxation constants, namely magnetization transfer (Sect. 6.3.3) and contrast agents (Sect. 6.3.4).

For diagnostic usefulness, the signal level should be well above the noise level. This is described by the signal-to-noise ratio. In Sect. 6.4 a detailed theoretical treatment of this signal-to-noise ratio (SNR) will be presented. The difference between high-field and low-field MRI systems will get special attention in Sect. 6.4.3. The theory of Sect. 6.4 brings us to a position from which the dependence of the signal-to-noise ratio on the magnetic field strength, $B_0$, the RF coil properties, and the scan parameters can be studied (Sect. 6.5). In Sect. 6.6 some examples are given.

## 6.2 Contrast in MR Images

Contrast is the difference between two signal levels in adjacent tissue regions. It is a measure of the visibility of tissue borders and lesions. Let us first consider the contrast obtained in a SE sequence (see Fig. 6.1). After the 90° pulse the longitudinal magnetization is zero and starts to relax back according to (2.8) during the time TE/2. Its value grows to $1-\exp(\mathrm{TE}/2T_1)$. The 180° pulse then inverts the longitudinal relaxation after which it again relaxes back during the time TR – TE/2 until the next 90° pulse brings the longitudinal magnetization again to zero. Just before this next 90° pulse the longitudinal magnetization becomes

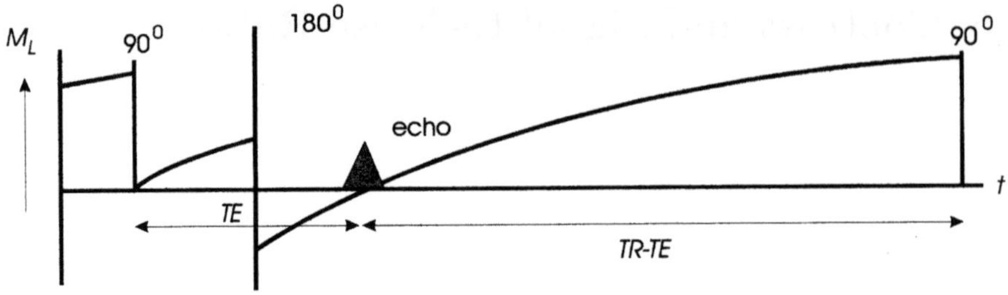


**Fig. 6.1.** Longitudinal magnetization as a function of time in a Spin-Echo sequence

$$M_\mathrm{L}(\mathrm{TR}) = M_0\,\{1 - (2 - \exp(-\mathrm{TE}/2T_1))\exp(-(2\mathrm{TR} - \mathrm{TE})/2T_1)\}\,. \quad (6.1)$$

After excitation this longitudinal magnetization is transformed into transverse magnetization, which is measured as an echo at time $t = \mathrm{TE}$ with magnitude:

$$\begin{aligned} M_\mathrm{T}(\mathrm{TE}) &= M_0\,\{1 - 2\exp(-(2\mathrm{TR} - \mathrm{TE})/2T_1) + \exp(-\mathrm{TR}/T_1)\} \\ &\quad \exp(-\mathrm{TE}/T_2) = M_0\,F_\mathrm{SE}. \end{aligned} \quad (6.2a)$$

When $\mathrm{TE} \ll \mathrm{TR}$, (6.2a) reduces to the contrast equation given by (2.10b):

$$M_\mathrm{T}(\mathrm{TE}) \cong M_0(1 - \exp(-\mathrm{TR}/T_1))\exp(-\mathrm{TE}/T_2) = M_0\,F_\mathrm{SE}. \quad (6.3a)$$

$M_0$ is proportional to the spin density, $\rho$. Equations (6.2a) and (6.3a) describe the dependence of the signal level, $M_\mathrm{T}$, on $M_0$, the proton density, and the relaxation times for an SE sequence. When TR is reduced the signal reduces and when TE is reduced the signal increases.

Equation (6.2a) can also be used for TSE when in the $T_2$-decay term the echo time TE is replaced by $\mathrm{TE_{eff}}$, the time delay between excitation and acquisition of the lowest-order $k$ profiles (see Sect. 3.2.1). This yields

$$\begin{aligned} M_\mathrm{T}(\mathrm{TE_{eff}}) &= M_0\{1 - 2\exp(-(2\mathrm{TR} - \mathrm{TE}')/2T_1) + \exp(-\mathrm{TR}/T_1)\} \\ &\quad \exp(-\mathrm{TE_{eff}}/T_2) = M_0\,F_\mathrm{TSE}, \end{aligned} \quad (6.2b)$$

where $\mathrm{TE}' = (F_\mathrm{T} + \frac{1}{2})\mathrm{TE}$ and $F_\mathrm{T}$ is the turbo factor. For TSE also $\mathrm{TR} \gg \mathrm{TE}'$, so (6.2b) is simplified to

$$M_\mathrm{T}(\mathrm{TE_{eff}}) \cong M_0(1 - \exp(-\mathrm{TR}/T_1))\exp(-\mathrm{TE_{eff}}/T_2) = M_0\,F_\mathrm{TSE}. \quad (6.3b)$$

Equation (6.3a) also applies to FE sequences with a 90° flip angle when $T_2$ is replaced by $T_2^*$ (see Sect. 2.5), so for $\mathrm{TR} \gg \mathrm{TE}$

$$M_\mathrm{T}(\mathrm{TE}) = M_0(1 - \exp(-\mathrm{TR}/T_1))\exp(-\mathrm{TE}/T_2^*) = M_0 F_\mathrm{FE}. \quad (6.3c)$$

Finally (6.3c) can also be used for EPI when $\alpha = 90°$ and TE is again replaced by $\mathrm{TE_{eff}}$:

$$M_\mathrm{T}(\mathrm{TE}) = M_0(1 - \exp(-\mathrm{TR}/T_1))\exp(-\mathrm{TE_{eff}}/T_2^*) = M_0 F_\mathrm{EPI}. \quad (6.3d)$$

When the flip angle is $\neq 90°$ we have to add the factor $\sin\alpha$ to (6.3c) and (6.3d).

The contrast equations for gradient-echo sequences have been calculated in the previous chapters. We found in (4.22) for N-FFE that

$$\begin{aligned} \langle M_{\mathrm{T}}^{+} \rangle &= M_0 \frac{\sin\alpha(1-E_1)}{C} \left\{ \frac{C + D\,E_2}{(D^2 - C^2)^{1/2}} - E_2 \right\} \exp(-\mathrm{TE}/T_2^*) \\ &= M_0 F_{\mathrm{N-FFE}}; \end{aligned} \tag{6.3e}$$

in (4.23) for $T_2$-FFE that

$$\begin{aligned} \langle M_{\mathrm{T}}^{-} \rangle &= M_0 \frac{\sin\alpha(1-E_1)E_2}{C} \left\{ 1 - \frac{C + D\,E_2}{(D^2 - C^2)^{1/2}} \right\} \exp(-\mathrm{TE}/T_2^*) \\ &= M_0 F_{T_2-\mathrm{FFE}}; \end{aligned} \tag{6.3f}$$

and in (4.16) for $T_1$-FFE that

$$\langle M_{\mathrm{T}} \rangle = M_0 \frac{\sin\alpha(1-E_1)}{1 - E_1\cos\alpha} \exp(-\mathrm{TE}/T_2^*) = M_0 F_{T_1-\mathrm{FFE}}. \tag{6.3g}$$

This latter equation also describes conventional FE sequences with arbitrary flip angles and $\mathrm{TR} \gg T_2^*$. In equations (6.3e–f), $E_1 = \exp(-\mathrm{TR}/T_1)$ and $E_2 = \exp(-\mathrm{TR}/T_2)$.

Contrast depends for all imaging methods in a more or less known way on the actual values of $T_1$ and $T_2$ and the density $M_0$, according to (6.3). In Sect. 6.3.1 of this chapter the global theory of these relaxation phenomena, the Bloembergen–Pound–Purcell (BPP) theory [1], will be outlined. Application to relaxation phenomena in tissue is treated in Sect. 6.3.2 on the basis of a paper by Fullerton [2], which also contains an extended bibliography. This treatment of relaxation phenomena in tissue also enables us to understand the effect of cross-relaxation, which can be used to enhance the contrast, a technique called Magnetization Transfer Contrast (MTC). The use of contrast agents will be described in Sect. 6.3.4.

## 6.3 The Physical Mechanism of Relaxation in Tissue

In this section the relaxation properties of homogeneous matter will be treated first. This is necessary to understand the magnetic resonance properties of tissue, which is a mixture of water and hydro-carbon macro molecules such as proteins. It is important to realize that the dominant mechanism determining the relaxation in MRI is the dipole–dipole interaction, both in static and in dynamic cases. This mechanism will be first described for homogeneous matter as the BPP theory [1]. When this theory is outlined, it must be applied to the special case of in-vivo tissue [2] in which the relaxation mechanism depends on the interaction of water and macro molecules.

### 6.3.1 The BPP Theory of Relaxation in Homogeneous Matter

The static dipole–dipole interaction will be discussed first. We consider a static water molecule in the main magnetic field (see Fig. 6.2). Depending on the orientation of the molecule, each of the protons "sees" a magnetic field, which is increased or decreased due to the magnetic moment of the other proton. The maximum field deviation appears to be $+/-10\,\mu$T, which results in a dephasing of both spins in a time equal to $11\,\mu$s. The dephasing signifies a signal decay with the same time constant $T_2$ and in a static situation $T_2$ is too short for MR acquisition. Such a static situation occurs in solids, which is the reason why bone structures are not imaged by MRI.

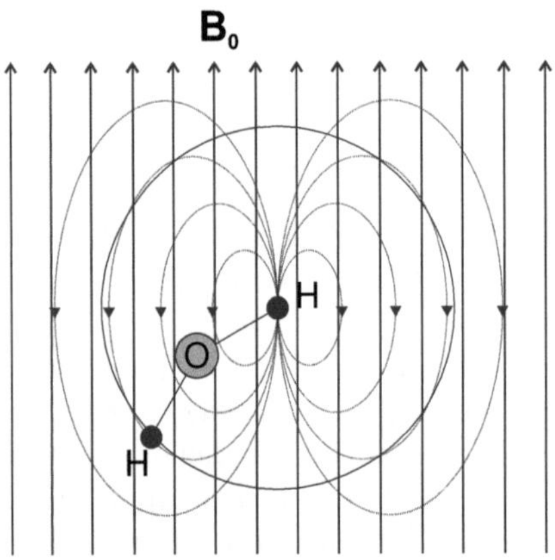


**Fig. 6.2.** The field of one of the spins of a hydrogen atom influences the magnetic field experienced by the other proton

Now in practice there is no static situation; matter is always in thermal agitation. This causes vibrational and rotational motion and the protons involved in this motion cause varying magnetic fields. In this situation the dephasing due to fields of other protons is averaged out and the dephasing is much slower than in the static case. In practice $T_2$ decay (spin–spin relaxation) in pure water takes seconds. In tissue, where the protons of the water are sometimes under the influence of the extra magnetic fields of the macro molecules so that dephasing can occur, the $T_2$ decay time decreases to tens of ms.

Thermal motion is characterized by a correlation time, $\tau_c$, which is the time that a certain situation is preserved for or the time in which the orientation of a molecule changes. For liquids this time is $10^{-12}$ s and for solids it is $10^{-5}$ s. Viscous fluids have correlation times of around $10^{-9}$ s. The frequency spectrum of the thermal motions is shown in Fig. 6.3. As is shown in Sect. 1.1.3 the rotating magnetic fields associated with molecules tumbling with the frequency $\omega_0 = \gamma B_0$ can excite or deexcite a neighbouring spin depending on the phase of the tumbling with respect to the phase of the spins. By this mechanism the spins can be de-excited and their energy transformed to a form of thermal energy. This is $T_1$ or spin–lattice relaxation.

This is an approximate classical description of spin exchange, with dissipation of spin energy in the thermal motion of the molecules. The actual spin–lattice relaxation rate ($= 1/T_1$) is proportional to the ratio of the number of molecules tumbling with a frequency in a small band around $\omega_0$ (the gyromagnetic frequency) to the total number of atoms present. For solids this fraction is small, for viscous fluids it is large (so $T_1$ goes through a minimum) and for liquids it becomes small again, as is demonstrated in Fig. 6.3 where $\omega_c$ is defined as $\tau_c^{-1}$. This result is shown in Fig. 6.4 and is known as the BPP theory [1].

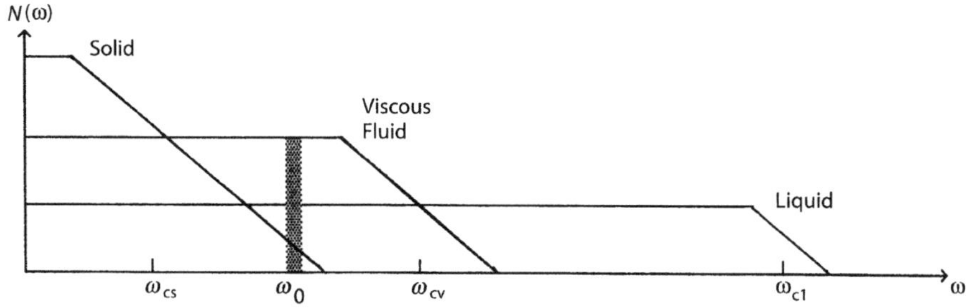


**Fig. 6.3.** Number of molecules tumbling at frequency $\omega$, note that $\omega_c \tau_c = 1$

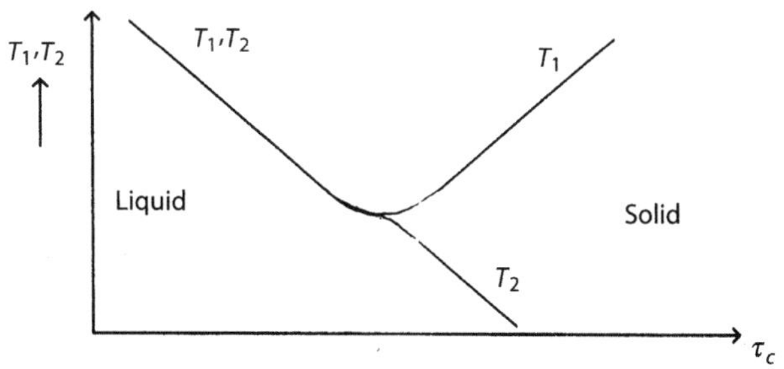


**Fig. 6.4.** The BPP theory of the relaxation time constants in homogeneous matter

It is furthermore observed in Fig. 6.3 that when $\omega_0 \ll \omega_c$ the value of $T_1$ is independent of $\omega_0$, and therefore independent of the magnetic field. However, when $\omega_0 \cong \omega_c$, so that the slope of the frequency spectrum becomes important, the relaxation rate becomes dependent on the magnetic field because the fraction of the molecules with thermal frequencies around $\omega_0$ changes rapidly with $\omega_0$. This makes $T_1$ dependent on $\omega_0$. A method to measure $T_1$ in tissue is described in image set VI-1, where a $T_1$ image is composed of a TSE image and a IR-TSE image.

The value of $T_2$ closely follows that of $T_1$ for liquids with a small correlation time $\tau_c$, so large $\omega_0$. For solids ($\tau_c \cong 10^{-5}$) the effect of the static dipole–dipole interaction (for which $T_2 \cong 11\,\mu s$) reduces the value of $T_2$ to values far below $T_1$. For tissue the value of $T_2$ is about ten times smaller than $T_1$. Without explaining the mathematics of the BPP theory (for which one can consult [1]), this qualitative description gives a sufficient understanding of relaxation in homogeneous matter.

### 6.3.2 Relaxation Effects in Tissue

#### 6.3.2.1 Fast Exchange

From the point of view of magnetic resonance imaging, tissue can be considered as consisting of 60% to 80% water in which macro molecules are suspended. The water is either free or is bound to the surface of the macro molecules by ionic or valence bonds. These bonds can be single, so the water molecule can still rotate, or double in which case the water is irrotationally bound. Free water has a very low relaxation rate (long $T_1$), because of the very low correlation time ($10^{-12}$ s, see Sect. 6.3.1). Bound water can have spin exchange with the "lattice" of the macro molecules and therefore can have a much higher spin–lattice relaxation rate. One can distinguish three fractions of water:

1. the fraction $f_w$ of free water with relaxation rate $1/T_{1w}$,
2. the fraction $f_r$ of rotationally bound water with relaxation rate $1/T_{1r}$, and
3. the fraction $f_i$ of irrotationally bound water with relaxation rate $1/T_{1i}$.

There is fast exchange between these three fractions, so each water molecule is from time to time a member of the different fractions. In the fast-exchange model the overall relaxation rate of tissue depends on the fractions of water mentioned above:

$$T_1^{-1} = f_w T_{1w}^{-1} + f_r T_{1r}^{-1} + f_i T_{1i}^{-1}, \tag{6.4}$$

where $T_{1w} \gg T_{1r} > T_{1i}$. From this model it is directly clear that the more free water ($f_r$ and $f_i$ become very small) in tissue the smaller the relaxation rate ($1/T_1$) or the larger $T_1$, because $T_{1w}$ is large. CSF, for example, has a large $T_1$ value. The specific $T_1$ values of a certain tissue depend both on the water content and on the surface properties of the macro molecules, which determine the number of water molecules they can bind. From (6.4) it follows that even small fractions $f_r$ and $f_i$ can have a large influence on the relaxation time $T_1$ of the tissue, because of the large relaxation rates of these fractions.

#### 6.3.2.2. Compartments and Slow Exchange

The mechanism of exchange of spins described in the previous paragraph is an exchange based on molecular motion on a microscopic scale, mainly by diffusion of free water. The condition of fast exchange is fulfilled when the exchange rate is fast compared with $T_1$ or $T_2$ relaxation. The radius of the fast-exchange region is of the order of the diffusion length $\lambda = \sqrt{2Dt}$ and is a few microns. On the scale of an entire voxel, the exchange is absent, so that tissue inhomogeneities can easily exist that are large in scale compared with the fast-exchange region and yet small compared to the size of the voxel. In such cases it is not always possible to assign a single $T_1$ or $T_2$ to the voxel. It is helpful in these cases to think of a subdivision of the voxel into compartments. Each compartment is homogeneous with respect to $T_1$ or $T_2$ and is equivalent to a single region of fast exchange (although it can be much larger). Between the compartments the conditions for fast exchange do not exist.

When such a voxel is subject to an imaging sequence, per compartment a different steady state will develop and the signal will be a complex function of the relaxation times of each compartment and of the exchange rate between the compartments. This means that (6.4) is no longer applicable. Instead, each component should be described with a continuity equation completed with terms for the exchange with other compartments, similar to (6.6) and (6.7) of the next section. These equations describe the exchange between two compartments: the spins bound to macromolecules and the spins of free water. It is shown that, in this case, two values of the $T_1$ relaxation time exist.

The most important compartments in tissue are (a) the microvascular network and (b) the interstitial space with parenchymal cells. Moreover, under normal conditions the relaxation rates in these compartments are sufficiently equal to allow the description per voxel with a single $T_1$ and $T_2$. A trivial exception, of course, is the partial volume effect across the border of two tissues.

An important case of multicompartmental behaviour of the voxel response is when the concentration of a contrast agent in the microvascular compartment differs strongly from that in the interstitium. This situation exists in the brain when the blood–brain barrier is intact; moreover, it exists in most other tissue early after bolus injection (first pass). In this latter situation, the observed signal enhancement is not only the result of the changing $T_1$ of the vascular compartment but also of the (slow) exchange of spins with the vascular compartment and the interstitium – see Sect. 6.3.2.1 and [3].

### 6.3.3 Magnetization Transfer

The insight gained in the previous section can be used to appreciate a physical method to influence the value of the equilibrium magnetization and also of the spin–lattice relaxation time $T_1$ of a tissue [4]. Under normal conditions the fraction of free water, which has a certain longitudinal magnetization, can exchange its spin with the fraction of water bound to the macro molecules (we only consider one single type of bound state), where part of this longitudinal magnetization is transferred to the molecule and dissipated in some state of thermal agitation. However, the remaining longitudinal magnetization returns in the free-water fraction and the $T_1$ value of the tissue depends on this equilibrium. Note that the spins of bound protons cannot be measured because of their fast dephasing (short $T_2$) due to "static" dipole–dipole interaction as discussed in Sect. 6.2.1.

The bound-water fraction, however, has a large range of resonant frequencies $\delta\omega$, again due to short $T_2$:

$$\exp(-t/T_2) \circ\!\!\overset{\mathfrak{F}}{\longrightarrow}\!\!\circ \frac{T_2}{1+\mathrm{j}\delta\omega T_2}. \tag{6.5}$$

From this equation it follows that the resonant line width is given by $\delta\omega T_2 = 1$, showing that $\omega$ is large when $T_2$ is small. This can be used to saturate the bound water protons without saturating the free water by merely choosing an excitation frequency outside the linewidth of the free water but within the much wider bandwidth of the bound protons (bound pool saturation). As we know, transverse magnetization of the bound protons is quickly destroyed by static dipole-dipole interaction. So now when water molecules are for some time bound to a macro molecule and are excited by a signal with a frequency somewhere within their wide linewidth, but outside the narrow free-water peak, their longitudinal magnetization is transferred into transverse magnetization, which is dephased immediately. When these water molecules return to the free-water pool, they never carry any longitudinal magnetization. This means that the longitudinal magnetization of the free-water pool is smaller than in the regular case without excitation of the bound pool. Application of RF power to excite the bound water will change the equilibrium situation in free water. These effects will be illustrated with a simplified model [4].

We consider a tissue with an overall spin–lattice relaxation time $T_1$ and an equilibrium magnetization $M_0$. Some of the water atoms are bound to the surface of macro molecules. The relaxation of the longitudinal magnetization of the free-water pool ($M_{\mathrm{LF}}$) is determined by its intrinsic relaxation properties, characterized by a relaxation time $T_{1\mathrm{F}}$, and the difference between what is lost to the bound pool with magnetization $M_{\mathrm{LB}}$, characterized by the rate constant $1/\tau_{\mathrm{F}}$, and what is regained from the bound pool (rate $1/\tau_{\mathrm{B}}$):

$$\frac{\mathrm{d}M_{\mathrm{LF}}}{\mathrm{d}t} = \frac{M_{\mathrm{F0}} - M_{\mathrm{LF}}}{T_{1\mathrm{F}}} - \frac{M_{\mathrm{LF}}}{\tau_{\mathrm{F}}} + \frac{M_{\mathrm{LB}}}{\tau_{\mathrm{B}}}. \tag{6.6}$$

For the bound pool a similar relation exists:

$$\frac{\mathrm{d}M_{\mathrm{LB}}}{\mathrm{d}t} = \frac{M_{\mathrm{B0}} - M_{\mathrm{LB}}}{T_{1\mathrm{B}}} - \frac{M_{\mathrm{LB}}}{\tau_{\mathrm{B}}} + \frac{M_{\mathrm{LF}}}{\tau_{\mathrm{F}}}. \tag{6.7}$$

The longitudinal relaxation time of the tissue under consideration (containing both the bound and free pools of water), the "measured" $T_1$ of the tissue, can be found by solving both equations simultaneously to be

$$2T_1^{-1} = -\mathfrak{T}^{-1} \pm \left[\mathfrak{T}^{-2} + 4\left((\tau_F\tau_B)^{-1} - (T_{1F}^{-1} + \tau_F^{-1})(T_{1B}^{-1} + \tau_B^{-1})\right)\right]^{1/2}, \quad (6.8)$$

where $\mathfrak{T}^{-1} = 1/T_{1F} + 1/\tau_F + 1/T_{1B} + 1/\tau_B$. We find two relaxation constants, one small and one large, and it is the combination of the effect of both relaxation constants we name "the" relaxation of the magnetization of the tissue. This equation will not be further pursued, but it illustrates that in a tissue containing a bound pool of protons a simple exponential decay cannot in general be expected. Usually the short $T_1$ is too short to be observed and the observable relaxation in regular tissue is a simple mono-exponential decay with time constant $T_1$. The equilibrium magnetization of the tissue under consideration can be found by taking $\mathrm{d}M/\mathrm{d}t$ equal to zero in (6.6) and (6.7), from which the equilibrium magnetization $M_0$ of the tissue can be expressed in the relaxation rates and transfer rates. Note that only the equilibrium value of $M_{LF}$ is of interest since excitation in the bound pool cannot be detected.

When now the bound pool is completely saturated this means that there can be no longitudinal relaxation in the bound pool, so $M_{LB} = 0$. Then (6.6) reduces to

$$\frac{\mathrm{d}M_{LF}}{\mathrm{d}t} = \frac{M_{F0} - M_{LF}}{T_{1F}} - \frac{M_{LF}}{\tau_F}, \quad (6.9)$$

which yields an equilibrium magnetization $M_{LF}$ equal to:

$$M_{LF} = M_{F0}\frac{\tau_F}{\tau_F + T_{1F}}, \quad (6.10)$$

and a relaxation time $T_{1sat}$:

$$T_{1sat}^{-1} = T_{1F}^{-1} + \tau_F^{-1}. \quad (6.11)$$

$T_{1sat}$ is the value of the (single) longitudinal relaxation time, observed under conditions of complete bound pool saturation. Since $M_{LF} < M_{F0}$ and $T_{1sat} < T_1$, it follows that the longitudinal equilibrium magnetization is reduced with respect to that of the free pool and that the relaxation rate can be enhanced by magnetization transfer and therefore the contrast can be influenced. The longitudinal equilibrium magnetization is also reduced with respect to the equilibrium under normal conditions. The consequences of these effects in multi-slice imaging, where many excitations may cause MT effects, are studied in image set VI-4.

The spin–lattice relaxation process under conditions of complete bound pool saturation is described by (6.6) with $M_{LB}$ equal to zero. This equation can be solved in the same way as (2.8) with the boundary condition $M_{LF}$ $(t = 0) = M_0$ (no saturation of the bound pool before $t = 0$). Under influence of saturation of the bound pool from $t = 0$ onwards one finds

$$M_{LF} = M_{F0}\frac{\tau_F}{\tau_F + T_{1F}} + \left\{M_0 - M_{F0}\frac{\tau_F}{\tau_F + T_{1F}}\right\}\exp(-t/T_{1sat}), \quad (6.12)$$

which proves that a new equilibrium is established at a lower level of magnetization and with a time constant $T_{1sat}$ [5]. The magnitude of $\tau_F$ is of the same order as the magnitude as $T_{1F}$ and both are comparable with the normal tissue value of $T_1$. As a result $T_{1sat}$ is somewhat lower than $T_1$ and $M_{sat}$ is lower than $M_0$. Examples of the resulting contrast changes in regular scans with MTC are shown in Table 6.1. It is shown that the effect of MTC in blood is much less pronounced than in tissue, because of the larger free-water pool.

**Table 6.1.** Typical values of attenuation in tissues using MT imaging

| *Tissue* | *Attenuation by MT* |
|---|---|
| Adipose tissue, bone marrow, CSF, bile and *in vivo* blood | $< 5\%$ |
| skin | 80% |
| hyaline cartilage | 70–75% |
| skeletal muscle | 60–80% |
| cardiac muscle | 50–70% |
| white brain matter | 42–69% |
| grey brain matter | 39–52% |
| liver | 35–40% |
| kidney | 25–35% |
| in vitro blood | 15–25% |

Bound pool saturation (causing “Magnetization Transfer”) can be incorporated in normal sequences to change the contrast or to enhance contrast differences. During an imaging sequence the bound pool must be saturated by RF pulses, which must be repeated at a rate that is fast compared with $T_{1b}$. The value of $T_{1b}$ is not well known, but can be taken to be not less than 30 ms. With ideal bound saturation pulses, saturation can be reached nearly completely, when interleaved with a normal sequence at regular intervals, for example once per excitation pulse.

There are two types of bound pool saturation pulses (somewhat erroneously named “Magnetization Transfer Pulses”). The first type employs the differences in bandwidth of the free-water protons and the bound-water and macromolecular protons, as described earlier – see (6.5). Narrow-bandwidth shaped RF pulses (e.g. sinc-Gauss pulses with bandwidth $\Delta B$) are applied at a frequency offset $\Delta f$ away from the resonance frequency ($\Delta f > \gamma \Delta B/2$). The resulting saturation of the bound pool improves with RF flip angle and with decreasing $\Delta f$. Both, however, are limited: the power is limited by safety considerations, and the minimum offset is limited by possible excitation of the free water pool [6]. A practical compromise is a 700° flip angle pulse at a frequency offset of 1.5 kHz. Note that, in multi-slice imaging, also Magnetization Transfer can appear for a particular slice when adjacent slices are excited. Although the flip angles applied are always less than 180°, Magnetization Transfer contrast can be visible in the images, especially in TSE scans.

The second type of bound pool saturation pulses employs the difference in $T_2$ relaxation time between bound and free protons. These pulses consist of on-resonance composite pulses with zero overall flip angle. The most commonly used composite pulse is a $90^\circ_x$, $90^\circ_{-x}$, $90^\circ_{-x}$, $90^\circ_x$ pulse with an overall length of about 2 ms. This composite pulse will have almost no effect on the free pool. However, the spins of the bound pool, with their very short $T_2$ relaxation time of less than 0.1 ms, will not be refocussed and therefore they will lose their longitudinal magnetization. The bandwidth of the composite (binomial) pulses can be influenced by introducing more components [7]. In practice one must be careful not to excite fat. The power dissipation of these pulses is smaller than that of off-resonance pulses, and so it is easier to comply with the RF safety regulations.

Applications of MTC (see [8] for a survey) are found in angiographic imaging [9], where use is made of the fact that blood remains unaffected by MT, while other tissue produces less signal, improving the vascular contrast (see Table 6.1). Other applications are improving the contrast of Gadolinium-enhanced imaging of brain tumours [10] and Multiple Sclerosis [11]. In the spine, MTC can be used to improve the contrast between spinal fluids and the nerve roots [12] – see image set VI-8. Finally, MTC has assisted in the important area of fundamental research of tissue characterisation [13].

### 6.3.4 Contrast Agents

In MRI, contrast agents are applied to alter the properties of the tissue, in this case to alter the relaxation times $T_1$ and $T_2$. This is different from the working of contrast agents in X-ray radiography, where the contrast agent changes the absorption of the X-rays. We have shown in the previous sections that the relaxation mechanisms in tissue depend on the interaction of the water with macro molecules via bound states. Varying magnetic fields caused by thermal motion induce exchange of spins with the protons of the macro molecule. These varying fields are enhanced with contrast agents, so the exchange mechanism is accelerated ($T_1$ effect).

Also the value of $T_2^*$ depends on the variation of the local magnetic field (see Sect. 3.3.2.1) and contrast agents which vary the susceptibility will therefore influence the contrast in an image. For this application, superparamagentic and ferromagnetic contrast agents are of interest. A thorough treatment of contrast agents is beyond the scope of this textbook, for further study we refer the reader to the abundant literature on this topic [14].

We concentrate on discussing paramagnetic contrast agents. These have a widespread clinical use. Contrast enhanced MR images based on the use of these agents can give specific information on the carcinogenic nature of solid tumors, they can show the extent of inflammation, they more recently are used to enhance the contrast in MR angiography (Sect. 7.5.2.2) and they can be used to find functional information (Sect. 7.6.1).

Paramagnetic atoms have a number of unpaired electrons and therefore electron spin, for example $Fe^{3+}$, $Mn^{2+}$, and $Gd^{3+}$. The magnetic field of an electron spin is 657 times stronger than the field of a proton spin, so the interacting field is much stronger. $Gd^{3+}$ even has seven unpaired spins in its valence shell. There are more transition metals and lanthanide metals with unpaired spins, but to be effective as a relaxation agent the electron spin-relaxation time must match to the Larmor frequency, which is met better for the three mentioned examples (about $10^{-8}$–$10^{-10}$ s, all others have $10^{-11}$–$10^{-12}$).

Unfortunately all the metal ions mentioned are toxic and cannot be used as such. Therefore they are bound to chelates at the cost of the minimum distance between the water molecules and the dipole of the unpaired electron spins. Also, part of the unpaired spins of the metal ion is now paired with the chelate molecule. $Gd^{3+}$ retains a number of unpaired spins and is, for example, bound to DTPA (diethylene triamine pentaacetic acid) which is highly stable. In Gd-DPTA the $Gd^{3+}$ ions are still in close contact to the water molecules of the tissue.

The paramagnetic contrast agents reduce both $T_1$ and $T_2$, depending on the concentration. From the theory it follows that the influence of the relaxivity $R$ of a contrast agent with concentration $C$ on the $T_1$ and $T_2$ values of tissue can be written as

$$T_1^{-1}(\text{observed}) = T_1^{-1} + R_1 C, \tag{6.13}$$

$$T_2^{-1}(\text{observed}) = T_2^{-1} + R_2 C, \tag{6.14}$$

where $R_1$ and $R_2$ are constants. For an agent like Gd-DPTA at 1.5 T the values of both relaxivities are roughly $4.5\,\text{s}^{-1}$ $(\text{mmol/kg})^{-1}$. Since $T_1^{-1}$ is very small, a low concentration of the contrast agent influences the spin–spin relaxation appreciably. At large concentrations the $T_2$ relaxation is also influenced.

The influence of the contrast agents on images must be judged on the basis of the contrast equations for the different sequences as mentioned in Sect. 6.2, because only then does it become clear whether a change in $T_1$ and or $T_2$ really has an effect. When the contrast agent resides in the vascular space only, the disturbance of the homogeneity of the local field is large and the reduction of $T_2^*$ can be the dominating effect. This is, for example, the case during the first pass of a Gd-DPTA bolus of the brain. The redistribution to the interstitial space has a time constant of about 5 min. After that the contrast medium is cleared out of the body through the kindneys with a time constant of 100 min. The best time between the administration of the contrast medium and its observation depends on our interest. Typically the maximum interstitial concentration is reached between 1 and 10 minutes. The study of uptake phenomena requires dynamic imaging, based on fast multi-slice imaging with a frame rate of 1–2 per minutes per slice. First-pass studies (for example, to study perfusion in the myocard) require still faster imaging. An example of a dynamic study of contrast-agency uptake is shown in image set VI-3.

For a discussion on "susceptibility agents" (ferromagnetic and superparamagnetic contrast agents) influencing $T_2^*$, we refer the reader to the literature [15]. Also the influence of these contrast agents depends on the sequences used and can be studied using the contrast equations mentioned in Sect. 6.2.

## 6.4 Signal-to-Noise Ratio (SNR)

The trend in MR imaging is towards fast scans and high resolution. During the efforts to achieve this we meet a fundamental restriction, the signal-to-noise ratio. The signal in a pixel is due to the total magnetic moment of the spins in a voxel. It is proportional to the volume of the voxel considered and is therefore small for small voxels (high resolution). The signal also depends on the history of the magnetization in the voxel: in $T_2$-weighted conventional scans (spin echo, SE, and field echo, FE, with long TR) excitation starts each time from close to the equilibrium situation. However, in short-TR, fast scans (FFE), where the magnetization cannot relax back to equilibrium between two excitation pulses (dynamic equilibrium), the total magnetization along the $z$ axis available for excitation (and thus the resulting signal) is reduced (see Sect. 6.2 and Chap. 4).

The random noise in an MR system is caused by ohmic losses in the receiving circuit and by the noise figure of the pre-amplifier. The loss in the receiving circuit has two components: first, of course, the ohmic losses in the RF coil itself; second, the eddy-current losses in the patient, which are inductively coupled to the RF coil [16]. In a well-designed high-field system the latter component must be dominant. Furthermore, the connecting circuit between the coil and the pre-amplifier must have low losses, which is generally achieved by integrating the pre-amplifier into the coil. Finally, the pre-amplifier itself must have a low noise figure.

In this section, all effects relevant to the signal-to-noise ratio will be described and included in a general equation for this ratio. The equation is complicated and, therefore, it will be deduced in steps. In Sect. 6.4.1 the general expression for the signal-to-noise ratio in a single acquisition is derived, some parameters in this expression are, however, not measurable quantities, so the expression is not yet of practical use. Before solving this problem we shall first calculate, in Sect. 6.4.2, the resistance induced in the receiving circuit by the eddy currents in the patient (patient loading) for a simple geometry. This resistance will be shown to increase quadratically with frequency, which explains the essential difference between low-field and high-field MR systems (see Sect. 6.4.3).

In Sect. 6.5 the signal-to-noise ratio, as deduced in Sect. 6.4, will be expressed in measurable quantities characterizing the scan properties (number of measurements, number of averages, etc; see Sect. 6.5.1), the resonant input circuit (quality factor $Q$, and effective volume, $V_{\mathrm{eff}}$), and noise figure of the receiver (Sect. 6.5.2), and the influence of relaxation in the scans considered (Sect. 6.5.3).

In Sect. 6.6 the full equation for the signal-to-noise ratio will finally be discussed and applied to some practical cases, for example the signal-to-noise ratio (referred to as SNR) as a function of the main magnetic field strength.

In Sect. 6.7 the influence of non-homogeneous sampling of the $\vec{k}$ plane, for example in half matrix scans, when not all profiles are acquired, or in spiral imaging will be studied.

### 6.4.1 Fundamental Expression for the SNR

The RF coil, including the circuit that tunes and matches it to the preamplifier, can in the neighborhood of resonance always be described as a series resonant LCR circuit. At resonance its impedance is $R$.

Assume an RF coil carrying a current of $I$. The field of this coil at a certain position, $\vec{r}$, is $B_{\mathrm{RF}}(\vec{r})$. The coil sensitivity at that point is then $B_{\mathrm{RF}}(\vec{r})/I = \beta_1(\vec{r})$ (actually $\beta_1$ should be a vector; we shall, however, consider mainly homogeneous field regions with constant $\beta_1$ and are therefore interested in the absolute values only). Conversely, using the reciprocity theorem for electromagnetic fields, the voltage (denoted $S$ to express that this is the signal) induced by a transverse magnetization $M_{\mathrm{T}}\,(\vec{R})$ in that coil can be written as

$$S(t) = \frac{\delta}{\delta t}\left[M_{\mathrm{T}}\left(\vec{r},t\right)\right]\beta_1\left(\vec{r}\right)\mathrm{d}V\left(\vec{r}\right), \tag{6.15}$$

where $\mathrm{d}V(\vec{r})$ is the volume of the voxel considered. In a stationary situation the time dependence is periodic, so the amplitude of the signal is

$$S = \omega_0|M_{\mathrm{T}}(\vec{r})|\,|\beta_1(\vec{r})|\mathrm{d}V(\vec{r}), \tag{6.16}$$

where $\omega_0$ is the local Larmor frequency, expressed as

$$\omega_0 = \gamma B_0 \tag{6.17}$$

(and for practical purposes the gradient field and the RF field can be neglected here). Note that we actually have to take the amplitude of the signal and not the effective value of the amplitude.

The noise voltage induced in the receiving coil is due to the ohmic losses of the coil itself and the losses due to patient loading of the coil. If we think of the coil as a series LCR circuit, the resistance $R$ will be the sum of the coil resistance, $R_{\mathrm{c}}$ and the resistance induced by the patient conduction losses, $R_{\mathrm{p}}$. So $R = R_{\mathrm{c}} + R_{\mathrm{p}}$. The noise voltage over the resonant circuit, $V_N$, is obtained from the well known Nyquist equation:

$$V_N = (4\kappa T(R_{\mathrm{c}} + R_{\mathrm{p}})\delta f)^{1/2}, \tag{6.18}$$

where $\kappa$ is Boltzmann's constant, $T$ is the absolute temperature of the resistive object and $\delta f$ is the bandwidth of the receiver. The signal-to-noise ratio

SNR of a voxel at position $\vec{r}$ is now easily found by dividing (6.16) by (6.18), which yields

$$\mathrm{SNR_s} = \frac{\omega_0 |M_\mathrm{T}(\vec{r})|\,|\beta_1(\vec{r})| \mathrm{d}V(\vec{r})}{(4\kappa T(R_\mathrm{c} + R_\mathrm{p})\delta f)^{1/2}}. \tag{6.19}$$

This is the fundamental expression for $\mathrm{SNR_s}$ of a single sample (denoted by the index s). The absolute bars are used for modulus images. However, in its present form (6.19) is not useful, since the values of $\beta_1$ (the coil sensitivity), $R_\mathrm{c}$ and $R_\mathrm{p}$ must first be related to measurable quantities. Furthermore, only one measurement of an isolated voxel is considered. In an imaging sequence, the object is measured many times. After a Fourier transform, the SNR of a voxel can be found from (6.19) after accounting for the number of samples of the object taken during the entire scan.

It is always useful to look at the dimensions of the quantities, which occur in an equation, especially when electromagnetic quantities are considered. We shall express all quantities in the fundamental SI units of meter [m] for length, kilogram [kg] for mass, second [s] for time, ampere [A] for electric current, and degrees Kelvin [K] for absolute temperature.

- The magnetization $M$ has the same dimension as the magnetic field strength $H$ and is expressed in [A/m]. The magnetic susceptibility, defined by $M = \chi_\mathrm{p} H$, is therefore dimensionless.
- $\delta f$ and $\omega_0$ in [$\mathrm{s}^{-1}$],
- $\beta_1(\vec{r})$ is expressed in T/Amp [$\mathrm{kg/s^2A}$], for a good coil it is constant over the imaging region (body coil, head coil, knee coil, etc),
- $\kappa T = 4 \times 10^{-21}$ Joule [$\mathrm{m^2kg/s^2}$] at room temperature (293 K).
- $R$ is expressed in $\Omega$ [$\mathrm{m^2kg/s^3A^2}$].

With these dimensions it is easy to show that SNR is dimensionless, as it should be. This exercise is, as always, a good check on the calculation.

### 6.4.2 Patient Loading of the Receiving Circuit

We shall start with finding an expression for the patient loading using a simplified geometry. In Fig. 6.5 a region with constant $B_\mathrm{RF}$ (the RF magnetic field) is assumed and a cylindrical conductive region serves as a simple model for eddy currents in the patient.

The voltage induced in a cylindrical path (d$r$) is proportional to the time rate of change of the enclosed flux, $\Phi$:

$$U = \frac{\delta\Phi}{\delta t} = -\mathrm{j}\omega_0 \pi r^2 B_\mathrm{RF}. \tag{6.20}$$

In a conducting medium this voltage gives rise to a current (eddy current). If we assign a conductivity $\sigma$ to the material, the conductance of the cylindrical ring drawn in Fig. 6.6 is

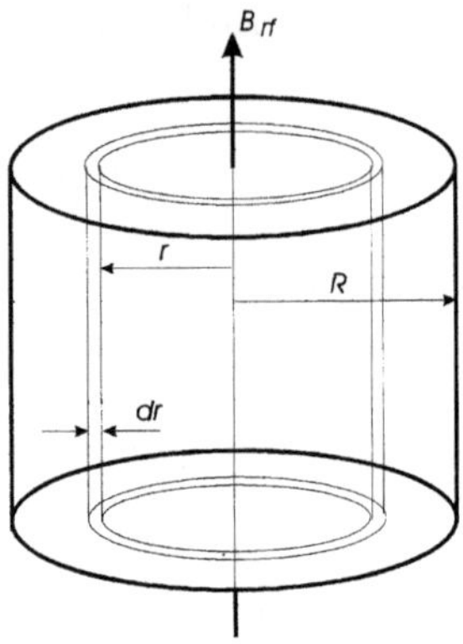


**Fig. 6.5.** Model for patient loading of the input circuit

$$\mathrm{d}G(r) = \frac{h\mathrm{d}r}{2\pi r}\sigma. \tag{6.21}$$

The power $\mathrm{d}P$ dissipated in the material is given by

$$\mathrm{d}P(r) = 1/2U^2(\mathrm{r})\mathrm{d}G(r). \tag{6.22}$$

The total power dissipated in the material is easily found by integrating over all values of $r$ between 0 and $R$:

$$P = \frac{1}{16}\pi R^4 h\omega_0^2\sigma B_{\mathrm{RF}} = \frac{1}{16\pi h}(\mathrm{Volume})^2\sigma\omega_0^2 B_{\mathrm{RF}}^2. \tag{6.23}$$

It is left to the reader to show that if we considered a sphere with radius $R$, (6.23) would read

$$P = \frac{1}{15}\pi R^5\sigma\omega_0^2 B_{\mathrm{RF}}^2 = \frac{1}{80R}(\mathrm{Volume})^2\sigma\omega_0^2 B_{\mathrm{RF}}^2. \tag{6.24}$$

Notify that we consider a constant RF field in the rotating frame of reference, which is a circular polarized RF field in the system reference frame. Such a field can be excited with a quadrature coil. If $B_{\mathrm{RF}}$ originates from a linear polarized field in the system reference frame there is also the counter rotating component (see Sect. 11.1.3) the dissipated power is twice as high.

Let us explore this equation a little. Suppose that there are, say, $N$ small spheres with radius $R_N$ in the same homogeneous field (the human body also has compartments), and with the same total volume. In that case (6.24) becomes

$$P = \frac{1}{15}\pi\left(\sum R_N^5\right)\sigma\omega_0^2 B_{\mathrm{RF}}^2. \tag{6.24a}$$

If ten small spheres are assumed to have equal radii and a total volume equal to the original volume, then it is easy to show that the power dissipated is about a fifth of the power dissipated in a homogeneous sphere of the same volume. The electrical conductivity of the patient is strongly different between tissues. So it is not unrealistic to describe the total RF eddy current pattern in the patient to be the result of a number of relatively mutually isolated compartments. The loading of the RF coil depends on the precise geometrical form of the compartments in the patient, determining the possible eddy-current paths. Furthermore there is anisotropy in the RF loading of the receiving coil, due to the elliptical cross section of the patient, an effect which is responsible for the fact that, when quadrature coils are used, the improvement over a linear coil is less than 3 dB [17].

Equations (6.22) and (6.23) show that, independent of the actual conductivity distribution in the patient, the dissipated power is proportional to $\omega_0^2 B_{\mathrm{RF}}^2$. For the purpose of describing noise, this power can be thought of as being dissipated in a series resistance $R_{\mathrm{p}}$ in the receiving circuit. This induced resistance $R_{\mathrm{p}}$ increases quadratically with increasing magnetic field strength. We come to this point later in more detail. First we consider the situation that $B_{\mathrm{RF}}$ is equal to $\beta_1$, which means that the current in the coil must be 1 A. In that case the power dissipated in the induced resistance is

$$P = R_{\mathrm{p}}.$$

Using this result in (6.24a) yields the resistance describing the patient loading of the receiving coil

$$R_{\mathrm{p}} = \frac{2}{15}\pi \left(\sum R_N^5\right) \sigma\omega_0^2\beta_1^2 = C\omega_0^2\beta_1^2, \tag{6.25}$$

where $\beta_1$ is the coil sensitivity at the position of the object, as defined before.

### 6.4.3 Low-Field and High-Field Systems

We shall now discuss an essential difference between high-field and low-field systems [18]. We therefore come back to (6.19) and realize that $R_{\mathrm{c}}$ is, due to the skin effect in the copper wires used in the coil, proportional to $\omega_0^{1/2}$. Introducing this and (6.25) into (6.19) we find

$$\mathrm{SNR_s} = \frac{\omega_0 |M_{\mathrm{T}}(\vec{r})|\,|\beta_1(\vec{r})|\mathrm{d}V(\vec{r})}{(4\kappa T(R_{\mathrm{c}} + R_{\mathrm{p}})\delta f)^{1/2}} = \frac{\omega_0 |M_{\mathrm{T}}(\vec{r})|\,|\beta_1(\vec{r})|\mathrm{d}V(\vec{r})}{(4\kappa T(A\omega_0^{1/2} + C\omega_0^2\beta_1^2)\delta f)^{1/2}}. \tag{6.26}$$

Since a homogeneous field is assumed, $\beta_1$ is taken to be constant. From this equation it is clear that at "low" frequencies the resistance $R_{\mathrm{c}}$ dominates. In that case loading by the patient can be neglected and SNR is proportional to $\beta_1$. So at low frequencies one should try to apply coils with a high sensitivity. For example, in an MR system with transverse field (permanent H-shaped magnet where the patient is located between the poles) one can apply a solenoid coil with many windings, so that $\beta_1$ is high (but be careful that the unfolded length of the coil does not surpass $\lambda/4$ of the wavelength used in the system). In practice, saddle coils with more than one winding have been applied in low-frequency systems of 0.1 T and below.

However, in the ideal system the loading of the RF coil by the patient will be more important than the coil's own resistance (the system should not add an appreciable part of the noise). In all high-field MR systems this is generally the case. As can be seen from (6.26) the SNR is then independent of the coil sensitivity $\beta_1$. From (6.26) the (vague) boundary between a "low" and a "high" field is given by $\omega_\mathrm{b}$, for which:

$$A\omega_\mathrm{b}^{1/2} = C\beta_1^2\omega_\mathrm{b}^2, \quad \text{or} \quad \omega_\mathrm{b} \propto \beta_1^{-4/3}, \tag{6.27}$$

and this boundary depends on the coil sensitivity. This latter form means a warning: the sensitivity of a coil should not be too low because one may enter the low-field regime, where other design rules must be followed. This is a real danger when, for example, the RF screen around the coil is too close, which causes $\beta_1$ to be low. This will be explained with a simple model. The sensitivity of a saddle coil with radius $r$ can be calculated to be (see [1.2], pp. 267–9)

$$\beta_1 \simeq \frac{7.5}{10^4\, r}\left[\frac{\mathrm{T}}{\mathrm{A}}\right]. \tag{6.28}$$

An RF screen with radius $a$ around the coil causes a "mirror" coil in the reverse direction around the screen with radius $2a - r$. The sensitivity of the total system is now

$$\beta_1 = \frac{7.5}{10^4}\left(\frac{1}{r} - \frac{1}{2a - r}\right). \tag{6.29}$$

This means that if the radius of the coil is 0.30 m and the RF screen has a radius of 0.34 m the sensitivity is lowered by a factor of five, which increases the boundary frequency between high-field and low-field systems considerably. For a well-designed MR system the patient losses should dominate the circuit losses, which can be influenced by proper design.

## 6.5 Practical Expression for the SNR

We shall now return to the expression for the signal-to-noise ratio as given by (6.19) in order to cast this equation into a more useful form. We use here the expressions deduced in Sect. 2.3.4 for the quality of the RF coil, $Q$, and the effective volume of the RF coil, $V_\mathrm{eff}$. The quality of a series resonant circuit is defined by:

$$Q = \frac{\omega L}{R_\mathrm{c} + R_\mathrm{p}}. \tag{6.30}$$

The value of the self-inductance of the coil, $L$, in this equation is easily found from the expressions for the total stored energy, $E_\mathrm{St}$, in a magnetic field generated by a current $I$ in the coil:

$$E_\mathrm{St} = \frac{1}{2} L\, I^2 = \frac{1}{2} \iiint \frac{B_\mathrm{RF}^2(\vec{r})}{\mu_0} \mathrm{d}\vec{r}^3 \equiv \frac{1}{2}\frac{B_\mathrm{RF}^2}{\mu_0} V_\mathrm{eff}. \tag{6.31}$$

Here $B_{\mathrm{RF}}$ stands for the value of the RF magnetic field in the region of interest of the coil (for example, within a body or head birdcage coil). The last form in (6.31) defines $V_{\mathrm{eff}}$. It means that we take the RF magnetic field to be constant over a volume $V_{\mathrm{eff}}$ so that the identity for $E_{\mathrm{St}}$ holds. $V_{\mathrm{eff}}$ is not the actual volume of, for example, a birdcage coil, but it can be assumed to be proportional. Dividing (6.31) by $I^2$ (and using the definition of the coil sensitivity: $\beta_1 = B_{\mathrm{RF}}/I$), we find

$$L = \frac{\beta_1^2}{\mu_0} V_{\mathrm{eff}}. \tag{6.32}$$

Using (6.30) and (6.32) we can eliminate $\beta_1$ and $R_{\mathrm{c}} + R_{\mathrm{p}}$ from (6.19), which yields

$$\mathrm{SNR_s} = |M_{\mathrm{T}}(\vec{r})| \mathrm{d}V(\vec{r}) \left( \frac{\omega_0 \mu_0 Q}{4\kappa T\, V_{\mathrm{eff}}\, \delta f} \right)^{1/2}. \tag{6.33}$$

This expression shows that the SNR of the signal from a voxel is not only proportional to the local transverse magnetization and the volume of the voxel but also to the quotient $Q/V_{\mathrm{eff}}$, describing the input coil cicuit. When the pixel volume and the effective volume are reduced proportionally it is clear that the SNR decreases with $(d\mathrm{V}(\vec{r}))^{1/2}$. Finally we see that the SNR is inversely proportional to the square root of the bandwidth of the input circuit. This bandwidth is therefore taken to be as small as possible but the sampling theorem requires that it should not be smaller than $1/t_{\mathrm{s}}$, the inverse of the sampling time (see Sect. 2.4.1.1). In further calculations we shall use

$$\delta f = 1/t_{\mathrm{s}},$$

since this yields the highest possible signal-to-noise ratio by using the highest possible sampling time.

### 6.5.1 Introduction of the Scanning Parameters

Until now we have the signal-to-noise ratio for a single sample of a voxel. During an imaging sequence a signal from this voxel is measured many times: for a 2D scan, $N_{\mathrm{m}}$ samples per profile and $N_{\mathrm{p}}$ profiles are measured. The signals from the different samples add in amplitude, because they are correlated, and the signal is proportional to $N_{\mathrm{m}}, N_{\mathrm{p}}$. However, the noise amplitude increases as well. For uncorrelated noise the noise powers add, so its amplitude rises with $(N_{\mathrm{m}}, N_{\mathrm{p}})^{1/2}$. Therefore the SNR for a complete 2D scan rises by a factor $(N_{\mathrm{m}}, N_{\mathrm{p}})^{1/2}$ above the SNR of a single voxel.

Sometimes the signal-to-noise ratio of a scan is marginal and the operator will decide to do the same measurement several times. Also, in this case the SNR improves by $(N_{\mathrm{a}})^{1/2}$, where $N_{\mathrm{a}}$ is the number of samples acquired (NSA) for each point in $k$-space.

Three-dimensional scans require a further increase of the number of profiles. The increase is equal to the number of partitions envisioned $N_{\mathrm{s}}$ in the third direction. The SNR improves with $(N_s)^{1/2}$. This is the reason why 3D measurements may usually be assumed to have a high SNR so that thinner slices can be used.

The use of quadrature coils improves the signal by a factor of 2, but the noise increases by $2^{1/2}$. One can see this by realizing that a quadrature coil actually consists of two coils generating RF magnetic fields, which are perpendicular and which are 90° shifted in phase. So for quadrature coils we introduce the factor $N_c^{1/2}$, $N_c$ being 1 for a linear coil and 2 for a quadrature coil. A word of caution: as has been explained earlier, depending on the inhomogeneity of the conductivity in the patient the actual improvement is usually less than $2^{1/2}$ (see Sect. 6.3.2).

Multi-channel coils [1.23] are constructed for the study of a large field of view without the need to cope with a single coil with large $V_{\text{eff}}$, which lowers the SNR. Since each coil "looks" at another part of the field of view, each gives a signal from a different region so their contributions do not add for a single pixel. However, the signals always come from coils with a small effective volume and this means that we must use this small effective volume in our expression for SNR. This is the advantage of using multi-channel coils [19].

In an approach that is a hybrid between a 3D scan and a multiple-slice scan several slices can be excited by a single excitation pulse [20]. By changing the phase of the slices with respect to each other, one can distinguish during the signal-processing phase the signals from the different slices. Compared with single-slice excitation this gives an improvement (within the same scan time) of $(N_s)^{1/2}$, where $N_s$ is again the number of slices (as in the case of 3D).

We conclude this section with the resulting equation for the signal-to-noise ratio for a voxel in a complete scan:

$$\text{SNR} = |M_{\text{T}}(\vec{r})|\text{d}V(\vec{r}) \left(\frac{\omega_0 \mu_0 \, Q t_{\text{s}}}{4\kappa T \, V_{\text{eff}}}\right)^{1/2} (N_{\text{m}} \, N_{\text{p}} \, N_{\text{s}} \, N_{\text{a}} \, N_{\text{c}})^{1/2}. \tag{6.34}$$

### 6.5.2 Influence of the Receiver on the SNR

Let us assume that the input circuit has a power damping of $\delta$ (= power in/power out, so $\delta > 1$) and that the noise figure of the receiver is $F_{\text{r}}$. We first consider a passive four-pole with power damping $\delta$.The general definition of the noise figure $F_{\text{i}}$ of the four-pole is (note that $S$ and $N$ are amplitudes):

$$(\text{SNR})^2_{\text{in}} = (\text{SNR})^2_{\text{out}} \, F_{\text{i}}. \tag{6.35}$$

Since the available noise power at the output terminal of the passive four-pole is equal to the available power at the input $N_{\text{out}} = N_{\text{in}}$. We therefore have

$$F_{\text{i}} = S_{\text{out}}/S_{\text{in}} = \delta.$$

This passive four-pole is followed by a receiver with a noise figure $F_r$. The noise figure of the total receiver (passive four-pole followed by pre-amplifier) can be found with the Friis equation:

$$F_t = F_i + \frac{F_r - 1}{g_1},$$

where in our case $F_i = \delta$ and $g_i = 1/\delta$. So for the total noise figure of the receiver plus input circuit we find

$$F_t = \delta F_r. \tag{6.36}$$

Introducing this into (6.35) and realizing that $(\mathrm{SNR})_{\mathrm{in}}$ is given by (6.34) we find that the SNR is deteriorated by the receiver plus input circuit by a factor $(\delta F_r)^{1/2}$. $F_r$ is mainly determined by the input transistor. The damping $\delta$ must be kept as low as possible. This is why the input stage of the pre-amplifier must be integrated into the coil. If we express, as is usually done, these factors $\delta$ and $F_r$ in decibels, the factor $(\delta F_r)^{1/2}$ multiplying the input signal-to-noise ratio becomes

$$10^{-(\delta+F_r)/20}, \tag{6.37}$$

and so the full equation now reads

$$\begin{aligned} \mathrm{SNR} &= |M_T(\vec{r})|\mathrm{d}V(\vec{r}) \left(\frac{\omega_0 \mu_0 Q t_s}{4\kappa T V_{\mathrm{eff}}}\right)^{1/2} (N_m N_p N_s N_a N_c)^{1/2} \\ &\quad \times 10^{-(\delta+F_r)/20}, \end{aligned} \tag{6.38}$$

which now contains all system parameter and scan parameters (the dependence on $B_0$ is hidden in $M_T$ and $\omega_0$). Note that the SNR is proportional to the voxel volume but inversely proportional to the square root of $V_{\mathrm{eff}}$.

### 6.5.3 Influence of Relaxation on the SNR

The influence of relaxation during the repetition and echo times is hidden in $M_T(\vec{r})$. Expressions for $M_T(\vec{r})$ are given in Sect. 6.2. $M_T(\vec{r})$ is always proportional to the equilibrium magnetization $M_0$, which is given by the relation

$$M_0(\vec{r}) = \chi_p(\vec{r})\mathrm{H}_0 = \chi_p(\vec{r})B_0/\mu_0, \tag{6.39}$$

where the magnetic susceptibility, $\chi_p(\vec{r})$, is a dimensionless quantity depending on the nuclear spins and is position dependent.

After a 90° pulse in a conventional SE or FE sequence with $\mathrm{TR} \gg T_1$, the complete magnetization $M_0$ is rotated in the transverse plane, so $|M_T(\vec{r})|$ is $M_0(\vec{r})$. The echo is formed TE seconds after the excitation and during this time $T_2$ decay (for SE) or $T_2^*$ decay (for FE) takes place, which means that the result must be multiplied by $\exp(-\mathrm{TE}/T_2)$ or by $\exp(-\mathrm{TE}/T_2^*)$. The resulting expression now becomes (we shall drop the $\vec{r}$ dependence during the rest of our treatment)

$$\mathrm{SNR} = \left(\frac{B_0^3 \gamma \chi_\mathrm{p}^2 Q t_\mathrm{s} \mathrm{d}V^2}{4\kappa T \mu_0 V_\mathrm{eff}}\right)^{1/2} (N_\mathrm{m}\, N_\mathrm{p}\, N_\mathrm{s}\, N_\mathrm{a}\, N_\mathrm{c})^{1/2} 10^{-(\delta+F_\mathrm{r})/20} \exp\left(-\frac{\mathrm{TE}}{T_2}\right), \tag{6.40}$$

where for spin echo only the $T_2$ must be taken into account whereas for field echo the magnetic field inhomogeneity and the varying susceptibility must also be accounted for by using $T_2^*$.

For imaging sequences with short $\mathrm{TR}(< T_1)$ the factor $F$ as given in (6.2) can be used. For an N-FFE sequence, for example, (6.3e) is used and the signal-to-noise ratio is given by

$$\mathrm{SNR} = \left(\frac{B_0^3 \gamma \chi_\mathrm{p}^2 Q t_\mathrm{s} \mathrm{d}V^2}{4\kappa T \mu_0 V_\mathrm{eff}}\right)^{1/2} (N_\mathrm{m}\, N_\mathrm{p}\, N_\mathrm{s}\, N_\mathrm{a}\, N_\mathrm{c})^{1/2} 10^{-(\delta+F_\mathrm{r})/20} F_{\text{N-FFE}}. \tag{6.41}$$

For the other sequences the complete equation for the signal-to-noise ratio is obtained in a similar way, using (6.2) and (6.3). Since $T_1$ and $T_2$ depend on the tissue imaged, one can only speak of the SNR for a certain tissue.

## 6.6 Application to Practical Situations

In this section the complete equation for the SNR will be given and applied to practical cases. Since there has been a long dispute about the frequency dependence of the SNR [18], it is interesting to study this dependence when using, as much as possible, the measured values of $Q$ and $V_\mathrm{eff}$ for (for example) the head coil at 1.5 T and 0.5 T and introduce some realistic guesses of the frequency dependence for this coil, manufactured with the optimal technology, in low-frequency regions [19]. Also, the measuring gradient can be lower in low-frequency regions (sampling time $t_\mathrm{s}$ may be elongated), where the absolute water–fat shift is lower and the absolute homogeneity of the main magnetic field is better. Finally we must also introduce the magnetic field dependence of $T_1$, as described in Sect. 6.3.

The full equation for the signal-to-noise ratio reads

$$\mathrm{SNR} = \left[\frac{B_0^3 \gamma \chi_\mathrm{p}^2 (t_\mathrm{s} \mathrm{d}V^2 N_\mathrm{m}\, N_\mathrm{p}\, N_\mathrm{s}\, N_\mathrm{a})}{4\kappa T \mu_0}\right]^{1/2} \times \left\{\sqrt{\frac{N_\mathrm{c} Q}{V_\mathrm{eff}}} 10^{-(\delta+F_\mathrm{r})/20}\right\}_\mathrm{c} F. \tag{6.43}$$

The factor $F$ is for the different imaging methods given by (6.3a)–(6.3g).

The terms in (6.43) are rearranged in such a way that under the first square root we find the magnetic field strength of the main field, some physical constants and {between brackets} the scan parameters. The brackets with index c contain the properties of the RF circuit. The last factor describes the influence of the relaxation properties of the tissue during the scan considered. With (6.43) the SNR (or the trend of the SNR) of most scan types maybe evaluated. In the following we shall give some examples of this evaluation.

As has been discussed earlier, (6.43) contains three terms that must be considered to be functions of the frequency. The first one is the $Q$ factor of the RF coil. It is well known that at higher frequencies it becomes increasingly difficult to obtain a high $Q$, since the RF losses increase (see (6.25)). We shall consider in our example a head coil. It has a measured effective volume of $0.04\,\mathrm{m}^3$. From measurements we know that at 1.5 T (64 MHz) the loaded $Q$ of a quadrature head coil (including patient loading) is around 40. At 0.5 T

Parameters used to obtain Fig. 6.6.

I. General parameters

$$\chi_\mathrm{p} = 4 \times 10^{-9};\ \mu_0 = 4\pi 10^{-7};\ \gamma = 2\pi 42.6 \times 10^6;\ \kappa T = 4 \times 10^{-21}$$

where $\kappa$ denotes Boltzmann's constant.

II. Scan parameters

$$t_\mathrm{s}(B_0) = 20\frac{10^{-6}}{B_0^{0.3}};\ \mathrm{d}V = 10^{-9}\,\mathrm{m}^3;\ N_\mathrm{m} = 256;\ N_\mathrm{p} = 256;\ N_\mathrm{s} = 1;$$

$$N_\mathrm{a} = 1.$$

$$\mathrm{TR} = 2000;\ 1000;\ 500;\ \mathrm{TE} = 12;\ \alpha = \frac{\pi}{2}.\quad (\mathrm{TR},\ \mathrm{TE\ in\ ms})$$

$$\text{Scan time} = \mathrm{TR}\ N_\mathrm{p} N_\mathrm{s} N_\mathrm{a} 10^{-3}\ 60^{-1} = 8.533\,\mathrm{min}$$

III. Properties of the RF circuit

$$Q(B_0) = 40\frac{1.5}{B_0};\ V = 0.04\,\mathrm{m}^3;\ \delta = 0.5;\ F_\mathrm{r} = 1;\ F_\mathrm{t} = 10^{-(\delta+F_\mathrm{r})/20}.$$

$$N_\mathrm{c} = 1$$

IV. Relaxation properties

$$T_1(B_0) = 850\,B_0^{0.4};\ T_2 = 50.\quad (T_1,\ T_2\ \text{in ms})$$

V. Signal-to-noise ratio

$$\mathrm{SNR}(B_0) = \left(\frac{B_0^3 \gamma \chi_\mathrm{p}^2 t_\mathrm{s}(B_0) \mathrm{d}V^2 N_\mathrm{m} N_\mathrm{p} N_\mathrm{s} N_\mathrm{a} N_\mathrm{c}}{4\mu_0 \kappa T}\right)^{1/2} \left(\frac{Q(B_0)}{V_\mathrm{eff}}\right)^{1/2} F_\mathrm{t}$$
$$\times \frac{[1-\exp(-\mathrm{TR}/T_1(B_0))]\sin\alpha}{1-\cos\alpha\exp(-\mathrm{TR}/T_1(B_0))} \exp\left(\frac{-\mathrm{TE}}{T_2}\right).$$

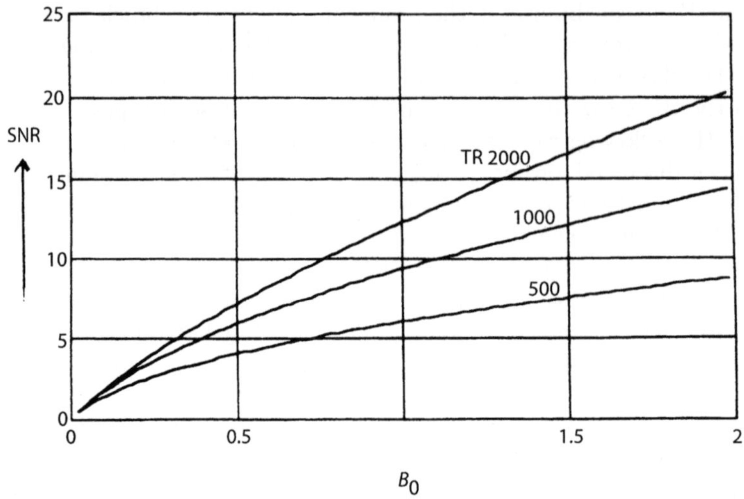


**Fig. 6.6.** Signal-to-noise ratio of grey matter as a function of the main magnetic field, $B_0$, expressed in Tesla, for a Spin-Echo sequence

(21 MHz) the quality $Q$ is 120 and at 0.15 T a loop array [19] has a $Q$ factor of 300. Therefore we write as a good approximation

$$Q(B_0) = 40\frac{1.5}{B_0}. \tag{6.44}$$

Note that for coils with other configurations (such as low-frequency coils) the frequency behaviour may be different.

The sampling frequency $t_s^{-1}$ is also a function of $B_0$. For a constant water–fat shift we can take $t_s(B_0) \propto 1/B_0$. In practice this relationship cannot be used to the full extent, since this elongates the acquisition time beyond the restrictions set by the sequence. So we choose

$$t_s(B_0) = \frac{20 \times 10^{-6}}{B_0^{0.3}} s, \tag{6.45}$$

so that the sampling time increases from 18 $\mu$s at 1.5 T to 35 $\mu$s at 0.15 T.

Finally for $T_1(B_0)$ we take (see [1.2], p. 23)

$$\mathrm{T}_1(B_0) = 850\, B_0^{0.4}\,\mathrm{ms},$$

which yields a value of 1000 at 1.5 T and of 400 at 0.1.5 T. For $T_2$ we take 50 ms. It will be clear that all choices made are rather arbitrary, but serve the illustration of our point.

We shall now use (6.43) to study the SNR for a few different cases. In all cases we take the pixel volume to be 1 mm$^3$ ($10^{-9}$ m$^3$); since the SNR is linearly dependent on the pixel volume it is easy to scale it to a desired value for diagnostic imaging. There is no practical requirement on the SNR for a diagnostic image, it very much depends on the object imaged. In general it

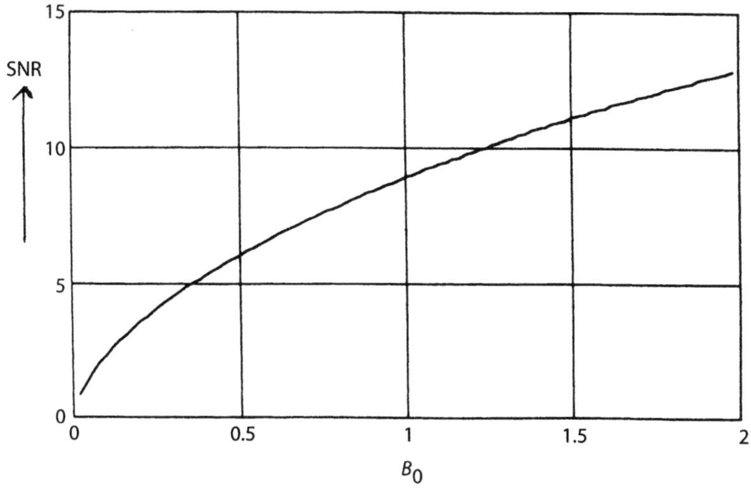


**Fig. 6.7.** The signal-to-noise ratio of grey matter as function of the main magnetic field, $B_0$ for 3D $T_1$-FFE. The parameters are as in Fig. 6.6, except for: TR $= 9$ ms, TE $= 6$ ms, $\alpha = \pi/12$, $N_s = 128$, $t_s(B_0) = 10 \times 10^{-6}/B_0^{0.3}$, the scan time is 5 min

is thought that a value between 5 and 10 is a acceptable; see image set VI-5. A sequence of images in cine may even have a lower minimum.

The first example is a 2D SE sequence. In this case $N_s = N_a = 1$. TR is taken to be 2000, 1000, and 500 msec and the excitation flip angle is 90°. The signal is received by a linear coil, so $N_c = 1$. All assumed parameters necessary for solving (6.43) are shown in the text box below Fig. 6.6 and the resulting graph representing SNR as a function of $B_0$ is shown in Fig. 6.6 for three TR values. It is shown in this figure that, under the assumptions made, the SNR as a function of $B_0$ is close to a linear function, especially for the highest TR value.

In Fig. 6.7 the SNR of a 3D $T_1$-FFE, measured during the steady state, is shown; TR $= 9$ and TE $= 6$ ms, $N_c = 2$, and $\alpha = \pi/12$. The sampling time had to be decreased so that the acquisition time for a single profile fits between the excitations. We have halved this time, which means that the input bandwidth of the receiver must be increased. As has been explained earlier, this lowers the SNR.

A 2D FFE scan is used for inflow measurements, which show the inflowing fresh blood as a very bright image (see Sect. 7.5.2.1). The SNR of the blood is much higher than that of the stationary tissue, because it is not saturated when it enters the slice to be imaged. This effect is taken into account by giving the blood a much larger TR than the stationary tissue. In Fig. 6.8 the signal-to-noise ratio for the stationary tissue and for the inflowing blood is shown. For good contrast the flip angle $\alpha$ was increased to $\pi/3$.

As has been shown in this section, from (6.43) it is straightforward to study the signal-to-noise ratio of applied imaging methods. For example

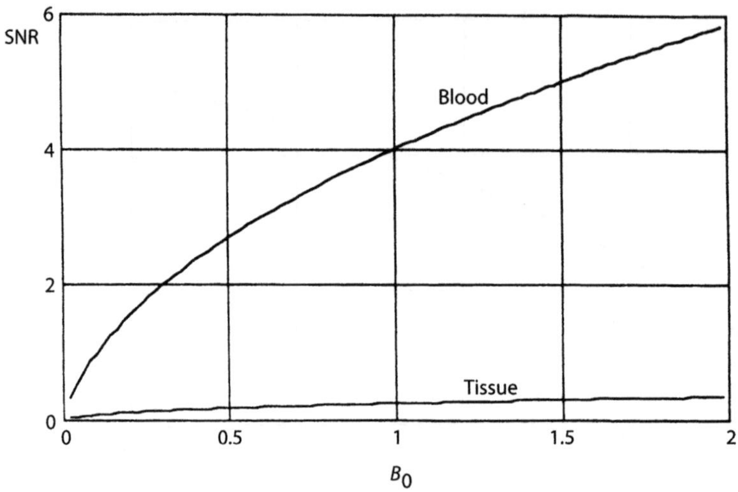


**Fig. 6.8.** The signal-to-noise ratio for inflowing blood (see Sect. 7.5.2.1) in multi-slice 2D $T_1$-TFE; $N_s = 1$, TR = 9 ms, TE = 6 ms, $T_2^* = 20$ ms, $t_s(B_0) = 10^{-5}/B^{0.3}$, $\alpha = \pi/3$. Other parameters as in Fig. 6.7

(6.43) can be integrated into the software of MRI systems to evaluate the influence of changing the parameters of the preset procedures.

## 6.7 SNR for Non-uniform Sampling of the $\vec{k}$ Plane

In many sampling methods not all Cartesian points in the $\vec{k}$ plane are acquired. This section describes the consequence in SNR for some methods where sampling is non-uniform.

### 6.7.1 One-Sided Partial Scans

One example of non-uniform sampling is a "half-matrix" acquisition, which is sometimes required so as to reduce the scan time. For example, suppose only 62.5% of the profiles are acquired, as shown in Fig. 6.9. In Turbo Spin Echo or Turbo Field Echo, a half-matrix can be used in conjunction with the profile order (see Sects. 3.2.1 and 5.5) to place the central profiles at the desired time within the echo train (TSE) or the shot (TFE). For instance, a TSE single-shot method with a "linear" profile in Fig. 6.9 could start with a profile at the upper edge of the partly sampled $\vec{k}$ plane, in which case the effective echo time would be short (in fact shorter than half the duration of the entire echo train).

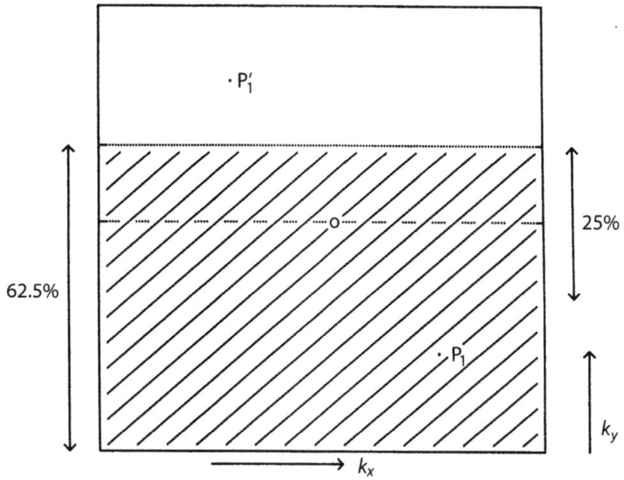


**Fig. 6.9.** One-sided partly sampled $\vec{k}$ plane (half matrix)

Another example is the acquisition of only part of each profile (a "half echo" acquisition – see Fig. 6.10). This can be used to obtain a very short echo time TE where, for example, before TE only 12.5% of the samples per profile are acquired. Note that in this technique the area under the compensating lobe of the read gradient is also small.

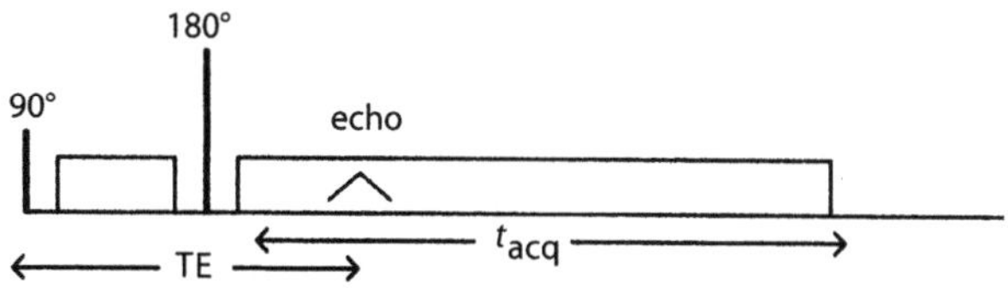


**Fig. 6.10.** "Half-echo" sequence

The consequence of either half-matrix or half echo acquisition is a lower signal-to-noise ratio when compared with a scan with full coverage of the $\vec{k}$ plane. The factor by which the signal-to-noise ratio is lowered depends of the fraction of the full $\vec{k}$ plane that is sampled, the "half-scan fraction" (HSF), as well as on the reconstruction technique that is used. To illustrate this, three reconstruction techniques are compared:

(a) *Immediate reconstruction.* The sudden amplitude change in the $\vec{k}$ plane will introduce heavy and unacceptable ringing in the image. The method is impractical and given only for comparison.
(b) *Ramping down* of the intensity of the scanned profiles near the unsampled region of the $\vec{k}$ plane, for instance as

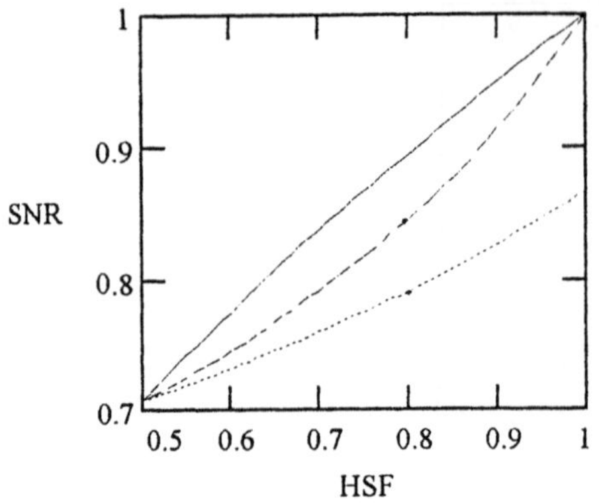


**Fig. 6.11.** Comparison of the SNR of some reconstruction strategies for half scan data (see text). Upper trace: case (a), middle trace: case (c) and lower trace: case (b)

$$
\begin{aligned}
s'(k_y) &= s(k_y)\, f(k_y), \text{ with} \\
&f(k_y) = 1 \qquad\qquad\qquad\qquad k_y/k_{y\,\mathrm{max}} < 1 - 2\mathrm{HSF} \\
&f(k_y) = \frac{(k_y/k_{y\,\mathrm{max}} + 1 - 2\mathrm{HSF})}{2(1-2\mathrm{HSF})} \quad 1-2\mathrm{HSF} < k_y/k_{y\,\mathrm{max}} < 2\mathrm{HSF}-1 \\
&f(k_y) = 0 \qquad\qquad\qquad\qquad k_y/k_{y\,\mathrm{max}} > 2\mathrm{HSF} - 1
\end{aligned}
$$

The ramping adequately suppresses the image ringing at the cost of some loss of signal to noise.

(c) *Missing samples* can be obtained from the samples at the other side of the center of the $\vec{k}$ plane (see in Fig. 6.9 the points $P_1$ and $P_1'$) since the signals in $P_1$ and $P_1'$ are complex conjugates of each other. When all echoes have the same size, ringing is suppressed effectively.

The relative value of SNR for these three examples (SNR of the full scan is 1) is compared in Fig. 6.11 for the case that all echoes have identical amplitudes as in conventional sequences. In technique (a) above, simply $\mathrm{SNR}_a = \sqrt{\mathrm{HSF}}$, as can be seen from (6.34) by replacing $N_p$ with $\mathrm{HSF}N_\mathrm{p}$. In technique (b), the signal amplitude is exactly halved and the noise power is multiplied by the factor $\sum_{k_y} f^2(k_y) = (2 - \mathrm{HSF})/3$, so that $\mathrm{SNR}_b = 1/[2\sqrt{(2-\mathrm{HSF})/3}]$. In technique (c), the signal amplitude does not change. The noise in the mirrored sample pairs is correlated, so that for these pairs their amplitudes have to be added. This multiplies the total noise power by the factor

$$(1+1)^2(1-\mathrm{HSF}) + (1)^2(2\mathrm{HSF}-1) = 3 - 2\,\mathrm{HSF},$$

and in this case $\mathrm{SNR}_\mathrm{c} = 1/\sqrt{3-2\mathrm{HSF}}$.

Comparison of the strategies shows that suppression of ringing goes at the cost of some SNR, while strategy (c) performs better than strategy (b).

So far we have assumed all echoes to be equal; such as can be expected in conventional SE. In TSE and TFE, where the echoes are not equal, strategy (c) does not remove all ringing (the echo amplitude varies over the $\vec{k}$ space), and one has to accept the use of strategy (b). It is also clear that, for strategy (b) in the case of unequal echoes, the point spread function is influenced by both the function $f(k_y)$ and the variation of the echoes.

### 6.7.2 Non-uniform Sampling with Non-linear Trajectories

Another imaging sequence in which the $\vec{k}$ plane is sample in an inhomogeneous way and not on the cartesian points is spiral imaging (see Sect. 3.6.2). Now we can follow the same reasoning as above. For the gridding process the sampled signals are multiplied with $D^{-1}(k_x, k_y)$ and so are the noise amplitudes. The total signal obtained is

$$S_{\mathrm{T}} = \sum_{k_x} \sum_{k_y} S(k_x, k_y) D^{-1}(k_x, k_y). \tag{6.46}$$

The overall noise amplitude $N$ is given by

$$N_{\mathrm{T}} = N_{\mathrm{s}} \left\{ \sum_{k_x} \sum_{k_y} \left\{ \frac{1}{D(k_x, k_y)} \right\}^2 \right\}^{1/2} . \tag{6.47}$$

The signal-to-noise ratio is now obtained from the quotient of (6.46) and (6.47).

To give a simple example of an inhomogeneously sampled $\vec{k}$ plane, assume that in the center part of the $\vec{k}$ plane 25% of the area is sampled with double density. After gridding, the total signal is $S\, N_{\mathrm{a}}\, N_{\mathrm{p}}$. The noise, however, becomes $0.875\ N_{\mathrm{a}}\, N_{\mathrm{p}}\, N_{\mathrm{s}}$. The signal-to-noise ratio increases by 12.5% due to the extra sample points in the center of the $\vec{k}$ plane. For spiral imaging, $D(k_x, k_y)$ is obtained from (6.31), which must be solved by numerical means.

### 6.7.3 Reduced Matrix Acquisition

In the interest of reducing reconstruction time, the reconstruction matrix that can be selected has a size that equals a power of 2; e.g. $512 \times 512$ or $256 \times 256$. It is not always convenient to select a number of $k_y$ values acquired that is equal to these reconstruction matrix sizes. A common strategy in that case is to sample all $|k_y|$ values below a reduced maximum $k'_{y,\mathrm{max}}$, lower than the corresponding $k_{y,\mathrm{max}}$ of the full scan. This type of scan is called a reduced matrix scan and obviously can be completed in a shorter scan time by a factor RMF, named the "reduced matrix factor", which equals $k'_{y,\mathrm{max}}/k_{y,\mathrm{max}}$. In the reconstruction strategy for reduced matrix scans, we usually give all samples for the remainder $k_y$ lines a zero value, called "zero padding". Without further measures, the consequence would be that the reconstructed image shows ringing ("Gibbs ringing"), but ringing is often suppressed by a low pass filter applied to the data before reconstruction at the expense of spatial resolution.

In reduced matrix images, the line-spread function has changed for lines perpendicular to the phase-encode gradient. The estimate of SNR in these images therefore depends on the size of the detail that is considered. The noise is always proportional to $\sqrt{\mathrm{RMF}}$. For a large detail, the signal value in the detail does not change with RMF, so that SNR increases as $\sqrt{\mathrm{RMF}^{-1}}$. For a small detail (smaller than one pixel), the signal decreases proportional to RMF and SNR decreases as $\sqrt{\mathrm{RMF}}$.

### 6.7.4 Other Partial-Scan Methods

In Cartesian data sets, the partial-scan methods are by no means restricted to omission of profiles with just the highest $k_y$ values, such as discussed in Sects. 6.7.1 and 6.7.3. An interesting alternative is the omission of profiles that are selected on the basis of prior knowledge of the object, for instance on the structure of edges. The reconstruction strategy then consists of adding into the missing profiles estimates of the sample values. [2.11], [2.23]. The estimates are free of noise, so that the performance of the resulting images cannot be judged on the basis of their SNRs only. Image Set II-14 gives some examples.

Another interesting option is the omission of $k_y$-profiles during acquisition with multi-element receive coils. In Sect. 2.4.1.1 it is shown that the field of view is inversely proportional to the distance of the sample points, $\delta k_x$, see (2.32):

$$\gamma G_x t_s = \delta k_x = \frac{1}{\mathrm{FOV}} . \tag{6.48}$$

This also holds for the $k_y$ direction. So when only half of the $k_y$ profiles are acquired, so that the distance between the profiles is doubled, the field of view is halved. This is shown in Fig. 6.12, where the image of a circular object is shown in the gray area. The upper and lower parts of the cylindrical object are aliased in the image. For instance, pixel $A$, $B$ contains the signal from object voxel $B$ superimposed on that of object voxel $A$.

When the measurements are acquired simultaneously with two receiving coils with sensitivities $\beta_{1,2}(x, y)$ that are different functions of position, using (6.16) we have for the pixel $A, B'$ as the signal $S_1(A, B')$ of the first coil:

$$S_1(A, B') = \omega_0 \mathrm{d}V(\beta_1(A) m_\mathrm{T}(A) + \beta_1(B) m_\mathrm{T}(B)) , \tag{6.49a}$$

and the second coil yields for this pixel a signal equal to

$$S_2(A, B') = \omega_0 \mathrm{d}V(\beta_2(A) m_\mathrm{T}(A) + \beta_2(B) m_\mathrm{T}(B)) . \tag{6.49b}$$

When $\beta_{1,2}(x, y)$ is known, the transverse magnetizations $m_\mathrm{T}(A)$ and $m_\mathrm{T}(B)$ of the voxels $A$ and $B$ can be determined from these two simultaneous measurements. With this knowledge the complete image can be "unfolded". The

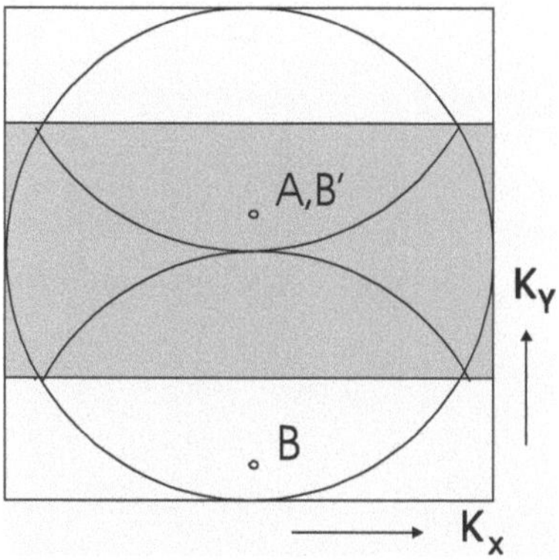


**Fig. 6.12.** Aliased image of a circular object, when the FOV is equal to the radius of the object. Voxel $B$ is imaged in the same pixel as voxel $A$

advantage of this method is that the scan time is reduced by a factor of two. The sensitivities of the coils must be acquired in a separate experiment. A requirement for the design of the coils is that these sensitivities are sufficiently different. When the coils are also fully decoupled, the noise of the signals per coil is not correlated, so the SNR is exactly $1/\sqrt{2}$ of its value in a scan in which for all profiles the coil signals are collected. This method is named SENSE (SENSitivity Encoding). It can be extended by using more coils, in which case higher reduction factors are possible, as is shown in image set II-13 and discussed more formally in the text with that image set.

In dynamic imaging, the acquisition of some profiles of later scans can be omitted and replaced by values that were acquired in earlier scans out of the same dynamic series. This technique is called "keyhole imaging" [2.12]. SNR will be constant, but systematic errors will occur in the late images because the substitute profiles are not exactly descriptive of the actual state of the subject, for instance after arrival of the contrast agent.

# Image Sets
# Chapter 6

## VI-1 Measurement of $T_1$

(1) Knowledge of the relaxation times of tissue is useful to obtain an estimate of the contrast in MR images. More importantly, it may be used clinically as a basis for quantitative diagnosis. The lack of published material on clinical use of relaxation times is probably related to the difficulty of obtaining such data in a sufficiently reproducible way in a clinically acceptable time. In this image set, an example of a fast and accurate approach for measurement of $T_1$ in a transverse slice through the head of a volunteer is given. $T_2$ values of the head of another volunteer are listed under image set II-2.
Our measurement of $T_1$ is based on a scan that mixes two sequences to observe the signals in two sufficiently different states of relaxation. The value of $T_1$ is calculated from the ratio of these signals (two-point method). In this type of calculation one has to assume that the $T_1$ relaxation is mono exponential.

Parameters: $B_0 = 1.5\,\text{T}$; FOV $= 180\,\text{mm}$; matrix $= 128 \times 128$; $d = 3\,\text{mm}$; TE $= 14\,\text{ms}$

| image no. | 1 | 2 | 3 |
|---|---|---|---|
| scan method | TSE | IR-TSE | - |
| TR (ms) | 1170 | 5000 | - |
| TI (ms) | - | 330 | - |
| Turbo factor | 8 | 8 | - |
| NSA | 2 | 2 | - |
| image type | M | M | $T_1$ |

The image types are indicated as M (modulus) or $T_1$ (calculated $T_1$). The scan method used was a "single-slice" method, in which the inversion pulse and the refocussing pulse were not slice-selective. Image 3 was calculated from the ratio of the pixel values in images 1 and 2.

(2) The choices of TR and TI in the scans is based on the mathematical optimisation of the error in $T_1$ for the range of TR from 500 to 1500 ms [21]. The choice of slice thickness and FOV is aimed at a voxel shape that is more nearly cubical than the voxel size used normally for imaging. With this choice, partial volume effects in the calculated $T_1$ image are decreased.

(3) The standard error in $T_1$ per voxel, calculated from the noise measured in the SE and IR images, is 7 ms at $T_1 = 1000\,\text{ms}$. The total error in $T_1$ of course must be larger because of the finite accuracy of flip angle and slice profile. Repeatability of the $T_1$ values in a volunteer was better than 20 ms. Calibration of the method on an EEG phantom [22] showed an absolute accuracy better than 4%.

(4) Values of $T_1$ of brain tissue can be read from the calculated image. Per tissue a small ROI was taken, in the visual centre of the tissue area inspected, away from the tissue boundary in order to reduce partial volume. The $T_1$ values obtained are tabulated and compared with the average over a number of healthy volunteers, as collected in [23] Thus:

| | $T_1$ values (ms) | |
|---|---|---|
| | this measurement | [23] |
| frontal white matter | 583 | 556 |
| occipital white matter | 604 | 588 |
| genus | 558 | 546 |
| splenium | 583 | 572 |
| putamen | 820 | 826 |
| head of caudate nucleus | 1020 | 928 |
| grey matter | 1014 | 999 |

(5) The values obtained in this measurement correspond within 10% to those in [23]. This correspondence is satisfactory in view of the rather large interindividual variation in $T_1$ reported there. The description of the details of the method used illustrates the type of care that is likely to be needed to obtain clinically meaningful values of $T_1$.

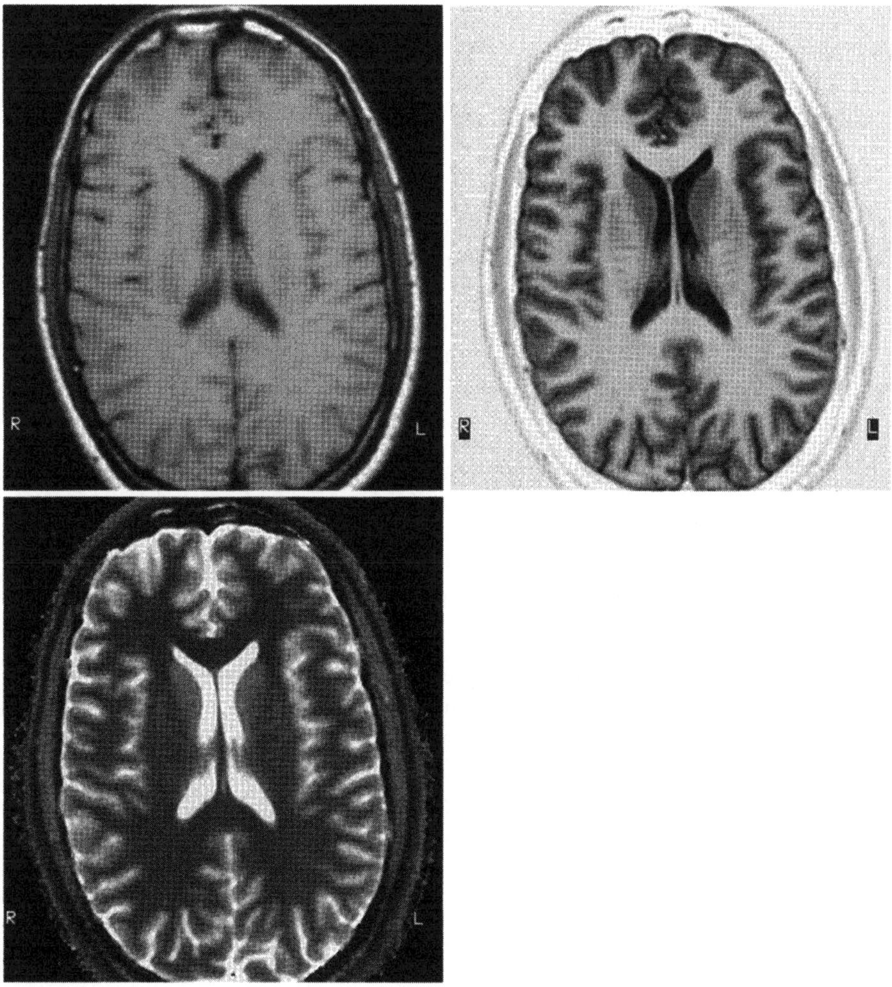


1 2 3

**Image Set VI-1**

## VI-2 Contrast Enhancement in SE and in Liquor-Suppressed TSE

(1) The clinical use of Gd based contrast agents in MR imaging is widespread, especially for inspection of the central nervous system. Typically scans are used that offer a $T_1$-weighted contrast. In this way, use is made of the $T_1$ shortening induced by the contrast agent (see Sect. 6.4.3). In this image set, the use of Gd DTPA in a patient with a brain tumour is compared in images of two scans (Courtesy of dr Prevo of Medisch Spectrum Twente). One of the images is offering $T_1$-weighted and one $T_2$-weighted contrast. For comparison, images are added that were taken before administration of the contrast agent. All scans were obtained with a multi-slice technique with 19 slices per scan.

Parameters: $B_0 = 1.5\,\text{T}$; FOV $= 230\,\text{mm}$; matrix $= 190 \times 256$; $d = 6\,\text{mm}$

| image no. | 1 | 2 | 3 | 4 |
|---|---|---|---|---|
| Gd-DTPA | no | no | yes | yes |
| scan method | TSE | SE | SE | IR-TSE |
| TI (ms) | | | | 2400 |
| TR/TE (ms) | 2400/100 | 600/18 | 600/18 | 8000/120 |
| NSA | 4 | 2 | 2 | 2 |
| scan time | 3′26″ | 3′21″ | 3′21″ | 5′36″ |

(2) The timing of TI and TR in the IR-TSE scan offers $T_2$ contrast for brain tissue as well as nulling of the signal for CSF. When used in combination with SE, this technique is also known as FLAIR (fluid attenuation by inversion recovery [24]). The use of the technique in a Turbo-Spin-Echo method is interesting because of the relatively short scan time that results. (See image set VII-9 for similar images and a discussion of inflow of CSF during the inversion time)

(3) The enhancement of the tumour region after Gd-DTPA administration is clearly seen in image 3 in $T_1$ weighting. Oedema in and around the tumour cannot be seen in images 2 and 3, but shows up in images 1 and 4 because these are $T_2$ weighted. Image 4 shows both the oedema and the Gd-DTPA enhancement, although the enhancement is more subtle than that in image 3.

(4) Image 4 may be of clinical value because of the combined display of edema and contrast-medium enhancement. The suppression of the CSF may add further to this value because it allows differentiation between oedema and CSF space (as in the oedematous region at the left of the occipital part of the septum). The relatively low signal-to-noise ratio of image 4 (23 for white matter, compared with 51 in image 3) is a disadvantage.

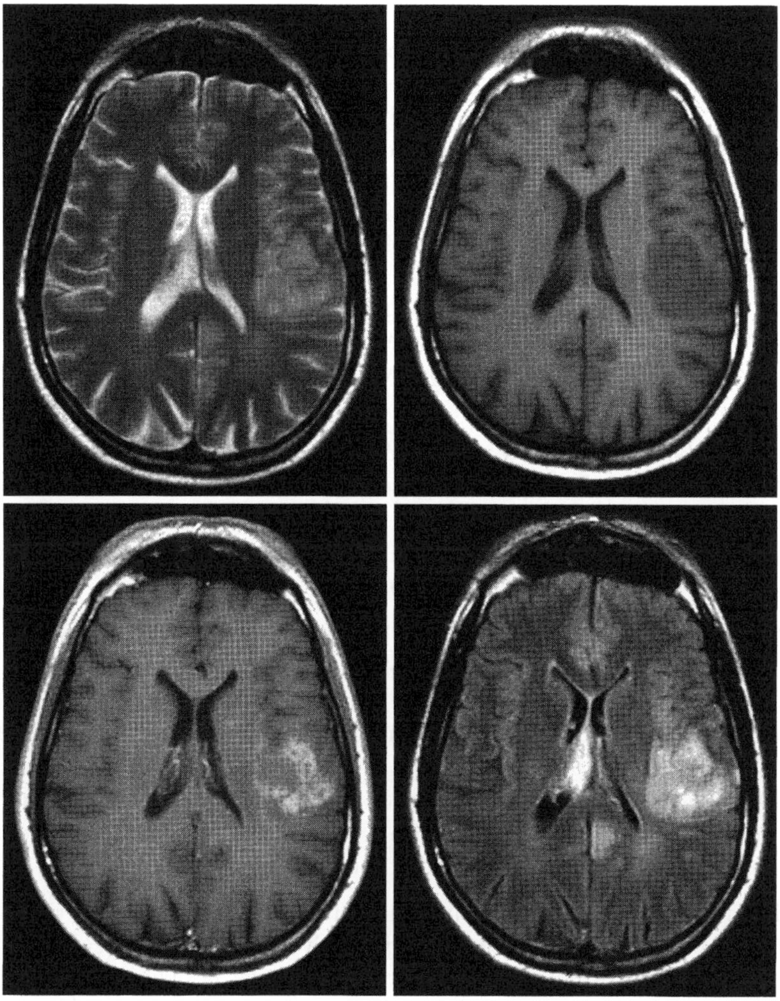

1 2 3 4

**Image Set VI-2**

# VI-3 Dynamic Behaviour of Contrast Agent Distribution

(1) Interest in the dynamic behavior of contrast agents is growing. Its observation with MRI may offer the basis for assessment of the speed of uptake of contrast agents by enhancing lesions, with the perspective of increasing the specificity of the diagnosis. Various publications on the use of this approach for differentiation between benign and malignant breast tumours have appeared [25], [26]. In this image set, images are shown of a dynamic study of a patient with a breast tumour. (Courtesy of Dr Dornseiffen, Medisch Spectrum Twente). The imaging technique was a dynamic series of 3D $T_1$-FFE scans.

Parameters: $B_0 = 1.5\,\mathrm{T}$; FOV $= 320\,\mathrm{mm}$; matrix $= 180 \times 256$; $d = 6\,\mathrm{mm}$; TR/TE/$\alpha = 19/5.5/45$. 36 slices per scan were acquired. A bilateral breast surface coil was used.

| | | | | | | |
|---|---|---|---|---|---|---|
| image no. | 1 | 2 | 3 | 4 | 5 | 6 |
| time after injection (s) | 0 | 40 | 70 | 100 | 150 | 250 |

(2) After completion of scan 1, a Gd-DTPA bolus was injected intravenously. The timing of the dynamic series was minimized between images 2, 3, and 4 in order to allow proper observation of a fast increase in the contrast medium enhancement of tumour tissue. The time per scan could be reduced to 30 s with the use of a half scan technique and by restricting to profiles that in the $\vec{k}$ plane have $(k_y, k_z)$ values inside an ellipse. This elliptic reduction of the coverage of the $\vec{k}$ plane does reduce scan time by $\pi/4$ without reducing $k_{y,\max}$ or $k_{z,\max}$ so that the scan resolution remains the same.

(3) The images show a relatively large invasive ductal carcinoma in the left breast. In image 6, the enhancement curve is given for the lesion (curve 2) as well as for a region of normal glandular tissue (curve 1). The enhancement of the lesion is much more rapid as well as stronger than that of normal tissue. For the case shown, the histological diagnosis of this patient showed a malignancy.

(4) The speed of the enhancement is possibly not well-enough resolved with the temporal resolution of the dynamic scan that was used. Faster scanning would have been possible by reduction of scan resolution, by combination of a small FOV and a small matrix so that one breast only is covered, or by reduction of the number of slices. All these alternatives, however, have a compromise character.

(5) The use of dynamic imaging of the breast after bolus injection is shown. Rapid enhancement may be highly specific of malignant tumours, but sufficient statistical proof of this still is lacking.

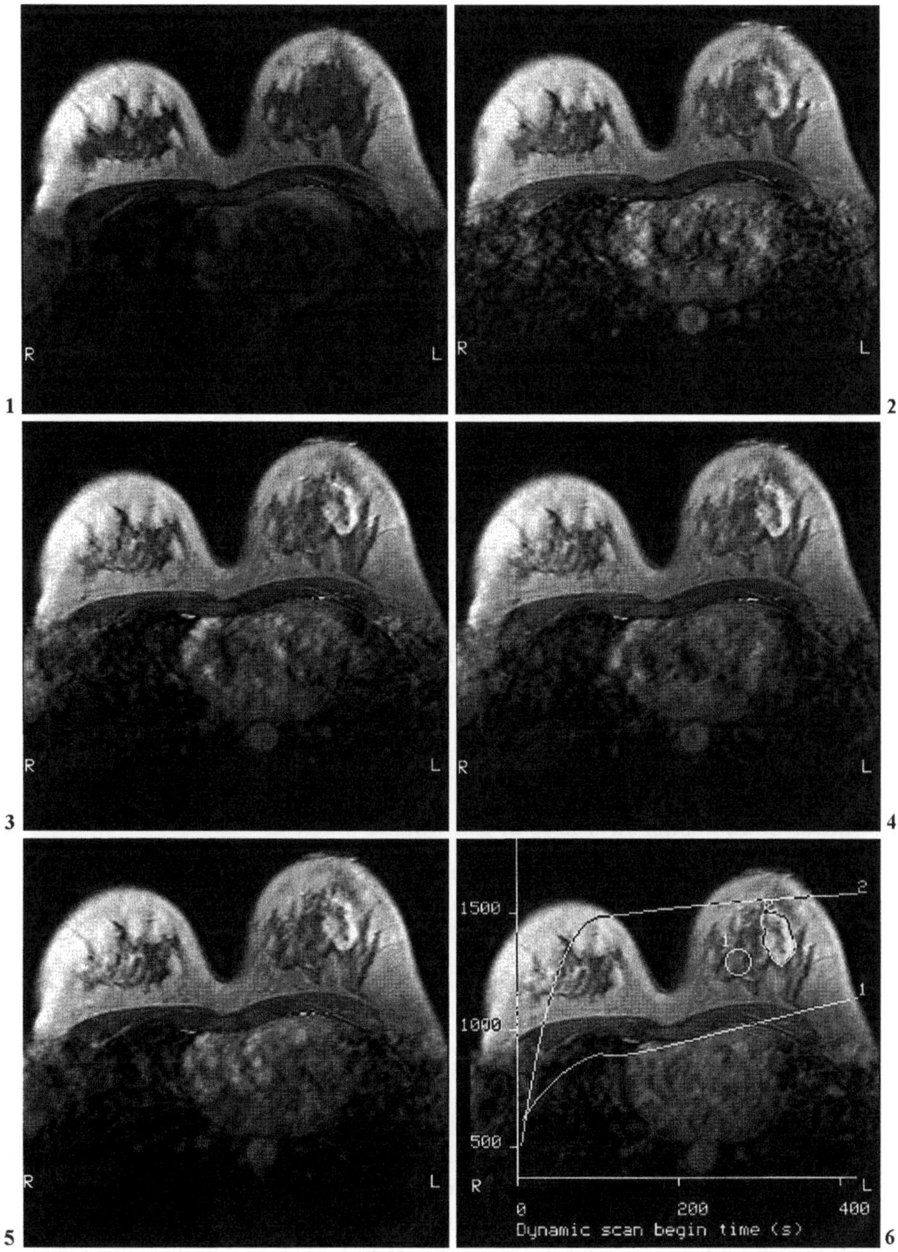


**Image Set VI-3**

## VI-4 Effects of Multi-slice Imaging on Magnetization Transfer in SE

(1) The occurrence of each slice-selective RF pulse has an effect on the saturation of the longitudinal magnetization in macromolecular protons in a wide region of space around the slice [27]. This saturation in turn can be transferred to the spins of the visible (water-bound) protons, via a mechanism called magnetization transfer (Sect. 6.3.3). In multiple-slice imaging, magnetization transfer will be the cause of a relation between the contrast in each MR image and the multi-slice scheme used.

This image set is chosen to inspect the magnitude of such a relation. All images are made in Spin Echo, using a multi-slice technique, but the number of slices per TR varied.

Parameters: $B_0 = 1.5\,\mathrm{T}$; FOV = 230 mm; matrix = 256 × 256; $d = 6\,\mathrm{mm}$

| image no. | 1 | 2 | 3 | 4 | 5 | 6 |
|---|---|---|---|---|---|---|
| TR/TE | 2000/25 | | | 425/25 | | |
| no. of slices per TR | 1 | 41 | 41 | 1 | 9 | 9 |
| MT pulses | | | yes | | | yes |
| NSA | 1 | 1 | 1 | 4 | 4 | 4 |

(2) In images 3 and 6, magnetization-transfer (MT) pulses are added to the scan method. These pulses are given before each excitation. Each MT pulse is a spatially non-selective ("hard") 1$\underline{2}$1 pulse, with a pass band of +/− 1000 Hz around the resonance frequency. This is sufficient to avoid direct excitation of water-bound spins anywhere in the volume imaged.

(3) The influence of the extent of magnetization transfer can be inspected by comparison of the SNR between images for some tissues

ratio of SNR values per tissue type

| images | 2:1 | 3:1 | 5:4 | 6:4 |
|---|---|---|---|---|
| CSF | 1 | 1 | 1 | 1 |
| grey | 0.81 | 0.49 | 0.85 | 0.74 |
| white | 0.80 | 0.39 | 0.84 | 0.62 |
| fat | 0.90 | 0.64 | 1.0 | 0.90 |

The comparison of images 1 and 3 shows a sharp decrease of signal from brain tissue compared to CSF upon introduction of MT pulses. The contrast between white and grey is strongly increased; the contrast between white matter and CSF is reversed. Comparison of images 1 and 2 also shows similar albeit smaller differences in contrast. This indicates that the RF pulses addressing the other slices in the multi-slice method have an influence similar to that of the MT pulses.

(4) From theory (see Sect. 6.3.3), it is known that magnetization transfer has two effects. It shortens $T_1$ and it causes a lowering of the equilibrium

magnetization. The long TR used in image 1 to 3 makes that the contrast change in these images reflects predominantly the second effect, a lowering of the equilibrium magnetization.

In images 4, 5, and 6 the signal strengths of grey and white matter and of CSF vary in the same direction as in images 1, 2, and 3. However, the changes in contrast are not as large. The much shorter value of TR in these images creates saturation by direct excitation because $T_1$ of the tissue is much longer than TR. The $T_1$ shortening from magnetization transfer now decreases the saturation; which gives an increase in signal that counteracts the signal decrease imposed by the lowering of equilibrium magnetization.

(5) Precise adjustment of the contrast weighting in MR requires not only the control of the timing of the sequence but also the control of the multi-slice logic. Magnetization transfer from the excitation of other slices in a multi-slice sequence influences the contrast. For contrast and signal-to-noise it is rewarding to not use more slices than needed.

1,2

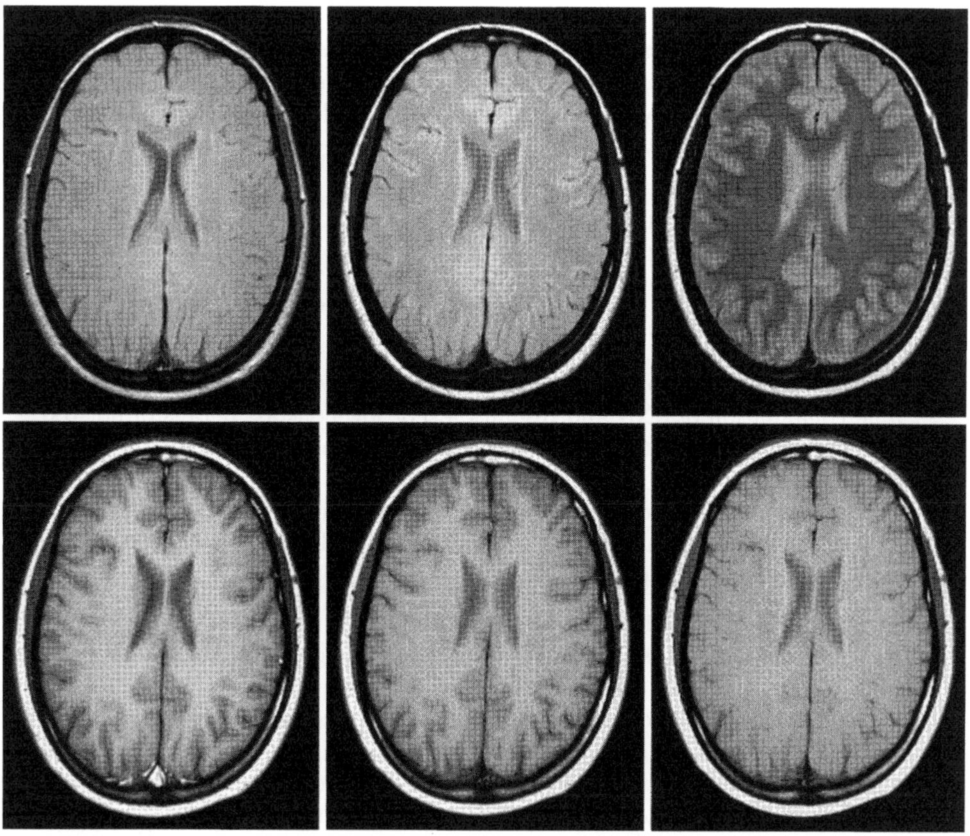

3

4,5

6

**Image Set VI-4**

## VI-5 Signal-to-Noise Ratio Depending on Choice of Pixel Size

(1) For scans that are identical except for the pixel size, the signal-to-noise ratio (SNR) is proportional to this size. The choice of a large pixel size (and large SNR) leads to poor spatial resolution and vice versa (Sect. 6.6). In the mathematical argument, no indication can be found of the optimum compromise between SNR and resolution. The image with this optimal strategy should have the most desirable combination of SNR and resolution, corresponding to the best "detail visibility"; but this term is defined only in the "perceptive fields" for the desired detail in the brain of the observer.

In the image set a transversal SE scan of the head is repeated with different FOVs.
Parameters: $B_0 = 1.5\,\mathrm{T}$; matrix $= 256 \times 256$; $d = 4\,\mathrm{mm}$; TR/TE $= 300/15$.

| image no. | 1 | 2 | 3 | 4 | 5 | 6 | 7 |
|---|---|---|---|---|---|---|---|
| FOV (mm) | 320 | 280 | 240 | 200 | 180 | 155 | 135 |

(2) The noise was measured for each of the images as the standard deviation of the pixel values in a ROI outside the head. Its value should be proportional to $(\mathrm{FOV})^2$. The observed values of the SNR for white matter are inspected for this proportionality:

| image no. | 1 | 2 | 3 | 4 | 5 | 6 | 7 |
|---|---|---|---|---|---|---|---|
| SNR | 44.8 | 35.3 | 26.3 | 19.0 | 15.8 | 11.3 | 8.7 |
| $\mathrm{SNR}/(\mathrm{FOV})^2$ | 4.38 | 4.50 | 4.57 | 4.75 | 4.88 | 4.70 | 4.75 |

The expected proportionality holds satisfactorily over the entire range on FOVs.

(3) Visual inspection of the images gives a subjective clue for the most desirable combination of SNR and resolution. For low-contrast details of not-too small size; e.g., the internal capsule, image 4 or 5 appears to give the best information.
For smaller and sharper details such as the horns of the ventricles, image 5 or 6 offers the best readability. In these images SNR has a value between 15 (image 5) and 11 (image 6).

(4) A generalization of these subjective findings is that scans should not be designed with an excessively high SNR. In exchange of a reduction of such SNR the resolution of the image may be improved.
There are other image characteristics that have to be brought in balance as well. Typical pairs of such characteristics are contrast versus SNR, slice thickness versus resolution, etc. For a typical MR scan like the one shown in this image set, with $T_1$ weighting and a with slice thickness of 3 mm, a value of SNR between 10 and 20 is close to the perceptive optimum.

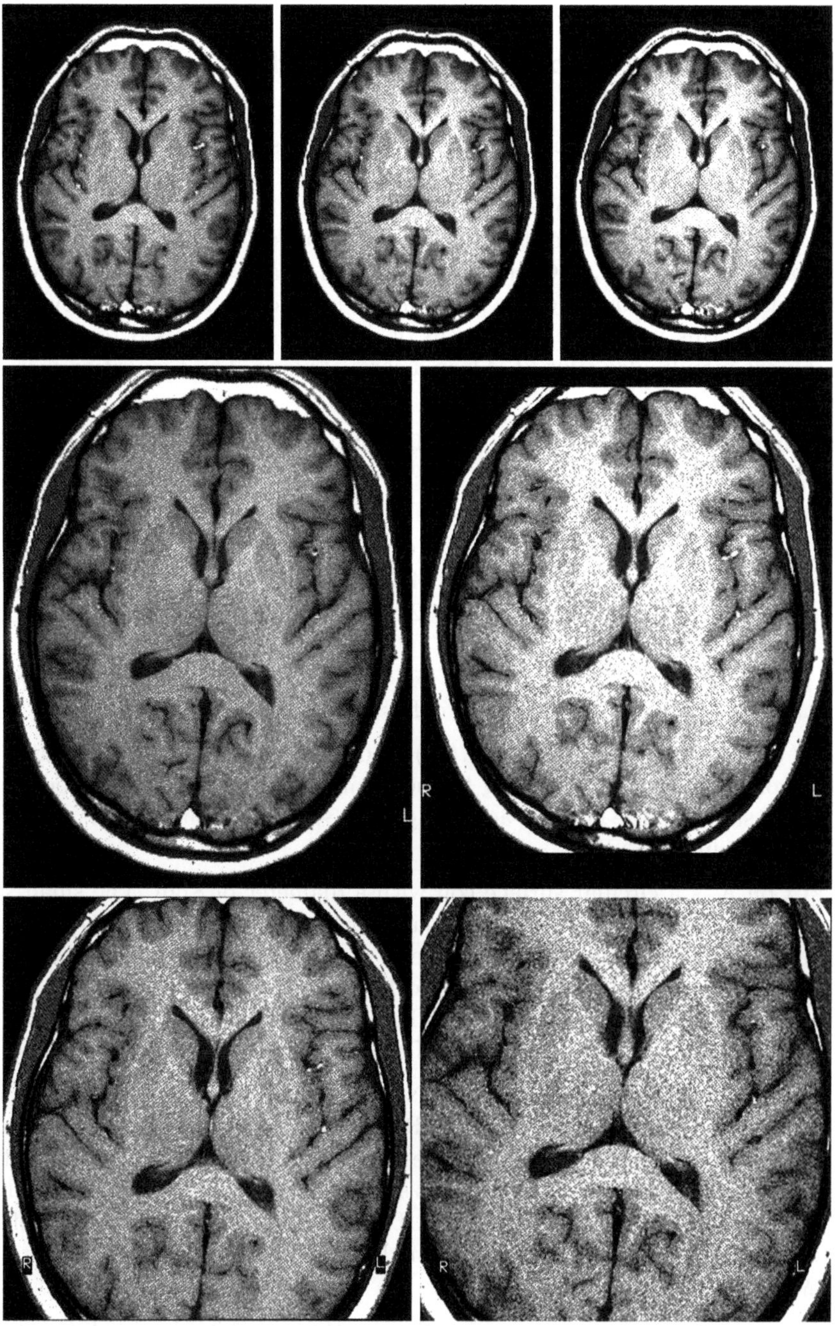


1,2

3

4

5

6

7

**Image Set VI-5**

## VI-6 Signal-to-Noise Ratio Depending on Choice of Surface Coil

(1) The choice of the receiver coil influences the signal-to-noise ratio (SNR) of the image. The properties of the coil of interest are its effective volume $V_{\text{eff}}$ and its quality factor $Q$, as defined in Sect. 6.5. The SNR in the centre of the coil is proportional to the square root of $Q/V_{\text{eff}}$ (6.43). A small coil usually has a small $V_{\text{eff}}$. On the other hand, a small coil has a rapid drop-off of sensitivity outside its centre, causing an unwelcome inhomogeneity in the image. In this image set, three coils of different size are used to image the temporo mandibular joint (TMJ) in a coronal plane. The acquisition method used is TSE;

Parameters: $B_0 = 1.5\,\text{T}$; FOV $= 160\,\text{mm}$; matrix $= 256 \times 256$; $d = 3\,\text{mm}$; TR/TE $= 4000/70$

| | | | |
|---|---|---|---|
| image no. | 1 | 2 | 3 |
| coil type | A | B | C |
| $Q$ | 30 | 26 | 46 |
| $V_{\text{eff}}$ ($\text{m}^3$) | 0.002 | 0.030 | 0.028 |
| $N_{\text{c}}$ | 1 | 1 | 2 |
| $\sqrt{N_{\text{c}}Q/V_{\text{eff}}}$ | 122 | 29 | 57 |

$N_{\text{c}}$ is the number of coils used, as defined in Sect. 6.5.1.

For the surface coils, $V_{\text{eff}}$ (Sect. 6.5) is defined at 3 cm from the plane of the coil; roughly corresponding to the position of the TMJ. The product of factors shown in the last line according to (6.34) is a measure for the expected relative value of SNR.

(2) Coil A is a circular surface coil with a diameter of 8 cm. The small size of the coil causes drop off of the signal near the extremes of the FOV. Coil B is a circular surface coil with a diameter of 17 cm. Over the FOV, the signal does not drop off visibly. Coils A and B were placed immediately on top of the joint. Coil C is a quadrature head coil. Note the relatively small value of $V_{\text{eff}}$ of this coil, which is the result of careful design.

(3) A relatively large value of the SNR near the TMJ in image 1 is clearly visible in the image. (The value of SNR in these images could not be measured because no area free of detail was present). Images 2 and 3 have roughly the same visual impression of SNR. This subjective evaluation of SNR is in line with the listed values of $\sqrt{N_{\text{c}}Q/V_{\text{eff}}}$. The relatively strong signal drop-off in image 1 has no negative influence on the readability of the image near the TMJ.

(4) In conclusion, for areas located close to the skin, like the TMJ, the inhomogeneous, high-SNR image obtained with a small surface coil is preferred above the more homogeneous but noisier image obtained with the head coil. A large surface coil is not useful for imaging of the head.

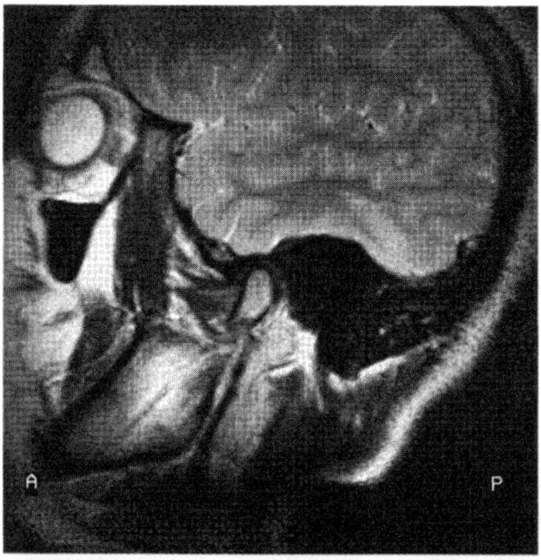

1

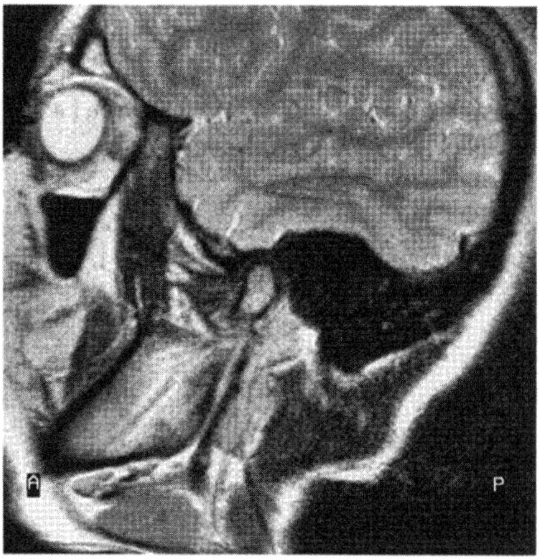

2

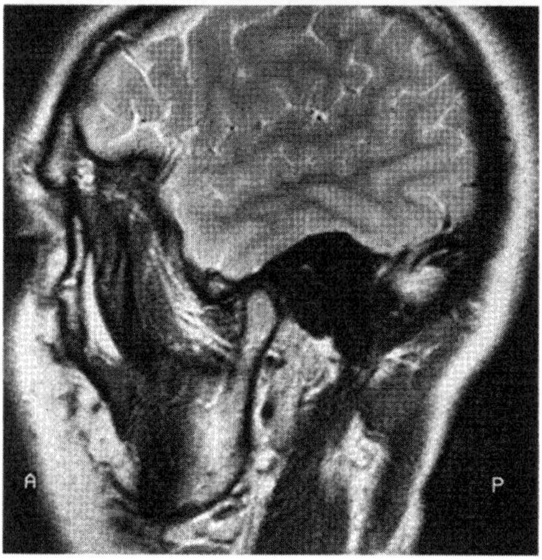

3

**Image Set VI-6**

## VI-7 Signal-to-Noise Ratio and Field Strength

(1) In this image set, images are collected that allow for some acquisition methods the observation of the relation of the signal-to-noise ratio to field strength.

The observed values are compared with the theoretical predictions of (6.43). The images are obtained from one volunteer, scanned in three MR systems, operating at 0.5 T, 1.0 T, and 1.5 T. The systems were all from one manufacturer, so that the system hardware was of comparable design. Moreover, the sequences used per field strength in each of the acquisition methods were completely equal.

Parameters: FOV = 230; matrix = 256 × 256; $d = 6$ mm; TR/TE = 2700/80; NSA = 2. For other parameters, see image set III-7.

| image no. | 1 | 2 | 3 | 4 | 5 | 6 | 7 | 8 | 9 |
|---|---|---|---|---|---|---|---|---|---|
| field strength (T) | 0.5 | 0.5 | 0.5 | 1.0 | 1.0 | 1.0 | 1.5 | 1.5 | 1.5 |
| image method | TSE | SE-EPI | GRASE | TSE | SE-EPI | GRASE | TSE | SE-EPI | GRASE |
| $Q$ of receive coil | | 119 | | | 59 | | | 40 | |
| $V_{\text{eff}}$ of receive coil ($m^3$) | | 0.028 | | | 0.028 | | | 0.028 | |

(2) The coils used are quadrature head coils of equal dimensions. As a consequence, the effective volume of the coils is equal. The quality factor $Q$, however, is a function of field strength, in agreement with (6.44).

(3) The observed and theoretical values of the SNR of white matter for the images, when normalized per acquisition method to their value at 0.5 T, are

| image no. | 1 | 2 | 3 | 4 | 5 | 6 | 7 | 8 | 9 |
|---|---|---|---|---|---|---|---|---|---|
| relative SNR (observed) | 1 | 1 | 1 | 1.7 | 1.9 | 2.0 | 2.6 | 2.4 | 2.7 |
| relative SNR theory | 1 | 1 | 1 | 1.9 | 1.9 | 1.9 | 2.7 | 2.7 | 2.7 |

(4) For each of the acquisition methods, the observed relative value of the SNR at 1.0 T is roughly constant ($1.9 \pm 0.1$). This is true as well at 1.5 T where the relative SNR = $2.5 \pm 0.2$. The SNR is predicted by (6.43) assuming constant $T_2$ and neglecting the influence of $T_1$. This last assumption is allowed because of the long TR used in the scans. The theoretical rate of increase of SNR with field strength is roughly equal to the observed rate. Some deviations still exist. Apart from observational errors they could be due to field-strength dependance of tissue properties, such as $T_2$, $T_2^*$, or magnetization transfer (see image set VI-4).

(5) For a range of $T_2$-weighted imaging methods, the experimental values of SNR are in rough agreement with the theoretical prediction.

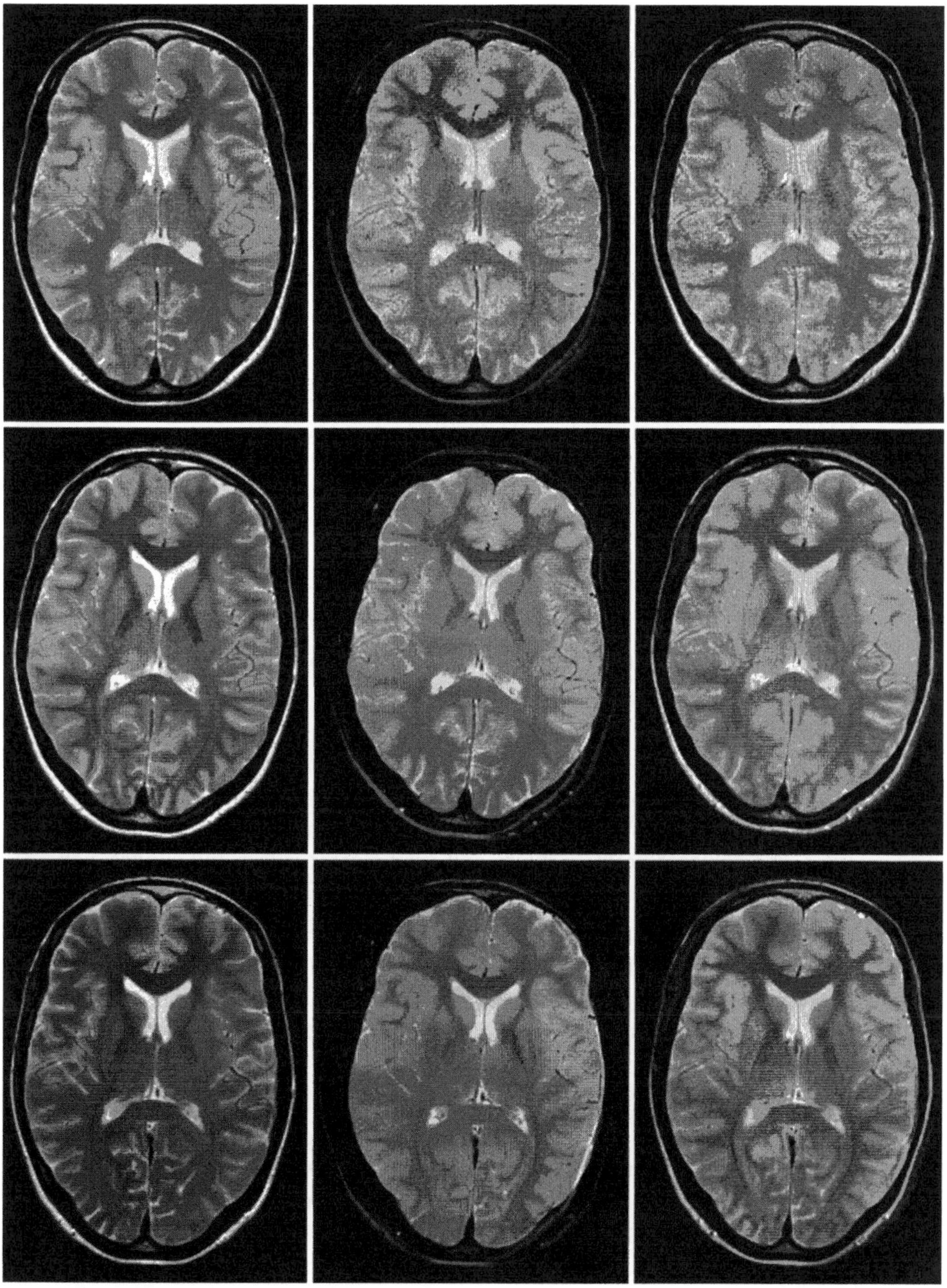

1,2 3

4,5 6

7,8 9

**Image Set VI-7**

## VI-8 The Influence of Magnetization Transfer in $T_1$-FFE Imaging of the Brain

(1) FFE scans are typically used with short TR and TE values. The contrast in these scans can be varied between wide margins depending on the technique factors used. These technique factors include TR, TE, spoiling and flip angle as well as magnetization preparation prepulses (see Sect. 4.5.3.3 and Image sets V-1 and V-2). In addition magnetization transfer (MT) pulses (Sect. 6.3.3) can be used. Such pulses are typically given once per TR and require about 5 ms. In many cases that is not a prohibitively long increase in TR, so that MT can be seen as a prepulse compatible with FFE imaging. In this image set, the influences on the image contrast of flip angle, spoiling and MT pulses are compared in 3D $T_1$-FFE imaging.

Parameters: $B_0 = 1.5\,\mathrm{T}$; FOV = 230, matrix = 256 × 256; $d = 4\,\mathrm{mm}$, TR/TE = 26/2.6; partitions = 30, scan time = 3'20"

| image no. | 1 | 2 | 3 | 4 | 5 | 6 |
|---|---|---|---|---|---|---|
| flip angle (degrees) | 50 | 50 | 20 | 20 | 6 | 6 |
| MT prepulses | no | yes | no | yes | no | yes |

(2) At large flip angles (50°; images 1 and 2), addition of the MT pulse does not bring any advantage; it merely lowers the signal-to-noise ratio in the brain parenchyma.

When the flip angle is lower (20°, image 3) the image is still $T_1$-weighted, the ventricles are dark as expected and the signal-to-noise ratio is much better than that of image 1. When adding MT prepulses to this scan, as shown in image 4, the contrast between white and gray matter is lost and suppression of the soft tissue signal is sufficient to bring up the CSF signal as intermediate gray.

At a flip angle of 6°, the $T_1$-FFE image is proton density weighted (image 5). At this flip angle the addition of MT prepulses is advantageous (image 6). The soft tissue contrast is enhanced and the image shows a strong bright contrast of CSF.

(3) The use of MT prepulses as a contrast parameter creates an image with a contrast that resembles that of $T_2$-weighted Spin Echo images. Although the contrast obtained is a magnetization transfer contrast, it can be useful for imaging in regions where $T_2$-weighted Spin Echo sequences show flow voids (e.g., the cervical spine; see Image Set VII-8).

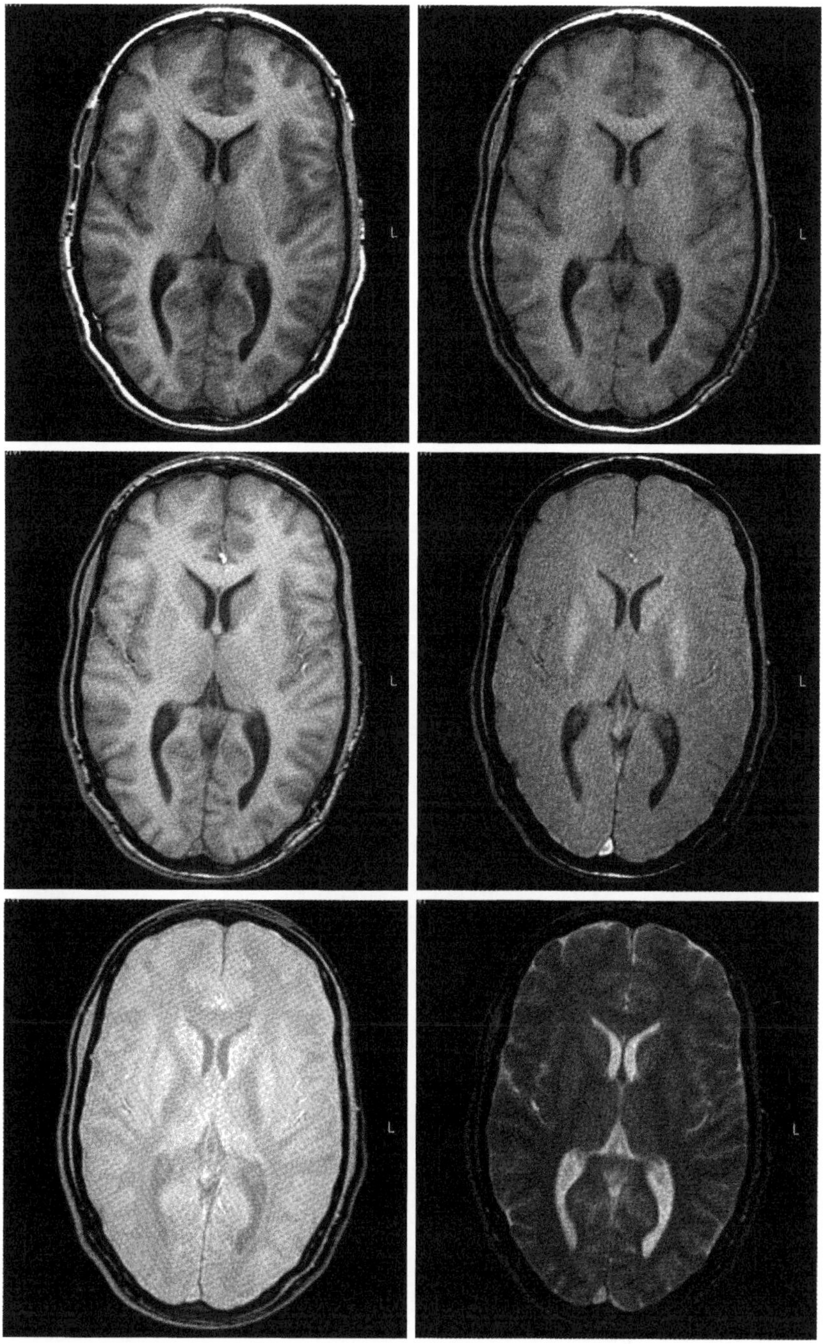


**Image Set VI-8**

# 7. Motion and Flow

## 7.1 Introduction

So far in this textbook only stationary tissue is considered. However, patient motion during acquisition causes inconsistencies in phase and amplitude of the (transverse) magnetization, which lead to blurring and ghosts [1]. Methods to minimize these artifacts will be discussed in this chapter. Also, motion itself may be the object of an MRI study, as is for example the case in the study of cardiac motion [2]. Methods used to study moving structures and to avoid the artifacts caused by these moving structures will be surveyed in Sect. 7.2.

A special case of motion in a patient is the flow of blood or CSF. Like the moving structures, flow also causes artifacts. To minimize these artifacts, their mechanism must be known. However, it is also possible to study the properties of flow by visualizing the flow paths (MR angiography) or by measuring the velocity of flow (quantitative flow measurements) [3]. There are two different groups of methods for this "flow imaging". The first one depends on the phase difference of the (transverse) magnetization of flowing blood with respect to that of stationary tissue (phase-contrast methods). The second group of flow imaging methods is based on the difference in magnitude of the (transverse) magnetization between flowing and stationary tissue (modulus contrast methods).

The phase shift of flowing tissue with respect to stationary tissue is due to its motion during an imaging sequence. Flowing blood will experience during the sequence a different gradient field from stationary tissue and therefore the phase of its magnetization is different. The magnitude of this phase difference will be discussed for a number of gradient waveforms in Sect. 7.3, as a preparation for the discussion of flow-artifact suppression and phase-contrast "flow imaging".

Section 7.4 is devoted to flow-artifact suppression in practical imaging sequences. In Sect. 7.5 flow-imaging methods will be described. Section 7.5.1 deals with methods based on the phase shift between flowing and stationary tissue (phase-contrast methods). Methods based on the difference in magnitude of the transverse magnetization (modulus contrast methods) are described in Sect. 7.5.2. The group of modulus contrast methods comprises "inflow", "contrast enhanced" and "time-of-flight" methods. Finally this chapter is closed with a discussion of MR diffusion in Sect. 7.7 and perfusion measurements in Sect. 7.6. In this chapter, the gradient waveforms are assumed to have negligible rise time. This somewhat unrealistic approximation is used to allow the mathematics to be more straightforward.

## 7.2 Moving Structures, Artifacts, and Imaging Methods

In this section four types of motion of structures will be discussed: patient motion (involuntary motion, see image set VII-4, or intended motion for the study of joint motion), cardiac motion, respiratory motion, and peristaltic motion. When substantial motion occurs within one scan, so that the profiles are measured in a changing geometry, blurring and/or ghosts occur in the image. Methods to minimize these artifacts are based on ECG triggering or gating, respiratory gating, or the application of fast imaging methods. Some methods require reordering of the data after completion of the scan and before reconstruction (retrospective gating).

Apart from causing artifacts, motion during the acquisition process can also be used to obtain information on the motion itself. Knowledge of the motion of the heart wall and the functionality of the heart (for example, ejection fraction) are of diagnostic importance, see Chaps. 5 and 6 of [2]. Since many acquisition methods last too long to freeze the motion of the heart, methods must be designed to compose an image from measurements taken in the same heart phase of a number of heart beats Chap. 2 of [2]. ECG triggering is then necessary (see Sect. 1.3.5).

Also for other types of motion, for example in orthopedic studies of the motion of the knee or other joints [4, 5], the precise development of the motion as a function of time is the object of the MRI study.

Respiratory motion causes blurring and ghosts in the image, which must be counteracted. Respiratory motion is rather slow and using a sensor (see Sect. 1.3.5) the respiration phase can be monitored and accounted for in the imaging sequence. Respiratory motion must also be monitored during cardiac studies to correct for the part of the motion of the heart that is caused by respiration [6]. Peristaltic motion cannot be anticipated or monitored and its effect can only be avoided by using very fast imaging methods which reduce the chance of having such motion during the scan.

### 7.2.1 Cardiac Motion

A way to cope with cardiac motion is to trigger the measuring sequence by the cardiac rhythm using an electrocardiogram [7]. The profiles of a single image made with a conventional imaging method (for example SE with TR equal to the length of the R–R interval) (see Fig. 1.36) are acquired in the same heart phase (= delay after the previous R peak). Since only one profile is measured the rest of the R–R interval can be used for the acquisition of data from other slices. In this way a multi-slice scan is obtained, in which each slice is acquired at a different heart phase, Chap. 2 of [2].

An extension of this method is to repeat the scan with a cyclic permutation of the heart phase of the slices per heart beat. So in one R–R interval one may acquire a profile of the slices 4, 5, 6, ..., $n$, 1, 2, 3 and in the next of slices 5, 6, ..., $n$, 1, 2, 3, 4, and so on, see Chap. 2 of [2]. In that case one ends up with $n$ slices each in $n$ heart phases. Of course the total scan time is now also n times longer. It is then possible to make a cinematic display of the heart motion for each of these slices and to observe the heart motion. The volume of the heart can be calculated from the n cross sections of the blood pool. The ejection fraction can be calculated from the difference in volume during systole and diastole, see Chap. 5 of [2]. Furthermore, visualizing the motion of the heart wall is of importance for finding infarcted regions, see Chap. 6 of [2].

One can make use of an imaging sequence with a TR which is much shorter than the R–R interval. By triggering a series of these excitations after the R top and sorting the data in the correct way, one can obtain images of a single slice at a number of heart phases shifted by TR. A suitable method for this approach is FFE, for which sufficiently short TR (e.g., 15 ms) can be reached. The irregularity of the heart beat makes it necessary to interrupt the excitations shortly before the expected next R top (e.g., 100 ms). This interruption imposes a modulation of the steady-state magnetization in the images. Especially the first image after the R top is relatively bright, since there is no saturation yet. Moreover the heart phases just before the R top cannot be studied.

An improvement that removes these problems is to sample continuously during the heartbeats and rearrange the acquired profiles into time slots of which the phase is known by the simultaneously acquired and stored ECG, so that each image can be built up from profiles taken at the same heart phase. This is "retrospective" gating [8]. When sufficient images are acquired at different heart phases a cinematic display of the heart motion can be made.

The acquisition process can be accelerated by measuring more than one profile per slice per R–R interval. In that case one has to decide on the time resolution required (for example, 40 ms per heart phase, so about 20 phases per heart beat). The number of profiles per slice per heart beat depends on the method used and is limited by the gradient system and the noise requirements. Fast imaging sequences as described in Chaps. 3 and 5 of this book can be of great advantage in this approach.

In extremes this leads to methods in which all data for a single slice are collected in a time interval that is short compared to a single R–R interval, assuming that the heart motion can be neglected during that interval. The time available depends on the amount of motion that one is willing to accept. For low-resolution EPI scans such imaging is in principle possible.

When only a certain time interval of the cardiac R–R cyle is used for acquisition during a sequence that runs continuously, so that the steady state is not interrupted, one speaks of a gated sweep. This method is applied to reduce the artifacts due to pulsating flow [9] and is used mainly in flow imaging. An example is shown in image set VII-12, where the vessels in the upper leg are imaged.

For a cardiac scan with high spatial resolution, only a short time per heart beat can be allowed for data collection, since otherwise the motion spoils the resolution. For such scans a combination of triggering, segmentation, and fast imaging is used. Per R–R interval, as many profiles as possible are acquired, centered around a certain phase of the heart motion. One example is the effort to visualize the coronary arteries. For this purpose there is an available period of about 110 ms per heart beat during diastole, in which the heart is relatively quiet and the coronary flow is strong. Data is collected in these intervals. Visualizing coronaries using 3D TSE techniques or spiral imaging is described in [11, 12]. Image set VII-13 shows 3D EPI images with excellent visualization of the coronary arteries. Further improvement of the quality of cardiac scans is obtained by including control of respiration motion.

### 7.2.2 Respiratory Motion

In scans that last long in comparison with the characteristic times of respiration, respiratory motion gives rise to severe blurring and ghost artifacts (see image sets VII-6 and VII-7). Various organs, including the heart, show significant movement with respiration. The wall of the abdomen with its high-intensity subcutaneous fat layer is a prominent source of respiratory artifacts. A number of approaches that can be used to reduce these artifacts are given in the following.

#### 7.2.2.1 Ordering of Phase Encoding

The motion due to respiration can be observed with a sensor (pressure sensor or strain gauge). The profiles can be acquired in the relative motionless period after exhalation. This method is called respiratory gating. However, gating is not applied frequently because it makes inefficient use of time (see Fig. 7.1). A frequently used method of artifact reduction is based on the notion that the low-order profiles (low $k_y$ values), which determine the contrast and its overall distribution, would give the most intense and disturbing artifact when measured during movement. These profiles are collected accordingly in the motionless period and the high-order profiles in the periods with motion (see Fig. 7.1). This method is named ROPE (reordered phase encoding) or PEAR (phase encode artifact reduction) [13] (see image set VII-6).

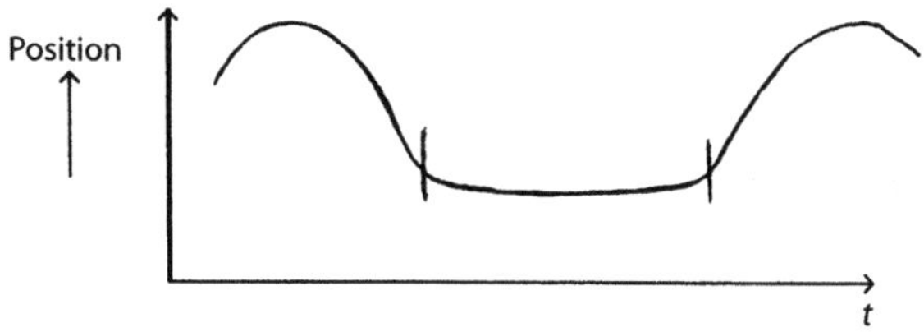


**Fig. 7.1.** Respiratory motion as a function of time

#### 7.2.2.2 Breath Hold

In many cases, the method of choice for suppressing artifacts due to both respiratory and peristaltic motion is to use breath holding in connection with sufficiently fast imaging methods whenever possible. In image set VII-7 an example is shown of a 3D-EPI image, showing the liver imaged within a breath hold. Some SNR is sacrificed, but a considerable reduction of motion artifacts is reached.

The combination of breath hold and cardiac triggered imaging can be used. One method available for this purpose is TFE. During each heart beat eight profiles are acquired during diastole, so the $128^2$ single-slice scan takes 16 heart beats. Increasing flip angles are used to make the response per profile more constant (see Sect. 5.2.2). This procedure must be repeated for the other slices, each acquired within a breath-holding period of sixteen heart beats [10].

#### 7.2.2.3 Respiratory Gating

The respiratory movement can for a large part be eliminated by respiratory gating and restriction of the acquisition to the relatively motionless end-expiratory phase of breathing (see Fig. 7.1). This technique can be combined with $T_2$ sequences in which a repetition time of a few seconds is already wanted. This technique is used in $T_2$-weighted abdomen imaging such as MRCP.

The best results are reached when the respiration sensing is based on navigator echoes (see Sect. 7.2.2.4). This technique is used in coronary imaging, see image set VII-13.

#### 7.2.2.4 Correction of Respiratory Movement Using Navigator Echoes

A method to correct for respiratory motion is based on the use of a "navigator echo". In the original version a $k_y = 0$ profile of a slice intersecting with a moving structure is measured at regular intervals during the scan. The Fourier-reconstructed (1D) projection shows the position of this moving structure – for example an abdomen wall – as a function of the time. With this knowledge, the slice position can be corrected during the scan (prospective navigator echo) [6]. The same knowledge can be used to correct for in plane motion by post processing of the raw data.

The application of a navigator echo is very important for the imaging of coronary arteries. Apart from cardiac motion itself, the heart moves up and down with the respiratory motion of the diaphragm. Although a slice can in principle be measured during a single breath hold, the result is not sufficient for clinical use. Multiple slices or 3D imaging with high spatial resolution are necessary for viewing small structures such as coronary arteries. Therefore a 1D-navigator echo is proposed to find the position of the diaphragm as a function of time.

A sagittal slice through the right half of the chest and the liver could be used to find the position of the diaphragm. However the diaphragm is convex and therefore its position cannot be determined accurately in a 1D-projection of this slice. Therefore instead of a sagittal slice, a cylindrical volume with a diameter of a few cm (a "pencil beam") in the FH direction is excited using a 2D excitation pulse (see Sect. 3.7 and image set III-8). This cylindrical volume includes a small part of the diaphragm, which is then well delineated in the 1D image. The heart follows the motion of the diaphragm to a certain extent, and usually a proportionality constant $C$ having a value between about 0.6 and 0.9, depending on the part of the heart considered, is used [14].

For imaging a coronary artery, double oblique volumes, from which slices (even curved slices) containing a large section of a coronary artery can be reconstructed, must be acquired. Due to respiratory motion the imaged volume moves in the FH direction. Assume that the volume is characterized by the unit vectors $\vec{s}, \vec{r}, \vec{p}$, where $s$ stands for selection direction, $r$ for read-out direction and $p$ means phase encode direction, and also assume that the relative displacement of the diaphragm in the FH direction is $\Delta s\vec{k}$, where $\vec{k}$ stands for the unit vector in the $z$ direction (the FH direction).

To be sure the same tissue is excited by each excitation, we must find its displacement in the $z$ direction. This displacement is given by $C\Delta s\vec{k}\vec{s}$ and to excite the same tissue the transmitter frequency must be shifted by

$$\Delta\omega = \gamma G_s C \Delta s(\vec{k}\cdot\vec{s}) = \gamma G_s C \Delta s \cos\alpha$$

where the in-product of the vectors $\vec{k}$ and $\vec{s}$ is equal to the cosine of the angle $\alpha$ between the FH direction and the selection direction of the slice. The in-plane displacement of the tissue is $\Delta s \sin\alpha$ in the direction of the intersection of the plane through $\vec{k}$ and $\vec{s}$ and the slice. When this direction makes an angle $\beta$ with the direction of $\vec{r}$, we have a displacement $\Delta r = \Delta s \sin\alpha \cos\beta$ in the read-out direction. This can be compensated for by shifting the detection frequency with $\Delta\omega_0 = \gamma G_r \Delta r$. In the phase direction the displacement is $\Delta p = \Delta s \sin\alpha \sin\beta$. This displacement can be compensated for by changing the phase of the receiver by $\Delta\phi = \gamma G_p \Delta p T_p$, where $T_p$ is the duration of the phase gradient. With these precautions, the position in the image is corrected for respiratory motion.

The navigator echo can be acquired between the R wave and the diastolic phase of the ECG, so that the interleaves of the scan to be measured within one heartbeat can be prospectively corrected for respiratory motion. Results are shown in [15] and [16] and an example of a coronary artery scan made with this method is shown in image set VII-13.

It is, of course, possible to use a navigator echo in a more dedicated position for imaging a certain part of the coronary tree, or to apply more than one navigator echo in different directions, in order to improve the accuracy. A special challenge is to excite only a small volume around the coronary artery to be imaged so as to reduce the field of view, which is normally determined by the dimensions of the chest. This can be done using two-dimensional pulses, which excite a cylindrical volume with small diameter, as discussed in Sect. 3.7. Similarly, it is also possible to excite a volume with rectangular cross-section by using another (for example a meandering) path through the $\vec{k}$ plane for RF pulses. This makes a small FOV, matched to the dimensions of coronary arteries, in principle possible.

### 7.2.3 Tagging

The precise motion of the left ventricle wall is of diagnostic importance, so methods are designed to measure this motion. The first condition is, of course, a cine display of the wall motion. However, the simple cine shows only the wall thickness as a function of time. There is also rotation and deformation, which is not visualized because of the lack of contrast in the heart wall. This can be provided using tagging.

The idea of tagging is to make a periodic grid-like pattern of saturated areas in the slice to be imaged and to study the deformation of this grid as a function of time [17]. One implementation of the method [64], [65] is shown in Fig. 7.2. It is based on $N$ equally spaced very short $(90/N)^\circ$ pulses in a large

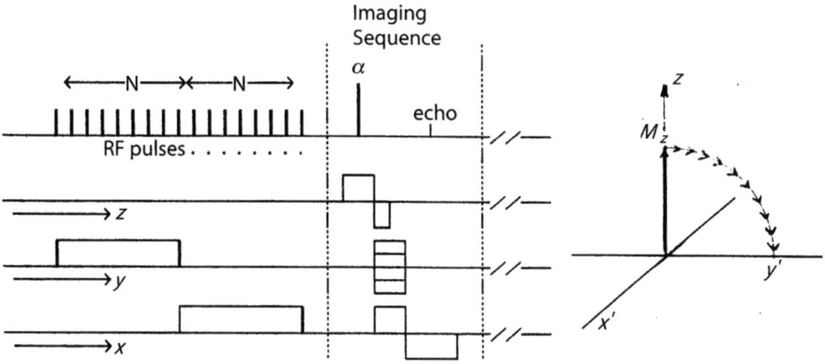


**Fig. 7.2.** Waveform for tagging; $x$ is the measuring direction, $y$ is the phase-encode direction, and $z$ is the selection direction. Between the tagging sequence and the imaging sequence a $z$ gradient is applied to spoil the transverse magnetization

gradient in the plane of the slice; for example, in the $x$ or $y$ direction. These short (large frequency bandwidth) pulses rotate the magnetization in a wide region containing the slice to be measured through $(90/N)^\circ$ around the $B_1$ field, notwithstanding the gradient field. Between the pulses the spins precess in the gradient field. Only the spins that precess by $2n\,180^\circ$ $(n = 0, 1, 2 \ldots)$ between two subsequent pulses will be rotated $90^\circ$ around the $B_1$ field after $N$ pulses (and will give no signal in the actual imaging sequence), the other spins will have rotated over a smaller angle or will still have longitudinal magnetization. Take for example those spins that between each pair of pulses precess by $(2n+1)180^\circ$: they remain practically unexcited by the tagging and will give a full signal in the actual imaging sequence. The precessing speed of the spins increases linearly with the distance in the direction of the gradient applied during the pulse train and a pattern of stripes will result. When the RF pulse trains are repeated in the two directions of the imaging plane we have a grid overlaying the image and we can follow the motion of this grid as a function of time or as a function of the heart phase (see image set VII-10).

Observation of the motion of tissue is also possible with phase contrast MR. It is used, for example, in wall motion analysis and can be combined with tagging, as discussed in [72]. The phase contrast method is discussed in Sect. 7.5 in connection with observation of flow.

This concludes the discussion of moving structures.

## 7.3 Phase Shift Due to Flow in Gradient Fields

Flow is a phenomenon of motion that deserves separate treatment. In a case of flow, most of the image is occupied by stationary tissue and the flow is restricted to the relatively small vessel lumens. Flow causes artifacts in the image and the understanding of the mechanism of these artifacts may open the way to suppressing them. Also, flowing blood may show up in the image outside the vessel in which it flows (misregistration). The reason for these artifacts are inconsistencies in phase or amplitude due to flow during the acquisition process. As a preparation for the discussion on flow artifacts in this section, phase shift due to flow will be discussed.

The influence of gradient fields on the phase of flowing tissue during the acquisition process can also be used to obtain information on the properties of the flow, such as the position, velocity, and acceleration of flowing blood. This information can be used to visualize the paths of the arteries and veins in projective views (angiography) or to measure the velocity and acceleration of the flow as a function of time and/or position (quantitative flow measurements). The flow-imaging methods based on phase shift are called "phase contrast methods".

For the discussion of phase contrast methods it is useful to first develop the basic mathematical tools, such as the relation between the motion and phase of the magnetization of blood when it is influenced by certain gradient waveforms [18]. Also the influence of gradient field inhomogeneities and eddy currents on the phase must be considered. This will be done for blood (or CSF) having both velocity and acceleration.

The position of blood, $\vec{r}(t)$, can in general be written as a Taylor series around an arbitrarily chosen expansion time $t_e$:

$$\begin{aligned}\vec{r}(t) &= \vec{r}(t_e) + (t - t_e)\left(\frac{d\vec{r}(t)}{dt}\right)_{t_e} + \frac{1}{2}(t - t_e)^2\left(\frac{d^2\vec{r}(t)}{dt^2}\right)_{t_e} \\ &\quad + \frac{1}{3!}(t - t_e)^3\left(\frac{d^3\vec{r}(t)}{dt^3}\right)_{t_e},\end{aligned} \tag{7.1}$$

where $(d\vec{r}(t_e)/dt)_{t_e}$ is the velocity at $t = t_e$, $(d^2\vec{r}(t)/dt^2)_{t_2}$ is the acceleration at $t = t_e$, and $(d^3\vec{r}(t)/dt^3)_{t_e}$, the third-order term, describes the change of the acceleration (sometimes named "jerk") at $t = t_e$. For ease of argument the discussions in this treatment will be restricted to flow with constant acceleration. The general case, including the higher-order terms, is treated in [19], where it is shown that the influence of terms of order higher than the second (acceleration) is negligible. So the position of the blood can now be described by the well-known equation of motion:

$$\vec{r}(t) = \vec{r}_e + \vec{\nu}_e(t - t_e) + \frac{1}{2}\vec{a}(t - t_e)^2, \tag{7.2}$$

where $\vec{r} = \vec{r}(t_e)$ is the position at $t = t_e$, $\vec{\nu} = \vec{\nu}(t_e)$ is the velocity at $t = t_e$, and $\vec{a}$ is the (constant) acceleration. We consider an experiment in which the motion and the gradients are in the $x$ direction, and the resulting phase is measured at the time $t_m$ (see Fig. 7.3), which is chosen as our expansion time, so $\vec{r}(t_m) = x(t_m) = x_m$. This can easily be extended to three dimensional situations.

The gradient is switched on at $t = T - \tau/2$ and switched off at $t = T + \tau/2$. We now calculate the phase of the spins with position $x_m$ at the time $t = t_m$. As usual we consider only the phase evolution due to gradient waveforms. For stationary spins we only have to consider the term with $x_m$ in (7.2), but for moving spins this is not allowed, since the position is a function of time. Therefore the complete form of (7.2) has to be inserted into the integral describing the phase at $t = t_m$:

$$\varphi(t_m) = \gamma \int_{T-\frac{\tau}{2}}^{T+\frac{\tau}{2}} x(t)G_x dt = \gamma \int_{T-\frac{\tau}{2}-t_m}^{T+\frac{\tau}{2}-t_m} (x_m + \nu_m t' + 1/2 \cdot a \cdot t'^2)G_x dt', \tag{7.3}$$

where $t' = t - t_m$. For the waveform of Fig. 7.3 this integral is easily evaluated and yields

$$\varphi(t_m) = \gamma x_m A + \nu_m \gamma A(T - t_m) + \gamma\frac{aA}{2}\left\{(T - t_m)^2 + \frac{\tau^2}{12}\right\}, \tag{7.4}$$

where $A$ is the area under the gradient of the strength-versus-time plot as given in Fig. 7.3: $A = G_x\tau$.

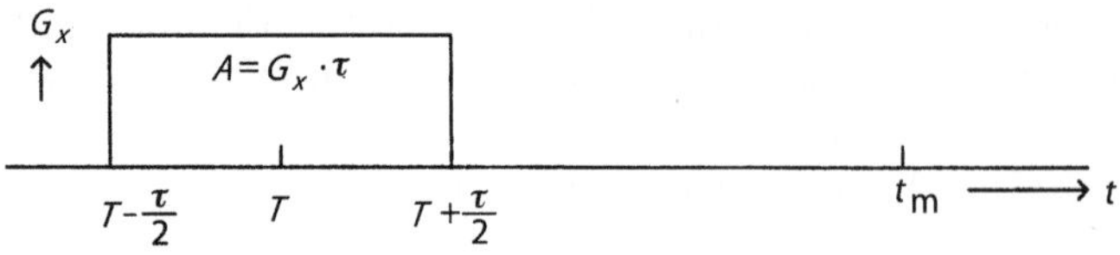


**Fig. 7.3.** Effect of the gradient waveform on moving spins

In the following part of this section the discussion is restricted to flow with constant velocity and the discussion of acceleration effects is postponed to Sect. 7.3.4. In that case the phase at time $t_m$ is a linear function of the velocity, $\nu(t_m) = \nu$:

$$\varphi(t_m) = \gamma x_m A + \gamma \nu A(T - t_m). \tag{7.4a}$$

This equation expresses the fact that $\varphi(t_m)$ is proportional to $\nu$. The proportionality factor, $A(T - t_m)$, is the first-order moment $M_1$ of the gradient at $t = t_m$. Equation (7.4a) can be rewritten by choosing an expansion time $t_e$ that differs from $t_m$. An interesting alternative is $t_e = T$, which yields:

$$\varphi(t_m) = \gamma x(T) A, \tag{7.4b}$$

where $x(T)$ is the position of the blood at $t = T$. It follows that $\varphi(t_m)$ is independent of $t_m$. Equation (7.4a) is easily understood physically when one realizes that at the time $t = T$ the spins (which are observed at position $x_m$) were at position $x(T) = x_m - \nu(t_m - T)$ and had obtained the gradient phase shift that is specific for that position.

For arbitrary gradient waveforms the area under the gradient-versus-time curve can be decomposed into rectangular gradients of length $\Delta\tau$, and their effect can be obtained by summing up the individual effects. So for the discussion of flowing blood in arbitrary gradient waveforms one can draw general conclusions, notwithstanding our restriction to simple rectangular waveforms.

### 7.3.1 Velocity Measurement Using a Bipolar Gradient

Based on the results of the preceding section, a simple tool can be designed to compose gradient waveforms with different properties for moving and stationary tissue. For example let us consider two equal-gradient areas with opposite polarity, as shown in Fig. 7.4 (a "bipolar-gradient" waveform). Restricting ourselves to flow with constant velocity, (7.4b), we see that the phase at $t_m$ due to the first gradient is "area times arm", $\gamma A x(T_1)$, and that the phase due to the second gradient is $-\gamma A x(T_2)$. So the total phase is given by

$$\varphi = \gamma \nu A x(T_1) - \gamma \nu A x(T_2) = -\gamma \nu A(T_2 - T_1) = -\gamma \nu A \tau', \tag{7.5}$$

and is independent of $t_m$. A bipolar gradient gives a phase shift that is proportional to the velocity $\nu$. The proportionality factor $A\tau'$ is the "first-order moment" of the bipolar gradient waveform: $m_1 = \int G(t) t dt = A\tau'$. It should

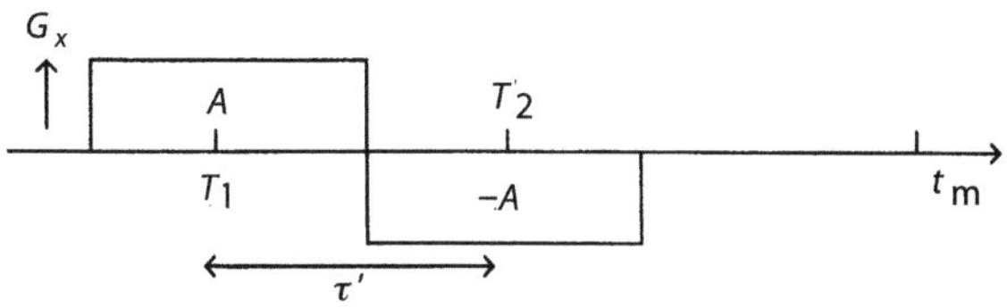


**Fig. 7.4.** "Bipolar" Gradient waveform

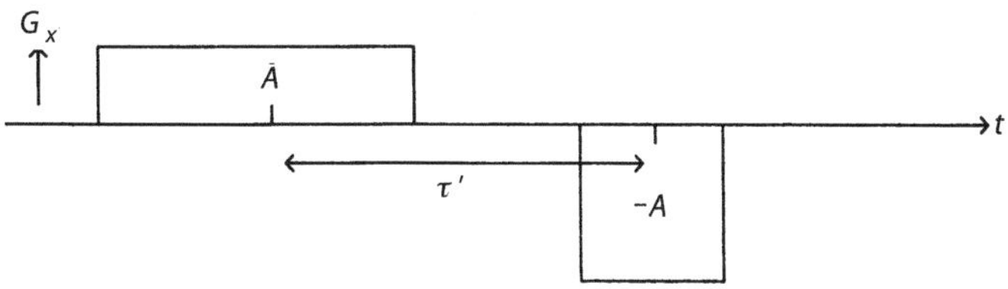


**Fig. 7.5.** General "bipolar" gradients

be realized that there is no reason for both gradient lobes to be adjacent. The time $\tau'$ (the "arm") is in this case the distance between the "centers of gravity" of the single gradient lobes (see Fig. 7.5). If they are some distance apart this merely means that the "arm" between the gradient lobes is longer. The areas of the gradient lobes must be equal so that the bipolar gradient waveform has no influence on the phase of spins in stationary tissue. This is expressed by the fact that $m_0 = \int G_x(t)\mathrm{d}t$, the zero-order moment, is zero.

The magnetization of spins that move with constant velocity through a bipolar gradient obtains a phase, $\varphi(t_\mathrm{m}) = \varphi$, that is proportional to the velocity and to $m_1$ (VENC: Velocity ENCoding). The maximum (minimum) velocity that can be distinguished uniquely occurs when $\varphi = \pm\pi$, which corresponds to a velocity $\nu = \pm\pi/\gamma A\tau'$ (see Fig. 7.6). For higher velocities aliasing occurs because $\varphi$ cannot be distinguished from $\varphi + 2n\pi$ $(n = \pm 1, \pm 2, \ldots)$.

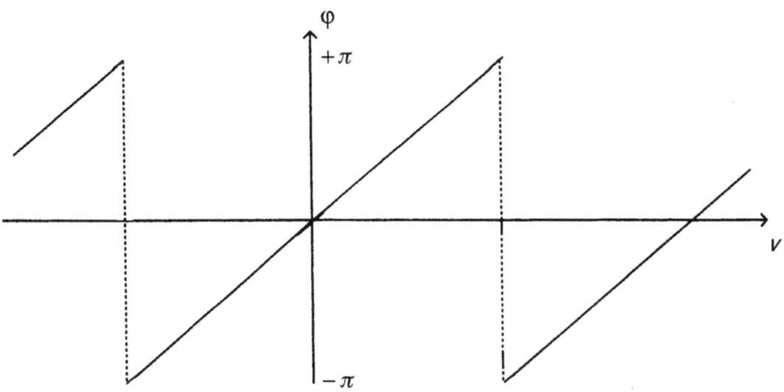


**Fig. 7.6.** The phase of magnetization of flowing tissue as a function of its (constant) velocity

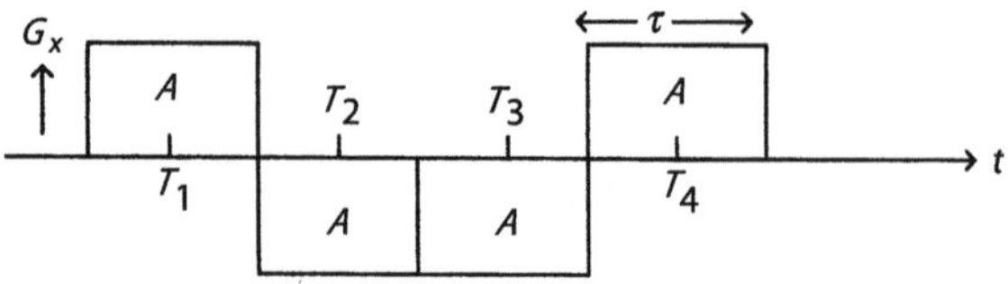


**Fig. 7.7.** Velocity-insensitive gradient waveform

This means that the sensitivity of the phase-versus-velocity curve has to be suitably adapted to the expected value of $\nu$ by choosing the product $A\tau'$. For low velocities (as is the case, for example, with perfusion phenomena) $m_1 = A\tau'$ must be very large. Practical design of sequences that measure flow velocity is described in Sect. 7.5.

### 7.3.2 Velocity-Insensitive Gradient Waveform

When two bipolar gradient waveforms with first-order moments that are equal in magnitude but reversed in polarity are applied to the spin system (see Fig. 7.7), the resulting phase is independent of the velocity and position since the moments of both bipolar gradient waveforms are equal and have opposite signs. This means that both the zero-order and the first-order moments, $m_0$ and $m_1$ of the back to back bipolar waveform, are zero. Again it is not necessary that the two "back-to-back bipolars" have zero distance, we have the same design freedom as mentioned in connection with single bipolar gradients. The "back-to-back" bipolar waveform can be used to design velocity-insensitive sequences. When the image resulting from a velocity-sensitive sequence is subtracted from the image of a velocity-insensitive sequence of the same object slice, only a signal due to flow is shown in the resulting subtracted image. In case of blood a projective view of the blood vessels (angiogram) can be made (phase-contrast angiography). We shall discuss examples in Sect. 7.5.

### 7.3.3 Flow with Acceleration

In this section the discussion will be extended to flow with constant acceleration. Now the full form of (7.4) must be used. The bipolar gradient method can still be used to measure the velocity, but since the velocity is a function of time now only the velocity at a certain moment can be measured. To prove this, (7.4) is applied successively to a gradient lobe with an area $A$ and its center at $T_1$ and one with area $-A$ and centered around $T_2$ (Fig. 7.4). When the expansion point is taken at the measuring time $t_{\mathrm{m}}$, the result is

$$\varphi(t_{\mathrm{m}}) = \gamma\nu(t_{\mathrm{m}})A(T_1 - T_2) + \gamma\frac{aA}{2}\left[(T_1 - t_{\mathrm{m}})^2 - (T_2 - t_{\mathrm{m}})^2\right]. \tag{7.6}$$

Obviously $\varphi(t_{\mathrm{m}})$ is a function of $\nu(t_{\mathrm{m}})$ and $t_{\mathrm{m}}$ as soon as a $\neq 0$. We can again change the expansion time. For an arbitrary time $t_{\mathrm{e}}$ the velocity $\nu(t_{\mathrm{m}})$ can be written as

$$\nu(t_{\mathrm{m}}) = \nu(t_{\mathrm{e}}) + a(t_{\mathrm{m}} - t_{\mathrm{e}}).$$

Introducing this into (7.6) yields

$$\varphi(t_{\mathrm{m}}) = \gamma\nu(t_{\mathrm{e}})A(T_1 - T_2) + \gamma a A\left[t_{\mathrm{e}}(T_2 - T_1) + \frac{1}{2}(T_1^2 - T_2^2)\right]. \tag{7.7}$$

The expansion time for which the phase due to motion is independent of the acceleration, $a$, is

$$t_{\mathrm{e}} = \frac{1}{2}(T_1 + T_2), \tag{7.8}$$

in which case

$$\varphi(t_{\mathrm{m}}) = \gamma\nu(t_{\mathrm{e}})A(T_1 - T_2). \tag{7.7b}$$

This means that for a bipolar gradient and an accelerating flow the velocity at the time $t_{\mathrm{e}}$ can be found from the measured phase according to (7.7b). This is an interesting result, because the acceleration sensitivity of bipolar gradient waveforms is frequently reported, but in reality only the velocity is measured correctly. The value obtained is that at the expansion time given by (7.8). For each arbitrary bipolar waveform this expansion time lies at the "center of gravity", $1/2\ (T_1 + T_2)$, of the bipolar gradient waveform [19]. For more complicated gradient waveforms an expansion time with the same properties can always be found, only its definition has to be extended, so as to encompass more gradient lobes. At any other expansion time the phase is related to both velocity and acceleration, so it is then impossible to separate the two contributions.

The theory described here can be extended to flow with higher-order components, such as jerk and the fourth order (see [19]). It follows that our conclusion is still approximately right: with a bipolar gradient the phase of the transverse magnetization of moving tissue can be related to the velocity at the center of gravity of the bipolar gradient; the acceleration has no influence on the phase and the higher-order terms usually have a negligible influence.

The acceleration must be measured with an independent method. This can, for instance, be done by using gradient waveforms with both the zero-order and first-order moments, $m_0$ and $m_1$, equal to zero. For the simple back-to-back bipolar gradient waveform, shown in Fig. 7.8, we find from (7.4)

$$\varphi = 4\gamma a A \cdot \tau^2, \tag{7.9}$$

where use has been made of $T_n = T_1 + (n-1)\tau$. The second-order moment of the back-to-back bipolar gradient waveform is $4A\tau^2$, and it determines the relation between the phase and the acceleration.

Sequences can also be made acceleration insensitive. The number of gradient lobes used for this goal must be kept as small as possible to save time in an acceleration-insensitive imaging sequence. For a position-independent phase (at the echo) a bipolar gradient (with one positive and one negative

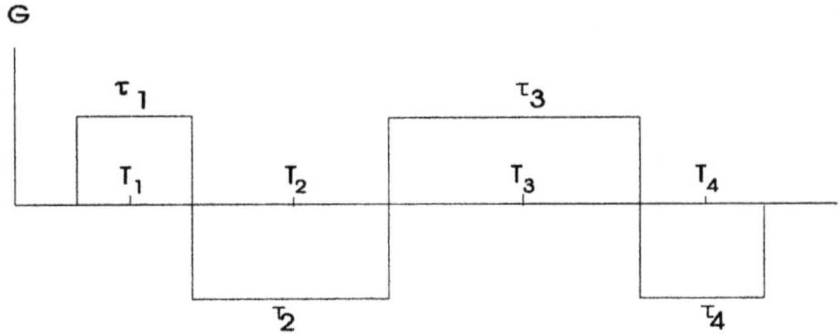


**Fig. 7.8.** Gradient waveform for acceleration insensitivity, the $\tau$'s stand for the length of the blocks

lobe) is needed (see Fig. 7.4). For a velocity-independent phase two back-to-back bipolar gradients are needed (three lobes; see Fig. 7.7). So for acceleration insensitivity, gradient waveforms with four lobes may be expected. This statement will only be proven formally because the equations do not lead to simple expressions and furthermore acceleration compensation is not frequently used, except in sequences with long TE; an example is the acceleration compensation of the second echo in a SE sequence.

Applying (7.4) to the gradient lobes sketched in Fig. 7.8, we can easily sum up the effects of the four gradient lobes and subsequently require that the coefficients of $x_0$, $\nu$, and $a$ are zero. This yields three equations in $\tau_1$, $\tau_2$, $\tau_3$, and $\tau_4$, in the first, second, and third power of the $\tau$s. This means that when one $\tau$ is chosen, three equations remain with three unknowns, which can be solved. It is no use performing these actions in this analytical treatment on phase-contrast flow imaging, because the equations must be separately solved numerically for each case. However, the numerical calculation is straightforward.

### 7.3.4 Influence of Field Inhomogeneities and Eddy Currents

In order to keep the equations simple, until now the influences of the inhomogeneity of the main magnetic field, $\delta B$, and the eddy current fields, $\delta B_{\mathrm{e}}(t)$, on the phase of flowing tissue have been disregarded. However, when these are taken into account the phase is correctly described by

$$\varphi(t) = \gamma \int [\delta B + \delta B_{\mathrm{e}}(t) + xG_x(t)]\mathrm{d}t = \varphi_B + \varphi_{\mathrm{e}} + \gamma \int xG_x(t)\mathrm{d}t. \quad (7.10)$$

The last term of this equation has been discussed in the previous sections and this has led to interesting and useful conclusions, which can be threatened by the effect of the first two terms. However, the first term can be canceled out through a method in which the phase is measured twice by imaging twice with different first-order moments and subtracting the phases per pixel. The phase difference between the two measurements no longer contains the term $\varphi_B$, which is equal in both measurements. Now the phase difference depends on the difference in the first-order moments of the bipolar gradient waveforms used in the two measurements and on the difference between the eddy-current fields associated with these different gradients. The eddy currents are not equal in the two measurements, so they compromise the subtraction method. The eddy-current problem can only be solved by making the eddy currents as small as possible by proper design of the gradient coils.

## 7.4 Flow Artifacts

Before embarking on MR flow imaging, we will first discuss some frequently encountered artifacts. Methods to avoid these artifacts are also discussed.

### 7.4.1 Ghosting Due to Pulsating Flow

Flowing blood (or CSF) causes artifacts in conventional SE sequences. During systole the refocusing pulse does not hit the spins that were excited by the 90° pulse, since because of the high velocity the blood moved out of the excited slice before the refocusing pulse was applied. So the signal from the blood during systole is zero. During diastole, however, the signal is high because the blood velocity is very low. So in SE imaging the blood gives a varying signal in the phase-encode direction, depending on the heart phase in which the profiles are measured (see image set VII-1). Signal amplitudes that vary from profile to profile will occur when the scan is not triggered. These signal variations will give rise to ghosts shifted in the phase-encode direction (see Sect. 2.6.1 and Fig. 3.5). A well-known example is an artery crossing the imaged layer, which can give a number of ghosts in the phase-encode direction (image set VII-3).

Of course this artifact can be avoided by applying ECG triggering so that all profiles are measured during the same heart phase. Also the method of presaturating the parallel slabs, which removes the transverse magnetization from the blood before it enters the imaged slice, can be used to avoid ghosts due to blood pulsation. In image set VII-9 a method is shown to make the signal of CSF very small, by inversion and by pausing the TSE scan until the longitudinal magnetization of CSF passes through zero (see also Sect. 2.7.1).

### 7.4.2 Flow Voids

Flow artifacts can be caused by a variety of factors. Strong spatial velocity gradients can occur, for example, behind a stenosis or near a bifurcation, or in regions of turbulent flow (image set VII-2). The intra-voxel dephasing caused by this large velocity spread can be such that the signal from the voxel is averaged out totally (flow void). This effect can be counteracted by using very short (effective) echo times during which the dephasing is insignificant, for example by using spiral imaging.

In image set VII-8 the CSF flow void due to outflow of CSF in a TSE image of the cervical spine could be avoided by taking the profile of the refocussing pulses wider than that of the excitation pulses. Another factor playing a role in the same image set is the inflow of CSF that is already saturated by excitation in another slice. Both inflow and outflow voids are discussed in image set VII-1.

### 7.4.3 Shift in Phase-Encoding Direction Due to Flow

A well-known artifact results when the images of the blood and the vessel in which it flows do not coincide: the blood is shifted with respect to the stationary tissue (see Fig. 7.9). We call this artifact "misregistration". This artifact is understood by realizing that the position in the phase-encode direction is determined by the subsequent phase-encode gradients (see Sects. 1.2.2 and 2.4.1.3). For stationary tissue this determination of position in the phase-encode direction is of course correct. However, for flowing blood the position is encoded when it is at the center of gravity $T$ of the phase-encode gradient. This gradient is unipolar (see Fig. 7.3) so that (7.4a) applies. The successive strengths of the phase-encode gradients are given by $G_{yn} = n\Delta G_y$, and the duration is $\tau_y$. During the $n$th phase-encode gradient the spins of the flowing blood, then at position $y(T)$, obtain a phase shift of

$$\varphi_{\text{ph}} = \gamma n \tau_y \Delta G_y y(T) = \gamma n \Delta A y(T), \tag{7.11}$$

where $\tau$ is the duration of the phase-encoding gradients (see Fig 7.3). So at the moment of the echo, when the blood is at $(x(\text{TE}), y(\text{TE}))$, it is observed in the image at $(x(\text{TE}), y(T)) = [x(\text{TE}), (y(\text{TE}) + \nu_y(T - \text{TE}))]$. Image set VII-5 demonstrates this misregistration artifact in the brain.

A method used to correct this artifact is to add a bipolar gradient to each phase-encoding gradient in such a way that the blood is characterized at the correct position during the echo. This will be discussed in Sect. 7.4.4.3.

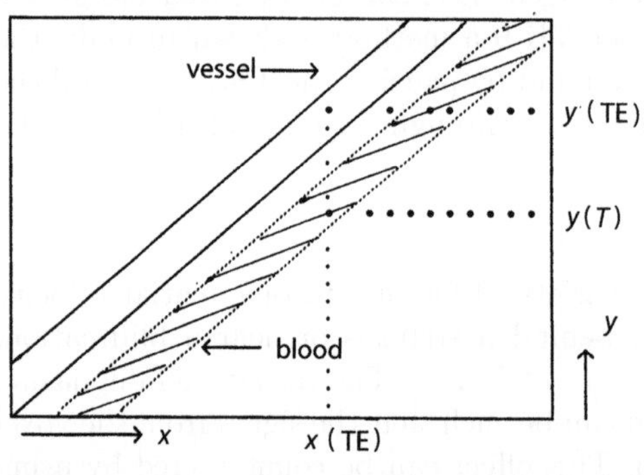


**Fig. 7.9.** Due to flow in the phase-encode direction, the blood is shifted in the $y$ direction

### 7.4.4 Velocity-Insensitive Imaging Sequences: Flow Compensation

For both quantitative flow measurements and phase-contrast angiography (PCA) we need further insight into the flow sensitivity of our standard imaging sequences [18]. Flow-insensitive sequences are usually made by choosing gradient waveforms that are not sensitive to velocity. Such sequences are of use in a variety of flow-imaging tasks (see, for example, image set VII-2). One reason is that subtraction of a flow-sensitive and a flow-insensitive image is frequently used.

#### 7.4.4.1. Selection Direction

In Sect. 2.3.2 it was shown that the selection-gradient waveform must have the form depicted by the full line in Fig. 2.2. We can assume that the spins are excited in the center of the pulse ($t = 0$) (see Sect. 2.3.2). When $t = t_r$ we have refocused the stationary spins. Between $t = 0$ and $t = t_r$ the gradient waveform is bipolar, so it is flow sensitive as given by (7.4). Velocity insensitivity is obtained by adding a reversed bipolar gradient as discussed in Sect. 7.3.3 and shown in Fig. 7.10 by the dotted bipolar gradient.

#### 7.4.4.2. Read-Out Direction

In the measuring direction the same phenomenon is met: at the time of the echo the zero-order moment of the gradient is zero. This holds both for SE (where after the refocusing pulse the sign of the phase must be changed) and FE sequences. When the gradient waveform is bipolar, the phase of the moving spins is velocity dependent. To obtain velocity insensitivity one must add a second bipolar gradient. How this is done is shown in Fig. 7.11 for FE and SE. Note that the modification of the FE sequence has created a new gradient echo (echo 0), before the first echo of the unmodified sequence. Note also that in the unmodified SE and FE sequences the second echo is velocity insensitive. Finally note that true velocity insensitivity is reached only at the center of the echo. For all other sampling points there is a small but finite velocity sensitivity. This leads to a small difference in magnification between image regions that move in the read-out direction and image regions at rest.

In the figure, only the formal solutions are given. In practice there are more considerations for making a useful velocity-insensitive sequence. For example the 180° pulse in a SE sequence is never exact, so a fid is also excited. In the velocity-insensitive sequence of Fig. 7.11d the echo of this fid refocusses at the center of the echo (zero-gradient area). This gives an artifact because the refocussed signal has not felt a preparation gradient, as echo 1 has. This means that as far as possible the complete extra bipolar gradient should be placed before the refocussing pulse (with reverse polarity). Furthermore, the

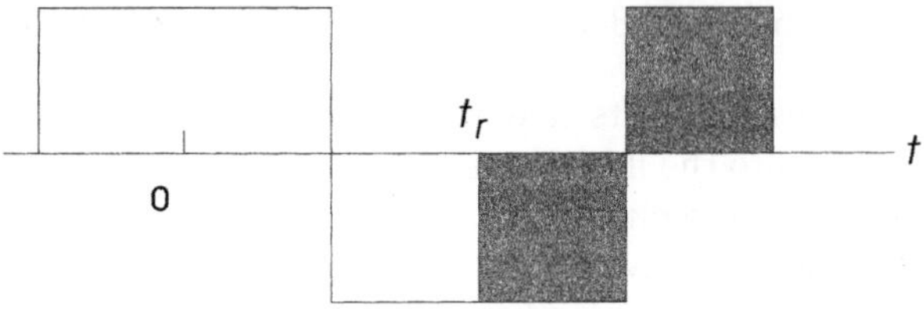


**Fig. 7.10.** Flow-insensitive selection gradient waveform

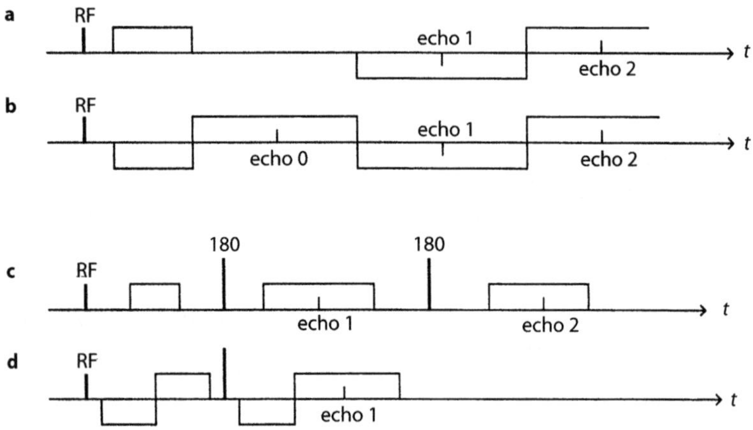


**Fig. 7.11.** **a** Normal and **b** velocity-insensitive FE in the read-out direction. In the normal scan the first echo is flow-sensitive and the second echo is flow-compensated. **c** Normal SE sequence and **d** an SE sequence that is velocity-insensitive for the first echo. In the normal SE the second echo is also velocity insensitive

read-out gradient should be kept as low as possible in view of the signal-to-noise ratio, but one also wants to achieve the lowest possible echo time. The gradient lobes before the 180° pulse can be taken as high as possible, so as to make them short. For this design job the simple rules deduced in the previous section (see (7.3) with $a = 0$) can be applied.

### 7.4.4.3. Phase-Encoding Direction: Correction for Misregistration

In the phase-encode direction, flow does not lead to a phase error, but, as discussed in Sect. 7.4.3, to misregistration. Correction of this misregistration is obtained when the phases measured by the phase-encode gradients are corrected in such a way that it is as if they are measured at $t$ = TE. So instead of the measured phase increase due to the phase-encode gradient of (7.11), the shift in position during the time interval between the encode gradients and the echo time should be corrected for by adding the velocity term

$$\varphi_{\mathrm{TE}} = \gamma n \Delta A[y(T) + \nu_y(\mathrm{TE} - T)] = \varphi_{\mathrm{ph}} + \gamma n \Delta A \nu_y(\mathrm{TE} - T). \tag{7.12}$$

This means that at each phase-encoding step an extra phase equal to $\gamma n \Delta A \nu_y \times (\mathrm{TE} - T)$ must be added. This can be done by adding extra bipolar gradient waveforms to the phase-encode gradients. These bipolar gradient waveforms, which depend on the phase-encode step $n$ have an area per gradient lobe $A_{\mathrm{B}}(n)$ and an arm $\tau'$. Their first-order moment $m_1$ is found from (7.12) and (7.5) to be

$$A_{\mathrm{B}}(n)\tau' = n\Delta A(\mathrm{TE} - T), \tag{7.13}$$

and is independent of $\nu_y$. This latter fact makes it possible to to apply this method to flow with unknown velocity.

This concludes the discussion of flow-insensitive sequences. If no extra bipolar gradient waveforms are added, the first echo of the imaging sequence is sensitive to velocity. Needless to say that the extra gradient lobes take time and therefore increase the minimum echo time TE. In sequences where a small TE is required, partial velocity insensitivity can sometimes be realized.

## 7.5 Flow Imaging

The MRI methods used for the diagnosis of flow are collected in this textbook under the general name of "flow imaging" (FI). The discussion in this section will be restricted to the basic principles of flow imaging. For a detailed treatment of flow imaging and its applications we refer the reader to the available literature [3, 20–22].

The spins of flowing blood (or CSF) experience the different parts of the imaging sequence in a different way as compared with those of stationary tissue, due to their motion during the sequence. This is used in flow imaging. There are two main groups of methods.

- Phase-contrast (PC) methods. These methods are based on the phase difference between the transverse magnetization of blood (or CSF) and that of stationary tissue, which are at the same location during the time of the acquisition. The flowing blood has experienced the gradient fields earlier in the sequence than had the stationary tissue, and at that time it had another location (see Sect. 7.3).
- Modulus-contrast (MC) methods. This second group of methods makes use of differences in the magnitude of the transverse magnetization between moving spins and stationary spins, also arising from their different positions earlier in the measuring sequence. The first member of this group is the "inflow" method, depending on the excess signal strength of fresh in-flowing blood in a slice, in which the stationary tissue is saturated in a fast gradient-echo sequence (see Chap. 4). A new and important method is contrast enhanced MR angiography (CE-MRA). There are more members of the

group of MC methods, such as "time of flight" methods, which include various labelling techniques of blood flowing into or out of the slice [22].

The two objectives of flow imaging, projective views of the vessels (MR angiography, MRA) and quantitative flow measurements ($Q$ flow), are served with acquisition methods coming from both main groups, viz. the phase contrast methods and the modulus contrast methods, and all four combinations exist (see Table 7.1). Our discussion on flow imaging will follow the division in acquisition methods (PC and MC methods) in order to get a systematic theoretical treatment of all methods for flow imaging. The objective of each individual flow-imaging method, "angio" or "$Q$ flow", and the display of the results, will be stated and described in all separate cases.

**Table 7.1.** Acquisition methods and their use in flow imaging

| | Flow imaging[a] | |
|---|---|---|
| | Angio | $Q$ flow |
| Phase contrast | *** | ** |
| Modulus contrast | ***** | * |

[a] The number of *s shows the order of importance of the group of methods

### 7.5.1 Phase-Contrast Methods

As mentioned above the aim of phase-contrast methods is either to make angiograms to show the path of the arteries and veins (phase-contrast angiography or PCA) or to measure the velocity of flow at different positions (quantitative-flow measurements or $Q$ flow).

Phase-contrast flow imaging relies on subtraction methods. First an image that is flow-insensitive (in three dimensions) is made; the (complex) signal per pixel is given by $\vec{I}_\mathrm{i}(x, y)$. In the following the coordinates $x, y$ will not be repeated in the equations for the complex signal per pixel. Then images are sensitized for flow by changing the gradient waveforms, as discussed in the previous section, in one of the three directions $x$, $y$, or $z$. These images are characterized by $\vec{I}_{sx}$, $\vec{I}_{sy}$, and $\vec{I}_{sz}$, respectively. The flow sensitization is reached by addition in the sequence of bipolar gradient waveforms in the $x$, $y$ and $z$ directions. When flow-sensitive and flow-insensitive images are subtracted the difference image will be non-zero only at the location of the flow paths, where the magnetization has a different phase in each of the original images.

Spin-phase deviations due to inhomogeneities in the main magnetic field are compensated for by the subtraction. In principle the flow-insensitive image can also be replaced by three flow-sensitive images, $\vec{I}_{s'x}$, $\vec{I}_{x'y}$, and $\vec{I}_{s'z}$, with different (but known) flow sensitivity s'. These images are then subtracted from the corresponding images with sensitivity $s_1$. This, however, is time consuming, since six images are required. A better solution is based on four flow-sensitive images, $\vec{I}_{s\alpha}$, $\vec{I}_{s\beta}$, $\vec{I}_{s\gamma}$, and $\vec{I}_{s\delta}$, with flow sensitivity oriented in space as a regular tetrahedron (Hadamard). In the following it will be assumed that the subtraction scheme includes a flow-insensitive image and three flow-sensitive images along the three main directions. If flow is predominantly in one direction only two images are necessary.

### 7.5.1.1. Phase-Contrast Angiography

For phase-contrast angiography (PCA) the flow-insensitive image $\vec{I}_{\mathrm{i}}$ is subtracted from each of the flow sensitive images $\vec{I}_{sx}$, $\vec{I}_{sy}$, and $\vec{I}_{sz}$. In the subtracted images, pixels showing stationary tissue have zero signal (black background) because in the original images the signals are equal. In pixels showing flowing blood the signal of the subtracted images results from primary signals with different phase. The complex difference vector is obtained in each pixel (note that these difference images are less sensitive to aliasing due to phase wrap as shown in Fig. 7.6). Finally the three subtracted images are added by complex addition and the modulus is taken. The result is:

$$\left|\vec{I}_{\mathrm{PCA}}\right| = \left|\left(\vec{I}_{sx} - \vec{I}_{\mathrm{i}}\right) + \left(\vec{I}_{sy} - \vec{I}_{\mathrm{i}}\right) + \left(\vec{I}_{sz} - \vec{I}_{\mathrm{i}}\right)\right| . \tag{7.14}$$

PCA can be performed with multi-slice 2D or with 3D acquisition techniques and the result is a 3D data set containing the angiographic information as shown in Fig. 7.12a. As we shall see later, inflow methods lead to a similar 3D data set. These data sets can be visualized on a two-dimensional screen as shown in Fig. 7.12b,c,d. The images b, c, and d are the result of a maximum intensity projection (MIP) in different directions, as shown in Fig. 7.12a, which means that along a line in the projection direction, only the maximum intensity is imaged (see image set VII-11). When the direction of the MIP is continuously changed, a movie shows the vessels from all directions. It is, of course, also possible to use simple subtraction schemes, where only one velocity sensitized and the velocity-insensitive image are used.

### 7.5.1.2. Quantitative Flow Based on Phase Contrast

The subtraction of flow-insensitive and flow-sensitive images can be used to measure the flow velocity and to form quantitative velocity maps ($Q$ flow) [Historic Introduction 22].

Quantitative flow measurements can be described with the same formalism as also used for PCA. One flow-insensitive image and three images, flow sensitized in one of the three directions, are acquired. For each pixel the

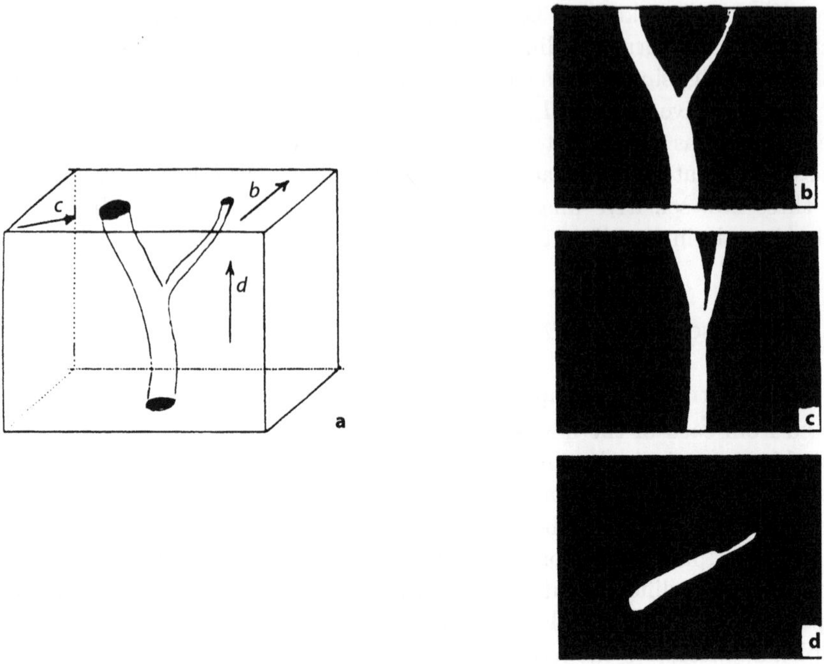


**Fig. 7.12.** Multi-2D or 3D image to be displayed in three directions with maximum intensity projection. Continuously changing the direction in one plane makes cine possible. **a** 3D image. **b–d** 2D images in the directions shown in **a**

phase of the flow-insensitive image is subtracted from the phase of the flow-sensitized images. The method can be characterized by

$$\vec{I}_{\text{Qflow}} = \vec{\imath}[\varphi(\vec{I}_{sx}) - \varphi(\vec{I}_{\text{i}})] + \vec{\jmath}[\varphi(\vec{I}_{sy}) - \varphi(\vec{I}_{\text{i}})] + \vec{k}[\varphi(\vec{I}_{sz}) - \varphi(\vec{I}_{\text{i}})], \quad (7.15)$$

where $\varphi(\vec{I})$ means the phase of the vector $\vec{I}$, which is calculated in each pixel. It will be clear that with the method described by (7.15), three-dimensional velocity plots can be obtained [23]. In a two-dimensional projection one can obtain in this way a map in which the velocity vector is shown in each pixel. A very original way of displaying the result is to simulate little signal voids in the vessels (bubbles) that move along with the blood [24].

The velocity vector is found from (7.6) and (7.15) to be

$$\vec{I}_{\text{Qflow}} = \vec{\imath}A_x\tau_x'\nu_x + \vec{\jmath}A_y\tau_y'\nu_y + \vec{k}A_z\tau_z'\nu_z, \quad (7.16)$$

where $A_x$, $A_y$, and $A_z$ are the (bipolar) gradient area differences between the flow-sensitized and flow-insensitive sequences. $A\tau'$ is the "first-order moment" of the bipolar gradient difference between both acquisitions. The magnitude of the phase can also be translated into a grey level which can be shown as an image. This grey level is proportional to the modulus of the velocity vector in each pixel. As an example we consider a gradient-echo sequence with a short

TR that is flow sensitive in the measuring direction (SE cannot be used, due to outflow between the excitation and refocusing pulses). After subtraction of a flow-insensitive image, the moving spins in the subtraction image will have a phase

$$\varphi(\mathrm{TE}) = \nu_x A_x \tau'_x, \tag{7.17}$$

where $\nu_x$ is the velocity in the measuring direction, $A_x$ now is the normalized area, and $\tau'$ is the "arm" of the bipolar gradient waveform difference of both sequences between excitation and echo.

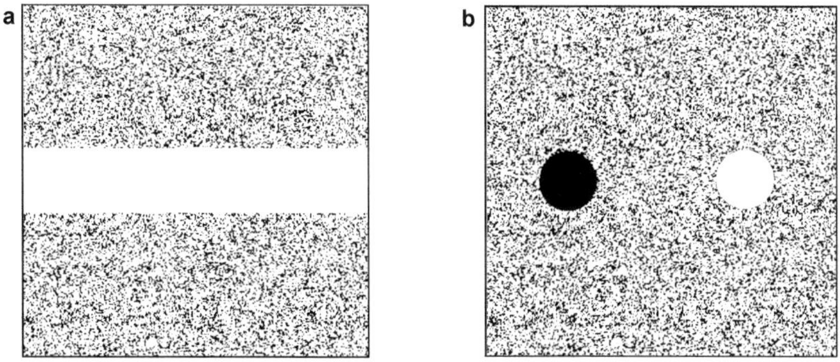

**Fig. 7.13.** **a** Subtraction PCA image of a vessel running in the measuring direction. Noise outside the vessel is due to the uncorrelated noise of the two images. **b** Image of vessels with opposite flow velocity perpendicular to a slice. Forward velocity is black, backward velocity is white, and static tissue is grey when noise is filtered

In the resulting subtraction image the grey level is taken to be proportional to the phase in a pixel, and so the velocity can be calculated as a function of the position from the measured grey level (note that velocities could also be measured in the two in-plane directions by interchanging the read-out and phase-encode directions). A typical image is shown in Fig. 7.13a in which flow is characterized by a definite grey level proportional to the velocity (beware of aliasing; see Fig. 7.6), and outside the flow area the image is noisy. This noise is due to the fact that the noise in the two subtracted images is uncorrelated and yields a random phase. This noise can be filtered out on the basis of the absolute value of the difference signal.

In Fig. 7.13b an example is shown of flow perpendicular to the selected slice. The image is the result of subtracting a scan in which the selection gradient is flow sensitive and one with a flow-insensitive selection gradient (see Fig. 7.10). The noise is filtered out on the basis of the length of the difference vector.

A practical example of measuring velocity perpendicular to the slice is found in the study of blood flow in a coronary artery [25]. The slices are taken perpendicular to the arteries as shown in Fig. 7.13b. The imaging method used is TFE, with which eight profiles are measured during a 110 ms period during diastole when the heart is relatively motionless. Two images are made, one flow insensitive and one flow sensitive in the selection direction. After subtraction the velocity of the blood flow can be determined. Practical values of the velocity depend on the artery considered and are of the order of 0.1 m/s.

With the method of Fig. 7.13a the in-plane flow can be measured and in this way the velocity as a function of the distance of a stenosis can be measured. This yields insight into the severity of the stenosis [26].

For clinical evaluation of flow, frequently the arterial flux to an organ is important. This parameter can be reached by taking the integral of the arterial flow velocities over the cross section of the vessel. The slices are taken perpendicular to the arteries as in Fig. 7.13b. In many cases the flow pulsates with the cardiac rhythm. In such cases it is necessary to trigger the PCA scans and to obtain flow values in a sufficient number of different times in the cardiac cycle. This type of studies shows for example promise for the evaluation of kidney stenotic vessels [74].

### 7.5.2 Modulus Contrast Methods

In this section the flow-imaging methods that are not based on phase difference but on modulus difference between flowing and stationary matter will be briefly discussed. The most important modulus-contrast angiography methods are the "inflow" and "contrast enhanced" methods. The "inflow" method is based on the inflow of fresh blood during an FFE sequence, which will be discussed in Sect. 7.5.2.1. The contrast enhanced method is described in Sect. 7.5.2.2. There are further modulus-contrast angiography methods, based on the use of RF pulses, which are added to the sequence for the purpose of labeling the blood. This will be described in Sect. 7.5.2.3. In Sect. 7.5.2.4 the "black-blood" method, based on the outflow of blood from the selected slice between excitation and refocusing in a SE scan, is discussed. Finally modulus-contrast quantitative flow measurement methods, which are also called "time-of-flight methods", are described in Sect. 7.5.2.6.

#### 7.5.2.1. Inflow Angiography

Since its introduction in 1982 [77] MR inflow angiography has been tried in a number of clinical problems. Its relevance is shown for vascular problems. In areas with sufficient vascular diameter and sufficient flow, such as the carotids, the method has gained clinical acceptance [75]. When pulsatile flow must be studied we can restrict the acquisition to the cardiac phase with high flow [76]. Image set VII-12 illustrates this technique for the upper legs.

In FFE sequences, with relatively high flip angles (30–60°) and with TR $\leq T_2$, stationary tissue is saturated by the repeated excitations. This gives a low signal from stationary tissue (see for example Fig. 4.22). Consider now a single-slice FFE sequence that is used to image a slice more or less perpendicular to the local blood vessels. The blood flows "fresh" (not excited earlier, so with the full longitudinal magnetization) into the excited layer and gives its maximum signal. So there is a contrast difference between the signal of the flowing blood and the suppressed signal of the static tissue, which can be visualized. The requirement that the blood is "fresh" is clearly that

$$\nu_z \mathrm{TR} > d, \tag{7.18}$$

where $\nu_z$ is the velocity of blood in the selection direction, $d$ is the slice thickness, and TR is the repetition time of the FFE sequence.

When this requirement is not fulfilled the blood will be excited a few times during its passage through the slice, thereby gradually loosing its excess signal, so that the angiographic contrast will decrease. This happens for example when one wishes to visualize the coronary arteries using inflow methods. The time available for imaging is a window of 110 ms during diastole [9]. During this time eight TFE profiles with TR = 12 ms and slice thickness 5 mm can be measured, and it is not probable that a steady state can build up in that time (110 ms). Furthermore, since the blood flow has a velocity of around 0.1 m/s in one TR period of 12 msec the blood advances only $12.10^{-4}$ m, or about 1 mm. At the downstream end of the slice the blood is excited five times and will be partly saturated. This problem does not occur when the number of excitations per heart beat is low, a condition that can be obtained in EPI imaging.

In regions where arteries and veins have opposite flow directions one can distinguish between the two by presaturating (REST, regional saturation) all the space on the upstream side of the arteries (or of the veins) with respect to the slice to be imaged. In that case only the veins (or arteries) show up in the image. One can repeat the single-slice scan for a number of contiguous slices and the result is a three-dimensional data set (see Fig. 7.14). Visualization of vessels in this data set as an angiogram occurs in the same way as discussed in connection with PCA (see Fig. 7.12).

Instead of acquiring a stack of single slices one can also apply 3D acquisition techniques. As in the multi-single-slice technique the stationary tissue is saturated by the repeated excitations with flip angle $\alpha$ and repetition time TR (see Fig. 7.14). However, in this case one must be careful that 3D acquisition is done in a volume that is thick in the $z$ direction and requires much acquisition time, so the blood experiences multiple excitations and may become saturated during its trip through the imaged volume. When the slice thickness $d$ is now replaced by the thickness of the imaged volume, (7.18) will not be satisfied. Therefore in comparison with multi-2D methods a smaller flip angle is used and as an extra measure the flip angle can be made to vary over space, so that on the upstream side of the of the excitation volume it has a relatively low value (Tilted Optimized non Saturating Excitation, TONE. This provision will prevent some of the saturation of the blood and distribute the brightness more evenly over the volume. Another method use to avoid a decrease in the blood intensity in the $z$ direction is to image several small

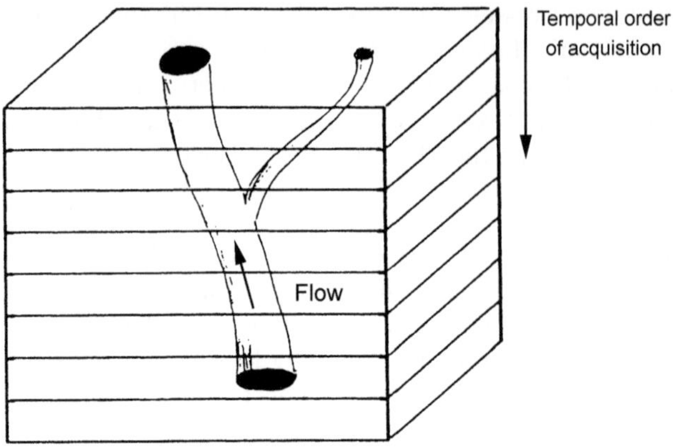


**Fig. 7.14.** Multiple-slice inflow image. The direction of flow is down-up

adjacent 3D regions (multi-chunk 3D or MOTSA, Multiple Overlapping Thin Slab Acquisition).

Another method is "outflow refreshment" [27]. The aim is again to image a number of slices more or less perpendicular to the vessels to be imaged. In this moving-slice method a gradient-echo sequence is used in which the slice selection rapidly advances over the slices to be imaged in the direction opposite to the blood flow. The slices are partially overlapping (overcontiguous) so that the static tissue in a slice is partially saturated by the excitation of the previous slice and gives a relatively low signal, whereas the blood is always fresh and gives a maximum signal and therefore maximum contrast. One profile per slice is acquired per excitation and this process is repeated until enough profiles are measured for reconstruction. When the heart beat is used for triggering for imaging the coronaries, one profile of all slices per heart beat is obtained. The motion of the heart deforms the arteries but they are imaged in a continuous way over their full length, so that malfunction due to stenosis becomes visible. The advantage of this method is that the full R–R interval is used.

### 7.5.2.2. Contrast-Enhanced MR Angiography

The application of contrast agents in MR angiography is rapidly attracting more attention because of its simplicity in comparison with other types of MRA [28–30]. Contrast-enhanced MR angiography (CE-MRA) relies on the shortening of the $T_1$ relaxation time of blood due to the application of a Gadolinium-based contrast agent, and it is relatively robust against artifacts. Contrary to inflow methods, motion plays only a secondary role (there are almost no saturation effects) and causes only mild artifacts. CE-MRA is also sensitive in cases of low blood velocity.

An arterial angiogram should not contain venous structures and must therefore be acquired before the contrast agent reaches the veins. The delay time between arterial and venous enhancement then determines the maximum available acquisition time. This delay time can be quite short, requiring very fast acquisition methods based on the application of powerful gradient systems, for example in the case of intercranial vessels and carotids. On the other hand, during this first pass a high concentration of the contrast agent can be realized by bolus injection. In peripheral CE-MRA, there is more time for acquisition (see image set VII-14).

CE-MRA imaging methods are usually based on spoiled gradient recalled sequences ($T_1$-FFE, see Sect. 4.5.3.3). The transverse magnetization at the echo of these sequences is given by (4.16):

$$M_{\mathrm{T}}(x,y) = M_0(x,y)\frac{\sin\alpha(1-E_1)\exp\left(-\frac{\mathrm{TE}}{T_2^*}\right)}{(1-E_1\cos\alpha)}, \tag{7.19}$$

where $E_1 = \exp(-\mathrm{TR}/T_1)$, TE is the echo time and $\alpha$ the flip angle. This equation formally holds for the steady state only (see Chap. 5). The duration of the transient, however, is very short and the steady state is established after a few RF pulses. This is due to the fact that, for CE-MRA, relatively large flip angles may be used. In Sect. 5.2.1 it has been shown that, assuming that spoiling is effective during the transient, the apparent relaxation time (see (5.6) and Fig. 5.1) becomes very short for large flip angles, so that the steady state is already reached after a few pulses. These statements will be illustrated with a numerical example.

Let us assume that the $T_1$ of blood is decreased to about 50 ms and its $T_2$ to about 30 ms (we shall show later that these values are realistic), and that the surrounding tissue has the relaxation parameters $T_1 = 700\,\mathrm{ms}$ and $T_2 = 50\,\mathrm{ms}$. The contrast between blood and background tissue is shown in Fig. 7.15, where the relative contrast of blood minus background is plotted against the flip angle. From this figure it is clear that flip angles $> 20°$ can be used for a large contrast. A large flip angle causes a very fast approach to the steady state. Using (5.6) for a flip angle of 0.5 radians, we find $T_{1\mathrm{app}}$ for both blood and background tissue to be about 25 ms, so that with a TR of 7 ms the steady state is reached after the fourth pulse. Of course, the assumption that spoiling is already effective from the first pulse is formally not true (see Sect. 4.2.7, which describes how spoiling is derived for the steady state only), but the result is shown to be good enough for our argument.

At the present time the available contrast agents are predominantly Gd chelates. Their effect on the relaxation times can be calculated with (6.13) and (6.14). The values of the constants in these equations are for most currently available contrast agents: $R_1 \cong 4\,\mathrm{s}^{-1}\,\mathrm{mM}^{-1}$ and $R_2 \cong 6\,\mathrm{s}^{-1}\,\mathrm{mM}^{-1}$. The quantity mM describes the concentration of contrast agent per ml: $1\,\mathrm{M} = 1\,\mathrm{mol/liter}$. The concentration of most commercial Gd chelates is $c_{\mathrm{ch}} = 0.5\,\mathrm{M}$.

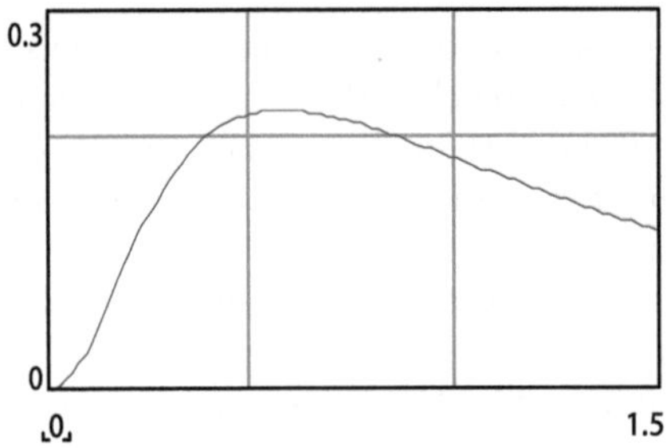


**Fig. 7.15.** Steady-state contrast $C(T_1) = \{M_T(\text{blood}) - M_T(\text{background})\}$ between contrast enhanced blood ($T_1 = 48$ ms and $T_2 = 28$ ms) and stationary tissue ($T_1 = 700$ ms and $T_2 = 50$ ms) as a function of the flip angle $\alpha$ (in radians), for TR = 7 ms and TE = 4 ms

We now have to calculate the concentration $C$ of the contrast agent in the blood during the first pass. Assume that the applied dose is $D_{\text{appl}}$ (mmol) and that the duration of the injection is $t_{\text{inj}}$ $s$. The injection speed is then $S_{\text{inj}} = D_{\text{appl}}/(c_{\text{ch}} t_{\text{inj}})$ ml/s. When this injected dose arrives in the heart, it is diluted with the average blood flow through the heart (the cardiac output), which is given by $\Phi_{\text{card}} \cong 100$ ml/s. Therefore the concentration during the first pass is given by $C = c_{\text{ch}} S_{\text{inj}}/\Phi_{\text{card}}$. We can now calculate the value of $T_1$ and $T_2$ of the blood when the dose is $D_{\text{appl}} = 10$ mmol in 20 s, which means $S_{\text{inj}} = 1$ ml/s. The concentration in the blood is $C = 5$ mM.

Using (6.13) and (6.14) and assuming the spin–lattice relaxation time of blood $T_1 = 1000$ ms and the spin-spin relaxation time $T_2 = 200$ ms, we find for the contrast-enhanced values that $T_{1\text{C}} = 48$ ms and $T_{2\text{C}} = 28$ ms, the values used in Fig. 7.15.

The timing of the injection of contrast agents with respect to the scan used is very important. At first the time of arrival of the first-pass bolus must be known to decide on the start of the angiographic scan [31]. This time can be measured with a fast dynamic single-slice scan following an injection. A method to increase the time resolution of this scan is to update the low-order profiles of the $\vec{k}$ plane more frequently than the high-order profiles [29]. The dose used for the injection is followed with saline to minimize the problem of residual contrast agents during the actual scan.

The duration of the injection should be matched to the duration of the scan used. Going back to the example of Fig. 7.15, the duration of a 3D $T_1$-FFE with a matrix of $256 \times 180 \times 16$ is 20 s.

The advantages of CE-MRA are the simplicity of the scan methods, leaving a large freedom of scan setup and reduced scan time compared with phase contrast angiography. Triggering is not always used, since the contrast also remains unchanged during periods of slow flow (diastole). However, changes in the diameter of the artery and measurement during the arrival of the bolus can lead to artifacts (see Sect. 7.5.2.6). The image quality is excellent (see the contrast calculation of Fig. 7.15) and can even be enhanced with subtraction methods.

Although the cost of a scan is increased, due to the contrast agent, CE angiography is rapidly finding application in clinical practice.

### 7.5.2.3. MR Angiography Based on Magnetization Preparation

With pre-pulses, as mentioned already in Sect. 2.7, the contrast between blood and surrounding tissue can be either established or improved. The contrast of inflow angiography, for example, can be improved with the application of magnetization transfer pulses (see Sect. 6.3.3), described in Sect. 7.5.2.3.1. Also, angiograms can be made using a subtraction method in which the inflowing blood is alternatively labelled with an inversion recovery pulse. In the resulting subtraction image, only the vessels containing labelled blood are visible – see Sect. 7.5.2.3.2. In other methods, use is made of the long $T_2$ relaxation time of oxygenated blood, which can be selectively addressed with a composite pre-pulse (a $T_2$ preparation sequence). This is discussed in Sect. 7.5.2.3.3.

Although the methods mentioned in this section are possibly not as important as the methods mentioned in Sects. 7.5.2.1 and 7.5.2.2, they illustrate the large number of possibilities for obtaining a desired contrast.

#### *7.5.2.3.1. MTC Pre-pulses*

In Table 6.1 it is shown that the Magnetization Transfer effect is very small in *in vivo* blood (due to the small relatively macromolecular content) and appreciable in the background tissue. This effect can be used to improve the contrast between the inflowing fresh blood and the saturated background, which obtains a lower $M_{\text{LF}}$ – see [32] and Sect. 6.3.3.

The MTC pulses, however, can also saturate the inflowing blood when this arrives from a region where the main magnetic field is no longer sufficiently homogeneous. To avoid this, a small gradient can be introduced, which makes the MTC pre-pulses further off resonance in the region from where the blood is washed into the region of interest [32].

#### *7.5.2.3.2. Arterial Spin Labelling*

The blood can be "labelled by inversion or saturation in a certain volume and detected in another slice. Detection occurs using a subtraction method of labelled and unlabelled images, as is shown in Fig. 7.16. This can be displayed as an angiogram. Of course, geometries other than that depicted in

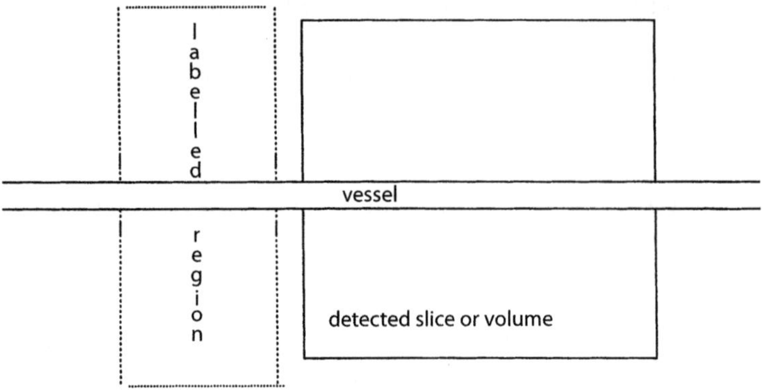


**Fig. 7.16.** Labelling is done outside the imaged slice

Fig. 7.16 can also be used. The imaging can be done with a variety of methods, including EPI, TSE, FFE, or TFE, each described in earlier chapters [33].

A practical implementation is described as "selective inversion recovery" (SIR) angiography [34]. The pulse sequence starts with an inversion pulse of the hyperbolic secant type [2.6]. When a selection gradient is switched on, this pulse inverts the magnetization in a homogeneous volume of defined width $W$. Without a selection gradient, the inversion pulse inverts all spins within reach of the excitation coil. After a time TI, the volume of interest with width $Z_e$, within the region defined by $W$ (see Fig. 7.17), is excited, and subsequently a 2D-FFE acquisition starts.

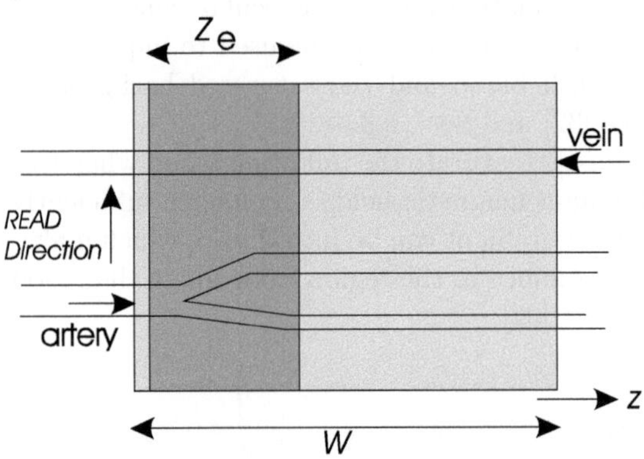


**Fig. 7.17.** Geometric situation in a SIR sequence. With a selection gradient, the inversion pulse excites a region with width $W$; without selection, all spins within reach of the transmission coil are inverted. So the arteries are inverted on alternative excitations. The veins and the stationary tissue within $Z_e$ are inverted in each excitation

By switching the selection gradient of the inversion pulse on and off on alternate measurements, two images are obtained. Only the region with washed in arterial blood is visible on the subtraction image; the venous blood and the stationary tissue are not visible thereon because both of the latter are always inverted (see Fig. 7.17). Phase encoding is in the $z$ direction and the read direction is in the transverse plane so that a projection image in a direction perpendicular to the $z$ and read directions can be formed. To avoid intra voxel dephasing due to turbulent flow, the TE of the FFE acquisition is made very short by using half-echo acquisition.

The scan method can be made faster by applying several excitation pulses after one inversion. This can be used, for example, for acquiring projection images in different directions in a quiet period of the heart cycle, or to obtain time-resolved images during the heart cycle without increasing the total acquisition time beyond acceptable limits. It is then required that the strength of the signals for the different images is equal. This can be achieved by varying the flip angle of the excitation pulses in such a way that they bring the same amount of longitudinal magnetization into the horizontal plane by variation of the flip angle. Before each excitation pulse, a spoiling gradient destroys the transverse magnetization and so, as a first approximation, we can consider longitudinal magnetization only (see also Sect. 5.2.1). Between two excitation pulses $n$ and $n+1$, the longitudinal magnetization relaxes back to its equilibrium value according to (2.8) and (5.1):

$$M_{z,n+1}^{-} = M_0 + (M_{z,n}^{-} \cos\alpha_n - M_0)\mathrm{e}^{-\frac{\mathrm{TP}}{T_1}}, \tag{7.20}$$

where TP is the time between the excitation pulses and $\alpha_n$ is the flip angle of the $n$th pulse. The measured signal of the region washed in blood in the difference image after the $n$th pulse, $S_n$, should be equal to the corresponding signal after the $(n+1)$th pulse:

$$\left(M_{z,n+1}^{-}(\mathrm{U}) - M_{z,n+1}^{-}(\mathrm{I})\right)\sin\alpha_{n+1} = \left(M_{z,n}^{-}(\mathrm{U}) - M_{z,n}^{-}(\mathrm{I})\right)\sin\alpha_n \tag{7.21}$$

where U means uninverted and I means inverted. Together with (7.20) this yields [34]:

$$\sin\alpha_{n+1} = \exp\left(\frac{TP}{T_1}\right)\tan\alpha_n \tag{7.22}$$

For a 4-excitation pulse sequence with maximum constant signal, we find that $\theta_1 = 29°$, $\theta_2 = 34°$, $\theta_3 = 44°$ and $\theta_4 = 90°$. Of course, these decreased flip angles result in a decreased signal-to-noise ratio.

In more recent literature, the method of selective inversion recovery and the forming of the difference image is referred to as selective targeting with alternating radiofrequency (STAR, [35]). A further discussion is given in Sect. 7.6 on perfusion.

There are more types of inflow labelling applied in angiography. It is difficult to be complete here, and so we restrict ourselves to two more types: double inversion pulses and $T_{1\rho}$ pre-pulses.

Double inversion pulses, where the first is non-selective and the second is selective, are applied in an "inverse" labelling scheme. The result of these pulses is that, in the slice to be imaged, the spins are not inverted, whereas outside this slice there is inversion. After some time the blood in the slice is replaced by blood from outside the slice and when this time TI is chosen correctly the inflowing blood will have a longitudinal magnetization equal to zero, resulting in a black blood image [36].

A $T_{1\rho}$ pre-pulse sequence consists of a $90^\circ_y$ pulse to bring the magnetization along (for example) the $x'$-axis, followed by a long "lock" pulse, exciting a $B_1$ field along the $x'$ axis. This lock pulse prevents $T_2$ dephasing of the spins. The spins are after some time brought back in the longitudinal direction with a 90° "flip-up" pulse. During the locking pulse there is interaction of the free water proton spins with the spins of the macromolecules, causing $T_{1\rho}$ decay depending on the relative amount of macromolecules and their properties, somewhat similar to the processes governing MTC. This prepulse was originally applied at very low fieldstrength, where the inhomogeneities of the main magnetic field are unimportant [37]. At high fieldstrength with large absolute values of field inhomogeneities, a composite $T_{1\rho}$ prepulse yields better locking [38]. A comparison of the different magnetization preparation schemes for improved blood-wall contrast in cardiac imaging is given in [39].

*7.5.2.3.3. Flow-Independent Angiography, $T_2$ Preparation Pulses*

Contrary to inflow or phase contrast angiographic methods, Flow Independent Angiography (FIA) does not rely on flow or, like CE-MRA, a contrast agent. The parameter that distinguishes oxygenated blood from most other tissues is its long $T_2$ relaxation time [40]. This parameter is addressed with a $T_2$ preparation pulse before the acquisition sequence, which selectively emphasises oxygenated blood with respect to other relevant tissue, like (partly) deoxygenated blood, muscle, fat and bone marrow. Since FIA does not rely on motion, it can depict vessels in any direction with slow and even with retrograde flow. An important application area of FIA could be the peripheral vascular system, the method, however, is still in an experimental phase.

A $T_2$ preparation sequence consists of a 90° pulse followed by a number of 180° pulses, after which the remaining transverse magnetization is brought back in the longitudinal direction by a 90° pulse (see in Fig. 7.18). During the $T_2$ preparation sequence there is $T_2$ decay, which is responsible for $T_2$ weighting in the image acquired following the preparation pulse. A special property of the $T_2$ preparation sequence is that the $T_2$ of the partly deoxygenated blood in the veins depends heavily on the time between the 180° pulses, $\tau_{180}$ [40].

This dependence of $T_2$ on $\tau_{180}$ is caused by dephasing due to diffusion in the varying fields of (paramagnetic) deoxygenated blood cells. When $\tau_{180}$ is increased from 5 ms to 40 ms, the $T_2$ of 30% deoxygenated blood lowers from

180 ms to 80 ms. This decrease is important for the contrast between arterial and venous blood; the effect can also be used directly for making arterial angiograms using TSE, as is shown in image set VII-15.

To improve the contrast between blood and muscle, the $T_2$-preparation pulse can be combined with an inversion pulse [41], as shown in Fig. 7.18. In this way the acquisition can start, for example, at the time that the longitudinal magnetization of muscle passes through zero. More generally stated, there are more free parameters to be chosen for optimal contrast in different situations.

An analytical expression for the contrast can be found by calculating the steady-state magnetization in a way similar to the method used in Sect. 5.2.1. As an example, we consider the sequence with only one excitation pulse with flip angle $\alpha_1$. The longitudinal magnetization before this pulse is assumed to be $M_{\mathrm{zss}}$, where the subscript ss stands for "steady state". After the excitation pulse the longitudinal magnetization is $M_{\mathrm{zss}} \cos\alpha_1$. This magnetization relaxes back to equilibrium during the time TR − TE − TI, according to (5.1). The resulting magnetization at $t = \mathrm{TR}$ is equal to $M_{zI} = M_0 + (M_{\mathrm{zss}} \cos\alpha_1 - M_0)\exp\{-(\mathrm{TR} - \mathrm{TE} - \mathrm{TI})/T_1\}$. This magnetization is inverted by the inversion pulse and is subsequently subjected to $T_2$ relaxation by the $T_2$-preparation sequence, according to $\exp(-\mathrm{TE}/T_2)$. After the $T_2$ preparation sequence, the magnetization is brought into the $-z$ direction by the second 90° pulse. Then $T_1$ relaxation occurs during the time interval TI, according to (5.1) again.

In a steady-state situation it may be assumed that the resulting magnetization after the $(n+1)$th excitation pulse is equal to $M_{\mathrm{zss}}$ after the $n$th pulse. It is left to the reader to complete this calculation. The result is:

$$M_{\mathrm{zss}} = \frac{M_0(1 - E_1 - E_1 E_2 + E_1 E_A)}{1 + E_1 E_A \cos\alpha_1} \tag{7.23}$$

where $E_I = \exp(-\mathrm{TI}/T_1)$, $E_2 = \exp(-\mathrm{TE}/T_2)$, $E_1 = \exp(-\mathrm{TR}/T_1)$, and $E_A = \exp(-\mathrm{TE}/T_2 + \mathrm{TE}/T_1)$. This equation can be used to find the optimum parameter setting for the different situations in which FI angiograms

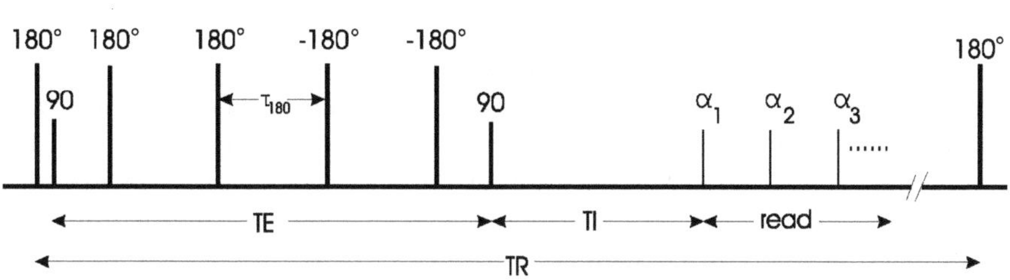


**Fig. 7.18.** $T_2$-Prep-IR pre-pulse. The duration of the $T_2$-Prep part is TE. Due to the initial 180° inversion pulse, the last 90° pulse brings the ($T_2$-coded) magnetization in the opposite direction to the main magnetic field. During TI, the remaining muscle signal relaxes to zero. The signal of the oxygenated blood is measured in the read sequences (only excitation pulses $\alpha_1, \alpha_2, \alpha_3 \ldots$ are shown)

are made. Equation (7.23) can easily be extended to include a sequence of excitation pulses [41].

The read-out sequence can be used to image a single thick slice, which shows a projection of the vessel of interest in a direction perpendicular to the slice, as discussed in Sect. 7.5.2.3. However it can be of interest to use 3D imaging, which makes possible MIPs (see Fig. 7.12) in different directions. Also, the contrast and the SNR are better. A disadvantage, however, is the longer scan time. To reduce scan time, segmented data acquisition can be applied by using several sequential excitations with increasing flip angle, so as to guarantee constant signal levels in the segments, as follows from (7.22). The read-out sequence can be a segmented EPI or spiral imaging, both in such a way that for further scan-time reduction the volume of $\vec{k}$ space is partially sampled – for example as spherical and completed as a cubic with zero padding. The 3D data sets can be used for MIPs in any direction or for segmentation of the desired structures by post processing.

An example of FIA is shown in image set VII-15. In the approach used there, strong $T_2$-weighting is used to obtain suppression of all tissue except blood.

#### 7.5.2.4. Black-Blood Angiography

The "black-blood" method is sometimes proposed as a supplement to the "brightblood" inflow method. The extent of a stenosis cannot reliably be determined by the latter method because of the "flow void" in FFE methods, which occurs behind a stenoses due to turbulent flow (inter-voxel dephasing). This flow void cannot be distinguished from the saturated signal of the static tissue, so the stenosis is overestimated. However, in the black-blood method the stenosis shows up in its correct dimensions.

In many cases, black blood images can be simply obtained by the use of a SE sequence. A slice is excited by the 90° excitation pulse and the spins are refocused by the 180° pulse (see Sect. 2.4). However, the blood will move completely out of the selected slice when

$$\nu_z \mathrm{TE}/2 > d,$$

where $d$ is the thickness of the imaged slice and TE is the echo time. In this case it will give no signal, creating contrast with static tissue that gives normal contrast (black blood) [42]. Most conventional SE scans can be used for blackblood angiography. Minimization of the scan duration can for instance be obtained with GRASE or SE-EPI. The signal-to-noise ratio achievable for black-blood angiography is limited.

A more powerful method for black-blood angiography is based on the black-blood pre-pulse, described in Sect. 2.7.2. This pre-pulse is usually combined with SE or TSE read out sequences, so that the effects of the black-blood pre-pulse and that of outflow combine. Image set VII-13 shows an example.

#### 7.5.2.5. Artifacts in Modulus Contrast Angiography

An artifact that is specific to modulus contrast angiography is due to the variation of the vessel diameter (about 10–15%) caused by pulsatile flow. Another artifact arises when a scan is performed during the increase or decay of the concentration of the contrast agent. Finally, turbulent flow during part of the heart cycle may also give rise to artifacts. In order to get some insight into these artifacts, we have developed a schematic model, which at least explains what can happen without trying to make a more exact model. Such a model is in principle possible, but it requires complicated simulations in practical cases [43–44].

We consider two situations: one with in-slice flow and one with flow perpendicular the slice. For the in-slice flow we assume that the vessel is straight and crosses the FOV. At first we consider a very thin vessel and construct the "image" in the $\vec{k}$ plane. This "image" of a thin vessel, $y = ax + b$, is calculated via (2.27) for the transverse magnetization in the object plane $m(x, y) = M_0\delta(y - ax - b)$ to be

$$\begin{aligned} M_\mathrm{T}(k_x, k_y) &= \iint\limits_{x,y} M_0\delta(y - ax - b)\exp\left(-j(k_x x + k_y y)\right)\mathrm{d}x\,\mathrm{d}y \\ &= M_0 \int_x \exp\left(-j(k_x + ak_y)x + bk_y)\right)\mathrm{d}x \\ &= M_0\exp(-jbk_y)\frac{\sin\left((k_x + ak_y)FOV/2\right)}{k_x + ak_y}\,. \end{aligned} \tag{7.24}$$

This equation shows us that the signal in the $\vec{k}$ plane has its maximum value when

$$k_y = -\frac{k_x}{a}\,. \tag{7.25}$$

Actually, (7.24) describes a "line spread" function in the $\vec{k}$ plane. This $\vec{k}$ plane "image" of any straight line in the object slice always goes through the origin – see Fig. 7.19. The intensity of the line in its longitudinal direction is modulated according to $\exp(-jbk_y)$.

A thick vessel can be considered as consisting of a number of parallel thin subvessels. Their "images" in the $\vec{k}$-plane coincide along one single line through the origin, however the modulation of the signal along this line for a certain sub-vessel, that passes the $y$ axis at $y = d$, is given by $\exp(-jdk_y)$, where $b < d < c$. This means that in the $k_y$ direction we have a superposition of spatial frequencies, which add to a sinc function similar to (2.18). When the diameter of the vessel increases, the width of the sinc function decreases, and vice versa. This sinc function can easily be evaluated.

When the $\vec{k}$ plane is scanned, the formation of artifacts depends on the direction of the vessel image in the $k$-plane in comparison with the phase direction. Assume, for example, that a vessel is in the $x$ direction ($a = 0$).

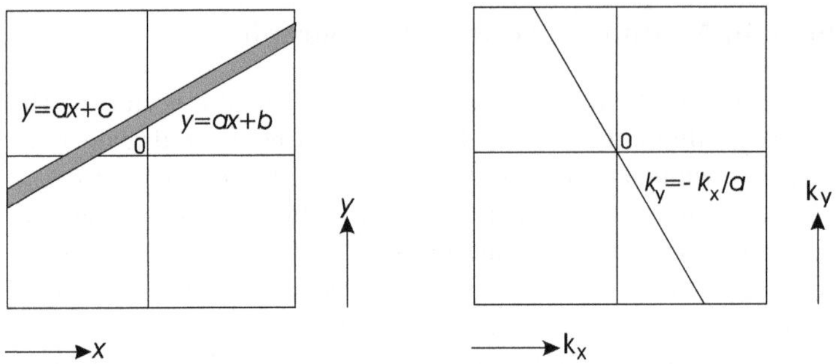


**Fig. 7.19.** Object slice with a vessel between the lines $y = ax + b$ and $y = ax + c$. Both lines transform to the line $k_y = -k_x/a$. The modulation of the signal along this line depends on the values $b$ and $c$

The "image" in the $\vec{k}$ plane is in the $k_y$ direction. In that case the different parts of the $\vec{k}$ plane "image" are scanned at different times during the width modulation of the vessel. So the width of the sinc function to be scanned periodically changes, and ghost artifacts arise in the phase direction of the object plane. If, in contrast, the vessel is in the $y$ direction ($a$ is infinite), then the $\vec{k}$ plane "image" is in the $k_x$ direction (the read-out direction) and the signal is measured in the time of a single profile measurement. During that short time, the vessel diameter is nearly constant and no artifacts occur.

When a vessel is perpendicular to the slice, the $\vec{k}$ plane "image" is rotationally symmetric (via a 2D sinc function) and a width modulation of the vessel will lead to an opposite width modulation of the $\vec{k}$ plane image, which always causes artifacts in the phase direction.

Measurement during a changing contrast concentration causes a changing intensity in the vessel. During scanning the intensity of the signal changes while traversing the $\vec{k}$ plane in the $k_y$ direction, so that when a vessel is in the $x$ direction the image will be blurred and shows some ringing when a linear profile order is used (compare Fig. 5.6b). In the case of a centric profile order during increasing intensity due to contrast, the high frequencies will be overaccentuated, which results in an intensity profile across the imaged vessel that shows maxima at the edges of the vessel, similar to the slice profile shown in Fig. 4.25. A vessel in the $y$ direction will appear in the image with its momentary properties at the time that the $k = 0$ profile is measured.

Finally, we construct a model for a periodic flow void (due to turbulence occurring at maximum flow velocity during the heart cycle). Assume that this flow void is in the isocentre and that the vessel is very thin (extension to thicker vessels is obvious but complicated). The varying intensity of the vessel in the isocentre can be described by the original vessel from which in the isocenter a voxel with varying intensity is subtracted. The "image" of this voxel in the $\vec{k}$ plane is constant throughout the $\vec{k}$ plane but has a varying amplitude depending on the occurrence of turbulence. Such an object will be imaged with ghosts in the phase direction. So, for a vessel in the phase direction, these ghosts will overlap with the vessel.

In conclusion, we can state that it has been shown that the types of artifacts in modulus contrast angiography can be understood using simple arguments, based on the theory developed earlier in this book. However, the exact appearance of artifacts in practical situations may become very complicated, and their description requires complicated simulation models.

#### 7.5.2.6. Modulus-Contrast Quantitative Flow Measurements

These methods make use of a refocussing RF pulse which addresses a refocussing slice that is different from the excited slice in a SE sequence. Only flowing blood that has the correct velocity for it to move in the time delay between excitation and refocussing from the excited slice into the refocussing slice will show up in the image. In the example of Fig. 7.20a, the selected slice and the refocussing slice of an SE sequence are parallel.

In the time between the refocussing pulse and the echo, the blood moves further but this has no influence on the signal. This method can be extended by stepping down the refocussing plane along the vessel. Also the refocussing slice can be taken parallel to the vessel (see Fig. 7.20b). In that case the flow profile, which is frequently assumed to be parabolic (Poiseuille flow), can be shown and the blood velocity measured [19].

In one of these methods, a thick region perpendicular to the blood stream is pre-saturated and the inflow into this region of fresh blood from outside of it is studied. This can be done by imaging a thin slice within the pre-saturated thick region in which only the incoming unsaturated blood will give a signal (see Fig. 7.21). When we image this thin slice with a fast imaging method (an FFE, TFE, EPI, or spiral sequence), after making an assumption on the flow profile as a function of time (for example, a quadratic Poisseuille profile), we can calculate how much blood is transported through the saturated region

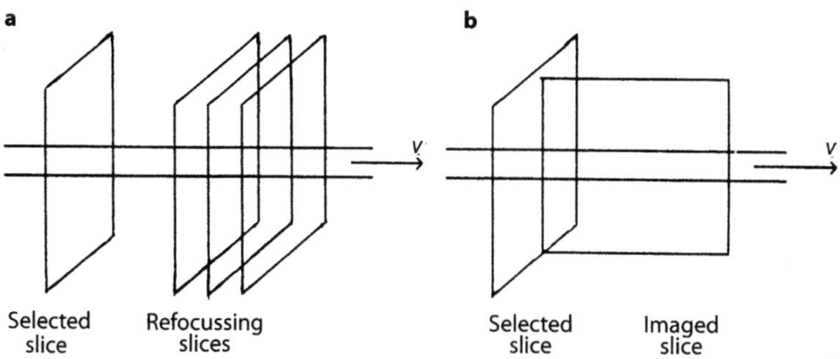


**Fig. 7.20.** Time-of-flight method for imaging flow. **a** Refocussing slices parallel to selected slice. **b** Refocussing slice parallel to blood vessel

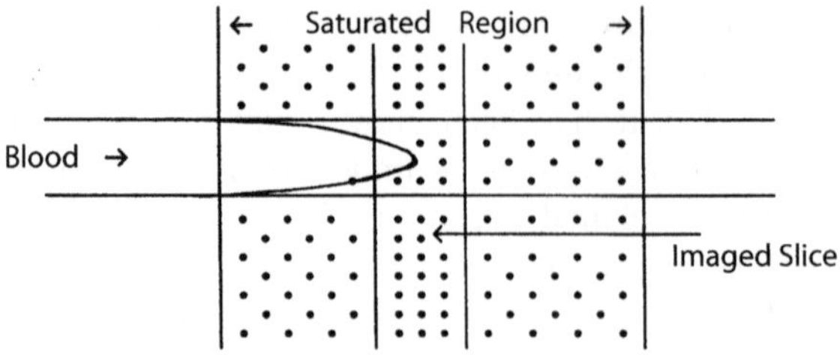


**Fig. 7.21.** Inflow of fresh blood in a saturated region

into the imaged slice. In combination with cardiac triggering and EPI imaging, this method has been applied to the transport of blood through coronary arteries with and without stress [45].

## 7.6 Perfusion

Perfusion is the transport process by which oxygen and other nutrients are delivered to the cells. It is a key factor in the functional evaluation of tissue, and various approaches have been suggested to obtain information on this process through the use of magnetic resonance. In these approaches, use is made of the motion of the blood in the capillary bed that pervades each tissue. When the blood is suitably labelled, perfusion can be observed by MR methods. Measurement of perfusion by the use of radioactive indicator solutions based on Positron Emission Tomography (PET) or Single Proton Emission Computed Tomography (SPECT) have a much longer history [46], and the theoretical framework for this type of measurement is well developed. Therefore, only its outline will be treated here.

We will first deal (in Sect. 7.6.1) with the analysis of the signal enhancement during the passage of a bolus of contrast agent. In Sect. 7.6.2 a discussion will be given of a more recent approach in which the blood is labelled with an MR technique.

### 7.6.1 MR Perfusion Imaging with Dynamic Bolus Studies

This subsection concentrates on methods suitable for assessment of brain perfusion. The restriction to this anatomical area simplifies the analysis of the dynamic bolus study because the blood–brain barrier will prevent leakage of the normal clinically available contrast agents, such as Gd-DTPA, so that the agent behaves as a non-diffusable indicator substance labelling the blood. (In animal research some attempts have been made to measure flow using Deuterated water (DHO), $H_2^{17}O$, and as a blood $T_2$ contrast agent, $^{19}F$; these methods will not be discussed in this book.) Information on brain perfusion is of special interest to the clinician because it is a relevant indicator of the status of the patient with brain ischemia or stroke.

For the analysis of perfusion in a certain tissue, the ideal arterial input function is reached when the arterial concentration of the contrast agent is high during a short time period and zero before and after. For a bolus of 1 mmol delivered in 1 s in an artery, with an arterial flow of $\Phi$ [ml/s], the arterial concentration $a$ is $1/\Phi$ [mmol/ml]. Such a "sharp" arterial input is shown by the upper trace in Fig. 7.22.

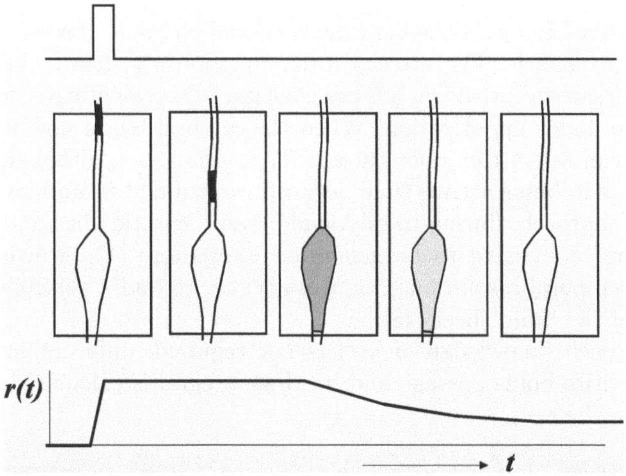


**Fig. 7.22.** Schematic representation of the capillary vascular bed in a voxel, with the distribution of a contrast agent at five instants in time after arrival of a sharp bolus of this agent; the temporal bolus shape is given in the upper trace, and the response is given in the lower trace

Figure 7.22 shows schematically, for a single voxel, how the capillary net could be distributed. At the arterial side (the upper side in the figure), its diameter is small and the flow is plugwise. At the venous (lower) side, the capillary net behaves as a mixing chamber. When the MR signal increases linearly with the amount of contrast agent in the voxel, the response function of the voxel, $r(t)$, indicates the fraction of the agent that is left in the voxel after the sharp input. This response function is shown in Fig. 7.22 as well: it is dispersed with respect to the input. Immediately after bolus arrival, it has a value 1 and after some delay it decays to zero.

When the artery in Fig. 7.22 transports all the blood that reaches the voxel, its flow defines the perfusion of the voxel. The perfusion rate is expressed in milliliters of blood per second per milliliter of tissue, and it has the dimension [1/s]. When a sufficiently sharp bolus is given with experimentally controlled duration $\Delta$ and arterial concentration $a$ [mmol/ml] in a tissue with perfusion rate $F$ [$s^{-1}$], the average concentration $c(t)$ [mmol/ml] of the agent in the voxel is related to $F$ as:

$$c(t) = Fr(t)a\Delta \qquad \text{(sharp bolus)}. \qquad (7.26)$$

$F$ can be found directly by observing $c(t)$ at an early time $t$ where $r(t)$ still is equal to unity. In practice however, sufficiently sharp boluses cannot be produced by intravenous injection. This means that, as well as the voxel concentration $c(t)$, the arterial concentration $a(t)$ also has to be measured. When this is done over the entire time of bolus passage, $F$ can still be obtained together with $r(t)$ by deconvolution of:

$$c(t) = Fr(t) \otimes a(t) \,. \tag{7.27}$$

Various attempts have been published in which $a(t)$ and $c(t)$ were observed and used to derive $r(t)$ and $F$. The arterial input function $a(t)$ has to be measured in a suitable artery in which the bolus shape is representative of its shape in the tissue under investigation. When the cerebral blood flow is to be measured, one can select the inner carotid for this purpose, although correction for the delay in bolus arrival time between carotid and brain may be necessary. In this approach, during the dynamic scan a carotid slice and the slices through the brain have to be generated alternately [47], which implies some loss of temporal resolution. One can also try to find a suitable arterial pixel in one of the brain slices [48].

In a simpler approach, knowledge of $a(t)$ is not required; only $c(t)$ is observed during the entire bolus passage and its time integral is calculated. From (7.26), it follows that:

$$\int c(t)\mathrm{d}t = V \int a(t)\mathrm{d}t \,, \tag{7.28}$$

where $V$ is the fractional blood volume in the voxel of interest. In (7.28) use has been made of the fact that $\int r(t)\mathrm{d}t = V/F$ (Central Volume Theorem, [49]). The obvious disadvantage is that the integral in (7.28) does not depend on the perfusion rate. Nevertheless, the approach is used clinically to inspect brain perfusion [50], the rationale being that perfusion and fractional blood volume frequently change together in pathological circumstances and so, although no flow is measured, the value of $V$ can be diagnostically determined.

Even in this simpler approach, some problems remain. Firstly, the tissue response is observable only during the first pass of the contrast agent. The return of the bolus in a second pass prevents the observation of the tail of the curve, and the value of the integral in (7.28) has to be estimated. This is usually done by assuming that the response has the shape of a gamma-variate function [46] (see also Fig. 7.23). Next, the calibration of $V$ requires knowledge of $\int a(t)\mathrm{d}t$, which is not observable – only the *relative* value of $V$ can be measured. This parameter is usually called rCBV (relative cerebral blood volume). This is not too much of a problem when this relative value is determined for each voxel and when the diagnosis is based on the image of these values.

The measurement of $c(t)$ is based on the enhancement of $T_2^*$ decay during bolus passage. The decay rate $R_2^*$ is pronounced because of the inhomogeneous distribution of the contrast agent (only intra-vascular) and the associ-

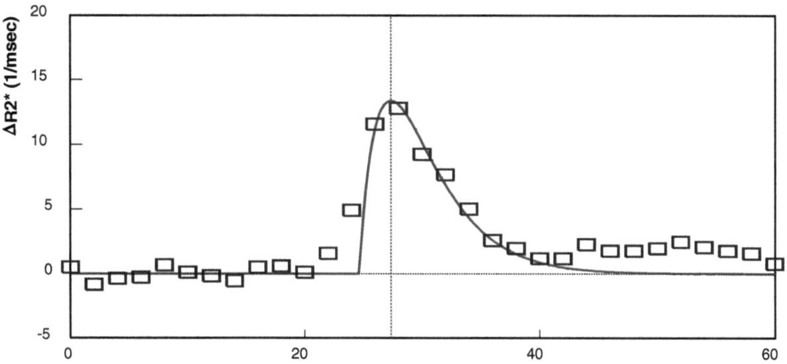


**Fig. 7.23.** Transient values of R2* observed using a GR-EPI method during bolus passage of a contrast agent through the brain, fitted to a gamma-variate curve: $\Delta R2^* = t^p \exp(-t/q)$

ated microscopic magnetic field inhomogeneity (susceptibility contrast). The exact relation between $\Delta R_2^*(t)$ and $c(t)$ is not easy to predict; it will depend on the size, direction and distribution of the capillary vessels in the tissue, as well as on the strength of the main field. The relation was studied through Monte Carlo calculations, in which the magnetic field deviations in the vessel bed were simulated, and by experiment [51]. It was shown that, in the range of concentrations of interest, $\Delta R_2^*$ varies approximately linearly both with $V$ and with $a$, which means that for a given bolus $c(t)$ is proportional to $\Delta R_2^*(t)$. The observed signal $S(t)$ during bolus passage relates to the native signal $S(0)$ and $c(t)$ according to:

$$c(t) \cong \frac{-1}{\mathrm{TE}} \ln\left(\frac{S(t)}{S(0)}\right) . \tag{7.29}$$

Sequences that are used commonly for perfusion imaging based on dynamic bolus study are FFE or Gradient Echo EPI sequences. A low flip angle is used to avoid influence of the $T_1$ change brought about by the contrast agent, and an echo time of about 30 ms is used to create a clear susceptibility contrast. Alternatively, a Turbo Spin Echo sequence can be used with a TE of about 100 ms.

Finally, MR perfusion imaging with dynamic bolus studies can be used without any further modelling to determine the bolus arrival time (or time to peak response). Images of such parameters can also be of diagnostic value, as delayed arrival is associated to reduced perfusion [52]. Examples of some images obtained by dynamic bolus tracking are shown in image set VII-16.

### 7.6.2 Arterial Spin Labelling

An interesting alternative for the measurement of the perfusion rate $F$ is given by the technique of arterial spin labelling. In this technique, the blood is labelled by changing the state of its longitudinal magnetization using prepulses in the imaging sequence. The labelling is applied to the arterial blood in a region upstream of the slice that is being imaged.

Arterial spin labelling can be said to create an endogenous indicator. The indicator substance is the free-water pool in the blood with altered spin state; it is diffusable through the interstitial and cellular compartments of the voxel. Before it arrives in the voxel, it has a lifetime that is given by the $T_1$ of blood. This situation resembles that of PET studies with short-lived radiolabelled diffusible indicators. The indicator (the labelled spins) enters the slice at a constant rate, exchanges with tissue at the capillary level, and the resulting change in magnetization reflects the ratio of the perfusion rate $F$ and the perfusion-free tissue relaxation rate $1/T_1$. The differential equation for the longitudinal tissue magnetization $M$ in the image voxel is [33]:

$$\frac{\mathrm{d}M}{\mathrm{d}t} = \frac{M_0 - M}{T_1} + F(M_a - M_v) \tag{7.30}$$

with $M_a$ the magnetization in the arteries, $M_v$ that in the veins and $M_0$ the magnetization of the brain tissue, including the blood fraction.

Before the blood reaches the veins, the exchange of magnetization is assumed to be complete, and $M_v$ can be replaced by $M/\lambda$, where $\lambda$ is the factor that takes into account the difference between the proton density in blood and that in the brain tissue. It is called the tissue–blood partition coefficient [53] and is usually assumed to have a value of 0.9. (Note that the observed signal may not be an adequate representation of the mean magnetization in the voxel, because the labelled arterial magnetization does not exchange immediately with that of the static tissue and complete exchange may not even be reached.)

The variable $M_a$ is constant and is imposed by the way the arteries are labelled. This means also that the product $FM_a$ is constant and can be lumped with $M_0$ in a new term $M_\infty^{\mathrm{sat}}$. So, (7.30) can be rewritten as:

$$\frac{\mathrm{d}M}{\mathrm{d}t} = \left(M_\infty^{\mathrm{sat}} - M\right)\left(\frac{1}{T_1} + \frac{F}{\lambda}\right), \tag{7.31}$$

where the bracketed factor is the flow-dependent apparent relaxation rate $1/T_{1\mathrm{app}}$, $M_\infty^{\mathrm{sat}}$ is the steady state magnetization and can be expressed [71] as function of the degree of inversion $\alpha = \frac{1}{2}\left(1 - \lambda\frac{M_a}{M_0}\right)$ of the arterial spins, and $\alpha$ is zero when there is no labelling and increases to unity when there is complete inversion. The steady-state magnetization $M_\infty^{\mathrm{sat}}$ now becomes:

$$M_\infty^{\mathrm{sat}} = T_{1\mathrm{app}}\left(\frac{M_0}{T_1} + FM_a\right) = M_0\left(1 - \frac{2\alpha F T_{1\mathrm{app}}}{\lambda}\right). \tag{7.32}$$

Equation (7.32) shows that $F$ is obtained directly and quantitatively from $M_\infty^{\mathrm{sat}}/M_0$ and $T_{1\mathrm{app}}$. The value of $T_{1\mathrm{app}}$ has to be determined separately; and to observe the ratio of $M_\infty^{\mathrm{sat}}$ and $M_0$, imaging has to be repeated twice, once without arterial labelling (the control scan) and once with labelling. As long as both sequences are the same, the ratio of the signals observed equals $M_\infty^{\text{sat}}/M_0$, because $T_{1\text{app}}$ does not depend on the labelling, as shown in (7.31).

To maximize the sensitivity of the technique, complete inversion is desirable ($\alpha = 1$). Even then, the values of $F$ to be observed are so low that $M_\infty^{\text{sat}}$ differs from $M_0$ by only a few per cent. This calls for a precise labelling technique, in which even slight crosstalk between the labelling procedure and the magnetization in the imaging slice has to be accounted for. Crosstalk can be reduced when use is made of a decoupled separate transmission coil to perform the labelling at a sufficient distance. In that case, an additional advantage stems from the possibility to use continuous low power RF radiation for "flow-induced adiabatic inversion" [53], in which technique complete inversion is reached reliably. In brain perfusion imaging, this could be done (for instance) at the level of the carotids [54], but the required provision of a double transmission coil is not normally available in the clinical MR system.

In the more usual system with only one transmission coil, it is possible to fold into the imaging sequence a high-quality inversion pre-pulse, given to a slab of tissue parallel to, and upstream of, the image slice. In that case crosstalk is much larger, and its major contribution is from magnetization transfer. Elimination of the influence of magnetization transfer is needed and can be accomplished by adding to the control scan a copy of the labelling RF pulse that is used to address a slab of tissue symmetric to the labelled slab but at the downstream side of the image slice. The use of this technique in combination with pulsed arterial labelling is called STAR (Signal Targeting with Alternating Radiofrequency) [55]. Various refinements in the sequence are needed to get rid of the remainder of the crosstalk that arises, for instance, from imperfect inversion slab profiles [56].

A somewhat different approach is used in the FAIR sequence (Flow-sensitive Alternating Inversion Recovery) [57]. Here, an inversion pulse is given to the image slice in both the control scan and the labelling scan. In the control scan this is a slice-selective RF pulse and in the labelling scan the same RF pulse is given without a slice-selection gradient. The advantage is that the distance between the labelled region and the image slice is minimal, and the lack of precision of the flanks of the inversion profile plays no role.

Even with sufficient control of crosstalk, the arterial spin labelling techniques suffer from the delay needed to allow the inverted spins to travel to the image slice and reach the capillaries, where they can exchange with the spins in the stationary tissue. In the the human brain, this is about one second [58]. The decay of the inverted blood during this time lowers the degree of inversion $\alpha$ and hence the sensitivity. Moreover, the dispersion in the travel time gives a further loss in $\alpha$ because, at a given image delay, only part of the inverted spins have reached the image slice. When the labelling region is removed further from the image slice, the travel time and its dispersion increase. This becomes heavily problematic when the transit time is unknown

– for instance when it is changed due to the presence of pathology such as a stroke.

All these problems together seriously limit the power of the arterial spin labelling technique in the human brain. Nevertheless, new refinements of the technique are being developed (see, for instance [70], from which image set VII-17 is taken), motivated by the attractiveness of its principle in which non-invasive imaging and quantitative information of flow can be combined. Recently a general kinetic model for quantitative perfusion imaging with arterial spin labeling, describing both the pulsed and continuous ASL methods, was published [86].

## 7.7 Diffusion

In analogy with flow, random thermal motion of spins in a gradient field causes a phase shift of their transverse magnetization with respect to static spins [59, 60]. For example, in a Spin-Echo sequence this means that the rephasing after the refocussing pulse is no longer complete, because of the changed position of the individual spins. The maximum transverse magnetization in the echo top is decreased due to the phase differences between the magnetization vectors of the individual spins.

This "attenuation" depends on the gradient history and the diffusion coefficient, $D$, of the protons in the tissue considered. This diffusion coefficient is in practice governed by the diffusion of water. In tissues the diffusion is restricted by barriers, confining the random walk to restricted volumes. When considered within a short time slot this restriction does not change the diffusion coefficient, but during the longer time periods involved in MRI acquisition the observed diffusion coefficient *is* restricted. In that case we speak of an *apparent* diffusion coefficient $D_{\mathrm{a}}$.

The restriction of random motion may be direction-dependent because of the structure of the tissue. In such cases the apparent diffusion will be anisotropic. Leaving out the subscript a, we can write the apparent diffusion coefficient as:

$$\overset{\Rightarrow}{D} = \begin{pmatrix} D_{xx} & D_{xy} & D_{xz} \\ D_{yx} & D_{yy} & D_{yz} \\ D_{zx} & D_{zy} & D_{zz} \end{pmatrix}.$$

To study the attenuation due to diffusion analytically, we refer to the Bloch equation for the precession of the transverse magnetization, $M_x + jM_y$, in a gradient field where according to (2.9)

$$\frac{\mathrm{d}M_{\mathrm{T}}}{\mathrm{d}t} = -j\gamma(\vec{G}\cdot\vec{r})M_{\mathrm{T}} - \frac{M_{\mathrm{T}}}{T_2}. \tag{7.33}$$

This is actually a continuity equation: the rate of change of the transverse magnetization, $\mathrm{d}M_{\mathrm{T}}/\mathrm{d}t$, is equal to the rate of change due to precession (the first form on the right-hand side) minus the loss of transverse magnetization due to $T_2$ relaxation.

The transport of a macroscopic quantity such as the transverse magnetization, due to anisotropic diffusion of the free water carrying this quantity, can be written as:

$$\vec{j}_{M_\mathrm{T}} = \vec{\vec{D}} \nabla M_\mathrm{T}, \tag{7.34}$$

where $\nabla$ is a column vector defined by: $\nabla^\mathrm{T} = \left(\vec{i}\frac{\mathrm{d}}{\mathrm{d}x}, \vec{j}\frac{\mathrm{d}}{\mathrm{d}y}, \vec{k}\frac{\mathrm{d}}{\mathrm{d}z}\right)^\mathrm{T}$, where the superscript T means "transposed".

When diffusion is to be taken into account in (7.33), we must add the net inflow of transverse magnetization into the volume element considered [61]:

$$\frac{\mathrm{d}M_\mathrm{T}(t)}{\mathrm{d}t} = -\left(j\gamma\vec{G}(t)\vec{r} + \frac{1}{T_2}\right) M_\mathrm{T}(t) + \nabla^\mathrm{T}(\vec{\vec{D}} \nabla M_\mathrm{T}). \tag{7.35}$$

Note that $M_\mathrm{T}$ is a complex number describing the amplitude and the phase of the transverse magnetization. Therefore the last term is also a complex number. The solution of (7.35) is found by first solving the continuity equation without the diffusion term (note the similarity with solving (2.14)):

$$\begin{aligned} M_\mathrm{T}(t) &= M_\mathrm{T}(0) \exp\left(-j\gamma\vec{r}\int_0^t \vec{G}(t')\mathrm{d}t' - \frac{t}{T_2}\right) \\ &= M_\mathrm{T}(0) \exp\left(-j\vec{k}(t)\vec{r} - \frac{t}{T_2}\right). \end{aligned}$$

Here we introduce $\vec{k}(t) = \gamma \int_0^t \vec{G}(t)\mathrm{d}t$, in accordance with (2.26). In order to solve (7.35), $M_\mathrm{T}(0)$ is now replaced by a function $A$ of $t$:

$$M_\mathrm{T}(t) = A(t) \exp\left(-j\vec{k}(t)\vec{r} - \frac{t}{T_2}\right). \tag{7.36}$$

Introducing this into (7.35) yields:

$$\frac{\mathrm{d}A(t)}{\mathrm{d}t} = \nabla^\mathrm{T}\left(\vec{\vec{D}} \nabla A(t)\right) = -\vec{k}^\mathrm{T}(t)\,\vec{\vec{D}}\,\vec{k}(t)A(t),$$

where we replaced $\nabla$ by $-j\vec{k}(t)$, which is also a column vector. The solution of (7.35) is then given by

$$\begin{aligned} M_\mathrm{T}(t) &= M_\mathrm{T}(0) \exp\left[-\int_0^t \left(\vec{k}^\mathrm{T}(t')\,\vec{\vec{D}}\,\vec{k}(t')\mathrm{d}t'\right)\right] \\ &\quad \exp\left(-j\vec{k}(t)\vec{r} - \frac{1}{T_2}\right). \end{aligned} \tag{7.37}$$

The first exponent describes the attenuation due to (apparent) anisotropic diffusion, and the second exponential describes (as usual) the precession and relaxation.

### 7.7.1 Measurement with Diffusion Sensitization in One Direction

Initially the one-dimensional solution will be considered [61]. In that case, $G_y$ and $G_z$ are zero, so that $\vec{k}^{\mathrm{T}}(t) = (\vec{i}k_x(t), 0, 0)$, and the first exponent in the right-hand side of (7.37) becomes:

$$D_{xx} \int_0^t k_x^2(t')\mathrm{d}t' = b_x D_{xx} \,. \tag{7.38}$$

Since the transverse magnetization acquired at the lowest values of $\vec{k}$ determines the contrast in an image, we shall evaluate (7.37) at the echo top when

$$\int_0^{\mathrm{TE}} k_x(t)\mathrm{d}t = 0. \tag{7.39}$$

In that case we can rewrite (7.37) as:

$$M_{\mathrm{T}}(\mathrm{TE}) = M_{\mathrm{T}}(0) \exp(-bD_{xx}) \exp\left(-\frac{\mathrm{TE}}{T_2}\right), \tag{7.40}$$

where

$$b_x = \int_0^{\mathrm{TE}} k_x^2(t)\mathrm{d}t = \gamma^2 \int_0^{\mathrm{TE}} \left(\int_0^{\tau} G_x(\tau')\mathrm{d}\tau'\right)^2 \mathrm{d}\tau \,. \tag{7.41}$$

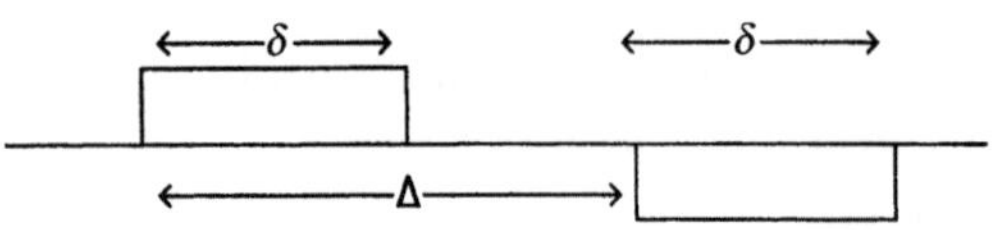


**Fig. 7.24.** Bipolar gradient waveform for diffusion sensitivity

The factor $b_x$ (the diffusion weighting factor) can be evaluated for a bipolar gradient as shown in Fig. 7.24, which satisfies the condition given in (7.39). The result is:

$$b_x = \gamma^2\delta^2 \left(\Delta - \frac{1}{3}\delta\right) G_x^2. \tag{7.42}$$

In the special case of adjacent gradient lobes, where $\Delta = \delta = \mathrm{TE}/2$, we find:

$$b_x = \frac{\gamma^2}{12}\mathrm{TE}^3 G_x^2. \tag{7.43}$$

The method of diffusion sensitization (diffusion weighting) with bipolar gradients is compatible with many imaging sequences. At first a Spin-Echo sequence will be considered. The two gradient lobes are inserted between the excitation pulse and the refocussing pulse and between the refocussing pulse and the acquisition gradient, respectively, as shown in Fig. 7.25.

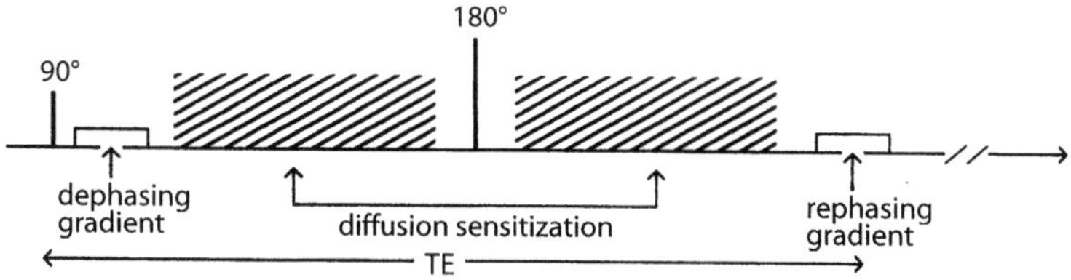


**Fig. 7.25.** Spin-Echo sequence with extra gradients for diffusion sensitization

A problem with diffusion measurements is that the diffusion sensitization gradients must be very high. Take, for example, the situation in which the attenuation due to diffusion is 10%, so $\exp(-bD)$ is 0.9. The diffusion coefficient of water at room temperature is $D = 2.2 \times 10^{-9}\,\mathrm{m^2/s}$. This means that $b$ must be $4.5 \times 10^8\,\mathrm{s/m^2}$. For an echo time of 60 ms we obtain from (7.43) for the gradient a value of about 18 mT/m. Since the gradient can be switched on only 80% of the time, because of the time needed for dephasing, rephasing and the refocussing pulse, we have $\delta = 24\,\mathrm{ms}$ and $\Delta = 28\,\mathrm{ms}$. In this case the gradient needed is found from (7.41) to be 23.3 mT/m. This illustrates a problem with diffusion imaging: very large gradients are needed and even so the effect is small, so that accurate methods are necessary to obtain an observable effect.

Clearly these methods consist of two measurements, one with diffusion-sensitizing gradients, yielding an image $M_{Tb}(x, y)$, and one without diffusion sensitizing gradients $M_{T0}(x, y)$. A diffusion-image is formed according to (7.40) by:

$$D(x, y) = \frac{1}{b} \ln \left( \frac{M_{Tb}(x, y)}{M_{T0}(x, y)} \right) . \tag{7.44}$$

In reality, the decay is steeper than that given by diffusion alone, because not only diffusion but also perfusion is the cause of signal decay, especially at low $b$ ($b < 150\,\mathrm{s/mm^2}$) [60]. To avoid the influence of perfusion in the calculation of $D$ from (7.44), it can be important to use in the denominator the data from a weakly diffusion-sensitised image.

The accuracy of large diffusion-gradient lobes is subject to very stringent requirements, because small errors (or eddy currents) in the large diffusion gradients will have a large influence compared with dephasing and rephasing gradients. However, modern systems that comply with these requirements are becoming available.

Another problem is the involuntary macroscopic motion of the patient. Because of the extreme motion sensitivity of the sequence, any small motion will lead to blurring and ghosting. Motion artifacts can be avoided by using a single-shot technique (see Fig.7.26). The same diffusion sensitization pulses are used, and subsequently the magnetization returns to the longitudinal direction by a 90° pulse at the time of the echo. The diffusion information is now present in the longitudinal magnetization, after which the excitation pulse (or

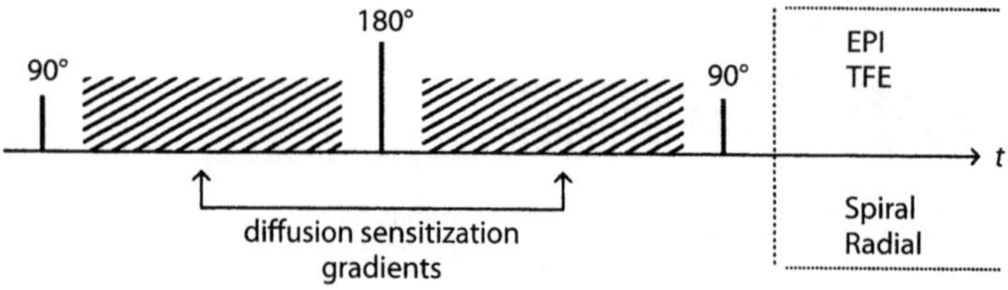


**Fig. 7.26.** Diffusion sensitization combined with fast imaging sequences

pulses) for a fast-imaging sequence can be applied. Such preparation pulses have already been mentioned in Sects. 2.7 and 5.3. Single-shot fast-imaging sequences, such as EPI or Spiral Imaging suffer, however, from susceptibility and resonance-offset effects. An example of single-shot diffusion-weighted images is given in image set VII-18. The compactness of the sequence allows the application of diffusion-weighted imaging in other organs, such as muscle or kidney.

An alternative approach to single shot imaging is multi-shot imaging with correction of motion before each shot. This approach will be described in the next session.

### 7.7.2 Diffusion Imaging of the Brain

"Diffusion images" (images in which the apparent diffusion coefficient is displayed) of the brain are important for the assessment of acute ischemia. The apparent diffusion coefficient decreases directly at the onset of ischemia. Navigator-guided diffusion measurements with sensitization of each of the three main directions can be made consecutively, with averaging of the resulting diffusion images. This is done to avoid misinterpretation of low diffusion coefficients, measured accidentally in a direction of low apparent diffusion, as a region with ischemia. Fortunately, this precaution is not always necessary.

It is possible to acquire diffusion images that directly display the average of the apparent diffusion coefficients in three directions. We shall discuss this method because it is a robust modern scan method that may find general application, and it utilizes many of the principles described earlier in this book [62]. An example of multi-shot diffusion-sensitized imaging with three orthogonal diffusion gradient directions and the resulting ADC image are given in image set VII-19.

There are three fundamental problems that hamper the acquisition of these images:

- the anisotropy of the diffusion coefficient, especially in white matter;
- susceptibility effects during read out in regions where the susceptibility varies; and
- the involuntary patient motion during the diffusion sensitisation gradients.

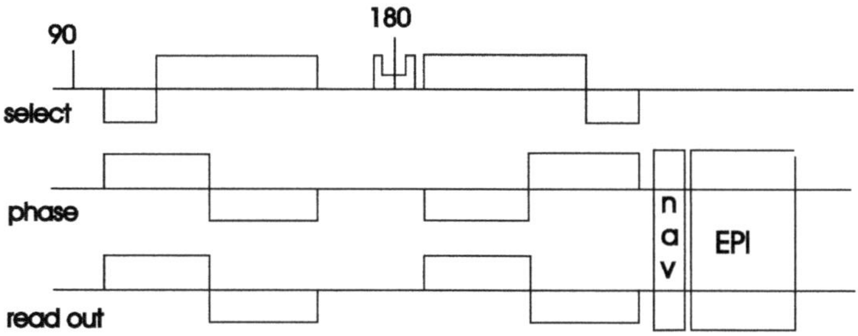


**Fig. 7.27.** Sketch of the simultaneous diffusion sensitizing gradients. The precise form of these gradients can be found in [63]. "nav" stands for navigator echo and "EPI" for the segmented EPI read out

The first problem is solved by programming the diffusion gradients to be sensitive to the average of the apparent diffusion coefficient along the three main axes (the trace of the diffusion matrix). To explain this method, we return to the solution of the diffusion–precession equation (7.36). The exponent, in describing the effect of diffusion, contains terms like

$$\int_0^t k_i(t')D_{ij}k_j(t')\mathrm{d}t = \Delta_{i,j}, \quad \text{where} \quad i,j = x,\, y \text{ or } z. \tag{7.45}$$

Since the average apparent diffusion coefficient ($D_{\mathrm{app}}$ or ADC) in the three main directions,

$$D_{\mathrm{app}} = D_{xx} + D_{yy} + D_{zz}$$

(the trace of the diffusion matrix), is to be measured, only the terms with diagonal elements of the diffusion matrix need to be retained in (7.45). Therefore the integrals containing the off diagonal terms, $\Delta_{i,j}$ with $i \neq j$, should be zero. A solution to this problem is proposed in [63] and applied in [62]. In this solution the diffusion-sensitizing gradients are applied simultaneously along the three main directions in such a way that the off-diagonal terms cancel and the diagonal terms are retained with the same weighting factor.

These simultaneous gradient lobes, for which the $\Delta_{i,j} = 0$ with $i \neq j$, are shown in Fig. 7.27. With this sensitization, twice the time needed for sensitizing one single direction is needed. The combined effect of diffusion in the three main directions is: $\exp(-b(D_{xx} + D_{yy} + D_{zz}))$, where $b = \int_0^t k_i^2(t')\mathrm{d}t'$, with $i = x, y, z$. This method is robust against errors due to diffusion anisotropy.

We now arrive at the second and third problem. In order to avoid degradation of the image quality due to susceptibility effects, it is usually proposed to use segmented EPI instead of single-shot EPI. In our case, with a read-out time of about 20 ms per shot, this segmentation introduces the third problem: image-quality degradation due to involuntary patient motion during the diffusion-sensitizing gradients between the different segments (shots). This motion results in extra phase shifts, which may be different for each segment (shot), corrupting the image quality. The solution is found in the application of navigator echoes, as discussed below.

Let us first consider motion for which the selected spins do not move out of the selected slice. Pure translation gives a phase shift, which is constant over the field of view. Rotations around the isocenter yield a linear phase shift over the FOV because the displacement increases linearly with distance from the center of rotation. Most motions are a combination of translation and rotation. The navigator echo now has to give two-dimensional information, and therefore differs from the navigator echo described in Sect. 7.2.2. for imaging coronary arteries. The 2D information can be obtained by any fast acquisition like Spiral or EPI acquisition of the center part of the $\vec{k}$ plane, making correction for these motions possible.

Depending on the motion, the phase at the end of the diffusion gradients has a non-zero value, which can be approximated by $\varphi = \varphi_0 + ax + by$. Because of this phase shift, the maximum of the signal in the $\vec{k}$ plane has a phase $\varphi_0$ and is shifted away from the centre of the $\vec{k}$ plane to $k_x = a$ and $k_y = b$. This can be seen directly by looking at the Fourier transform given in (2.23), which now reads:

$$\begin{aligned} M_{\mathrm{T}}(k_x, k_y) &= \iint\limits_{x,y} m(x,y)\mathrm{e}^{j(\varphi_0+ax+by)}\mathrm{e}^{-j(k_x x+k_y y)}\mathrm{d}x\,\mathrm{d}y \\ &= \iint\limits_{x,y} m(x,y)\mathrm{e}^{j\varphi_0}\mathrm{e}^{-j((k_x-a)x+(k_y-b)y)}\mathrm{d}x\,\mathrm{d}y, \end{aligned} \tag{7.46}$$

which shows the new center of the $\vec{k}$ plane shifted. From this shift, a and b can be calculated, and $\varphi_0$ follows from the phase of the magnetization in the center of the $\vec{k}$ plane.

A problem that cannot be solved directly arises from rotation of the excited spins out of the excited slice position. The indirect solution of the problem is based on the observation that such a motion causes large phase dispersion and therefore decreases the observed total transverse magnetization. In this way measurements of interleaves, in which the maximum signal in the $\vec{k}$ plane is lower than, say, 70% of the average value of the other interleaves, can be disregarded. This, however, makes oversampling in the phase-encoding direction necessary because of missing interleaves.

With the knowledge of $\varphi$, a and b, the measured profiles in the $\vec{k}$ plane can be brought to the correct position with respect to each other before reconstruction. Another option is to use the knowledge of the relative position of the acquired profiles in the $\vec{k}$ plane in a gridding procedure, as described in Sect. 3.6.1.2.

The result of this motion-corrected "anisotropic" diffusion sensitized method is a robust scan method for acute stroke patients [62].

But the same can be said of the combination of unidirectionally diffusion sensitized methods. Either way, the measurement of $D_{\mathrm{app}}$, the apparent diffusion coefficient is no longer an instrumental problem. It is easily added in a clinical brain examination and its use for the assessment of brain damage in acute stroke patients is actively explored ([62] and image set VII-19).

### 7.7.3 Q-space Imaging

In Sect. 7.7.1 the measuring method for the diffusion constant is described. As shown in Fig. 7.25, no restriction on the length of the gradient pulses was assumed. However that is formally only correct for a homogeneous material with free diffusion. In that case the self correlation function of a particle, $P(\vec{r}_0, \vec{r}_0 + \vec{R}, \Delta)$, that is the probability that a particle, which is in position $\boldsymbol{r}_0$ at time $t_0$, ends up in position $\boldsymbol{r}_0 + \boldsymbol{R}$ at time $t_0 + \Delta$, is a Gaussian function of $R$ and $\Delta$, see for example [78]:

$$P(\vec{r}_0, \vec{r}_0 + \vec{R}, \Delta) = \frac{1}{\sqrt{(4\pi D\Delta)^3}} \cdot \exp\left(-\frac{\vec{R}^2}{4D\Delta}\right). \tag{7.47}$$

The root mean square displacement $l_\mathrm{D}$ in the time interval $\Delta$ is named the diffusion length, given by $l_\mathrm{D} = \sqrt{(D\Delta)}$. This shows that the r.m.s. value of the displacement grows unrestricted with the square root of $\Delta$.

In tissue, however, we deal with inhomogeneous material with permeable cellular membranes, which separate the intra-cellular space from the extra cellular space. Furthermore the diffusion properties of the intra-cellular space are not equal to those of the extra cellular space, so in MRI experiments one always considers a mixture of the diffusion properties of both compartments. Now it is clear that, when the time $\Delta$ is long enough, the diffusion length $l_\mathrm{D}$ becomes of the order of the dimensions of the cells, or larger. In that case the diffusion is restricted and the self-correlation function is no longer Gaussian. Actually, we are then formally not allowed to measure a diffusion coefficient with the sequence suggested in Sect. 7.7.1 and shown in Fig. 7.25, since due to dephasing during a gradient pulse (7.41) and thus (7.42) and (7.43), in which unlimited space is assumed, are no longer valid. The “apparent diffusion coefficient”, $D_{app}$, that results from such measurements becomes, as a result of the spatial limits, a function of $\Delta$ and $\delta$, see [78–81]. Instead, in order to separate the dependence on $\Delta$ and $\delta$, one can use the short pulse gradient method (SGP), depicted in Fig. 7.28. The pulse time $\delta$ is assumed to be so short that the diffusion motion during this time interval may be neglected.

To describe the SGP method mathematically, we use the auto-correlation function $P(r_0,\ r_0 + R, \Delta)$ to find the root mean square motion during the interpulse time $\Delta$. If the protons did not move, the phase obtained during the first gradient pulse, being $\phi_0 = \gamma G r_0 \delta$, would be exactly rephased during the second gradient pulse, $\phi_1 = \gamma G r_1 \delta$. However, due to random motion the spins have a new position, according to $P(r_0, r_0 + R, \Delta)$, so that the signal change, due to the random motion becomes (see [78], Eq. (75)):

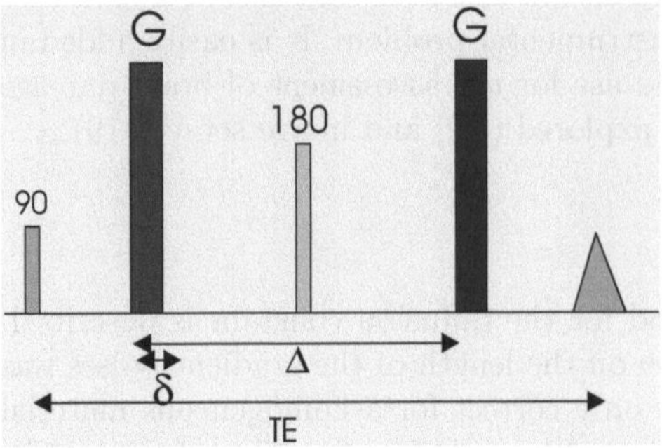


**Fig. 7.28.** Spin Echo Sequence with Short Gradient Pulses ($\Delta \gg \delta$)

$$
\begin{aligned}
E(G,\delta,\Delta) &= \frac{S(G\delta,\Delta)}{S(0,\Delta)} = \int_{r_1}\int_{r_0} \rho(r_0)P(r_0,r_o+R,\Delta)\mathrm{e}^{i(\phi_0-\phi_1)}\mathrm{d}r_0\mathrm{d}r_1 \\
&= \int_{r_1}\int_{r_0} \rho(r_0)P(r_0,r_0+R,\Delta)\mathrm{e}^{i\gamma G\delta R}\mathrm{d}r_0\mathrm{d}R \qquad (7.48)
\end{aligned}
$$

The integrations are over all starting positions, respectively final positions. The quotient of two measurements, one with and one without diffusion-sensitivity gradients is necessary to avoid the influence of $T_2$ relaxation, see (7.37). One can introduce a new function of only two variables, $P(R,\Delta)$, named the propagator or displacement distribution function, describing the probability that a distance $R$ is covered by any proton in the time $\Delta$. This function can be written, taking $r_1 - r_0 = R$, as

$$P(R,\Delta) = \int_{r_0} \rho(r_0)P(r_0,r_0+R,\Delta)\mathrm{d}r_0\ , \qquad (7.49)$$

so the signal change due to motion according to (7.48) is:

$$E(G,\delta,\Delta) = E(q,\Delta) = \int_R P(R,\Delta)\mathrm{e}^{iqR}\mathrm{d}R\ , \qquad (7.50)$$

where $q = \gamma G\delta$. This means that the propagator $P(R,\Delta)$ is the Fourier transform of the signal attenuation due to motion as a function of $q$ at constant $\Delta$, see for example [80]. This equation forms the basis for "$q$-space imaging", which enables us to determine the propagator $P(R,\Delta)$ by measuring $E(q,\Delta)$ as a function of $q$ for different values of $\Delta$. For free diffusion this then yields the Gaussian propagator, (7.47). The development of a theoretical model for restricted diffusion is beyond the scope of this book; we therefore refer to a number of papers [79–81] and the references mentioned in these papers. The experimental study of restricted diffusion is done by taking SPG measurements with constant $q$ and varying $\Delta$ or with constant $\Delta$ and varying $q$. Usually the attenuation described by (7.50) is equated to

$$E(q, \Delta) = \int\limits_R P(R, \Delta) e^{iqR} dR = \exp[-bD_{\text{app}}(q, \Delta)] \,, \tag{7.51}$$

where $b$ has its original value, given by (7.43).

The function $P(R, \Delta)$ not only describes restricted diffusion in tissue, but also restricted diffusion in porous materials or transport phenomena in general. We refer to an extended body of literature on these material studies [81–83], which may be useful also for the study of the structure of brain tissue, see Sect. 8.2.4.

For imaging free water protons, as is the case in MRI, the Short Gradient Pulse Method requires very high gradient pulses, which are not readily available in whole-body systems. Also they easily go beyond the IEC limits for the maximum gradient field and then also for $dB/dt$. Therefore, the study of restricted diffusion of free water in tissue is mostly done in vitro in small-bore systems. In the present clinical studies of the brain, the method with long gradient pulses of Fig. 7.25 is used in in vivo studies as an indicator for changes in the "apparent diffusion coefficient" of free water in tissue. This method yields clinical information, but cannot be used for the detailed study of the structure of the tissue.

Restricted gaseous diffusion in lungs is observable with MRI, using hyperpolarized Helium 3. At a diffusion time of 1.8 ms ($b = 0.67 \times 10^4\,\text{s/m}^2$) the alveolar diffusion coefficient of emphysematous lungs was found to be $0.67 \times 10^{-4}\,\text{m}^2/\text{s}$, three times larger that that in healthy volunteers [85].

Image Sets
Chapter 7

## VII-1 Inflow and Outflow Phenomena in Multi-slice-Triggered SE of the Abdomen

(1) In the multi-slice SE technique, per TR a number of slices are addressed sequentially. So, in cardiac-triggered multi-slice scans each image relates to a different cardiac phase. When such a scan is of transverse slices of the abdomen, in each slice the signal value at the cross section with the large vessels will depend on the blood velocity at the cardiac phase for that particular slice as well as on the magnetization carried by the bloodstream into the slice from slices excited earlier. The SE images in this set, taken from a healthy volunteer, are chosen to demonstrate these effects.

The scans used differ in the temporal order in which after the R top each slice is addressed.

Parameters: $B_0 = 1.5\,\mathrm{T}$; FOV = 300 mm; matrix = 256 × 256; $d = 5\,\mathrm{mm}$; slice gap = 1 mm. TR = 1 heart beat; TE = 15 ms; NSA = 1.

| | | | |
|---|---|---|---|
| scan no. | 1 | 2 | 3 |
| image no. | 1–8 | 9–16 | 17 |
| slice temporal order | down | up | |
| image type | | | *Q* flow |

Phase-encode artifact reduction (PEAR, see Sect. 7.2.2) was used to suppress respiration ghosting (see also image set VII-6). In each of scans 1 and 2, eight images were obtained. Scan 3 was of a single slice. It was measured with a quantitative-flow technique; see Sect. 7.5.1.2. In this scan two sequences were mixed; one sequence was flow sensitized in the FH direction and the other was flow insensitive.

(2) A small zoomed region of these images, containing the large vessels, is shown. The order in which the images per scan are presented is according to slice number and runs from slice 1 to slice 8. In our system slice 1 is the most caudal slice.
The temporal order of addressing the slices was either down or up. Hence stated by slice number, the down order for instance is 8, 7, 6, 5, 4, 3, 2, 1. The initial trigger delay (first slice addressed) was 10 ms. From slice to slice, the trigger delay increased with 33 ms, so that, for example, for scan 1 the longest trigger delay, that of slice 1, was 241 ms after the R top. Image 3 is a phase contrast quantitative flow image (Sect. 7.5.2). It was obtained by retrospective cardiac gating. A diagram showing the velocity of aorta and vena cava as function of the heart phase is shown in this image.

(3) *Outflow voids.* At large flow velocities, the spins excited in the large vessels flow out of the slice during the time between the excitation pulse and refocussing pulse (this is TE/2 or 7.5 ms in the scans of this image set). This outflow will result in loss of signal because of the lack of refocussing.

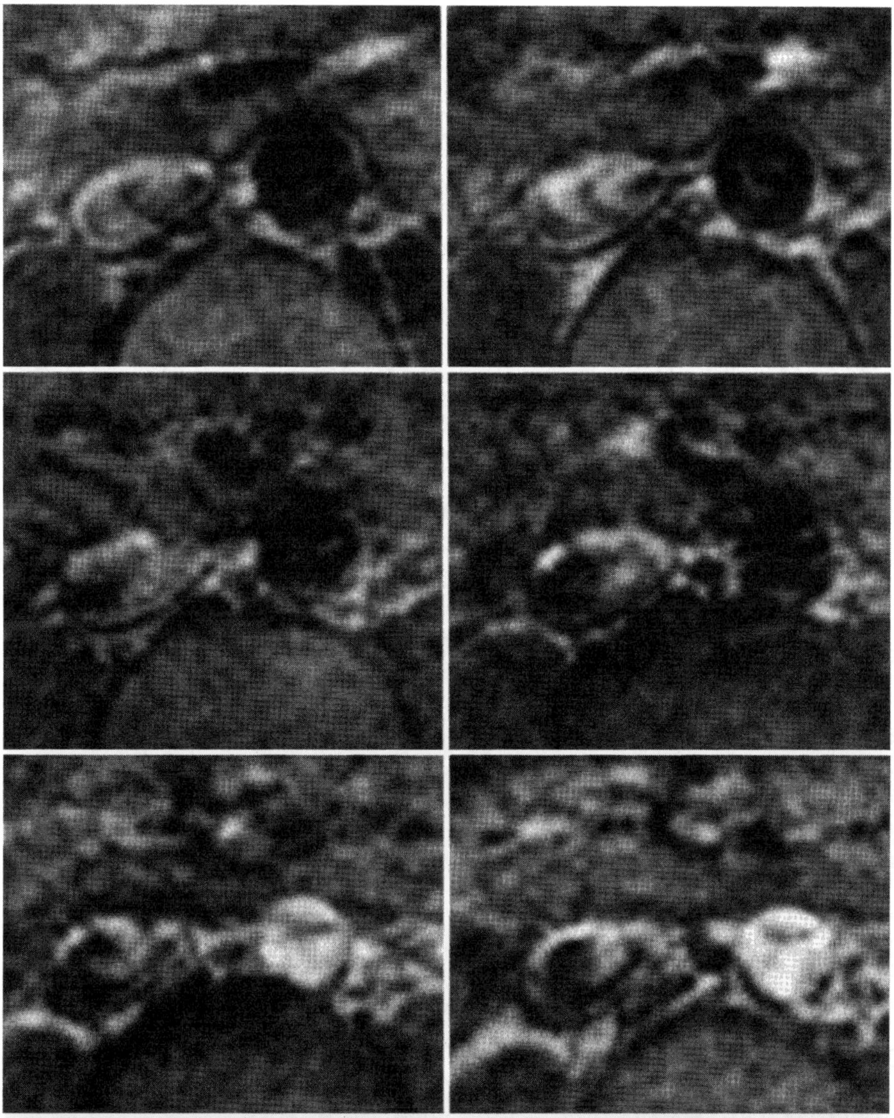


Image Set VII-1

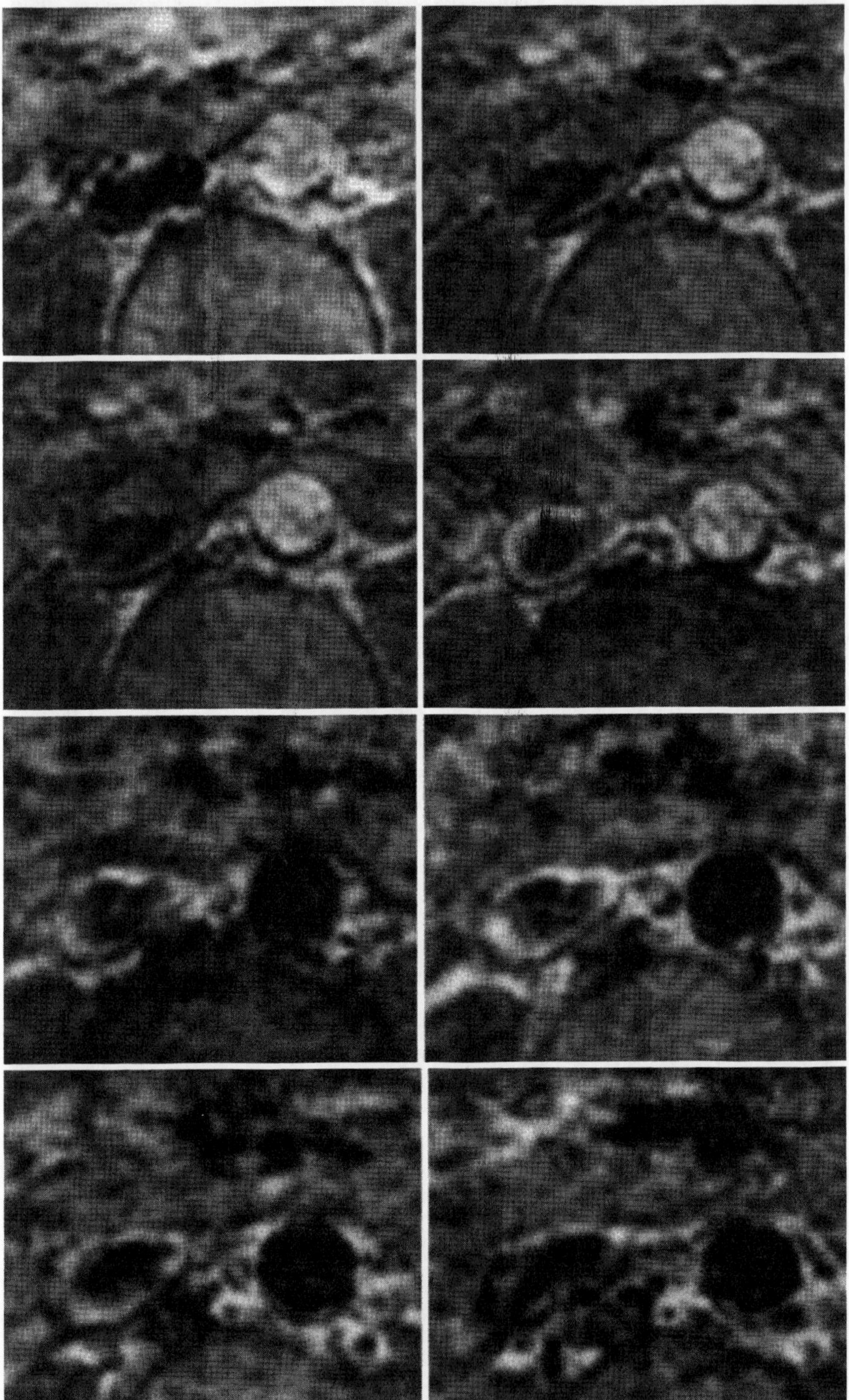

9 10 11 12 13 14 15 16

**Image Set VII-1.** (*Contd.*)

A complete outflow void will occur at a flow velocity $\nu > 2d/\mathrm{TE}$; see (7.19); corresponding to $\nu > 0.66\,\mathrm{m/sec}$. Image 3 shows that the flow velocity in the aorta at delays above 125 ms is large enough to create outflow voids. So, flow voids will occur in the aorta for the slices that are obtained at a trigger delay $> 125\,\mathrm{ms}$. In scan 1 this regards slices 4, 3, 2, and 1; in scan 2 slices 5, 6, 7, and 8. As expected, in these slices the aorta is dark.

(4) *Inflow enhancement.* In the images obtained at small flow velocities, when no outflow voids occur, the displacement per TR (one heart beat) of the blood is still much larger than the slice width. When through this mechanism the blood in a slice is replaced by fresh, unsaturated blood, the resulting signal will be relatively bright compared with the stationary tissue. This phenomenon is called inflow enhancement. It is visible in scan 1 as a bright aspect of the aorta in slice numbers 5–8 (most cranial slice positions).
In scan 2, slices 1 to 4, the aorta is less bright. This lack of brightness can be ascribed to signal reduction from partial saturation of the inflowing blood in those slices because this blood originates from upstream slices that were excited in the previous TR.

(5) *Inflow voids* will occur when during excitation in a certain slice, blood spins are found that are still strongly saturated such as will occur when blood arrives from an upstream slice that was already addressed in the same heart beat, because the travel of the excitation profile has the same velocity as the blood. This effect can be seen in the vena cava in scan 2 where the temporal slice order was up. The required blood velocity for a complete inflow void is equal to the ratio of centre to centre distance and the increase in trigger delay between adjacent slices. In the scans used this velocity equals 6 mm/33 ms or 0.18 m/s. Image 17 shows that the flow in the vena cava approximately had this velocity.

17

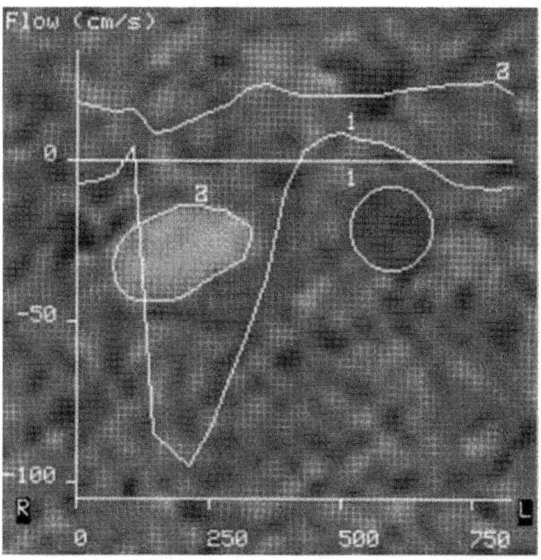


**Image Set VII-1.** (*Contd.*)

## VII-2 Signal Loss by Spin Dephasing Caused by Flow

(1) Partial or total loss of signal can occur in FFE imaging in voxels where at the echo time the spins have a significant dephasing. This phenomenon is called intra-voxel dephasing. Various causes exist and examples are given in image sets II-10 (main field inhomogeneity) and IV-3 (dephasing between water and fat). Intra-voxel dephasing can arise also as a consequence of strong flow divergence within the voxel, such as occurs in turbulent flow. This type of intra-voxel dephasing is called a flow void. Its occurrence depends on the flow sensitivity of the imaging method. A method with full flow compensation would never have a flow void. In such a method all first and higher order gradient moments (Sect. 7.3) have to be zero, which is never achievable. Velocity compensation (zero first order gradient moments) will already be reasonably effective in reducing the flow void. This is demonstrated in the present image set. Triggered single slice multiphase FFE images with controlled velocity sensitivity were made of the heart of a patient with a known defective mitral valve (courtesy of Dr Louwerenburg of Medisch Spectrum Twente). The heart is shown in an angulated axial slice, through the long axis, including left ventricle and left atrium. The atrium is dilated and is visible in the lower half of each image as a circular structure. Each image is of a separate scan and shows the heart in early systole at 127 ms after the R top. The images display a region of 130 × 130 mm. Phase encode direction was AP.

Parameters: $B_0 = 1.5\,\mathrm{T}$; FOV = 320 mm; matrix = 128 × 128; $d = 8\,\mathrm{mm}$; TE = 5.1 ms; TR = 25 ms; NSA = 2.

| image no. | 1 | 2 | 3 | 4 |
|---|---|---|---|---|
| VS | 0 | AP | LR | HF |

VS is the direction of the velocity sensitivity of each scan. It is controlled by the first gradient moment $M_1$ in each direction. In two directions $M_1$ is always zero and in the third direction $M_1$ is either zero (image 1) or equal to $M$ (image 2–4). In each of the images 2–4, $M$ corresponds to a numerical value of the velocity sensitivity of 0.8 m/s for 180° phase difference. With that, the gradient moment M is approximately equal to the one that is found in the read-out direction of a normal not flow compensated FFE scan with TE = 5.1 ms

(2) All images show a jet-shaped flow void emerging from the mitral valve and pointing into the left atrium in a dextero-posterior direction at about 30° downward. These flow voids correspond to a regurgitant jet flow resulting from the mitral valve defect. The mean velocity in the jet at its apex was estimated with a quantitative flow technique and was above 1 m/s at the heart phase shown in the images.

(3) The size and shape of the flow void vary per image. This is the result of the different velocity sensitivities used. In image 1 where VS is zero, the jet

has the smallest dimensions. This illustrates that at a time scale of 5.1 ms the tissue velocity in the jet is reasonably constant, so that in most of its area there is no phase dispersion on that time scale. In image 3 (VS = LR), the jet is longer and wider, indicating a large gradient in the left to right velocity component. To explain that the flow void is nearly complete in the first half of the length of the jet, the range of velocities in each voxel in that region must be equal to at least 0.8 m/s and this indicates that the voxels have to extend partly outside the jet. This is easily possible because of the slice width of 8 mm (in the HF direction). In image 2 (VS = AP), the flow void is less dark than in image 3. Apparently the maximum range of velocities in the AP direction is not sufficient to give complete extinction.
Image 4 (VS = CC) is somewhat of a surprise. Areas of nearly complete flow voids are found, including one at the ventricular side of the mitral valve. These regions must have a considerable gradient in the CC velocity component. An explanation can found by assuming a strong rotational component in the flow.

(4) Flow voids can occur in FFE sequences even when no turbulence is present. The size and shape of the void are no unique indicators of the flow pattern because they depend strongly on details of the imaging method, such as its flow sensitivity, slice thickness or pixel size.

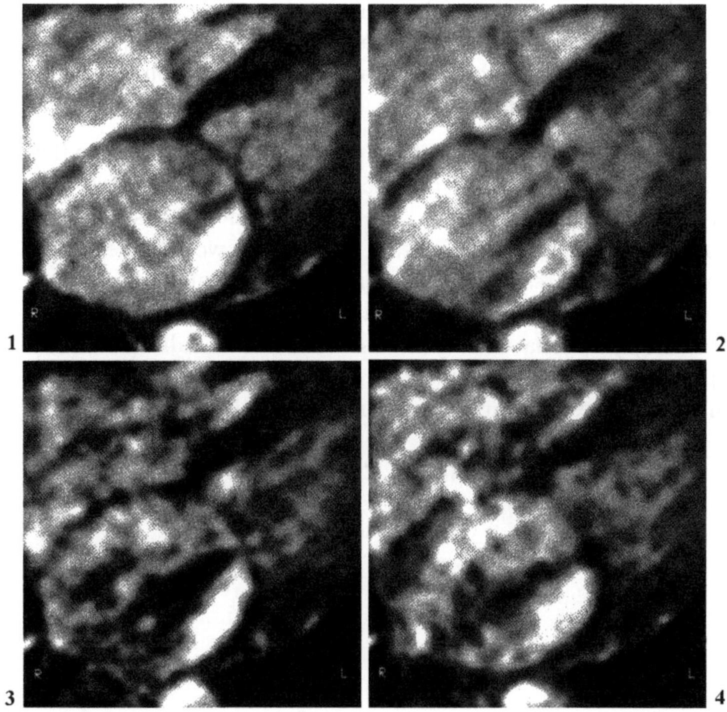


**Image Set VII-2**

## VII-3 Pulsatile Flow Ghosts in Non-triggered SE Imaging of the Abdomen

(1) In triggered transverse imaging of the abdomen, the pulsatile flow in the aorta causes a cardiac phase-dependent signal loss (see, e.g., image set VII-1 and Sect. 7.4.1.
In non-triggered imaging, this signal loss creates predominantly an amplitude modulation of the time-domain signals. This modulation causes a ghost artifact in the image. Section 2.6.1 describes the theory for this ghost mechanism. In this image set, some non-triggered and triggered transverse SE images of the abdomen are compared.

Parameters: $B_0$ = 1.5 T; FOV = 320 mm; matrix = 256 × 256; $d$ = 10 mm; TE = 20 ms; NSA = 1.

| | | | | | |
|---|---|---|---|---|---|
| image no. | 1 | 2 | 3 | 4 | 5 |
| cardiac triggering | no | no | no | no | yes |
| TR (ms) | 500 | 600 | 700 | 800 | RR |
| 1–TR/RR | 0.47 | 0.37 | 0.26 | 0.15 | 0 |

In the triggered scan, the repetition time equals the R-top interval RR, which for the volunteer scanned equalled 950 ms.

(2) The profile order in the scans for these images was linear. This means that no adaptation of the profile order to the respiration phase took place (see image set VII-6). This led to a strong respiration ghost. The phase-encode gradient was applied in the AP direction.

(3) Interference between TR and the cardiac-induced modulation of the flow in the vena cava induces as a rather circumscript ghost that is shifted with respect to the real position of this vessel. The signal from this vessel apparently was modulated as function of the heart phase. The well-defined ghost position in each image indicates a regular heart beat rate during the scan. The shift distance of the ghost is proportional to TR/RR – 1. When this expression equals 0.5, the vena cava ghost is shifted over half the field of view. This is approximately true for image 1. For the later images the same expression is smaller, corresponding to smaller shifts of the ghost.

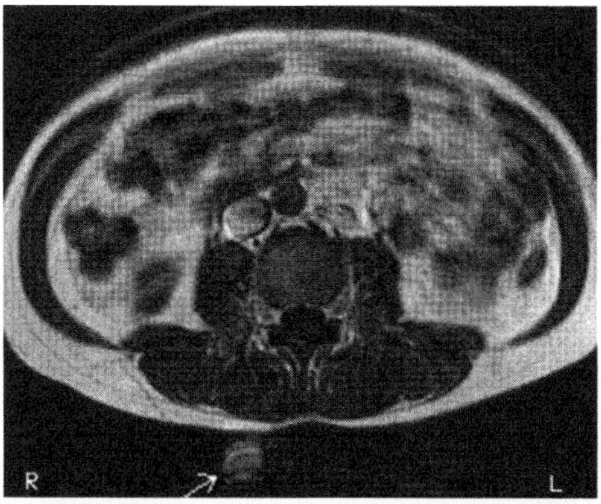


1

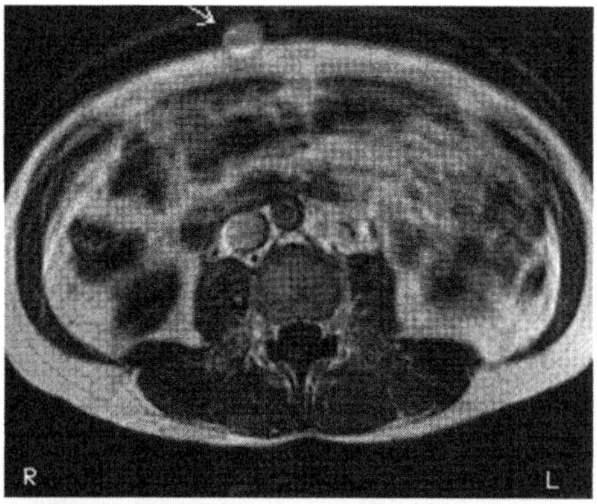


2

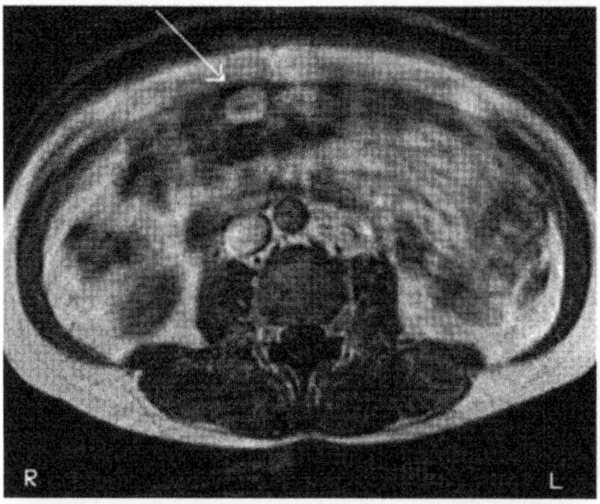


3

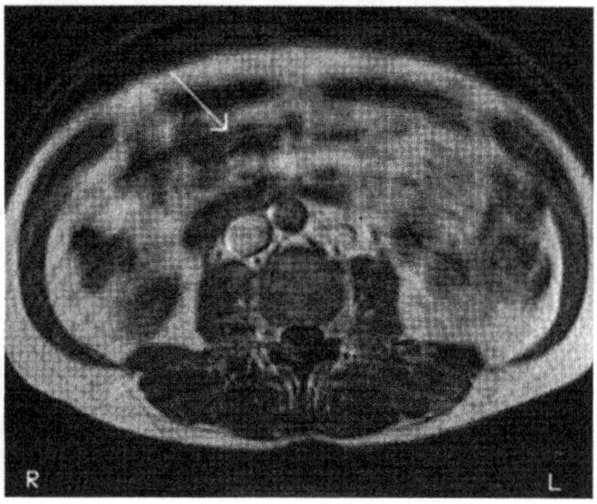


4

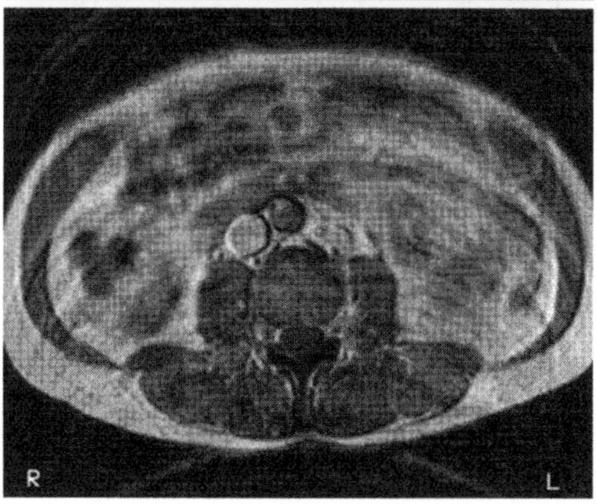


5

**Image Set VII-3**

## VII-4 Ringing from Step-Like Motion of the Foot

(1) Whereas periodic motion occurs in all patients; non-periodic movements caused, for example, by involuntary muscle force do occur in a significant number of cases and lead to degradation of the image. In this image set, such an event is simulated by transport of the patient table over 3 mm during the scan. The scan method was SE.

Parameters: $B_0 = 1.5\,\mathrm{T}$; FOV $= 220\,\mathrm{mm}$; matrix $= 205 \times 256$; $d = 4\,\mathrm{mm}$; TR/TE $= 600/20$; NSA $= 2$

| | | | |
|---|---|---|---|
| image no. | 1 | 2 | 3 |
| table shift | no | yes | yes |
| relative time of shift | – | 50% | 75% |

The time of the shift is given as a fraction of the scan time

(2) The profile order used in the scan was linear and the two signals used for averaging were obtained from excitations that were adjacent in time. As a result in the scan of image 2 the profile with $k_y = 0$ was recorded just before the moment of movement of the table. The phase-encode gradient was in the direction of the table movement, causing the movement artifact to be visible primarily in structures perpendicular to this direction.

(3) In the scan of image 2, the shift is equivalent to the addition of a discrete step in the phase of the positive $k_y$ values. The image defect is a ringing ghost extending from the image edges to both sides with clear maxima at the distance over which the table was moved.

(4) In the scan of image 3, the shift is equivalent to an addition of a phase step to the highest positive $k_y$ values only. The ringing artifact is now almost undamped, but of much lower amplitude. The damage to the image is visually less apparent.

1 2 3

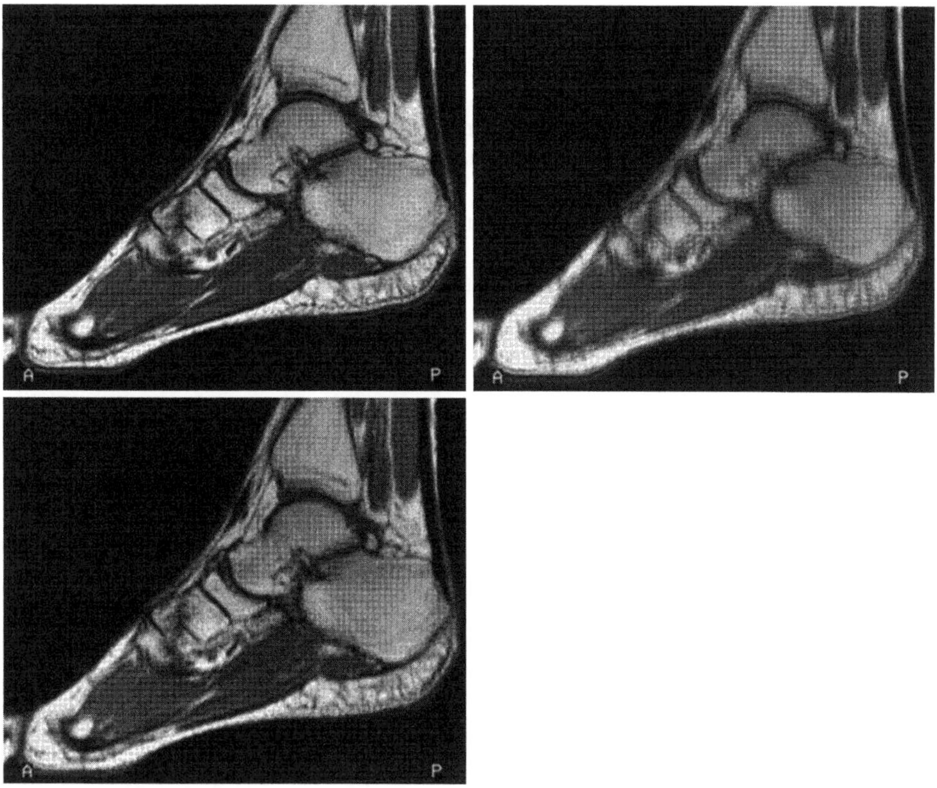


**Image Set VII-4**

## VII-5 Flow-Related Misregistration of Brain Vessels

(1) When tissue moves in the direction of the phase-encode gradient $G_y$, a misregistration artifact will occur, as described in Sect. 7.4.4.3 and shown in Fig. 7.9.
The misregistration is proportional to the tissue velocity $\nu$, the cosine of the angle $\phi$ of this velocity to the direction of the phase-encode gradient and the time delay between the centre of the phase-encode gradients $T$ and the echo time TE:

$$dy = \nu \cos\phi(T - \mathrm{TE})$$

In images of a curved vessel, $\phi$ is a function of place and so will be the size of the misregistration. The result is a distorted image of the vessel. This image set shows an example of the occurrence of this phenomenon in FE images of the cerebral arteries.

Parameters: $B_0 = 1.5\,\mathrm{T}$; FOV $= 320\,\mathrm{mm}$; matrix $= 205 \times 256$; $d = 8\,\mathrm{mm}$; TR $= 200\,\mathrm{ms}$.

| | | | |
|---|---|---|---|
| image no. | 1 | 2 | 3 |
| direction of $G_y$ | LR | AP | AP |
| TE (ms) | 5.6 | 5.6 | 12 |
| $T$ (ms) | 2.4 | 2.4 | 4.5 |

$T$ is the time of the centre of the phase encode gradient pulse; see formula.

(2) The images were adjusted to contain part of the anterior and the middle cerebral arteries. These parts are visualized from the point where they branch off from the carotid syphon. The anterior cerebral arteries flow towards the midplane and curve upward to the frontal brain. The middle cerebral arteries flow laterally, first with a gradual upward curvature and then a sharp downward curvature. The bright aspect of the arteries is due to inflow enhancement and to the use of an FE acquisition method.

(3) Comparison of images 1 and 2 shows that the shape of the curved section of the anterior cerebral arteries differs per image. In image 1, the distance between these arteries during most of their course is larger than that in image 2. All displacements in image 1 have to be in the LR direction, corresponding to the direction of $G_y$. This resulted in an outward displacement of the anterior cerebral arteries in this image. In image 2, the displacements are in the AP direction and they occur when there is flow in that direction. Result is that in this image the upwardly directed sections of the anterior cerebral arteries are displaced downward.
In image 3, the displacement is in the same direction as that in image 2, but it is larger, because of the longer TE used in this image. The upward sections of the anterior cerebral arteries have become shorter that those in image 2.

(4) Similar changes can be seen in the shape of the middle cerebral arteries. So for instance, the shape of their sharp downward curvature differs between the images.

(5) The original position of a displaced vessel is relatively dark. This effect is only faintly visible in the images, because the slice thickness used is much larger than the diameter of the artery.

(6) Significant displacement artifacts can occur in the arterial pattern of the brain, even at echo times as short as 6 ms. Usually the acquisition methods do not compensate for this artifact. The presence of misregistration often can not be discerned from the image. This was the case in the images of this set, because of the slice thickness used. Non-discernable displacement will occur also in the usual MR angiographic displays, such as a maximum-intensity display.

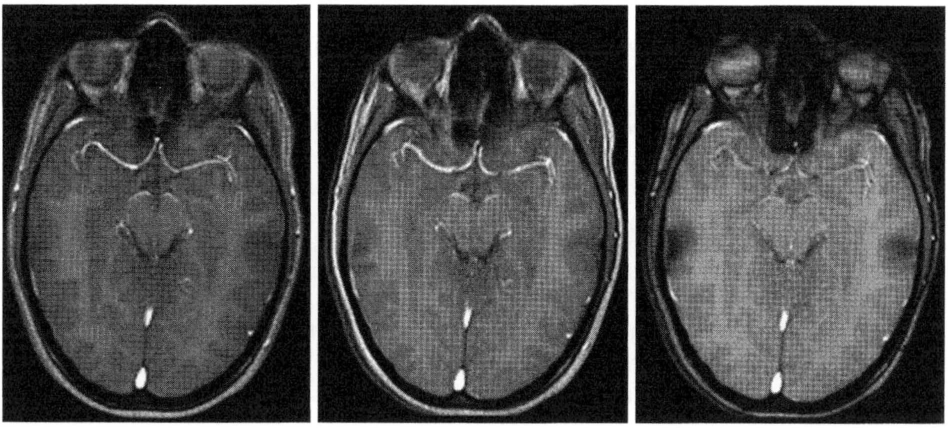

1 2 3

**Image Set VII-5**

## VII-6 Respiration Artifact in SE Imaging

(1) The clarity of $T_1$-weighted contrast of SE imaging is a desirable feature for imaging of the abdomen. Breath-hold imaging is not easy in this technique due to the long scan time, so means are introduced to overcome the respiration artifacts in SE abdomen images in different ways. The main cause of artifacts is the periodic movement from respiration. Image set VII-7 deals with imaging in a single breath-hold period. When breath holding cannot be used, increase of NSA is a simple but time-consuming way of dealing with the problem. Another classical way to cope with this artifact is the adaptation of the profile order to the respiration phase [13]. In this image set the performance of both methods are compared.

Parameters: $B_0 = 1.5\,\mathrm{T}$; FOV $= 320\,\mathrm{mm}$; matrix $= 256 \times 200$; $d = 8\,\mathrm{mm}$; TR/TE $= 300/20$.

| image no. | 1 | 2 | 3 | 4 | 5 | 6 | 7 |
|---|---|---|---|---|---|---|---|
| PEAR | no | no | no | no | yes | yes | yes |
| NSA | 1 | 2 | 4 | 8 | 1 | 2 | 4 |

(2) The method used for adaptation of the profile order in our system is called PEAR (phase encode artifact reduction). It requires the use of a respiration sensor. The system predicts the length of each new respiration cycle on basis of the observed mean cycle length and plans the $k_y$ value of the phase-encode gradient in that cycle. The expiration phase is reserved for excitations with low $k_y$ values.

(3) Without PEAR, the profile order is linear, so that the time periodicity of the respiration is translated in a $k_y$ periodicity. Section 2.6 describes the resulting artifact. For image 1, the shift of the ghost equals

$$\text{shift/FOV} = (\text{TR/respiration time}) - 1.$$

From image 1, the respiration period can be estimated to be about 15 TR or 4.5 seconds.

(4) When PEAR is not used and when NSA$>1$, each $k_y$ value is repeated for NSA successive excitations. This way of averaging is not unique. Alternatively one could, in a technique called serial averaging, delay the second round of $k_y$ values to after completion of the first round. However, in the averaging strategy used, the shift distance of the ghost increases with NSA as

$$\text{shift/FOV} = (\text{NSA}\times\text{TR/respiration time}) - 1.$$

For NSA $= 8$ the artifact suppression is successful; the shift distance of the artifact is about half a field of view (image 4).
Of course, this result is accidental; it would not apply for different TR or different respiration periods. However, also the intensity of the respiration

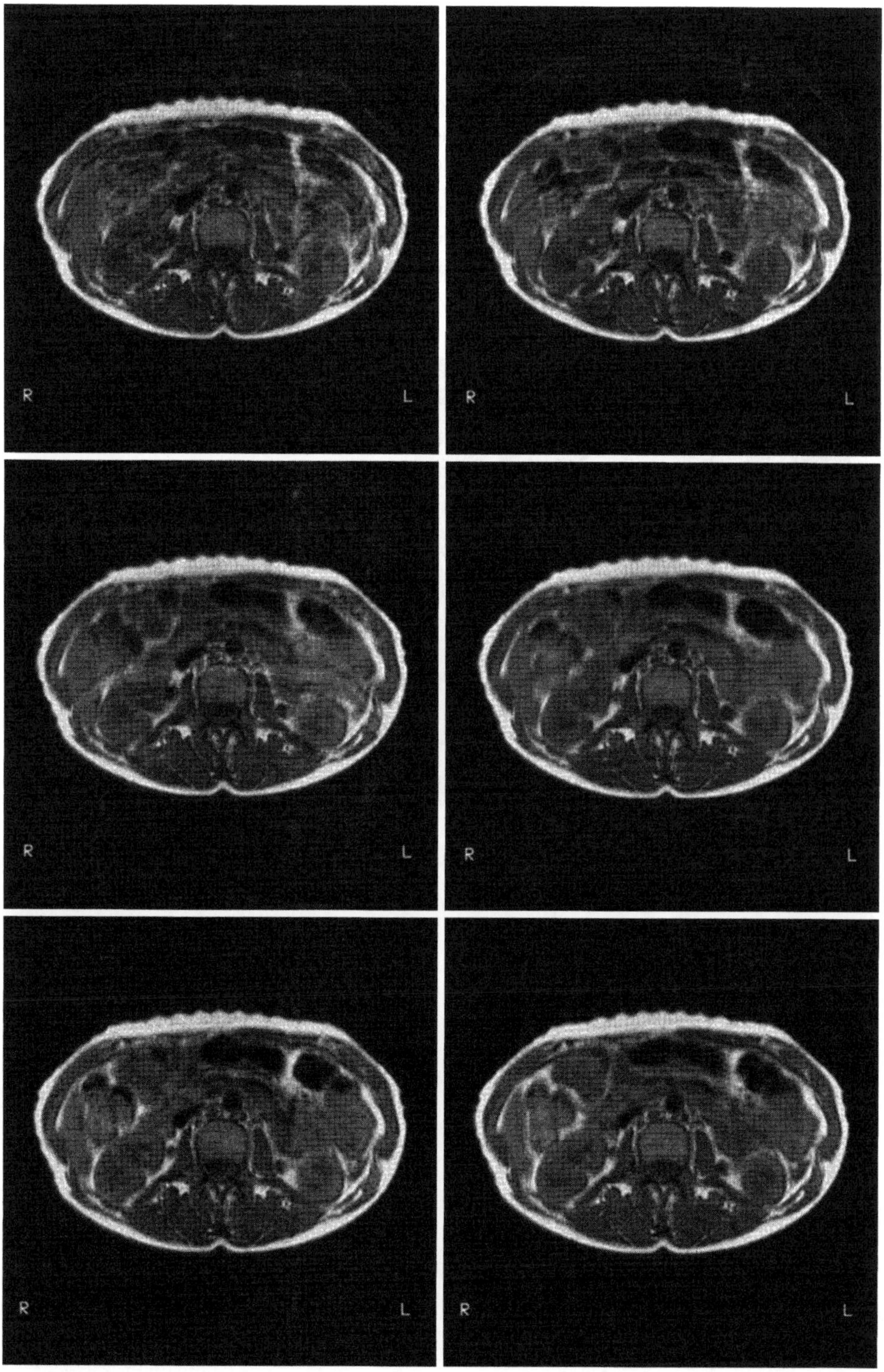


1 2 3 4 5 6

**Image Set VII-6**

artifact has diminished as a consequence of the incomplete periodicity of the respiration.

(5) When PEAR is used (images 5 to 7), the disturbance in the image is no longer present as a recognisable ghost; instead it gives the impression of increased image noise. This noise reduces with increase of NSA.

(6) The use of a phase-encode order that is ordered with respect to the respiration cycle as in PEAR is an attractive way to avoid respiration ghosts. Its use avoids the need for time-consuming scans with a large NSA, for instance, the overall impression of images 4 (NSA = 8) and 7 (NSA = 4) is equivalent.

7
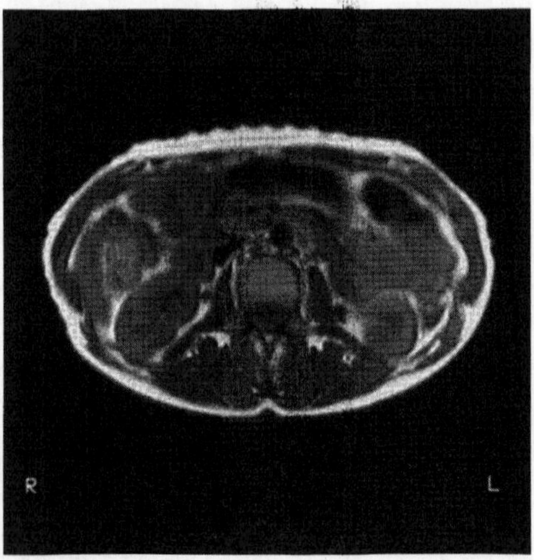


**Image Set VII-6.** (*Contd.*)

# VII-7 Respiration Artifact Level in SE and SE-EPI of the Liver

(1) In the previous image set, means were discussed to provide access to $T_1$-weighted SE imaging of the abdomen. In image sets V-1 and V-2, breath-hold imaging (Sect. 7.2.3) was suggested as the solution to the problem of respiration artifacts in TFE imaging. However, in these sets the contrast was that of a gradient-echo method. To obtain a contrast similar to $T_1$-weighted SE, the combination of SE-EPI imaging and a breath-hold technique can be used, as shown in this image set.

Parameters: $B_0 = 0.5\,\mathrm{T}$; FOV $= 375\,\mathrm{mm}$; matrix $= 200 \times 256$; $d = 10\,\mathrm{mm}$
Both methods are combined with a multi-slice technique for 7 slices

| | | |
|---|---|---|
| image no. | 1 | 2 |
| method | SE | SE-EPI |
| TR (ms) | 400 | 298 |
| TE (ms) | 15 | 14 |
| EPI factor | – | 7 |
| gradient echo spacing (ms) | – | 2.5 |
| scan time (s) | 197 | 17 |

(2) The contrast weighting of the SE and SE-EPI image is similar; the decrease in scan time is sufficient to allow breath-hold imaging in the SE-EPI scan.

(3) Although the EPI factor (the number of gradient echoes per excitation) is rather high; the corresponding shift in the phase-encode direction between water and fat (Sect. 3.3.2.2) still is acceptable, because of the low field strength and the short gradient echo spacing (see, eg, image set III-7).

(4) The signal-to-noise ratio of the SE-EPI image is much lower than that of the SE Image. The difference is caused by a short $t_{acq}$ (time per profile) used in the SE-EPI method, equivalent to a large bandwidth. For clinical readability, this difference in the SNR may nevertheless be acceptable in view of the absence of a respiration ghost in the breath-hold SE-EPI scan.

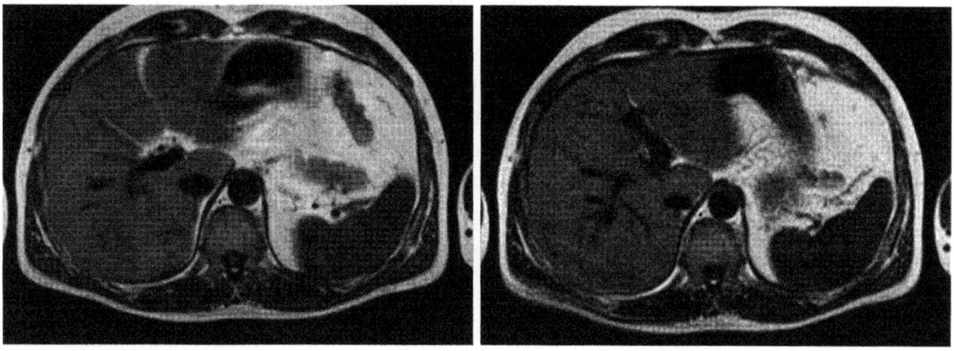

1 2

**Image Set VII-7**

## VII-8 Suppression of CSF Flow Voids in Turbo-Spin-Echo Images of the Cervical Spine

(1) Flow voids can occur in regions of CSF visible in $T_2$-weighted Turbo-Spin-Echo (TSE) imaging of the spinal canal. These flow voids can obscure the visualization of nerve roots and therefore reduce the clinical value of the scan. Transverse $T_2$-weighted TSE in particular is sensitive to this phenomenon because the transport of CSF then is mainly perpendicular to the slice direction. Various measures to suppress the flow void are possible. Image set III-3 deals with the possibility to use 3D TSE and reformatting. Image set VI-8 shows how $T_2$ weighting can be obtained in a 3D-$T_1$-FFE image. This image set compares some 2D-Multi Slice TSE images (zoomed) in which the suppression of flow voids is attempted by increasing the width of the spatial profile of the refocussing pulse.

Parameters: $B_0 = 1.5\,\text{T}$; FOV $= 190\,\text{mm}$; matrix $= 256 \times 256$; $d = 4\,\text{mm}$; TR/TE $= 3000/120$; NSA 8.

| | | | | |
|---|---|---|---|---|
| image no. | 1 | 2 | 3 | 4 |
| no. of slices per TR | 10 | 1 | 1 | 1 |
| ratio of width of spatial profiles (ref/exc) | 1 | 1 | 2 | 3 |

(2) Adjustment of the ratio of the width of the spatial profiles of the refocussing pulse (ref) and the excitation pulse (exc) was obtained experimentally by adjustment of the selection gradient for the refocussing pulse.

(3) Part of the flow voids can be attributed to "outflow". Outflow voids occur when the CSF is transported outside the region addressed by the refocussing pulses. The difference in the thickness of this region is cause of the reduction in the flow void level in image 4 compared to images 3 and 2.

(4) The difference in the flow-void level between images 1 and 2 indicates that another part of the flow void is due to "inflow". Inflow voids can occur in a multi-slice scheme by transport of CSF between slices. When CSF is transported between slices that are excited shortly after each other, for example in the same TR, the slice excited last will show a flow void. When only one slice per TR is excited, of course this inflow effect is not present (image 2).

(5) The scan used for successful suppression of the flow void in image 4 has little practical value, because it is a single-slice method. However, the method is not incompatible with a multi-slice technique. Clinical testing would be required to find out if such a technique would be of sufficient clinical value to replace normal multi-slice transverse scans.

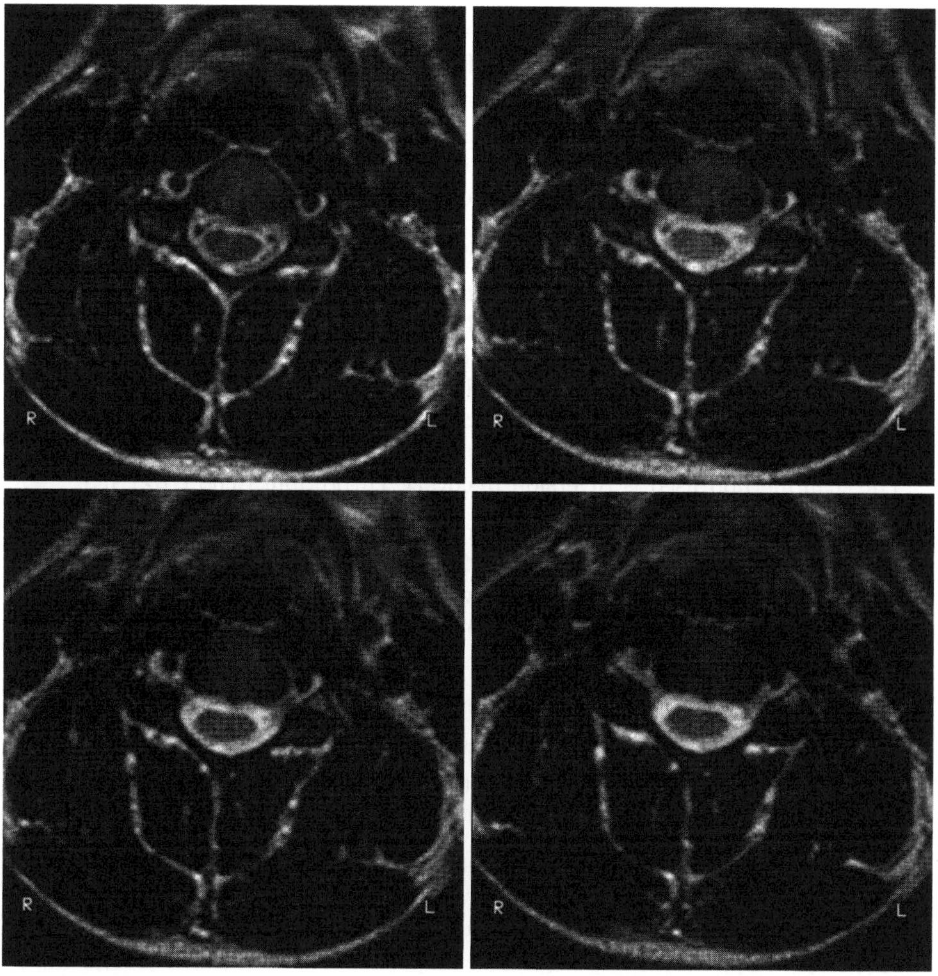


1 2 3 4

**Image Set VII-8**

## VII-9 Prevention of CSF Flow Artifacts in Liquid Suppressed IR-TSE

(1) Recently, the suppression of CSF by nulling of the magnetization of this after an inversion pulse was shown to be of clinical interest. This technique, in combination with SE was named FLAIR [6.25]. T2 weighting in combination with CSF nulling requires a very long TR (6–12 seconds) followed by a long TI (2000–3000 ms). The use of TSE and multi-slice for this purpose is attractive, because even with the required long TR, in that case imaging with the desired contrast can be realized in a scan time that remains acceptable for routine clinical work. When inversion of CSF is attempted in a multi-slice scan, the inversion pulse has to be slice selective; which can be problematic in the case of CSF flow. This image set illustrates the problem and suggests a solution. The scan method used is Inversion Recovery Turbo Spin Echo (IR-TSE).

Parameters: $B_0 = 1.5\,\mathrm{T}$; FOV = 200 mm; matrix = 256 × 256; $d = 6.6\,\mathrm{mm}$; TR/TI/TE = 8000/2400/120; NSA = 2. Per TR excitations were given to 10 slices separated by gaps of 6.6 mm. Two of these packages were scanned so that in total 19 adjacent slices were obtained in a scan time of 5:36 minutes.

| | | |
|---|---|---|
| image no. | 1 | 2 |
| ratio of width of RF profiles (inv/exc) | 1.0 | 2.0 |

(2) The ratio of the width of the RF profiles of the inversion pulse (inv) and the excitation pulse (exc) was obtained by adjustment of the selection gradient strength.

(3) Above the foramina of Monroe, a jet-like flow of CSF exists. This is visible in image 1, where the slice cuts through these jets. In the jet areas non-inverted CSF has reached the region of slice excitation. Per consequence bright spots appear in the image. These spots are accompanied by a ghost, related to the pulsatile nature of the jet movement. The presence of the ghost degrades the image.

(4) In image 2, the ghost is absent. The increased thickness of the inversion layer results in a fully effective inversion. The flow velocity apparently is not high enough to allow inflow of non-inverted spins from outside the inversion region into the excitation slice within the inversion delay time.

(5) The use of thicker than normal inversion layers avoids the presence of CSF flow artifacts as shown in this image set. This technique allows the clinical use of multiple-slice $T_2$-weighted liquor suppressed TSE imaging of the brain.

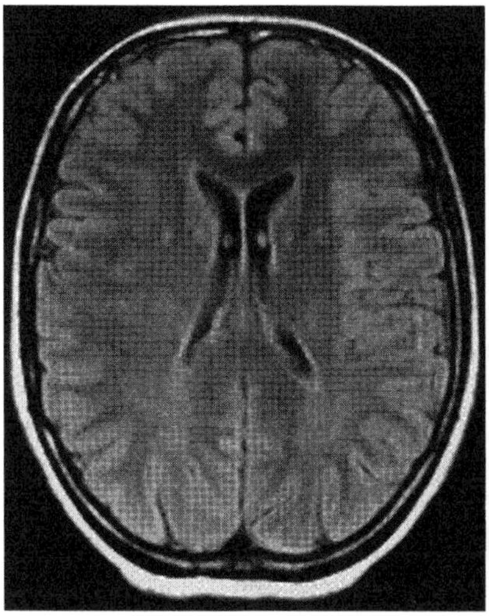

1

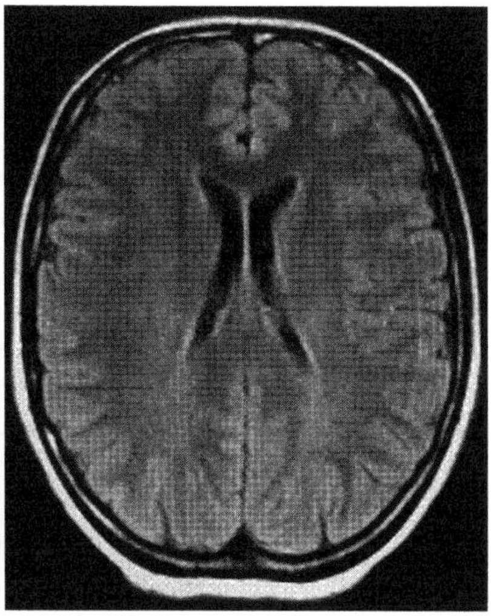

2

**Image Set VII-9**

# VII-10 Tagging of Spins in the Cardiac Muscle by Complementary Spatial Modulation of Magnetization (C-SPAMM)

Images courtesy of SE Fischer and P Boesiger, Biomedical Engineering and Medical Informatics, ETH Zürich, Switzerland.

(1) Tagging of part of the spins in the imaged slice can be added as a preparatory step in any imaging sequence. It creates a pattern in the image contrast that is the record of locations where the modulation was applied. The modulation depth in the tagged image is a function of the flip angle used for tagging and of the time between tagging and imaging. Section 7.2.3 describes a frequently used implementation of this technique, called spatial modulation of magnetization (SPAMM), after Axel [64]. Whereas that section addresses the use of a rather large number $N$ of $(90/N)°$ pulses for each set of tags, one can minimize this number to two 45° tagging pulses at the cost of sharpness of the tagging lines (see Fig. 8.13).
The technique has found application in imaging of the heart wall. When a regular pattern of modulation is "imprinted" shortly after the R top it will be deformed during systolic contraction. The deformation can give valuable information on the status of the heart wall. In this application it is important to find a sufficient modulation depth of the tagging at several moments during the entire RR interval. This is not easily reached. It is, however, possible to optimize the tagging technique for this special purpose [64, 65] and this image set demonstrates an example of this optimization.

(2) All images are obtained by SPAMM, but in the last three images this technique is modified to complementary SPAMM (C-SPAMM) [65]. In the images two perpendicular sets of tags are applied by pairs of $\beta°$ RF pulses. Tagging gradients, in the $x$ direction and the $y$ direction, are applied between the members of each RF pulse pair, and each pair is followed by a dephasing gradient in the $z$ direction. The imaging sequences used are multi-phase triggered $T_1$-FFE sequences.

Parameters: $B_0 = 1.5\,\text{T}$; TR/TE/$\alpha$ = 35 ms/3 ms/30°; 18 cardiac phases; 35 ms per phase.

| image no. | 1 | 2 | 3 | 4 | 5 | 6 |
|---|---|---|---|---|---|---|
| cardiac phase | 2 | 8 | 18 | 2 | 8 | 18 |
| tag angle $\beta$ | 45° | 45° | 45° | +/ − 90° | +/ − 90° | +/ − 90° |
| tag slice thickness (mm) | - | - | - | 7 | 7 | 7 |
| image slice thickness (mm) | 7 | 7 | 7 | 20 | 20 | 20 |

(3) Images 1 to 3 are obtained straightforwardly. In these images the effective flip angle from each pair of tagging pulses varies between 0 and 90°, the magnetization varies periodically between equilibrium and saturation. The

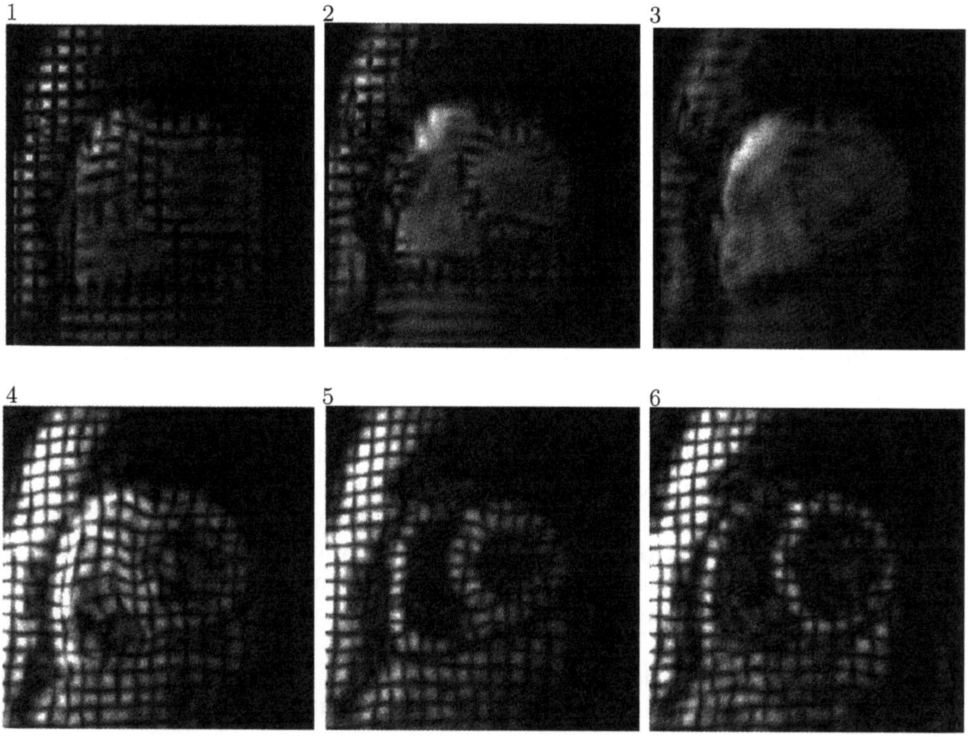


**Image Set VII-10**

tagging pulses are not slice selective and the subsequent excitation pulse defines the slice thickness.
Images 4 to 6, on the contrary, are the result of subtraction of two primitive images, one with all positive tagging flip angles and one in which the latest of the tagging pulses has a negative flip angle. Now the tagging pulses are slice selective and the excitation pulse covers a slab of tissue much thicker than the slice. Before subtraction, in the primitive images (not shown), the effective flip angles from each pair of tagging pulses vary between 0 and 180°; in these images the initial magnetization varies periodically from equilibrium to inversion. The difference between the tagging pattern in the first and second primitive images is that the positions of inversion are shifted over half a tag period.

(4) Images 1 to 3 demonstrate the well-known problem of tagging, namely that soon after the first few cardiac phases the tagging contrast decreases. The decrease is caused by a signal increase in the saturated regions through $T_1$ relaxation. On the other hand in images 4 to 6 the tag stripes remain black. This is a result of the subtraction that removes the signal contribution of regions where the tag angle was equal in the primitive images. The

subtraction at the same time adds constructively the signal in areas where in the first primitive image the tag angle was $\pm 180°$ (and in the second one $\mp 180°$), so that this pattern is retained in full contrast. Of course, in the subtracted images 4 to 6 the contrast-to-noise ratio decreases at the same rate as in images 1 to 3. However, the initial signal-to-noise ratio in the last images is better because a larger tagging angle could be employed. Moreover, the subtraction images are not disturbed by the signal of inflowing blood in the lumen of the heart. As a result the tagging pattern in images 4 to 6 is more conspicuous than that in images 1 to 3. This holds especially for the late cardiac phases.

(5) The two sub sets of images differ in yet another aspect. Images 1 to 3 show the part of the cardiac muscle that during the cardiac cycle passes through the excited plane. This latter plane itself is stationary with respect to the magnet coordinates. The tagging pattern that is shown therefore belongs to anatomical regions that change with cardiac phase. In images 4 to 6, however, the tagging pattern is defined in all three dimensions. Its position in space varies during the cardiac cycle, but it always remains in the thick slab excited by the imaging sequence. The movement pattern visible in the multi-phase scan represents the in-plane component of the true motion of the tagged anatomy.
The discussion and the images show that C-SPAMM is a viable and interesting modification of the original SPAMM technique [65].

# VII-11 Phase Contrast MR Angiography: Contrast Versus Velocity Sensitivity

(1) Phase contrast MR angiography is a technique that is capable of visualisation of small vessels. This scan technique is based on the acquisition of two or more datasets per slice with scans that have a different sensitivity for flow, followed by complex subtraction of the elementary data according to a suitable scheme; see Sect. 7.5. The resulting image depicts the spin phase, normalized to zero for stationary tissue. In the vessels, the phase is proportional to the flow velocity and to the flow sensitivity of the scan. This latter property is usually called "V encoding" or "Venc". It is defined so that in a phase contrast angiography image obtained with Venc = 100, a phase difference $\Delta\phi$ of 180 degrees results for spins that have a velocity of 100 cm per second.
The vessel contrast obtained depends on Venc. In this image sets some choices of this parameter are compared.
All scans are made with a 3D $T_1$-FFE acquisition method.

Parameters: $B_0 = 1.5\,\mathrm{T}$; FOV = 230 mm; matrix = 256 × 220; $d = 0.8\,\mathrm{mm}$; TR/TE/$\alpha$ = 10/9.9/30; NSA = 1.
50 transverse images covering a slab of 40 mm are obtained in a compound scan of 18 minutes. Each compound scan comprises four primitive scans; one with zero flow sensitivity and three with flow sensitivity in each of the three coordinate directions $x$, $y$ and $z$.

| image no. | 1 | 2 | 3 | 4 | 5 | 6 |
|---|---|---|---|---|---|---|
| image type | FFE-M | PCA-M | PCA-MIP | PCA-MIP | FFE-MIP | FFE-MIP |
| Venc (m/s) | 0.60 | 0.60 | 0.60 | 1.80 | 0.60 | 1.80 |
| $\Delta\phi\,(\nu{=}1)$ (degr) | 300 | 300 | 300 | 100 | 300 | 100 |

The modulus images are labelled as FFE-M. These images show the sum of the modulus of the data of all scans.
The phase contrast angiography (PCA) images are labeled as PCA-M (PCA-inodulus). These images show the modulus of the complex subtraction of the primitive images according to (7.14) in Sect. 7.5.1.1.
The maximum intensity projection images are labeled as MIP. The prefixes PCA and FFE are reminders that the data used to create these MIP images are the FFE-M and PCA-M data respectively.
The parameter $\Delta\phi$ ($\nu = 1$) is listed to obtain a better feel for the meaning of the parameter Venc. It is proportional to the inverse of Venc, and it is the spin phase (in degrees) obtained in tissue with a velocity of 1 m/s.

(2) The thickness of the region covered in the MIP images is not representative for what is used in normal clinical practice; however, it is sufficient to visualize the differences between the vessel contrasts obtained.

(3) The contrast in the FFE-M image (image 1) and the FFE-MIP images (image 5 and 6) is caused by inflow enhancement. This contrast is not dependent on Venc. The FFE-M image (image 1) shows a background of stationary tissue. This background is almost absent in the PCA-M image (image 2). The absence of background is a prerequisite for the retrieval of weak vessel signals by the MIP algorithm. As a result, PCA-MIP images can show small vessels, see for example image 3.

(4) The visualization in PCA-MIP images of vessels with a low velocity requires a suitable adjustment of Venc. Slow flow in the superior sagittal vein has influenced the brightness of this vessel between image 3 (Venc = 60) and image 4 (Venc = 180). Moreover, small vessels are less visible in image 4. The rule of thumb is that Venc should be about equal to the expected velocity in the vessels to be visualized. This results in large values of the spin phase in these vessels which of course is what is desired for a good contrast in the PCA-M images.

(5) Large vessels are visualized well in the FFE-MIPs shown in images 5 and 6. At the extreme cranial and caudal edge of the data volume this is true also for a number of smaller vessels. Here fresh inflowing blood has given sufficient signal in the small vessels to exceed the background. In most parts of these images however, the visualization of small vessels is poor, due to a lack of contrast between the vessel signal and the stationary tissue background in the FFE-M images. As expected, the vessel contrast in the FFE-MIP images is almost independent of Venc.

(6) Image 3 represents a properly adjusted PCA-MIP. It displays smaller vessels than images 4, 5 or 6. In this image, furthermore, the brightness of the vessels is more constant than in the other images. These are attractive features obtainable in phase contrast angiography. When the smallest vessels are not of prime interest FFE-MIP images can be used. Such images remain of interest also because they can be obtained with a shorter total scan duration.

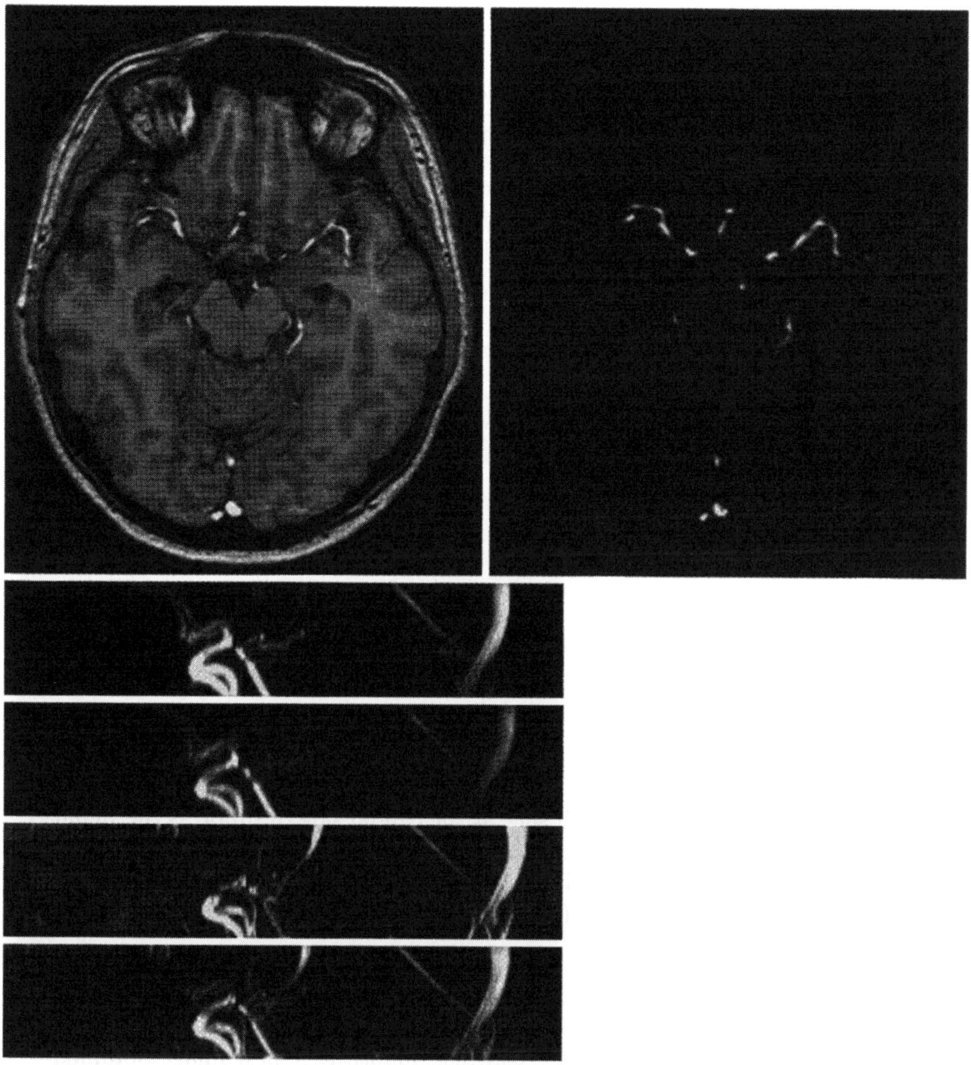

1 2 3 4 5 6

**Image Set VII-11**

## VII-12 Triggered and Gated Inflow MR Angiography

(1) Enhancement of the signal by inflow can be used for the projective angiographic imaging of blood vessels. The success of this technique, usually called inflow angiography, is based on numerous refinements in the scan. In this image set, examples are shown of two alternative methods for inflow angiography of the vessels in the upper leg:

Parameters: $B_0$ = 1.5 T; FOV = 325 mm; matrix = 166 × 256; $d$ = 5 mm; NSA = 1

| | | | | | |
|---|---|---|---|---|---|
| scan no. | 1 | | 2 | | 3 |
| image no. | 1 | 2 3 4 | 5 | 6 7 8 | 9 |
| cardiac synchronisation | triggering | | gating | | |
| acquisition method | $T_1$-TFE | | $T_1$-FFE | | $T_1$-TFE |
| TR (ms) | 15 | | 26 | | 18 |
| TE (ms) | 6.9 | | 6.9 | | 10 |
| flip angle (degrees) | 50 | | 50 | | 25 |
| scan time (minutes at cf = 65) | 9 | | 10 | | 5 |
| image type | M | MIP | M | MIP | q-flow |

(cf = cardiac frequency in beats per minute; MIP = maximum intensity projection)

Both angiographic scans (scan 1 and 2) are obtained with a transverse sequential single-slice technique. A saturated slab is positioned in a region that has its border 10 mm below each slice at the caudal side to suppress the signal from the veins. The slice thickness is 5 mm; the distance between adjacent slices is 4 mm. The overlap is included to obtain a smoother contour of the vessels in the projective view. Per scan 80 transverse images were collected.

(2) Image 1 and 5 are examples of the transverse images (image type modulus), obtained in scan 1 and 2. Images 2,3,4 and 6,7,8 are projective views of the entire data set. These views are obtained by maximum intensity projection (MIP). Image 9 is a phase-contrast quantitative-flow image obtained from a mixed scan with two sequences (Sect. 7.5.2). One sequence was flow sensitized in the FH direction and the other was flow insensitive.

(3) In scan 1, each R top of the ECG starts a shot of the TFE scan. This shot lasts 19 TR (285 ms) and has a linear profile order (Sect. 5.4). The shot is preceded by a slice-selective inversion pulse as well as by the regional saturation slab described above. The inversion pulse is designed to null the stationary tissue. Each shot starts 100 ms after the R top, so that the scan takes place in end systole; the arterial flow velocity then is high (image 9).
In scan 2, the R top is used to open a gate for data collection starting immediately after the R top. The sequence is continuous, but the data are acquired

during the gate only. The gate width is 390 ms, sufficient for 15 TR. The technique for image 2 is called gated sweep.

(4) The relatively long TR in scan 2 (FFE) is caused by the addition of a regional saturation pulse per TR, whereas in scan 1 (TFE) this pulse was given only once per heart beat. Both scans are designed to take about the same amount of time, so that per heart beat the data collection period of the FFE scan (gate width) had to be larger than that of the TFE scan (shot duration).

(5) The large flip angle has saturated the non-moving tissue in both scan 1 and scan 2. However, the saturation is more complete in the TFE technique, by virtue of the use of the inversion pulse. The MIP images in images 2, 3,4 and 6, 7, 8 reflect this as a difference in the visibility of background tissue structures; however, the visible background of the FFE based MIP images (images 6, 7, 8) does not hamper their readability.

(6) The suppression of the venous signal is more complete in the FFE technique. In images 2 and 4, the popliteal vein is visible as a grey structure, neighbouring the artery. In images 6 and 8, this structure is absent. This difference is due to the frequent repetition of the regional saturation slab in the FFE scan.

(7) In conclusion, both MR angiographic techniques are based on a number of provisions. Both scans result in clearly readable MIP images, obtained in about equal scan times. However, the gated sweep technique used for image 3 appears to be more attractive because of its superior suppression of the venous structures.

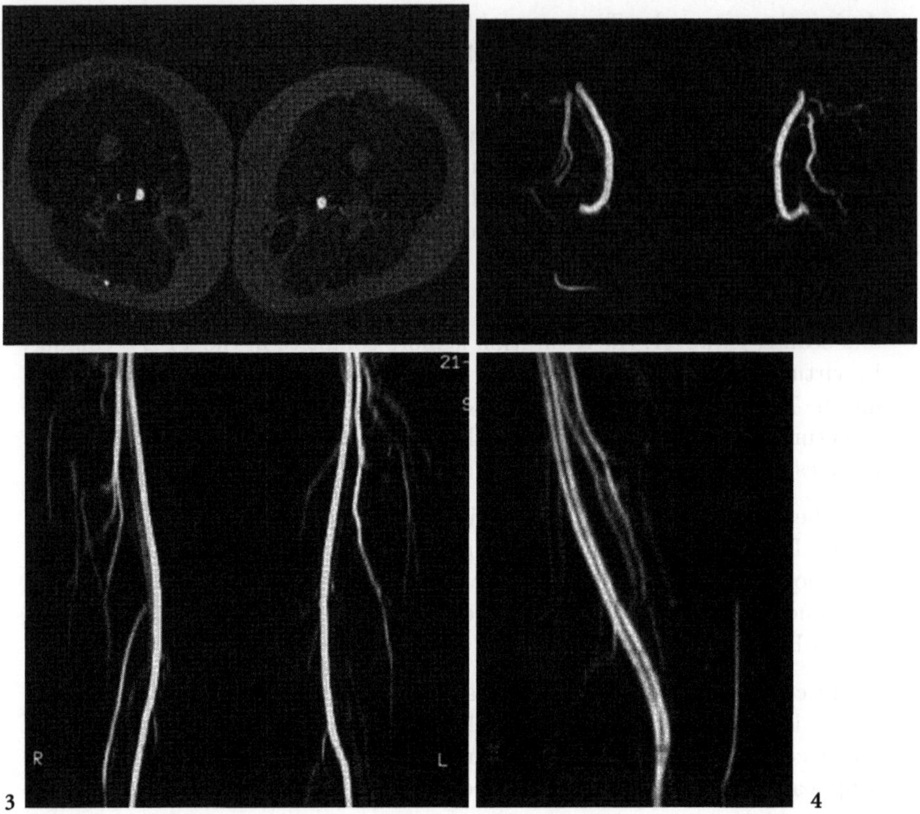


1 2 3 4

**Image Set VII-12**

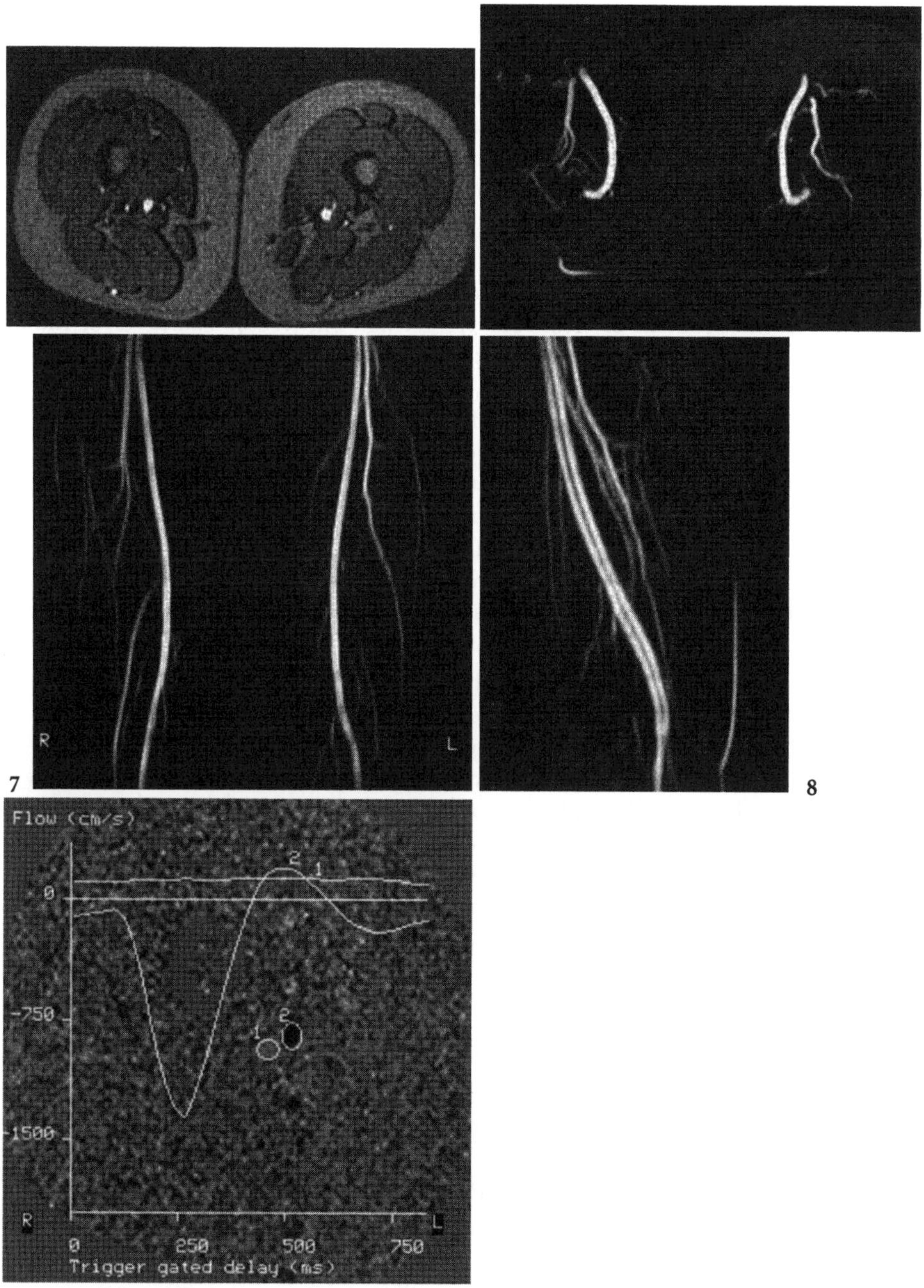


**Image Set VII-12.** (*Contd.*)

## VII-13 Imaging of the Coronary Arteries

(1) The images shown in the two earlier editions are replaced by more recent examples so that an up to date illustration is given of the status of the rapidly developing technique of coronary MR angiography. Coronary artery imaging has been promoted by the intense efforts of researchers and has had the benefit of a number of recent technical advances. Some of these are used in the images shown here.

- The use of a vector cardiogram to trigger the sequence [87] has strongly reduced the problem of erroneous triggering. The vector character of the signal is collected using a sufficient number of ECG electrodes (e.g. four). The vector direction of the magneto-hydrodynamic effect of flowing blood in the magnet field as well as that of the gradient-induced spikes differ from that of R-top, allowing its reliable detection.
- The use of navigator echoes for gating control over the respiratory movement of the heart. These navigator echoes are obtained by "pencil beam images" intersecting the diaphragm; see Sect. 7.2.2.4 for a description of the navigator principle and image set III-8 for the shape of the pencil beam. The use of navigator control allows the patient to continue breathing during the scan.

Image 1 is a representation of the dynamic sequence of the one-dimensional navigator images. The position of the diaphragm is obtained by online estimation (black markers). In a training period the acceptance window of diaphragm positions for data collection is determined (thin black lines). The occurrences of accepted diaphragm positions during the data collection period are indicated by the thick trace at the bottom of the image.

Images 2 and 3 show the normal-appearing left anterior descending coronary vessel of a patient suspected of coronary disease. The images are curved reformats of the corresponding 3D datasets, adjusted manually to follow the LAD. [All images are courtesy of Prof. Kyu Ok Choe, Yonsei University, Seoul, Korea.]

Parameters: $B_0 = 1.5\,\mathrm{T}$, FOV $= 360\,\mathrm{mm}$, matrix $= 512 \times 360$; $d = 3\,\mathrm{mm}$.

| | | |
|---|---|---|
| image no. | 2 | 3 |
| method | 3D N-TFE | M2D TSE |
| partitions/slices | 20 | 10 |
| prepulse | navigator, SPIR | navigator, black blood |
| TR, ms | 7.1 | 2 RR intervals |
| TE, ms | 2.0 | 26 |
| flip angle, degrees | 30 | 90 |
| turbo factor | n.a. | 24 |
| echo spacing, ms | n.a. | 5.6 |
| NSA | 2 | 2 |
| scan time | 3′12″ | 4′17″ |

n.a. = not applicable. Scan times are given for RR intervals of 0.857 s and a navigator efficiency of 100%. Its practical value is about 60%. In both scans, the slices overlap by half a slice thickness. In both scans, the images are obtained in end diastole. In the scan of image 3, the term M2D stands for an acquisition technique in which the data are collected sequentially per slice.

(2) The images show the main branch of the LAD with its bifurcations over a considerable length and in high resolution. The contrast in both images is complementary. Image 2 shows a bright blood image. The signal from pericardial fat is suppressed by the SPIR pulse (see Sect. 2.7.2 point 4). Image 3 shows this fat as a bright region, but there the arteries are displayed with low signal due to the use of the black-blood prepulse. The black-blood prepulse given in that scan consists of two inversion pulses given during systole (Sect. 2.7.2 point 9). The selective inversion resets the magnetization in a region 7 mm wide, so that almost all of the magnetization of the blood in the ventricle remains inverted. The image shows that the inverted blood has filled the complete length of the LAD.

(3) The pair of images is obtained in a clinical setting and in acceptable scan times. The complementary contrast is useful to rule out thrombi that may be poorly visible in the bright-blood image, as well as calcified stenosis that may remain low in signal in the black-blood image. The images demonstrate the clinical potential of this use of MR imaging.

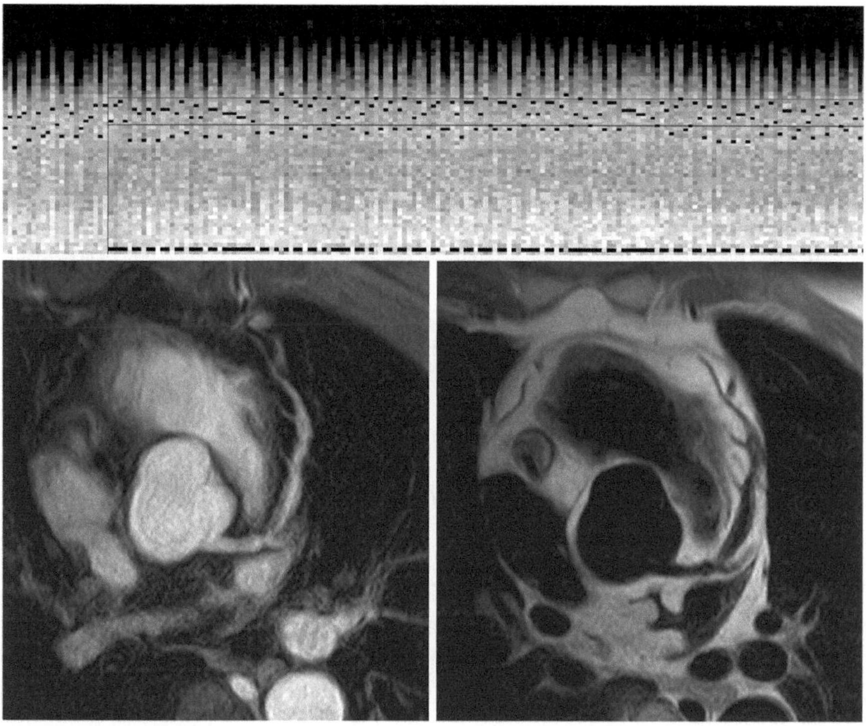

1

2 3

**Image Set VII-13**

## VII-14 Contrast-Enhanced MR Angiography of the Lower Extremities

Technique and images with permission of K.Y. Ho and J.M.A. van Engelshoven, University of Maastricht, The Netherlands.

(1) The use of contrast agents to enhance the value of MR angiography (MRA) is relatively new. Similar to X-ray angiography, this technique is usually based on imaging before venous return of the contrast agent has occurred. However, in MRA the acquisition time for a 3D angiographic scan complicates the problem of timing of the imaging run between bolus arrival and venous return. Contrast agent enhancement can be of considerable use in MRA. With respect to unenhanced MR angiography the potential advantages are twofold. First, the vessel contrast is not disturbed by flow voids that plague unenhanced MRA, especially near stenoses. Second, a mask image obtained before bolus arrival can be used for subtraction of the background, thereby improving the conspicuity of the vessel tree.
The arterial system below the abdominal aortic bifurcation is situated in a body region that is well suited for contrast agent enhanced MRA because the arterial phase of the contrast agent distribution is relatively long. However, this part of the arterial system is quite elongated and is more than two times the diameter of the homogeneous region of the imaging magnet and hence that of the maximum allowed field of view. When angiography of this entire system is desired, the problem of timing of the image increases considerably and the question rises how to reconcile the conflicting requirements of time, resolution and coverage and still obtain clinically useful angiographic data.
Images 1–4 show dynamic 2D MR $T_1$-FFE images of the arterial system above the knee and around the trifurcation of a healthy volunteer (31 years old), obtained after intravenous injection of 40 ml (20 mmol) of GD-DTPA dimeglumine at a rate of 0.3 ml/s. This technique of slow bolus injection is used also in the 3D MRA acquisitions shown in image 5 (healthy volunteer of 54 years).

Parameters for images 1–4: $B_0 = 1.5\,\mathrm{T}$; 3D $T_1$-FFE; 15 partitions; TR/TE/$\alpha$ = 8 ms/2.4 ms/30°; matrix = 90 × 128; FOV = 230 mm; $d = 2\,\mathrm{mm}$; time per scan 7 s.
Parameters for image 5 (merged MIPs of three subtracted scans): $B_0 = 1.5\,\mathrm{T}$; 3D $T_1$-FFE; 32 partitions; TR/TE/$\alpha$ = 11 ms/2.4 ms/40°; matrix 163 × 512; overlapping FOVs of 450 × 292 mm, $d = 2.7\,\mathrm{mm}$; time per scan 37 s.

| image no. | 1 | 2 | 3 | 4 | 5 |
|---|---|---|---|---|---|
| dynamic time | 42 | 63 | 178 | 227 s | 40, 80 and 120 s |

(2) The images 1 to 4 show the passage of the bolus as a function of time. At 42 s the bolus front arrives in the popliteal artery (image 1); at 63 s it has progressed to below the trifurcation (image 2); at 178 s the venous return of

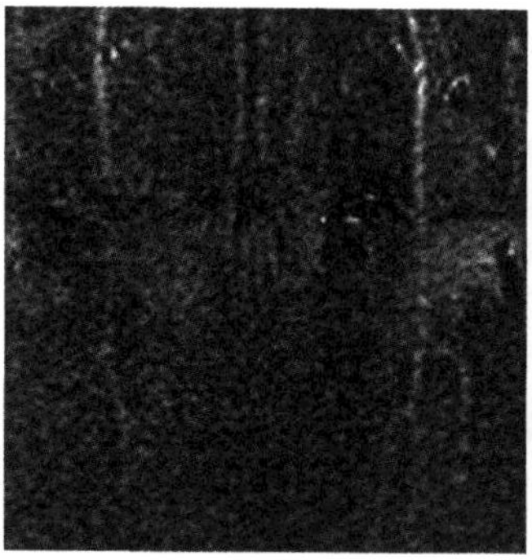

1

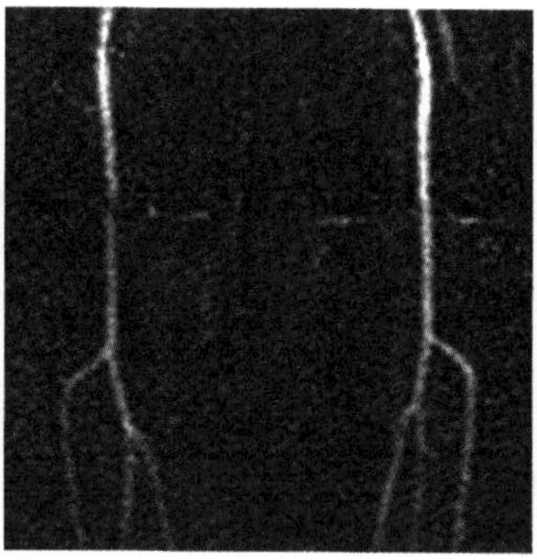

2

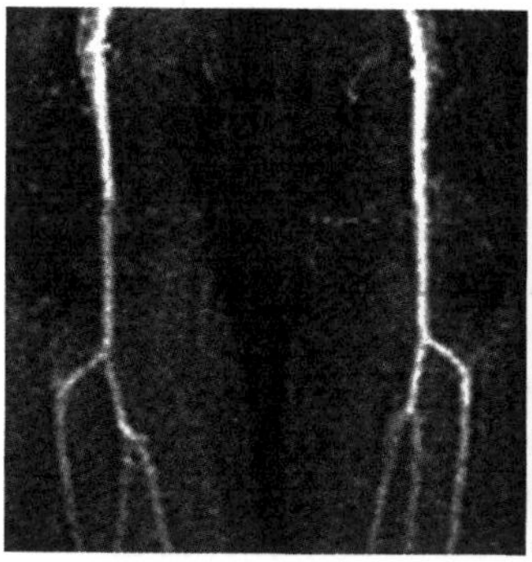

3

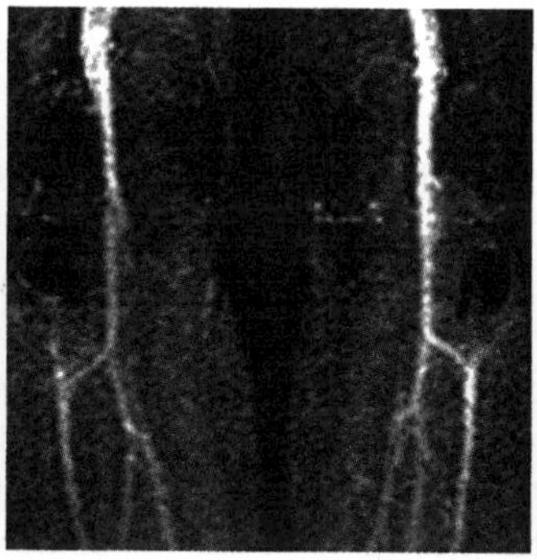

4

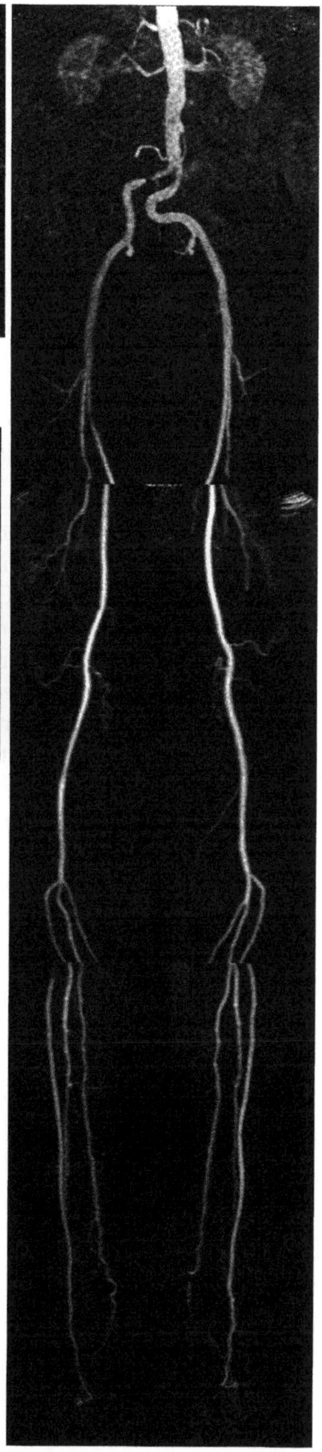

5

**Image Set VII-14**

the agent enhances the popliteal vein; at 227 s the contrast of the trifurcation area is diminished by soft tissue enhancement in the calve. The long interval between arterial and venous phase of the bolus passage (178–63 s) is due to the slow flow in the vascular system of the legs of the resting volunteer. This interval offers the window that can be used for angiographic imaging. The slow injection of the contrast medium that was used is optimal for this imaging task.

(3) The images show that the time window suitable for angiographic imaging shifts with position in the leg. This situation holds fairly typically for most patients as well. In the pelvis (not shown), the time window typically is between 30 and 80 s, in the region of the legs it is between 60 and 180 s. These time windows allow sequential acquisition of MR angiographic scans with scan durations up to 35 s at adjacent regions of the arterial tree of the lower extremities starting at 40 s (pelvis), 80 s (upper legs) and 120 s (lower legs) after injection.

(4) Sequential acquisition at three different positions with 5 s dead time between acquisitions requires special measures but is feasible. Image 6 shows coronal and sagittal MIPs of 3D contrast enhanced MR angiographic images of three adjacent regions of interest. The table travel to each position was programmed and the program was run twice; once before injection and once timed to the bolus injection. Use of the body coil to receive the signal obviated the need for further system adjustments during these runs. At the region of the pelvis, in the first part of the acquisition the volunteer held his breath as long as he could (about 25 to 30 seconds). His legs were stabilized by strapping the feet to a wooden rig and by fixation of the knees with sandbags. The total acquisition volume covered 88 mm × 292 mm × 1250 mm and this is sufficient to display the arterial system from the aortic bifurcation to the ankles without the need to adjust angulation and AP offsets for each of the scans separately.

(5) The favorable temporal behavior of the transport of the bolus in the arterio-venous system of the lower extremities allows contrast-enhanced 3D MR angiography with relatively little compromise. The protocol described above is used in clinical research [66].

# VII-15 Flow-Independent MR Angiography of Abdominal Aortic Aneurysms

Courtesy of D.W. Kaandorp, M.Sc., Eindhoven University of Technology and St Joseph Hospital, Veldhoven, The Netherlands.

(1) More and more MR angiographic problems are tackled with contrast-enhanced (CE) angiography (see e.g. Image Set VII-14). The abdominal aorta remains one of the areas for which CE angiography may not be called for. First, this is because of the rapid passage of the bolus through that anatomy combined with its early venous return so that the time window for CE angiography is short. Second, flow-independent angiography as an alternative has good merits for this imaging task, as will be demonstrated in this image set.

(2) Flow-independent angiography relies on the low transverse relaxation rate of blood (arterial blood has a $T_2$ of 180 ms and venous blood of 80–140 ms, depending on the oxygen saturation and the imaging technique used [40]). When suppression of fat is accomplished by a separate RF pulse and suppression of other tissues by the use of a long TE, the angiographic image can be obtained without subtraction. Moreover, in these images arterial blood, having the highest $T_2$, will have a distinctively brighter signal than venous blood. The acquisition method used in this image set is based on a long TE Turbo Spin Echo sequence that is fine-tuned for two further purposes. First, ghosting caused by flowing blood has to be removed from the image; second, modulation of the arterial signal by flow should be minimal. Compared are coronal MIP images of a patient with an abdominal aneurysm who volunteered his cooperation in the experiment. The images show the suppressed tissue background, the difference in arterial and venous signal and, in addition, the absence of ghosts. The flow-mediated signal loss in the aorta varies with echo spacing.

Parameters: $B_0 = 1.0\,\mathrm{T}$; 3D-TSE; $\mathrm{TE_{eff}} = 200\,\mathrm{ms}$; TR = 1 heart beat; cardiac frequency = 40/min; trigger delay 600 ms; matrix = 256 × 256; FOV = 350 mm; $d = 6\,\mathrm{mm}$; fat suppression with SPIR; NSA = 2. Phase Encode gradient order: ZY; refocusing pulse angle 160°; water–fat shift 0.3 pixel; read gradient head–feet.

| image no. | 1 | 2 | 3 |
|---|---|---|---|
| partitions retained | 26 | 19 | 15 |
| turbo factor | 52 | 38 | 30 |
| echo spacing $\Delta\mathrm{TE}$ (ms) | 7.5 | 10.3 | 12.9 |

(3) The absence of flow-induced ghosts is accomplished by making use of even echo rephasing in the TSE echo train. The uneven echoes are flow sensitive, while the even echoes are flow compensated. The $\vec{k}$-space modulation

of uneven and even echoes causes discrete ghosts, which can be handled in a controlled way. The strategy to suppress these ghosts [67] makes use of:

(a) choice of the temporal order of the looped preparation gradient values so that the $z$ gradient changes in the inner loop,
(b) collection of a number of $k_z$ gradients that is twice the number of partitions retained after reconstruction (a protection mechanism in our system against aliasing in the $z$ direction), and
(c) selection of the turbo factor so that the number of echoes equals the number of $k_z$ values.

With this combination, the first discrete flow ghost shifts in the $z$ direction and is moved over the maximum distance, ending up in a discarded partition.

(4) A large region of signal loss occurs in the aorta just above-the offspring of the renal artery. In this region the flow-mediated signal loss is seen to increase with echo spacing. This can be understood from (7.5) where it is shown that for a given gradient area, the phase shift of the uneven echoes is proportional to $\Delta$TE. The large gradient surfaces of the read gradient, which has its direction parallel to the aorta, make it likely that this phase shift is caused by longitudinal flow. For the largest $\Delta$TE used (image 3), the phase shift in regions with longitudinal flow approaches 180° and the signal loss is nearly complete. So, although the images are obtained in the diastolic phase and most of the aorta has low flow, the aortic signal loss shows that the renal artery has a steady flow and drains the aorta even in diastole. The pronounced flow void visible in most of the vena cava reflects the steady flow in that vessel.

(5) In addition to the lumen in the vessel tree, the images show an inhomogeneous high-intensity thrombus. The clear artifact-free presentation of thrombus, arterial blood and venous blood at separated brightness levels makes this type of scan a diagnostically valuable contribution in the assessment of abdominal aortic aneurysms. The images illustrate that the echo spacing is a critical parameter in the technique used for the images in this set. Short echo spacings are needed to avoid unwanted flow-mediated signal loss.

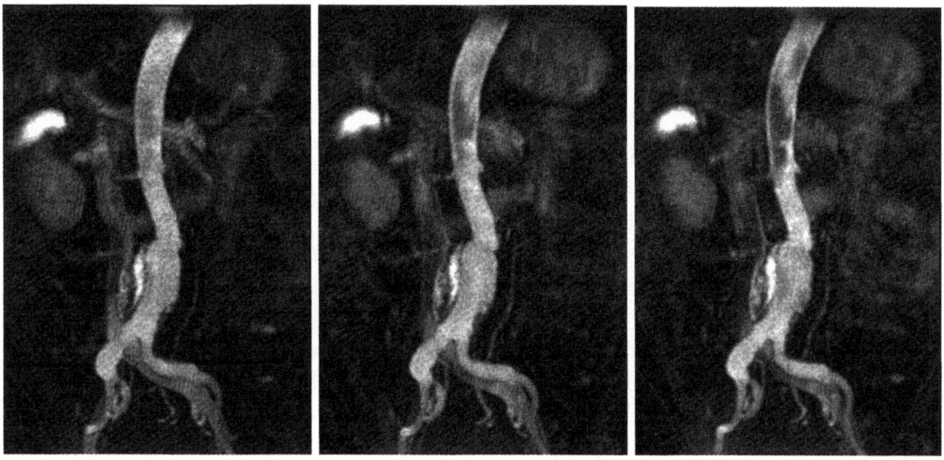

1 2 3

**Image Set VII-15**

# VII-16 Parameter Maps from Dynamic Scans after Bolus Injection of a Contrast Agent

(1) In well-perfused organs, such as brain, kidneys or the heart, dynamic scans after intravenous bolus injection of a contrast agent can be used to derive some diagnostically meaningful parameters. Examples are the area under the enhancing area in the time curve of the response, related to the relative blood volume (see Sect. 7.6) and the time of onset of the transient response of the organ.
The image set shows such images obtained in the brain of a volunteer, after intravenous injection with GD-DTPA dimeglumine. The scan technique used was a spoiled gradient echo sequence in a multi-slice multi-shot EPI method.

Parameters: $B_0 = 1.5\,\mathrm{T}$, TR/TE/$\alpha$ = 223/30/35°; EPI factor 9, matrix $89 \times 128$, $d = 6\,\mathrm{mm}$; dynamic scan time = 2.2 s per image. Bolus of 0.15 mmol/kg injected in 5 s.

| Image no. | 1 | 2 |
|---|---|---|
| | rCBV | time to arrival |

(2) The rCBV map (image 1, rCBV stands for relative cerebral blood volume) is obtained according to (7.29) and (7.28); the value of the integral estimated by curve fitting as shown in Fig. 7.23. The image shows a clear difference in signal strength between cortex and medulla. Although this type of image does not give absolute values of rCBV, the difference corresponds to the actual differences in the cerebral blood volume in these tissues of which the normal value is 8% in cortex and 4% in medulla. The low values of the cerebral blood volume explain the very strong bright signal contribution from the presence of macroscopic vessels, even when they are much smaller than the size of the rather coarse voxels in image 1. Both arterial and venous vessels are bright in the rCBV map. The dark aspect of the anterior ventricles corresponds to the absence of perfusion in that region. Some voxels are black, in these cases the curve-fitting algorithm did not converge.

(3) The time-to-arrival map (image 2) displays in gray values a map of the time of onset of the signal enhancement, so that early enhancement is bright. In the central area of the image some of the brighter regions correlate to areas in which the rCBV map shows vascularization. This correlation indicates an arterial contribution to the signal in both images. In the peripheral region of the brain, some of the dark gray areas of image 2 (late arrival) correspond to bright vascular areas in image 1. This can be interpreted as a venous contribution to the signal.
Image 2 shows little contrast, because the variation in the arrival time is small. From histogram analysis it was shown that over the entire slice shown its value is $28 \pm 4\,\mathrm{s}$. This narrow range of arrival times makes this type of image sensitive to the detection of disturbances in the cerebral blood flow.

Of course the arrival time is not a true descriptor of the local perfusion of the tissue that is imaged, rather it is influenced by the properties of the feeding arterial system. However, in situations of disturbed flow, the low cerebral blood flow often is caused by defects in this feeding system, so that in these cases the image can be used as a coarse qualitative indicator of cerebral blood flow.

(4) The combination of the rCBV map and the arrival time map can be relevant in the assessment of acute stroke [68].

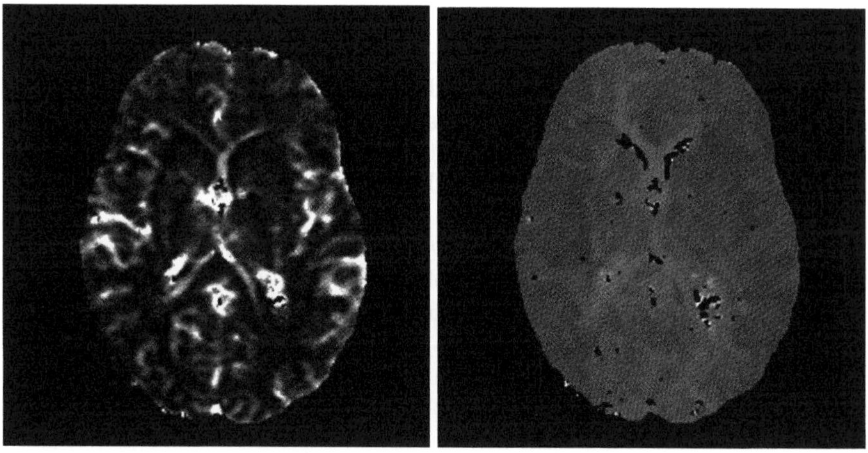

1 2

**Image Set VII-16**

## VII-17 Perfusion Imaging by Arterial Spin Labelling using TILT (Transfer Insensitive Labelling Technique)

Images are courtesy of Xavier Golay, Klaas Pruessmann, Matthias Stuber and Peter Boesiger of the Institute of Biomedical Engineering, ETH Zurich, Switzerland.

(1) In Sect 7.6.2, an introduction is given to arterial spin labeling techniques. In these techniques the effects of perfusion are visualized in a difference image from a slice that is scanned twice: one scan obtained during upstream arterial labeling and one control scan. The implementation is compatible with the hardware provisions of the typical clinical MR system (no separate transmit coil for labeling) and relies on labeling pulses that are intertwined in the imaging sequence. EPISTAR and FAIR can be seen as the parents of two families of methods that are suitable for such systems. In both families inversion pulses are used for the control as well as for the labeling scans. In EPISTAR-like methods the scans differ by upstream (labeling) and downstream (control) displacement of the inversion slab relative to the slice of interest; in FAIR-like methods the inversion pulses include the slice of interest, and the inversion slabs for control and labeling differ in width.
The inversion pulses always generate a certain degree of bound water saturation in the slice of interest, which leads to signal modulation by magnetization transfer. Direct signal modulation can also occur when the inversion slab has ill-defined shoulders or is too close to the slice of interest. The requirement that these modulations should be identical during control and labeling limits most of the pulsed arterial spin labeling methods to single-slice imaging. It may be clear that even then the control of all RF pulse profiles is a critical issue.

(2) The difficulties mentioned above are the result of the engineering details of the sequence actually used, and the transfer insensitive labeling technique (TILT) [69] combines solutions to some of them in an elegant way.
TILT belongs to the EPISTAR family, but instead of using two identical RF pulses with differently placed inversion slabs, it uses two different RF pulses for the same labeling slab, upstream of the slice of interest. The requirements now are:

(a) free water in the labeling slab has to be inverted when the one RF pulse is used but unchanged when using the other pulse, and
(b) the effect of both pulses on magnetization transfer in the slice of interest should be identical.

Inherent to the TILT approach is that the position of the slice of interest with respect to the labeling slab is not critical as long as it is outside its shoulder. This means that multi-slice imaging should be possible.

(3) Requirement (a) is reached by using a pair of two 90° pulses for inversion and a pair consisting of a +90° and a −90° pulse for control. Requirement

(b) is reached by separating the pulses of each pair by a time long enough for complete dephasing of the bound water. Where $T_2$ of bound water is well below 1 ms, a separation time of 500 to 1000 $\mu$s is amply sufficient. By virtue of this dephasing, both pulses will create an identical saturation of bound water in the slices of interest and in a multiple slice acquisition this identity will be true for all slices.

In [69] further details of the TILT sequence are discussed, including the means to obtain a short duration of the labeling pulses, a precise degree of inversion and a sharp definition of the labeling slab.

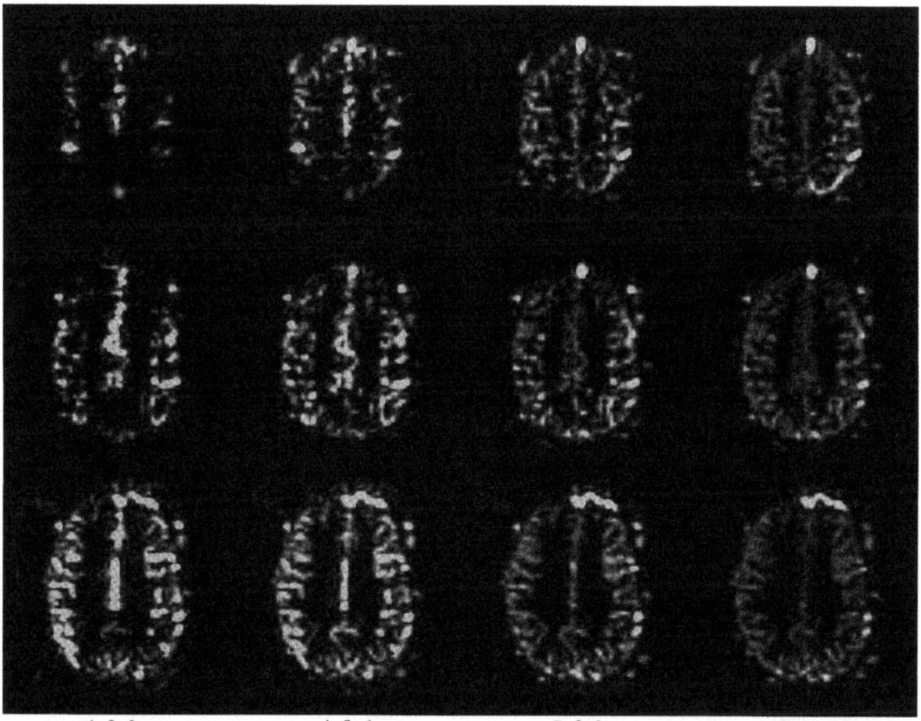

1, 2, 3 4, 5, 6 7, 8, 9 10, 11, 12

**Image Set VII-17**

(4) The images compare the influence of slice position in a mult-slice TILT method used with various labeling delays. Each image is the difference between a control and an inversion-labeled image. The labeling slab was 140 mm thick; slice saturation was performed in the imaging volume prior to the labeling. The saturation slab was 34 mm, covering all slices with a positive margin, of 4 mm on both sides. After labeling, multi-slice acquisition of three slices was obtained in a single shot SE-EPI sequence; TR/TE = 2000 ms/35 ms; matrix 128 × 57; FOV 430 × 190 mm; $d = 8$ mm; slice gap = 1 mm.

| image no. | 1 | 2 | 3 | 4 | 5 | 6 |
|---|---|---|---|---|---|---|
| labeling delay TL (ms) | 400 | 470 | 540 | 700 | 770 | 840 |
| gap to labeling slab (MM) | 23 | 14 | 5 | 23 | 14 | 5 |

| image no. | 7 | 8 | 9 | 10 | 11 | 12 |
|---|---|---|---|---|---|---|
| labeling delay TL (ms) | 1000 | 1070 | 1140 | 1300 | 1370 | 1440 |
| gap to labeling slab (mm) | 23 | 14 | 5 | 23 | 14 | 5 |

(5) At short delay (TL ≈ 400 ms), signal changes in the upper slices (images 1 and 2) are seen mainly in large arteries. In the lower slice (image 3), however, perfusion signal may be seen in arterioles closer to the brain parenchyma. Two effects may explain this observation:

- First, the slice acquisition order is top to bottom to avoid reduction of the perfusion signal of a slice by previous read-outs of the proximal slices. This results in a longer labeling delay TL for the lower slice.
- Second, the gap to the labeled slab is larger in the upper slices than in the lower one, and the same holds for the transport time of the labeled blood.

At larger delay (TL ≈ 700 ms; images 4–6), all images show a perfusion signal, but regions with large arteries still carry an excessive amount of signal, hampering the readability of these images for parenchymal flow.
At still larger delay (TL ≈ 1000 or ≈1300 ms; images 7–12) gradually the labeled blood in large arteries leaves the imaging volume and the image contrast is caused mainly by perfusion. Notwithstanding the decrease in perfusion signal between these two labeling delays, caused by $T_1$ decay of the labeled blood, the best presentation of the perfusion signal is obtained in images 10–12 at a delay TL ≈ 1300 ms.

(6) The image set demonstrates the possibility of implementation of the arterial spin labeling technique in a multi-slice acquisition. With that an important step towards clinical use of perfusion imaging by arterial spin labeling is reached [70].

The images in this set are borrowed from [69].

## VII-18 Diffusion-Weighted SE-EPI Imaging of the Brain

Diffusion-weighting is obtained by addition of strong bipolar gradient lobes in the sequence, as discussed in Sect. 7.7. Frequently a 180° refocusing pulse is placed between the lobes so that the diffusion-weighted signal is from the spin echo occurring after the second lobe. The read out of this echo, used often, is an EPI echo train. The echo time of the diffusion-weighted spin echo depends on the duration $\delta$ of the lobes, which in turn is related to the diffusion-weighting factor $b$ and the available maximum value of the gradient strength, as shown in (7.42).

The images shown are obtained with different gradient strengths, but all with a diffusion-weighting factor $b = 1000$. All images have the same scaling and are the average of three basis images, obtained with diffusion-weighted gradients in mutually perpendicular directions.

Parameters: $B_0$: 1.5 T; FOV = 230 mm; matrix = 128 × 89; single-shot half-scan SE-EPI; bandwidth = 1560 Hz/pixel.

| Image no. | 1 | 2 | 3 |
|---|---|---|---|
| Diffusion gradient strength, mT/m | 18 | 27 | 63 |
| TE, ms | 95 | 77 | 55 |

(2) The reduction of 20 ms in TE obtained between images 1 and 2 required a moderate rise of the gradient strength from 18 to 27 mT/m. Reduction by another 20 ms still is possible, but at the cost of an enormous increase of gradient strength to 63 mT/m. The gain of each of the reductions in TE is clearly visible in the images as increases in the signal-to-noise ratio.

(3) At given gradient strengths, lower $b$ values would of course allow still shorter TE, but this would not necessarily increase the diffusion-weighted contrast. Three arguments play a role:

- At given TE, the theoretical optimum diffusion-weighted contrast is at $bD = 1$, as can be shown easily by requiring in (7.40) that the double differential $\partial^2 M_T/\partial b \partial D = 0$. The diffusion coefficient of normal white matter $D_{\mathrm{wm}} = 7.5 \times 10^{-4}\,\mathrm{mm}^2/\mathrm{s}$ [88], and this leads to $b = 1333$.
- When TE is assumed to be a function of $b$, the theoretical optimum lowers because of the signal increase at short TE. Numerical solution shows that the size of that effect is about 15%.
- To avoid misreading in diffusion-weighted images ("$T_2$ shine through"), diffusion-weighted contrast should outweigh $T_2$ weighted contrast. In [89] it is shown by experiment that for old lunar infarcts shine through is absent at $b > 1000\,\mathrm{s/mm}^2$.

Accordingly, a diffusion-weighting factor $b$ of $1000\,\mathrm{s/mm}^2$ is typically found in literature on diffusion weighted brain imaging, see, for example, [88].

(4) At a maximum gradient strength of 27 T/m and at a diffusion-weighting factor $b = 1000\,\mathrm{s/mm}^2$, diffusion-weighted SE imaging is possible at an echo time of 77 ms, with an acceptable image quality and a nearly optimal diffusion-weighting.

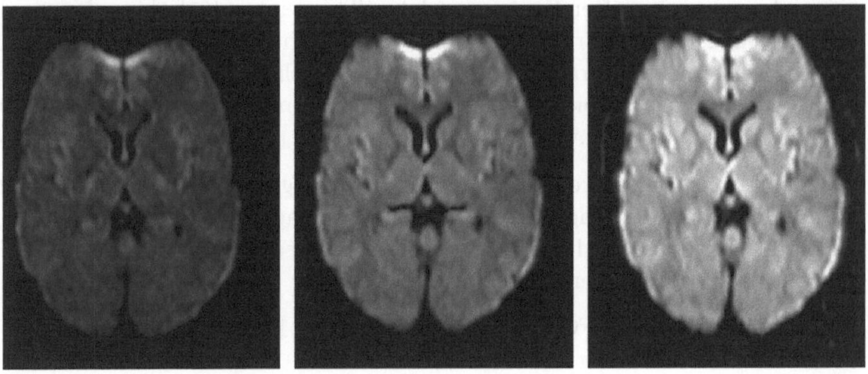

1 2 3

**Image Set VII-18**

# VII-19 Quantitative Diffusion Sensitized Imaging of the Brain

(1) Image Set VII-18 shows the influence of diffusion on the contrast in diffusion sensitized images. The contrast obtained was a qualitative indicator of the diffusion coefficients of the tissues depicted. More quantitative information can be reached by calculation of the diffusion coefficient from the ratio of two differently diffusion-sensitized images (Sect. 7.7). The parameter obtained is called the "apparent diffusion coefficient" (ADC). The adjective "apparent" is included because in the ratio method it is assumed that the diffusion process is not restricted by tissue septa and is isotropic. The presentation of the ADC as an image is called the ADC map.
The diffusion in some regions of the white matter has been shown to be considerably anisotropic and this aspect can confound the readability of ADC maps.

(2) In this image set images are compared with different directions of the diffusion-sensitizing gradient $G_D$ and different diffusion sensitivity $b$ (7.26).

Parameters: $B_0 = 1.5\,T$; cardiac triggered TSE; TR/TE1/TE2 = 2 beats/110 ms/130 ms; matrix = 256 × 256; FOV = 230 mm; $d = 7$ mm; NSA = 1; diffusion gradients (7.26), $\delta = 42$ ms and $\Delta = 50$ ms. The maximum gradient strength available was 20 mT/m.

| image no. | 1 | 2 | 3 | 4 | 5 | 6 | 7 |
|---|---|---|---|---|---|---|---|
| $b$ (s/mm$^2$) | 50 | 50 | 50 | 1250 | 1250 | 1250 | ADCt |
| Direction of $G_D$ | LR | AP | HF | LR | AP | HF | |

Image 7, labeled ADCt, is a map of the trace of the diffusion tensor, calculated from the ratios of image 4 over image 1, 5 over 2 and 6 over 3.

(3) Images 1 to 6 show the strongly anisotropic behavior of the diffusion coefficient of the white matter in the region of the corpus callosum. Especially in image 6 this tissue region is much brighter than in images 4 and 5. Its anisotropy clearly correlates with the direction of the nerves, which in this region are aligned in the transverse plane and mainly in the left–right direction. The biophysical cause of the anisotropy may be located in the myelinated nerve sheets that behave as effective barriers for diffusion perpendicular to the nerve direction.

(4) The ADCt map displayed in image 7 is free from the anisotropy contrast visible in images 1–6 (see Sect. 7.7.2). This is visible as a lack of contrast in the entire white matter.

(5) The ADCt map appears to present a useful summary of the diffusion properties of the brain tissue. It is of course not the only possible map. One can for instance think of an anisotropy index or of a representation of the

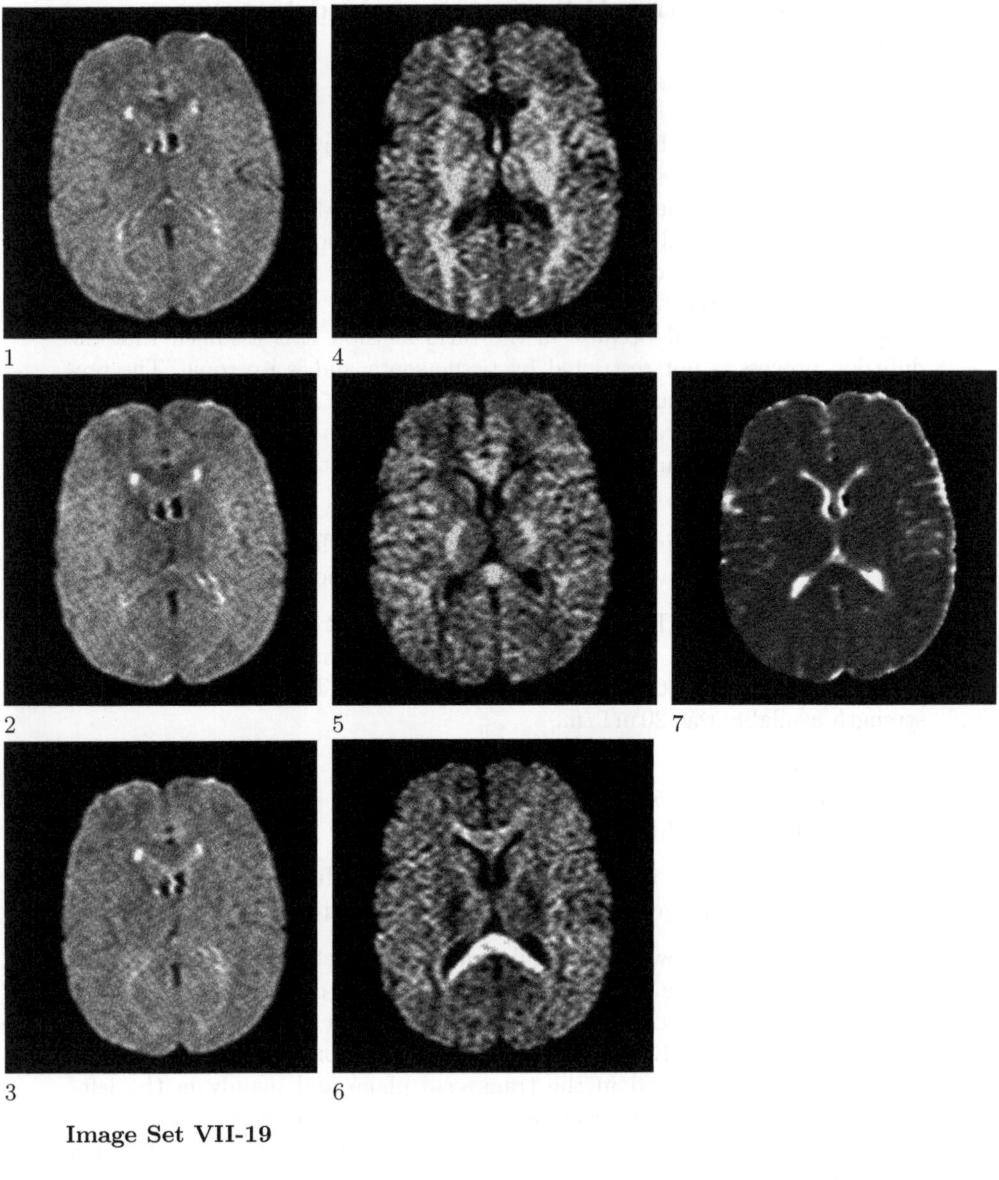


**Image Set VII-19**

anisotropy vector direction [71]. For the purpose of brain infarct evaluation, where diffusion imaging presently plays an important role [68] the use of an ADCt map gives sufficient information; it may even be sufficient to base diagnosis on simple (mono-directional) diffusion weighted images.

# 8. Partitioning of the Magnetization into Configurations

## 8.1 Introduction

In earlier chapters we considered the motion of the total magnetization vector under the influence of RF pulses and gradient fields. The $\vec{k}$ plane (or $\vec{k}$ space in the case of 3D scans) was used to describe the different scan methods. By finding the path through the $\vec{k}$ space, much of the properties of a scan method and its fundamental artifacts, like those due to $T_2$ or $T_2^*$ decay and resonance offset, could be explained. In this chapter it will be shown that additional information can be obtained by looking at the partitioning of the magnetization vector in "configurations" (or "base states" or "coherences"), characterized by their dephasing state.

The theory was developed for the description of NMR diffusion experiments long before the development of MRI [1, 2]. Recently it has been reintroduced [3] for the description of scan methods in which multiple RF excitation pulses, separated by periods with equal length TR and with equal zero order gradient moments $M_0 = \int_0^{\mathrm{TR}} \vec{G}(t)\mathrm{d}t$, are applied. Examples are the very fast "multi-excitation pulse sequences" such as BURST sequences.

It also will be shown that this theory can also be applied for the description of FFE sequences (see Chap. 4) or TSE sequences (see Chap. 3), which also have equidistant excitation or refocussing RF pulses. It will be shown that extra information on such scans may be obtained. Similar theoretical ideas can be applied to the detailed design of RF pulses, including the non-linearity of the Bloch equation. Examples of this application will be indicated at the end of the chapter.

### 8.1.1 Configurations and Phase Diagrams

To explain the general idea, we return to the Spin-Echo sequence, explained in Chap. 2 and particularly Fig. 2.4. For ease of argument we slightly change the situation by using an excitation pulse, which causes a rotation around the $y'$ axis and a refocussing pulse causing a rotation around the $x'$ axis. After the $90^\circ_y$ excitation pulse the longitudinal equilibrium magnetization $M_0$ is rotated into the $x'$ direction. Then the dephasing gradient causes position dependent dephasing of this transverse magnetization. The $180^\circ_x$ refocussing

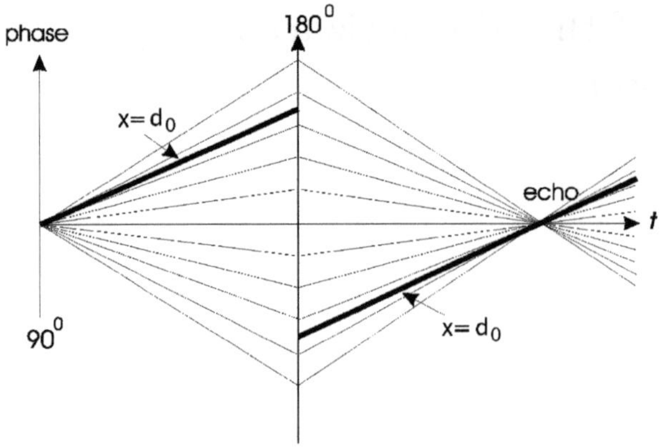


**Fig. 8.1.** Phase of transverse magnetization for some different isochromats. The 180° pulse reverses the phase. The phase of the isochromat $x = +1$ is accentuated

pulse rotates the magnetization vectors around the $x'$ axis so that their phases are inverted.

Due to the read-out gradient, the magnetization vectors of the different isochromats converge again to form an echo. Instead of using a figure like Fig. 2.4, we can also draw the phase evolution of the different isochromats as is shown in Fig. 8.1. Such a diagram is named a "Phase Diagram". The lines represent the phase evolution in the center of each voxel. The phase evolution of intra voxel spins yields a thin bundle of phase lines around a centre line (not shown in this picture).

We can simplify Fig. 8.1 by drawing only the phase evolution of one single isochromat. An isochromat is the collection of points in which the spins have equal phase evolution $\theta(t)$. For example, in case of an $x$ gradient, the lines $x =$ constant are isochromats. According to (2.24), the phase evolution $\theta_0(t)$ of a specific isochromat is just equal to the gradient moment times the distance $d_0$ of this isochromat from the isocenter, so that $\theta_0(t) = k(t)d_0 = \gamma d_0 \int G_x(t)dt$. Along the vertical axis the phase $\theta_0(t)$ is plotted, and along the horizontal axis we have the time $t$. We shall name this restricted $\theta_0(t) - t$ diagram also a "phase diagram".

When using such a phase diagram, we must always remember that only when all magnetization vectors have equal phase can they add together and form an echo. This only happens when the phases of all isochromats are zero, in other words when $k(t) = 0$. When $k \neq 0$ the different isochromats have different phases, according to $\theta(t) = k(t)d$, where $d$ is the distance to the isocenter. Therefore their magnetization vectors have different directions and compensate for each other. That means that only when $\theta(t)$ crosses the axis, where $\theta(t) = 0$, is an echo found.

The simplified phase diagram of our Spin-Echo sequence is shown in Fig. 8.2a, where only the accentuated isochromat $x = 1$ is drawn. The 180° re-

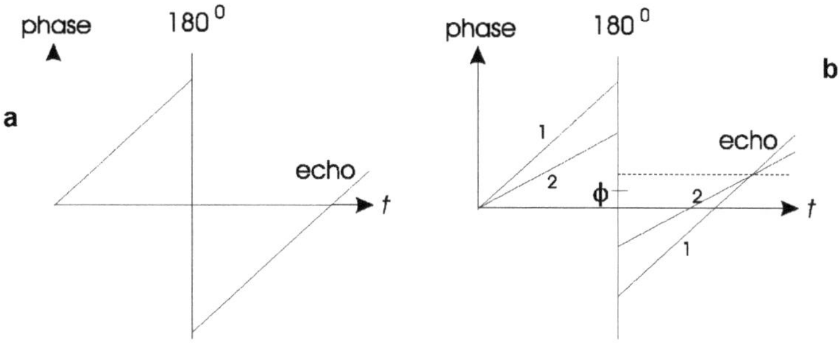


**Fig. 8.2.** Phase diagram for a Spin Echo. **a** The phase of the refocussing pulse is equal to that of the excitation pulse. **b** The refocussing pulse has a phase advance of $\phi$ radians. The echo now has a phase of $2\phi$ rads. For clarity a second isochromat is shown

focussing pulse changes the sign of the phase, as already discussed in Chap. 2. The value of the phase is assumed to be increasing for positive gradients applied, and so only after we created negative dephasing (as the result of inversion by an RF pulse) can an echo be generated, a spin echo.

A phase diagram even describes the situation in which the refocussing pulse causes a rotation around a line that makes an angle of $\phi$ radians with the $x'$ axis (the RF phase of the refocussing pulse deviates $\phi$ rads from the phase of an $x'$ refocussing pulse). This is shown in Fig. 8.2b. Phase reversal now occurs around the point $+\phi$ on the phase axis. After refocussing, the echo obtains a phase of $+2\phi$ with respect to the situation in Fig. 8.2a. It will be shown later that this latter phase shift (which is equal for all isochromats) can be described by using complex amplitudes for the description of the configurations and the echoes, in which case Fig. 8.2a can be used again to describe the situation of Fig. 8.2.b. This figure then only describes the dephasing state of the configurations due to precession, and the echoes (with complex amplitudes) converge only at zero phase on the $t$-axis.

For comparison, in Fig. 8.3 the evolution of the magnetization vector itself is shown. In this figure, part of Fig. 2.4 is repeated, but now for a $90^\circ_{y'}$ excitation pulse followed by a $180^\circ_\phi$ refocussing pulse that causes rotation around an axis with an angle $\phi$ with respect to the $x'$ axis, the situation described in Fig. 8.2b. Figure 8.3 shows that indeed the spin echo now has a phase equal to $2\phi$ with respect to the $x'$ axis. The echo has the same amplitude as in the case of zero phase difference ($\phi = 0$), and so all magnetization is used in this single echo and no other signals (configurations) are present.

It is left to the reader to verify that after a second $180^\circ_\phi$ refocussing pulse, the next spin echo coincides with the $+x'$ axis again. We have already discussed this situation in Sect. 3.2.2, in connection with TSE pulse sequences. The conclusion is that a change of the RF phase of the refocussing pulse only causes an overall rotation around the $z$ axis, and so only the overall phase of the echo is changed. The amplitude of transverse magnetization of the

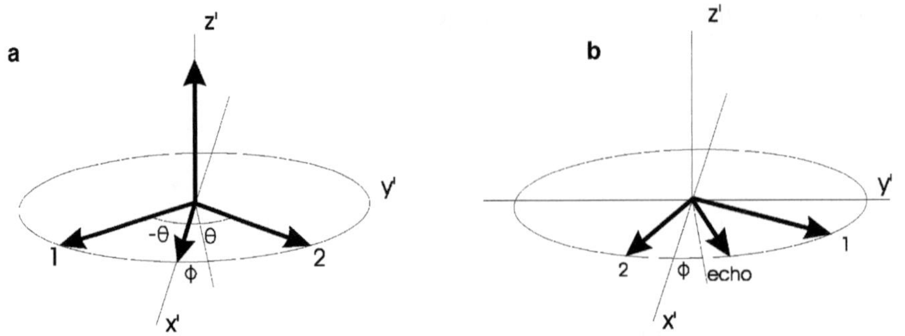


**Fig. 8.3. a** Magnetization vectors 1 and 2 of two isochromats after precessing away from the $x'$ axis during the dephasing period before the refocussing pulse. **b** Vectors 1 and 2 after the 180° pulse around the line that makes an angle $\phi$ with the $x'$ axis. After rephasing the echo has a phase shift of $2\phi$ with respect to the $x'$ axis

echo is complex, with phase $2\phi$. It will be shown later that the separation of dephasing of the magnetization (due to precession in a gradient field) and overall phase rotations of the echo amplitudes (due to the phase of the RF field) will appear automatically in the mathematical treatment.

A new situation arises when the flip angle of the second RF pulse is different from 180°. In this case, encoded longitudinal magnetization is also generated. This is demonstrated with the "Eight-Ball" echo sequence, introduced in Sect. 4.2.2 and shown again in Fig. 8.4c, in which the second pulse has a 90° flip angle around the $x'$ axis. We shall try to visualize the partitioning of the magnetization over the different dephasing states after the second pulse. In Fig. 8.4a the excitation, and in Fig. 8.4b the dephasing in the gradient field is shown for two different isochromats. Just before the second pulse, the phase of the magnetization vector of the isochromats is given by $\phi(\mathrm{TR})_{1,2} = k(t)d_{1,2}$ (neglecting the inhomogeneity of the main field). After the second pulse, a longitudinal component is present and the transverse magnetization is along the $x'$ axis (see Fig. 8.4d and also Fig. 4.4b). For both voxels 1 and 2 in Fig. 8.4e, the total transverse magnetization is split up into one component along the direction $+\theta(\mathrm{TR})_{1,2}$ and one component along the direction $-\theta(\mathrm{TR})_{1,2}$. This partitioning of the transverse magnetization of all isochromats in components along the $+\theta(\mathrm{TR})_{1,2}$ and the $-\theta(\mathrm{TR})_{1,2}$ directions forms two distinct "configurations" (one in the $(x', y')$ quadrant of Fig. 8.4e and one in the $(x', -y')$ quadrant). The term "configuration" is introduced by Hennig [3] and describes a certain state of dephasing due to precession.

The configuration formed by the isochromats with phase $-\theta(\mathrm{TR})_{1,2}$ rephases in the gradient field (see Fig. 8.4e) because the magnetization vectors move in the same direction and with the same speed as before the pulse. This configuration therefore forms the Eight-ball echo. The configuration with phase $+\theta(\mathrm{TR})_{1,2}$ continues to dephase. From Fig. 8.4d it is seen that

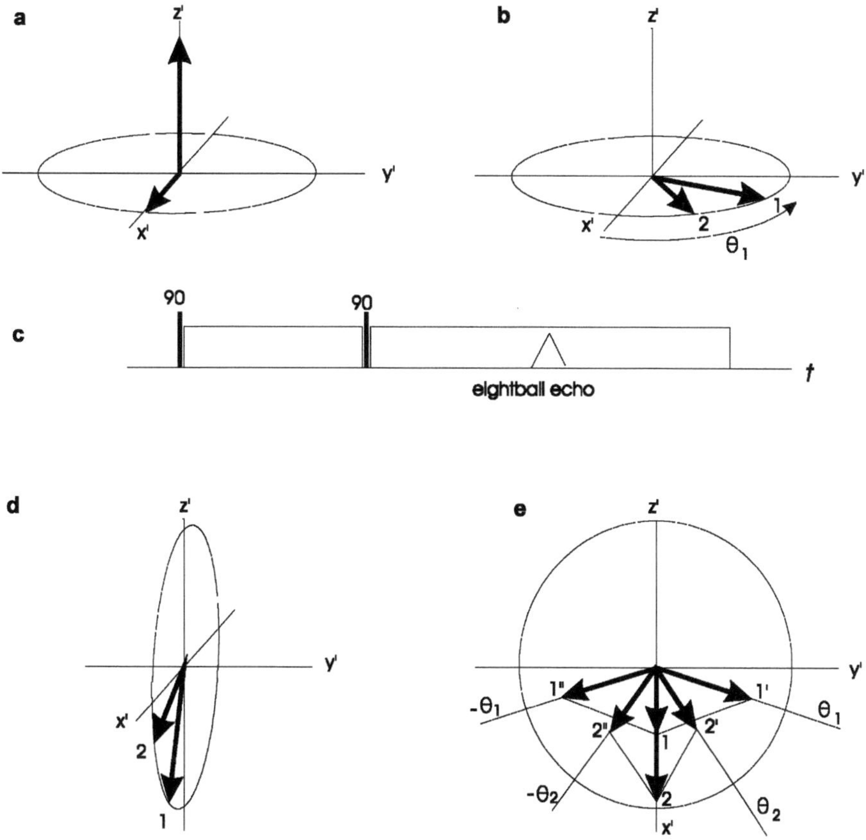


**Fig. 8.4.** **a** Magnetization vectors of all isochromats after the first pulse. **b** For two off-center voxels 1 and 2, the precession of the transverse magnetization $M_T$ in the gradient field over the angles $\theta_1$ and $\theta_2$ is shown. Voxel 2 is closest to the isocenter so $\theta_2 < \theta_1$. **c** Eight-ball sequence. **d** The magnetization just after the second 90° pulse. **e** The transverse components are decomposed along the directions $+\theta_{1,2}$ and $-\theta_{1,2}$ and precess further in a positive direction. Two configurations are formed, one (components $1'$ and $2'$) that dephases further in the gradient field and one (components $1''$ and $2''$) that rephases to form the eight-ball echo

the longitudinal magnetization of the voxels also depends on position (it is "coded"). However, this longitudinal configuration does not react on the gradient field and only suffers $T_1$ relaxation, which is neglected for the time being and therefore remains nearly constant during the read-out gradient. The dephasing and the echo formation after the second 90° pulse is elegantly displayed in the phase diagram, as shown in Fig. 8.5.

Since the longitudinal component is not affected by the gradient field, it is shown as a configuration with constant phase. This may seem rather artificial, but remember that a third pulse can reconvert the longitudinal magnetization into (position-coded) transverse magnetization and so with a

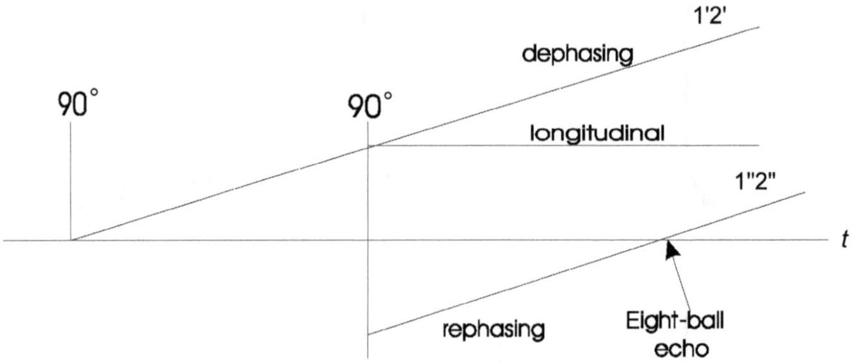


**Fig. 8.5.** Phase as a function of time due to two 90° pulses separated by precession periods with the gradients as shown in Fig. 8.4e

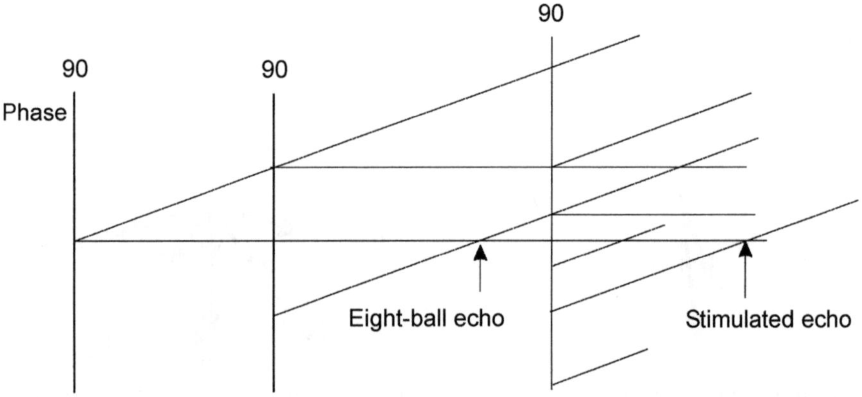


**Fig. 8.6.** Stimulated echo and Eight-ball echoes due to three 90° pulses. Due to the 90° excitation pulse, there is no FID after the second and third pulses

third pulse a new transverse configuration is generated. It is left to the reader to show that the phase diagram of a Stimulated Echo (see Sect. 4.2.3) is given by Fig. 8.6.

It now is clear that each configuration is split into three configurations by an RF pulse. So if we apply a sequence of equidistant RF pulses to a spin system, for example four equal small equidistant RF pulses, we obtain three echoes following the last pulse, as shown in Fig. 8.7. (Note that during the third and fourth pulse we already have echoes, but these cannot be acquired because of the presence of the RF pulses; see also similar situations in Figs. 4.9 and 4.11).

In all following figures, only the phase evolution due to the read-out gradient is considered. Moreover, its evolution between RF pulses is not necessarily described in detail, as the actual gradient waveforms in these periods are replaced by constant-gradient values with the same integral (namely

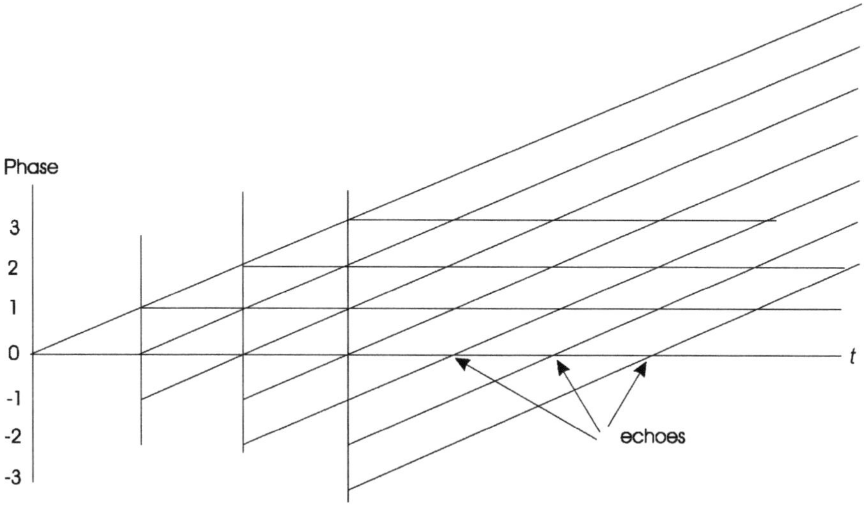


**Fig. 8.7.** Four RF pulses with flip angles smaller than 90° generate three echoes. Only configurations of the same order (the absolute value of the number along the vertical axis) interfere

$k(\mathrm{TR}) = \gamma \int_0^{\mathrm{TR}} G(t)\mathrm{d}t$), and so the phase evolution is linear in our examples. The "order" of a configuration is defined as the distance of its $k$ value to the $t$ axis at the time of the RF pulses, in units of $k(\mathrm{TR})$, and therefore it describes the dephasing state of the configurations. Only configurations of equal order (in absolute value) interfere by the action of the RF pulse resulting in a new mix of configurations of the same order. From Figs. 4.10 and 8.7 we conclude that the observed echoes can be a superposition of partial echoes caused by different configurations, characterized by different paths through the phase diagram. Similar reasoning on the basis of phase diagrams has already been used in the discussion of phase cycling ("spoiling") in $T_1$-FFE (Sect. 4.2.7).

The idea of using phase diagrams, which have been developed for the description of BURST-like sequences, can also be applied to describe fast imaging sequences in which many equidistant RF pulses are applied, such as Turbo Spin Echo or FFE. Multi-excitation pulse sequences such as BURST, QUEST, etc. will be described in terms of phase diagrams in Sect. 8.3. The mathematical basis for the application of phase diagrams only requires a reordering of the theory that we developed in Chap. 2 and will be described in the next section. Similar theoretical ideas can also be applied to the design of RF pulses, as will be shown in Sect. 8.4.

## 8.2 Theory of Configurations

In the previous section we have seen that several echoes are generated as a result of a number of equidistant excitation pulses. Each of these echoes is the result of the rephasing of a "configuration" with a negative dephasing, as shown in Figs. 8.5–8.7 and 8.2. In literature, other terms for "configuration" are also used, and we would mention "coherence", "base state" and "pathway" as three of those alternatives. The latter term illustrates the different paths through the phase diagram, as shown in Fig. 4.11. In this book we shall only use the term "configuration". Since the dephasing depends on voxel position and the time according to $\theta(x,t) = \gamma x \int G(t)\mathrm{d}t$, the dephasing can be characterized by $k = \theta(x,t)/x$ in units of $k = \gamma \int_0^{\mathrm{TR}} G(t)\mathrm{d}t$.

In an ideal Spin-Echo sequence, after the refocussing pulse there is only one single configuration ($k = -1$), which forms the spin echo. However, after the second 90° pulse of an Eight-Ball Echo sequence, we observe three configurations, as shown in Fig. 8.5, and this is generally true for any arbitrary flip angle, as is also shown in Fig. 4.10. One of these three configurations (with negative $k$) forms an echo.

We now consider a series of equidistant, hard RF pulses separated by precession periods. Such a series occurs, for instance, in a BURST sequence [4]. The gradients causing the precession (in the rotating frame of reference) are such that the zero order moment $k(\mathrm{TR})$ is equal in each inter-pulse spacing. Since there is no restriction on the detailed form of the gradient waveforms between the pulses, apart from the multi-excitation (BURST) sequences, all types of FFE sequences, as represented in Fig. 4.1, and TSE sequences (see Sect. 3.2) can be described with the present model [5].

Between the RF pulses the phase of each isochromat increases, depending on its position, due to precession – for ease of argument we assume a positive gradient field unless stated otherwise. During the precession periods the configurations do not interfere, the transverse configurations increase in order $k(n+1) = k(n) + 1$ ($n$ is pulse number) and the order of the longitudinal configurations remains constant. However, due to an RF pulse, the configurations of equal order (= absolute value $|k|$) are mixed.

In the following we shall further demonstrate these statements, already observed in the phase diagrams of the previous section.

### 8.2.1 Magnetization Expressed in Discrete Fourier Series

Our formalism is based on the partition of the magnetization into "configurations" [3], or "base states" [5], as is already proposed in [1] and [2]. The phase increment during the inter-pulse periods is:

$$\theta(\mathrm{TR}) = \gamma \int_0^{\mathrm{TR}} \vec{G}(t)\vec{r}\mathrm{d}t\,.$$

The effect of $N$ pulses and $N$ precession periods with phase increment $\theta(\mathrm{TR})$ may be expected to be an $N$th order Fourier series in $\theta(\mathrm{TR})$. At the $(N+1)$th pulse, each transverse configuration has experienced a certain (net) number $k$ ($|k| \leq N$) of phase increments $\theta(\mathrm{TR})$, which is the number of dephasing periods, $J$, minus the number of rephasing periods, $I$ (see also Sect. 4.2.7), so $k = J - I$.

The "order" of $k$ is given by the distance (in units $\theta(\mathrm{TR})$) from the $t$-axis in the phase diagrams. The longitudinal configurations will be labelled with the same number $k$, which describes the dephasing situation just before the RF pulse that caused the longitudinal configuration (or the dephasing state after reconversion to a transverse state). Therefore for $M_z$ after the $N$th pulse we can write:

$$M_z(N,\theta) = \sum_{k=-N}^{k=N} M_{z,k} \mathrm{e}^{jk\theta(\mathrm{TR})}, \tag{8.1a}$$

where the coefficients of the series have the property $M_{z,k} = M^*_{z,-k}$ because $M_z$ is always real. Note that $\theta(\mathrm{TR})$ is dependent on the position of the voxel considered.

For the transverse magnetization, this symmetry of the coefficients is not preserved: for one distinct value of $M_z$, the transverse magnetization vector may have many directions in the $x' - y'$ plane, and so its phase remains undetermined. Therefore:

$$M_T(N,\theta) = M_x(N,\theta) + jM_y(N,\theta) = \sum_{k=-N}^{k=N} F_k \mathrm{e}^{jk\theta}, \tag{8.1b}$$

where $F_{-k} \neq F^*_k$, to allow for overall phase differences.

### 8.2.2 Rotation

We now wish to find the matrices describing the action of an RF pulse and a precession period on each of the terms of the discrete Fourier series development of the magnetization (the configurations). As an intermediate step, we must first calculate the influence of rotation and precession on the components of the total magnetization vector, $M_z$ and $M_\mathrm{T}$. When this influence is known, (8.1) can be introduced in the rotation and precession matrices and the action of rotation and precession on the separate terms of the Fourier series can be studied. All we have to change in the mathematical treatment of the Bloch equations is to write the transverse magnetization in terms of $M_T = M_x + jM_y$ and $M^*_T = M_x - jM_y$. The relation between this and the previous formulation of Chap. 2 is thus:

$$\begin{aligned} \vec{M}_R &= \begin{pmatrix} M_T \\ M^*_T \\ M_z \end{pmatrix} = \overset{\Rightarrow}{S} \begin{pmatrix} M_x \\ M_y \\ M_z \end{pmatrix} = \overset{\Rightarrow}{S} \vec{M}, \quad \text{with} \quad \overset{\Rightarrow}{S} = \begin{pmatrix} 1 & j & 0 \\ 1 & -j & 0 \\ 0 & 0 & 1 \end{pmatrix}; \\ \vec{M} &= (\overset{\Rightarrow}{S})^{-1} M_R = (\overset{\Rightarrow}{S})^{-1} \begin{pmatrix} M_T \\ M^*_T \\ M_z \end{pmatrix}, \quad \text{with} \quad (\overset{\Rightarrow}{S})^{-1} = \frac{1}{2} \begin{pmatrix} 1 & 1 & 0 \\ -j & j & 0 \\ 0 & 0 & 2 \end{pmatrix}. \end{aligned} \tag{8.2}$$

We shall now describe a rotation of the magnetization vector by an RF pulse with flip angle $\alpha$ around the $x'$ axis. For the magnetization vector with Cartesian components, this rotation is given by (2.13). Such a rotation is easily translated into the components of the vector $\vec{M}_R$ given by:

$$\vec{M}_R^+ = \vec{\vec{S}}\,\vec{\vec{R}}_x(\vec{\vec{S}})^{-1}\vec{M}_R^- \quad \text{with} \quad \vec{\vec{R}}_x = \begin{pmatrix} 1 & 0 & 0 \\ 0 & \cos\alpha & -\sin\alpha \\ 0 & \sin\alpha & \cos\alpha \end{pmatrix}, \tag{8.3}$$

where $M_R^+$ is the magnetization vector after the pulse. For defining the directions of the rotation, we have used the "corkscrew" rule: a rotation around the $x'$ axis is positive when the corkscrew moves in the direction of the $x'$ axis, which means that a magnetization along the $z'$ axis is rotated away from the $z'$ axis into the direction of the $-y'$ axis. The result is found to be:

$$\vec{M}_R^+ = \begin{pmatrix} \frac{1}{2}(1+\cos\alpha) & \frac{1}{2}(1-\cos\alpha) & -j\sin\alpha \\ \frac{1}{2}(1-\cos\alpha) & \frac{1}{2}(1+\cos\alpha) & j\sin\alpha \\ \frac{-j}{2}\sin\alpha & \frac{j}{2}\sin\alpha & \cos\alpha \end{pmatrix} \vec{M}_R^- = \vec{\vec{\mathfrak{R}}}_{x,\alpha}\vec{M}_R^- \,. \tag{8.3a}$$

It is left as an exercise for the reader to find the matrices for rotations around the $y'$ axis using this same method.

Rotation through an angle $\phi$ around the $z$ axis is described by a rotation matrix with only diagonal elements, as should be expected:

$$\vec{M}_R^+ = \mathfrak{R}_{z,\phi}\vec{M}_R^-, \quad \text{with} \quad \mathfrak{R}_{z,\phi} = \begin{pmatrix} e^{j\phi} & 0 & 0 \\ 0 & e^{-j\phi} & 0 \\ 0 & 0 & 1 \end{pmatrix}. \tag{8.4}$$

Note that when $\phi$ is replaced by the precession angle $\theta$, this matrix also describes the effect of precession.

As a further example, we consider a rotation around an arbitrary axis in the $(x', y')$ plane. We assume this axis to make an angle $+\phi$ with the $x'$ axis. As shown in (4.5), the angle $\phi$ is the phase of the RF pulse causing the rotation. Now, the trick is that we first rotate the $(x', y')$ plane around the $z'$ axis over an angle $\phi$ (for a reference system the rotation is counted in the opposite direction, so we write $-\phi$) until the $x'$ axis coincides with the axis of rotation (direction of the $B_1$ field in the rotating reference system) – see (8.4). Then we perform the rotation $\alpha$ around the new $x$ axis – see (8.3a). Finally, we rotate the coordinate system back over $-\phi$ (in the equations we write $+\phi$) to the original situation. So the combined rotation matrix is:

$$\vec{\vec{\mathfrak{R}}}_{\phi,\alpha} = \vec{\vec{\mathfrak{R}}}_{z,\phi}\,\vec{\vec{\mathfrak{R}}}_{x,\alpha}\,\vec{\vec{\mathfrak{R}}}_{z,-\phi}$$

$$= \begin{pmatrix} \frac{1}{2}(1+\cos\alpha) & \frac{1}{2}(1-\cos\alpha)e^{2j\phi} & -j\sin\alpha e^{j\phi} \\ \frac{1}{2}(1-\cos\alpha)e^{-2j\phi} & \frac{1}{2}(1+\cos\alpha) & j\sin\alpha e^{-j\phi} \\ \frac{-j}{2}\sin\alpha e^{-j\phi} & \frac{j}{2}\sin\alpha e^{j\phi} & \cos\alpha \end{pmatrix}. \quad (8.5)$$

Similarly, for a rotation around the $y'$ axis ($\phi = 90°$), the rotation matrix for $M_R$ is given by:

$$\overset{\Rightarrow}{\mathfrak{R}}_{y,\alpha} = \begin{pmatrix} \frac{1}{2}(1+\cos\alpha) & -\frac{1}{2}(1-\cos\alpha) & \sin\alpha \\ -\frac{1}{2}(1-\cos\alpha) & \frac{1}{2}(1+\cos\alpha) & \sin\alpha \\ -\frac{1}{2}\sin\alpha & -\frac{1}{2}\sin\alpha & \cos\alpha \end{pmatrix}. \quad (8.6)$$

So far we have a good description of rotations caused by RF pulses around an arbitrary axis in the transverse plane. This means that we can describe the effect of "hard" RF pulses (see Sect. 2.3.1), in which a $z$ component of the effective magnetic field may be neglected. The matrix for a rotation with flip angle $\alpha$ around an arbitrary axis $\phi$ in the $(x', y')$ plane is defined as:

$$\begin{pmatrix} M_T^+ \\ M_T^{*+} \\ M_{z,l}^+ \end{pmatrix} = \overset{\Rightarrow}{\mathfrak{R}}_{\phi,\alpha} \begin{pmatrix} M_T^- \\ M_T^{*-} \\ M_{z,l} \end{pmatrix}. \quad (8.7)$$

In the design of slice-selective ("soft") RF pulses – see Sect. 2.3.2 – the selection gradient cannot be neglected with respect to the RF field, and later we shall show ways to cope with such pulses. However, we shall formally show that such pulses can in principle be described with the present theory. Consider the effective magnetic field due to both an RF field and the gradient field of the selection gradient in the rotating reference system and assume that it makes an angle $\psi$ with the $z$ axis, and that its transverse component makes an angle $\phi$ with the $x'$ axis (Eulerian angles – see Fig. 8.8). The rotation is now around this field with an angle $\theta$. To describe this rotation, we first rotate the frame of reference by an angle $\phi$ around the $z'$ axis, then by an angle $\psi$ around the new $y'$ axis until the new $z'$ axis coincides with the magnetic field vector, after which the precession can be calculated as the precession around the new $z'$ axis. We subsequently must transform back to the original reference system, and so:

$$\overset{\Rightarrow}{\mathfrak{R}}_{TOT} = \overset{\Rightarrow}{\mathfrak{R}}_{z,\phi}\overset{\Rightarrow}{\mathfrak{R}}_{y,\psi}\overset{\Rightarrow}{\mathfrak{R}}_{z,\theta}\overset{\Rightarrow}{\mathfrak{R}}_{y,-\psi}\overset{\Rightarrow}{\mathfrak{R}}_{z,-\phi}.$$

This yields an unwieldy expression, so we must try to find ways to simplify this calculation. One way to do this is to consider a soft pulse as a sequence of very short hard pulses separated by equal periods with precession in the field of the selection gradient. This idea will be described in Sect. 8.5.

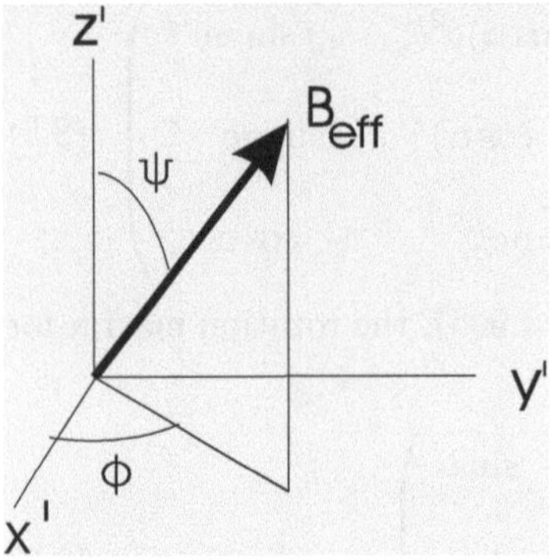


**Fig. 8.8.** The total magnetic field in the rotating frame of reference, consisting of a transversal component in the $(x', y')$ plane and a longitudinal component due to the selection gradient

### 8.2.3 Effect of Rotation and Precession on the Configurations

In the previous section we have deduced rotation matrices for the total magnetization vector. Actually, we went from a rotation matrix with real-terms $\overset{\Rightarrow}{R}$ working on the (real) components of the magnetization vector $M_x$, $M_y$ and $M_z$ – see, for example, (8.3) – to a matrix $\overset{\Rightarrow}{\mathfrak{R}}$, as per (8.4), with complex terms working on $M_\mathrm{T}$ $(= M_x + jM_y)$, $M_\mathrm{T}^*$ and $M_z$. The effect is that with (8.7) we now have two complex equations and one real equation working for $M_\mathrm{T}$, $M_\mathrm{T}^*$ and $M_z$. Since $M_\mathrm{T}$ and $M_\mathrm{T}^*$ describe the same quantity, this means that one of the complex equations is redundant.

So in a situation where dephasing due to a gradient field has not (yet) occurred, only the first and the third parts of (8.7), for respectively $M_x + jM_y$ and $M_z$, have to be solved. It is as if the transformation from $\vec{M}$ to $\vec{M}_R$ in (8.2) is a useless operation, and for non-dephased magnetization this is true. However, it will be shown that the benefit of this operation is that we are now in a position to calculate separately the effect of rotation and precession on each of the configurations of order $k$.

Introducing the discrete Fourier series of (8.1) into the rotation matrix, (8.7) yields:

$$\begin{pmatrix} \sum F_k^+ \mathrm{e}^{jk\theta} \\ \sum F_k^{*+} \mathrm{e}^{-jk\theta} \\ \sum M_{z,k}^+ \mathrm{e}^{jk\theta} \end{pmatrix} = \mathfrak{R} \begin{pmatrix} \sum F_k^- \mathrm{e}^{jk\theta} \\ \sum F_k^{*-} \mathrm{e}^{-jk\theta} \\ \sum M_{z,k}^- \mathrm{e}^{jk\theta} \end{pmatrix} . \tag{8.7a}$$

We use for the amplitudes of the configurations the notation $F_k^{\pm}(n)$, where $n$ is the pulse number, and + or − superscripts mean respectively after or before the RF pulse flanking the precession period. Collecting all terms with equal exponent $\exp(+jk\theta)$, we obtain:

$$\begin{pmatrix} F_k^+ \\ F_{-k}^{*+} \\ M_{zk}^+ \end{pmatrix} = \mathfrak{R} \begin{pmatrix} F_k^- \\ F_{-k}^{*-} \\ M_{zk}^- \end{pmatrix}, \tag{8.8}$$

where the common term $\exp(ik\theta)$ is omitted. This matrix equation describes the mixing of configurations with equal absolute values of $k$ (the order) due to a hard RF pulse. From now on, we only have to consider positive values of $k$ ($0 \le k < N$), since the negative values of $k$ yield the complex conjugates of the same equations – as is easily checked by writing the equations, for example, for $k = 1$ and $k = -1$. Note that this means that a different part of (8.8) is redundant than in the rotation equation for total magnetization, (8.7).

A special case is the situation $k = 0$ (no dephasing). Then the parts of (8.8) are redundant again, similar to (8.7), because there is no dephasing of a zero-order mode. When $k = 0$, the magnetization is rephased to form an echo and can then be described with a single transverse magnetization vector, which is the sum of the parallel magnetization vectors of all voxels. In our treatment for $k = 0$, we shall use the same rotation matrix as is used for the higher orders; however, we shall use only the first and third equations (8.7). Alternatively, in [3] the original rotation matrix $\overset{\Rightarrow}{R}$, as per (8.3), is used for $k = 0$.

It is important to investigate what happens in a precession period. The precession matrix describes a rotation around the $z$ axis over an angle $\theta(\mathrm{TR})$, and is shown in (8.4). The result of this operation is clearly seen thus:

$$\begin{pmatrix} e^{j\theta} & 0 & 0 \\ 0 & e^{-j\theta} & 0 \\ 0 & 0 & 1 \end{pmatrix} \begin{pmatrix} F_k^+(n)e^{jk\theta} \\ F_{-k}^{*+}(n)e^{jk\theta} \\ M_{z,k}^+(n)e^{jk\theta} \end{pmatrix} = \begin{pmatrix} F_k^+(n)e^{j(k+1)\theta} \\ F_{-k}^{*+}(n)e^{j(k-1)\theta} \\ M_{z,k}^+(n)e^{jk\theta} \end{pmatrix}$$

$$\Rightarrow \begin{pmatrix} F_{k+1}^-(n+1)e^{j(k+1)\theta} \\ F_{-(k-1)}^{*-}(n+1)e^{j(k-1)\theta} \\ M_{z,k}^-(n+1)e^{jk\theta} \end{pmatrix}. \tag{8.9}$$

The latter column vector shows the (complex) amplitudes of the Fourier series describing the magnetization just before pulse $(n + 1)$. We conclude that dephasing configurations increase by one order. However the rephasing configurations, $F_{-k}^*(n)$, decrease by one order (which is the absolute value of $k$), just as is shown in the phase diagrams of the previous section. Note that in the matrix calculation we always obtain the complex conjugate of the rephasing configuration. The longitudinal configurations do not change order and only suffer $T_1$ relaxation.

Equation (8.9) neglects relaxation. Its influence can be included in the normal way: for all non-zero longitudinal configurations, the spin–lattice relaxation during the precession periods is found by introducing (8.1a) into (2.10b) to give us:

$$M_{z,k}^{-}(n+1) = M_{z,k}^{+}(n)\mathrm{e}^{-\frac{\mathrm{TR}}{T_1}}. \tag{8.10a}$$

However, for the zero-order configuration we must take the restoration of the equilibrium magnetization, $M_0$, into account:

$$M_{z,0}^{-}(n+1) = M_{z,0}^{+}(n)\mathrm{e}^{-\frac{\mathrm{TR}}{T_1}} + (1-\mathrm{e}^{-\frac{\mathrm{TR}}{T_1}})M_0, \tag{8.10b}$$

which is the same equation as (2.10b). For the transverse configurations, spin–spin relaxation is described by the factor $\exp(-\mathrm{TR}/T_2)$, as usual.

The mathematical formalism for dealing with configurations resulting from a series of equidistant excitation pulses is now complete and so we can start with finding the effect of a first 90° excitation pulse starting from equilibrium, finding the input for the second pulse. At the start of the second pulse there is only one transverse configuration with $k = 1$. Then the rotation due to the second pulse is calculated. After the second pulse the only configurations are a rephasing transverse configuration with $k = -1$, a dephasing configuration with $k = 1$ and a longitudinal configuration with $k = 1$. When the first pulse does not convert all longitudinal magnetization into transverse magnetization, a new FID ($k = 0$) is excited by the second pulse. This process is repeated by the next precession period and at the end of the second precession period the second order is introduced and new second order configurations are caused by the mixing process of the third pulse. So the number of configurations increases as shown in the phase diagrams of Sect. 8.1.

### 8.2.3.1 Precession Matrix Including (Free) Diffusion

The precession matrix, as deduced in the previous section, can be extended so as to include diffusion. The result can be written as:

$$\begin{pmatrix} \mathrm{e}^{\mathrm{j}\theta-\tau/T_2-b_2\tau} & 0 & 0 \\ 0 & \mathrm{e}^{-\mathrm{j}\theta-\tau/T_2-b_2\tau} & 0 \\ 0 & 0 & \mathrm{e}^{-\tau/T_1-b_1\tau} \end{pmatrix}. \tag{8.11}$$

The precession angle $\theta$ is equal to $\mathrm{j}\gamma(G.r)\tau$, where $r$ is the distance to the isocenter in the direction of the gradient field, $\exp(-\tau/T_1)$ and $\exp(-\tau/T_2)$ describe the relaxation of the longitudinal and transverse magnetization during one inter-pulse delay, $\exp(-b_2\tau)$ stands for the attenuation of the transverse magnetization due to diffusion and $\exp(-b_1\tau)$ describes the effect of diffusion on the longitudinal magnetization. The values of $b_{1,2}$ are given by (7.41), which must be evaluated separately for the transverse and longitudinal configurations. Transverse configurations are recognized because of their increasing dephasing between the pulses, see Fig. 8.7. The amount of dephasing at the beginning of an interval is characterized by the number $n$, which is the distance to the horizontal axis in units $k_0 = \gamma \int_\tau G(t)\mathrm{d}t = \gamma G\tau$, where the integral is over the inter-pulse distance $\tau$. For a certain transverse configuration that starts after the RF pulse $m$ at a distance $n$ from the axis, we have $k(t) = nk_0 + \gamma Gt$. The value of $b_2$ as given by (7.41) can be found by taking $k^2(t)$ and integrating this function over the time $\tau$ between RF pulse $m$ and $m+1$. The result is:

$$b_2 = k_0^2 \left(n^2 + n + \frac{1}{3}\right)\tau \,. \tag{8.11a}$$

For the longitudinal magnetization the dephasing is constant, so $k(t) = nk_0$, and we find similarly from (7.41):

$$b_1 = k_0^2 n^2 \tau \,. \tag{8.11b}$$

This completes a full description of the influence of free diffusion on all multi pulse sequences, on which the configuration theory can be applied[1]. As will be shown later, the configuration theory can be applied to FFE, TSE and EPI sequences in general and therefore also the effect of diffusion in these sequences can be taken into account. A numerical analyses of the influence of diffusion on multi-pulse sequences is greatly simplified compared to the classical studies reported in refs. [1] and [2].

As an example we consider a CPMG sequence with ideal 180° refocusing pulses, see Fig. 3.9. The value of $n$ just after the pulses is always $-1$, which results in a decay constant for transverse magnetization, $b_2$, equal to $1/3k_0^2\tau$. At the $m$-th pulse the attenuation is therefore found to be proportional to $mb_2 = 1/3k_0^2 m\tau$ and therefore only proportional to the time $m\tau$ between excitation and the echo considered and not proportional to the third order of that time, as is the case with a spin echo. By choosing $k_0$ as small as possible the effect of the diffusion can be minimized and the effect of relaxation dominates. When the 180° pulses are not ideal, other configurations will arise and the situation must be described taking all configurations into account.

### 8.2.4 Use of the Theory of Configurations to Describe the Examples in Sect. 8.1

As an intermezzo, the present theory will now be applied to well-known simple "conventional" sequences containing several equidistant RF pulses and separated by periods in which only a (constant) $x$ gradient, $G_x$, is present. During these latter periods for each isochromat we have a certain precession angle $\theta = kx$ through which the spins precess around the $z'$ axis of the rotating frame of reference between the RF pulses. So for increasing distance from the isocenter, $x$, this precession angle increases. When many isochromats are considered, the phase angle $\theta$ can have all values between 0 and $2\pi$ for the different isochromats, which means that the total transverse magnetization vector becomes zero unless $k$ is zero.

[1] This work was done in cooperation with Dr. J. Groen and Ir. S.M.J.J.G. Nijsten.

To describe the effect of an RF pulse for each isochromat, a new reference system is defined that has in the transverse plane of the rotating frame of reference its axes in the direction $e^{j\theta}$ and $e^{-j\theta}$ (see Fig. 8.4, the components $1'$ and $2'$ for $\exp(+j\theta)$ and $1''$ and $2''$ for $\exp(-j\theta)$). The third axis remains, of course, in the $z'$ direction. The rotation matrix for that pulse now yields the relation between the old (before the pulse) and the new (after the pulse) partition of the magnetization along these three axes (see also Fig. 4.10). As a result, we obtain two new transverse configurations, with one along the positive $\theta$ axis and one along the $-\theta$ axis. The amplitude of the longitudinal configuration is also changed by the RF pulse.

In a following precession period, $\theta$ increases again and the dephasing also increases. The configuration along the $-\theta$ axis eventually reaches the value zero, as shown in Fig. 8.4. That is then true for all isochromats, and all the magnetization vectors add together to form an echo.

This situation will be demonstrated below for some of the examples already described in Sect. 8.1. From the examples it will be clear that the method can in principle also be used for calculating the magnitudes and phases of the echoes of TSE sequences, or Single-Shot TSE with reduced flip angle refocussing pulses [6]. A theoretical study of this sequence is given in [7]. Also, a study of FFE sequences is possible by applying the theory of configurations. In these cases, however, numerical methods are necessary, whereas here we wish to restrict ourselves to simple illustrative cases in which analytical methods are possible.

### 8.2.4.1. Multiple Spin-Echo Example

The first example is a Multiple Spin-Echo sequence (MSE), in which the refocussing pulses give a rotation around an axis under an angle $\phi$ with the $x'$ axis (see Fig. 8.2b). The question needs to be asked, "What is the amplitude and phase of the spin echoes?" We assume that the excitation pulse results in a rotation of the equilibrium magnetization, $M_0$, around the $y'$ axis, as given by (8.6) with $\alpha = \pi/2$:

$$\vec{M}_R^+ = \begin{pmatrix} +\frac{1}{2} & -\frac{1}{2} & 1 \\ -\frac{1}{2} & +\frac{1}{2} & 1 \\ -\frac{1}{2} & -\frac{1}{2} & 0 \end{pmatrix} \begin{pmatrix} 0 \\ 0 \\ M_0 \end{pmatrix} = \begin{pmatrix} M_0 \\ M_0 \\ 0 \end{pmatrix} = \begin{pmatrix} F_0^+(1) \\ F_0^{+*}(1) \\ M_z^+(1) \end{pmatrix}. \tag{8.12}$$

This transformation is applied to a situation with $k = 0$, the value applying before any dephasing has occurred, and so only the first and third equation in (8.12) should be used – the second equation is just the complex conjugate of the first and yields the same information. From (8.12) we find that after the first pulse $M_x + jM_y = F_0^+(1) = M_0$, which is real because the magnetization vector coincides with the $x'$ axis. This magnetization starts to dephase and ends up before the second pulse as a first-order term, with $k=1$ and thus $F_1^-(2) = F_0^+(1) = M_0$. The terms $F_{-1}^-(2)$ and $M_z$ are zero.

The action of the refocussing pulse (pulse 2) with phase $\phi$ is described by (8.5) with $\alpha = \pi$:

$$\begin{aligned} \vec{M}_R^+(2) &= \begin{pmatrix} 0 & e^{2j\phi} & 0 \\ e^{-2j\phi} & 0 & 0 \\ 0 & 0 & -1 \end{pmatrix} \begin{pmatrix} M_0 \\ 0 \\ 0 \end{pmatrix} \\ &= \begin{pmatrix} 0 \\ M_0 e^{-2j\phi} \\ 0 \end{pmatrix} = \begin{pmatrix} F_1^+(2) \\ F_{-1}^{*+}(2) \\ M_{z,1}^+(2) \end{pmatrix} . \end{aligned} \tag{8.13}$$

From this equation we see that after the refocussing pulse there is still only one non zero configuration, $F_{-1}^+(2)$, and this is a rephasing configuration with dephasing order $-1$, amplitude $M_0$ and phase $2\phi$. Note that we calculated the complex conjugate of this configuration, so for the magnetization vector $M_{T,-1}$ we must invert the phase. After such rephasing, the result is an echo, which clearly contains all magnetization available and therefore is the only configuration present. This result is already sketched in Figs. 8.2 and 8.3.

Now, in an MSE sequence the configuration that caused an echo dephases again before pulse 3. Actually we need to consider a new precession period between pulses 2 and 3, which is twice as long as the first precession period. This does not change the general picture with the discrete Fourier series given by (8.1), because this precession period can be considered as two equal precession periods separated by an RF pulse with zero flip angle. In such a precession period the order of the configurations increases by 2. Therefore the input configuration for the second 180° pulse (pulse 3) is $\vec{M}_{R,1}^-(3) = \left(M_0 e^{2j\phi}, 0, 0\right)^T$, where T means "transposed".

When the rotation matrix of (8.13) is applied again for an identical third refocussing pulse, the result is a configuration $F_{-1}^+(3) = M_0$ with phase zero. So when this configuration forms an echo, its phase is zero again. Extrapolation tells us that for every odd echo we find a phase equal to $2\phi$, while the even echoes have zero phase. This is in accordance with the discussion on artifacts in TSE sequences in Sect. 3.2.2 and Fig. 3.8.

With a similar calculation, the CPMG sequence that corrects for erroneous refocussing pulses in Turbo Spin Echo experiments [8, 9] can be illustrated. We assume that the refocussing pulses deviates somewhat from 180°, namely $\alpha = 180 - \delta$. The excitation pulse is taken as a rotation around the $y'$ axis. So just before the first refocussing pulse with rotation around the $x'$ axis, there is a magnetization $F_1^-(1) = M_0$. The result of this pulse is:

$$\vec{M}_R^+(2) = \begin{pmatrix} 0 & 1 & -j\delta \\ 1 & 0 & j\delta \\ -\frac{j}{2}\delta & \frac{j}{2}\delta & -1 \end{pmatrix} \begin{pmatrix} M_0 \\ 0 \\ 0 \end{pmatrix} = \begin{pmatrix} 0 \\ M_0 \\ -\frac{j}{2}M_0\delta \end{pmatrix}, \tag{8.14}$$

where we have only retained the terms of first order in $\delta$ so as to keep the calculation lucid. As is also shown in Fig. 3.7, there is now a transverse configuration $F_{-1}^+(2) = M_0$ and a new longitudinal configuration $M_{z,1}^-(3) = -(jM_0\delta)/2$. Precession between pulses 2 and 3 increases the order of the transverse configuration by +2, since the distance of the refocussing pulses is twice the distance between the excitation pulse and the first refocussing pulse. This double length can be seen as two single pulse distances separated by an RF pulse with flip angle of zero. So $F_{-1}^+(2)$ becomes $F_1^-(3) = M_0$ during the first precession period between the refocussing pulses. The effect of the second refocussing pulse (pulse 3) now becomes:

$$\vec{M}_R^+(3) = \begin{pmatrix} 0 & 1 & -j\delta \\ 1 & 0 & j\delta \\ -\frac{j}{2}\delta & \frac{j}{2}\delta & -1 \end{pmatrix} \begin{pmatrix} M_0 \\ 0 \\ -\frac{j}{2}M_0\delta \end{pmatrix} = \begin{pmatrix} 0 \\ M_0 \\ 0 \end{pmatrix}, \tag{8.15}$$

where, again, we have restricted ourselves to the first-order terms. We observe, however, that the longitudinal component is once more zero.

It is left to the reader to check that if we take the refocussing pulses as a rotation around the $y'$ axis (with the same phase as the excitation pulse), the longitudinal magnetization increases after each refocussing pulse. This proves the value of the CPMG requirement, as is also discussed in Sect. 3.2.2.

### 8.2.4.2. Eight-Ball Echo and Stimulated Echo Examples

We now turn our attention to the discussion of an Eight-Ball echo (see Sect. 4.2.2) and a stimulated echo (see Sect. 4.2.3). All pulses are taken as rotations around the $y'$ axis. The result of the first pulse is again $F_1^-(2) = M_0$, as in the case of SE. This is also the input for the second 90° pulse. The resulting magnetization after the second pulse is now calculated to be:

$$\vec{M}_R^+(2) = \begin{pmatrix} +\frac{1}{2} & -\frac{1}{2} & 1 \\ -\frac{1}{2} & +\frac{1}{2} & 1 \\ -\frac{1}{2} & -\frac{1}{2} & 0 \end{pmatrix} \begin{pmatrix} M_0 \\ 0 \\ 0 \end{pmatrix} = \begin{pmatrix} \frac{1}{2}M_0 \\ -\frac{1}{2}M_0 \\ -\frac{1}{2}M_0 \end{pmatrix}. \tag{8.16}$$

The output of the second pulse consists of three configurations: $F_1^+(2) = M_0/2$, $F_{-1}^+(2) = -M_0/2$, and $M_{z,1}^+(2) = -M_0/2$. It is now clear that the amplitude of the rephasing configuration, that forms the Eight-Ball echo,

$F^{+}_{-1}(2)$ is equal to $M_0/2$, as has been discussed extensively in Sect. 4.2.2 and Sect. 8.1.1.

The stimulated echo can be calculated when we take the $M^{+}_{z,1}(2)$ as the input for the third 90° pulse. This can only be done when no other configurations with the same order are present, in other words when

$$\vec{M}^{+}_{R}(3) = \begin{pmatrix} +\frac{1}{2} & -\frac{1}{2} & 1 \\ -\frac{1}{2} & +\frac{1}{2} & 1 \\ \frac{-1}{2} & \frac{-1}{2} & 0 \end{pmatrix} \begin{pmatrix} 0 \\ 0 \\ -\frac{1}{2}M_0 \end{pmatrix} = \begin{pmatrix} -\frac{1}{2}M_0 \\ -\frac{1}{2}M_0 \\ 0 \end{pmatrix}, \tag{8.17}$$

and the amplitude of the stimulated echo is $F^{+}_{-1}(3) = -M_0/2$ (see also Sect. 4.2.3).

When at the third pulse two configurations are present, which happens when the distance between the second and third pulse is twice the distance between the first and second pulse, there is interference of two configurations. This interference results in a lower amplitude of the echo, which is no longer a pure stimulated echo. Then:

$$\vec{M}^{+}_{R}(3) = \begin{pmatrix} +\frac{1}{2} & -\frac{1}{2} & 1 \\ -\frac{1}{2} & +\frac{1}{2} & 1 \\ \frac{-1}{2} & \frac{-1}{2} & 0 \end{pmatrix} \begin{pmatrix} -\frac{M_0}{2} \\ 0 \\ -\frac{M_0}{2} \end{pmatrix} = \begin{pmatrix} -\frac{3}{4}M_0 \\ -\frac{1}{4}M_0 \\ \frac{1}{4}M_0 \end{pmatrix}. \tag{8.18}$$

The amplitude of the echo is $M_0/4$, due to interference of $F^{-}_{1}(3)$ and $M^{-}_{z,1}(3)$.

Although the theory in this section did not teach us new facts, the successful description of the conventional sequences with phase diagrams shows that it is possible to apply the theory of configurations as an alternative to the classical theory described in earlier chapters.

## 8.3 Multi-excitation Pulse Sequences

Many fast-imaging sequences are already described in our book, particularly in Chap. 3 and Chap. 5 – we remember the TSE, EPI, Spiral Imaging and the different types of TFE. Still, there is always a need for faster imaging, for dynamic imaging, for real-time imaging, etc. The fastest sequence described up till now is EPI. In Sect. 3.3.1 an example of a very fast EPI is described, which requires less than 100 ms for a 74 × 128 "half-matrix" scan. This, however, requires a very strong gradient system, which is not (yet) currently available in whole-body MRI systems. Furthermore, in Sect. 3.3.2.3 it has been explained that at high gradient fields the concomitant "Maxwell" fields cause the actual value of the gradient field to deviate from the nominal value, which causes distortion.

Although multi-excitation pulse sequences also show complicated difficulties, they can be very fast and can still be operated within standard whole-body MRI systems. Therefore there is some promise that some of these sequences will be applied for very fast imaging. However, as has been shown with the example of TSE in Sect. 8.2.4, the theory of configurations developed for multi-pulse sequences can be used for the description of all sequences discussed earlier (Chaps. 3–5), in which equidistant pulses are applied, and also for FFE and TFE [5].

In this section we shall start with the description of sequences of $N$ equidistant, very short excitation pulses (called "DANTE excitation") applied simultaneously with a constant gradient field. This technique is named "BURST Imaging" [4]. When the configurations excited by these pulses are refocussed by a 180° pulse, the sequences are named "SE-BURST", or "DUFIS" (DANTE Ultra-Fast Imaging Sequence), and will be described in Sect. 8.3.1 (see also [10]). A problem with these sequences is that their signal, and therefore also their SNR, is low. This is partly due to the fact that the flip angle of the RF pulses is restricted to about $(90/N)°$, and so for a burst of 16 pulses the maximum flip angle is restricted to about 5.6°.

In Sect. 8.3.1.1 it is shown that a burst of equidistant pulses only excites thin equidistant stripes, the bulk of the spins (between these stripes) remaining nearly unexcited. This effect is already explained in the discussion of tagging in Sect. 7.2.3. A mathematical explanation follows from (2.19), where it is shown that the excitation profile of an RF pulse is the Fourier transform of the time dependence of this RF pulse. When this time dependence is a series of $\delta$ functions, the excited profile is also a series of $\delta$ functions, as shown by (2.31b), which means that small stripes are excited.

The question may now be raised: "How can all spins available in an object be used more efficiently?" The solution is found in varying the phase of the excitation pulses. Some proposals will be described in Sect. 8.3.1.2, and see also [11]. The sequence in which this phase cycling is applied is named OUFIS (Optimized Ultra-Fast Imaging Sequence) in [12]. A disadvantage is that the echoes also have different phase, which requires correction during post-processing. It is also possible to use only two phases, namely 0 and 180°, which at least does not introduce new phase shifts into the echoes. The sequence then is called "two-phase OUFIS" [12]. In Sect. 8.3.1.3 it is shown that BURST imaging principles can be combined with TSE to form a very fast sequence, which is relatively insensitive to magnetic field inhomogeneities and susceptibility changes [13].

When gradient reversal is used to refocus the configurations, we speak of GR-BURST (Gradient-Recalled BURST Sequences) or URGE (Ultra-Rapid Gradient Echo) [14]. These sequences are described in Sect. 8.3.1.4. In very fast segmented 3D-BURST imaging with short TR between the segments, the saturation of the spins, due to steady-state effects, can be minimized by using frequency shifts (FS-BURST) [15].

In Sect. 8.3.1.5, sequences with non-equidistant excitation pulses (although all inter-pulse distances are always an integer number times the shortest distance to make development in a discrete Fourier series possible) will be described. With a limited number of $N$ pulses $3^{N-1}$ echoes are in principle possible when all configurations are considered separately. However, the magnitude of the echoes in these sequences is very irregular, which requires correction during post-processing. When the first pulse is 90°, so that there is no transverse magnetization left during the next pulses to form a FID, this sequence is named QUEST (Quick Echo Split Technique) [16]. However, when the flip angle of the first pulse deviates from 90°, the sequence is named PREVIEW, in which also the (new) FIDs are used [17].

We will now answer the question, "How can we use the echoes of a multi-pulse sequence for imaging?". So far, we have assumed a homogeneous, unlimited object. A real, limited object can be obtained by multiplying the unlimited object profile (which is a constant) with the profile of the limited object. After Fourier transformation, this means that the Fourier transform of the unlimited homogeneous object (which is a $\delta$ function) is convolved with the Fourier transform of the limited object. Therefore each echo consists of the Fourier transform of the object. For example, if the object is homogeneous and limited, it can be described by a block function $U(x)$, with $-L/2 < x < L/2$, and so the echoes obtain the form $2\sin(Lt/2)/t$, similar to (2.35). When all subsequent echoes obtain the appropriate phase encode gradient, the $k$ profiles can be acquired and a 2D image can be obtained, as described in Sect. 2.4.

### 8.3.1 SE-BURST Imaging

To avoid fast gradient switching during a sequence of BURST pulses, in most cases a large gradient is switched on continuously. The RF pulses must be very short (broad frequency band) and therefore have large amplitudes, even for low flip angles. By these pulses the spins in a large volume are excited. The slice selection in the SE-BURST sequence is performed with the refocussing pulse.

The discussion of imaging with an SE-BURST sequence will be started by drawing a sequence and phase diagram for it, as shown in Fig. 8.9. In this sequence a selective refocussing pulse is used for slice selection. We distinguish primary echoes and mirror echoes [12]. The primary echoes follow directly after the pulse burst before and after the refocussing pulse, and they are heavily influenced by mutual (predominantly destructive) interference between the configurations. These primary echoes can of course be avoided by applying the 180° pulse at an earlier point in time so that they mix with the mirror echoes. The mirror echoes lie symmetrically with the excitation pulses on the other side of the 180° refocussing pulse and are less deteriorated by interference. The evolution of these mirror configurations is accentuated in the phase diagram shown in Fig. 8.9. In the figure, the primary echoes are

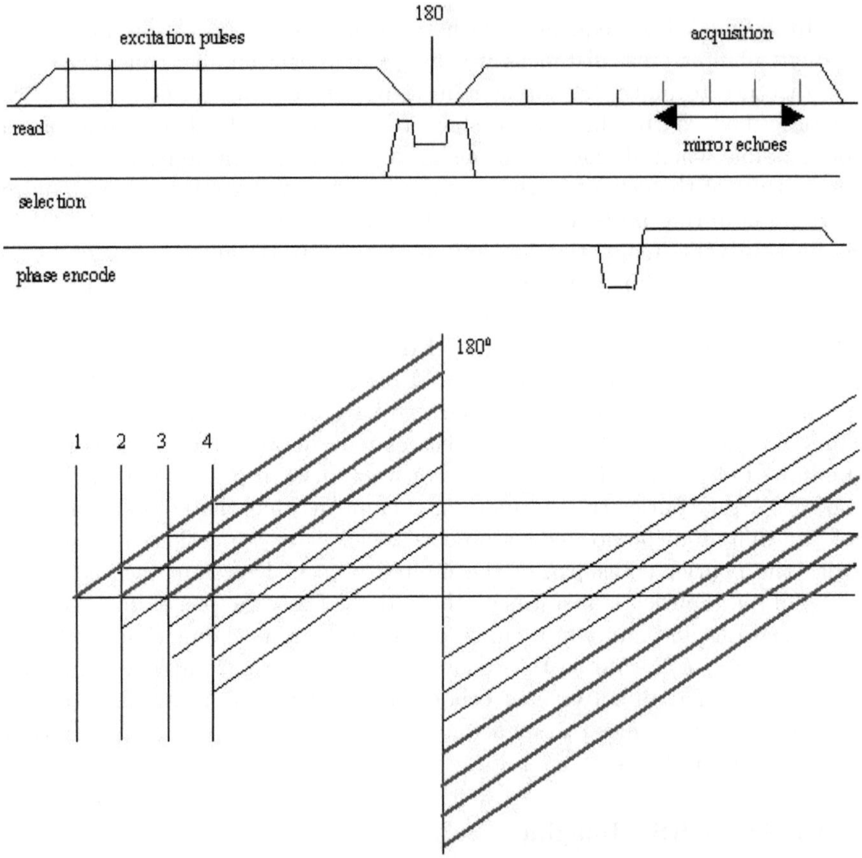


**Fig. 8.9.** Sequence diagram and phase diagram for a 4-pulse SE-BURST sequence. In the sequence diagram, the four primary echoes are not measured. The refocussing pulse is assumed to be exactly 180°, so that there are now no configurations after the refocussing pulse

not acquired. The mirror echoes experience different phase encoding, making imaging possible.

A burst of 64 pulses should have (by the $90/N$) rule) a maximum flip angle of 1.41°. When the read-out gradient is in the $x$ direction, a voxel on the line $x = 0$ experiences all pulses in series, so that a 90° flip angle is realized. Also, voxels on isochromats that have a precession phase angle of $2\pi n$ $(n = 0, 1, 2, \ldots)$ between the pulses experience the maximum excitation. The distance between excited stripes is therefore given by

$$\Delta x = 2\pi(\gamma G_x \tau)^{-1} \tag{8.19}$$

where $\tau$ is the time between the RF pulses. So the excitation profile consists of narrow stripes with optimum excitation, whereas the regions in between are almost unexcited, just as is the case with tagging (see Sect. 7.2.3). This makes the imaging process very inefficient and results in a low SNR for these BURST sequences.

Although the configurations leading to the mirror echoes are mixed with other configurations, we can still make a rough estimate of the order of their magnitude by neglecting this mixing. Each time a pulse is applied, the longitudinal magnetization is decreased by a factor $\cos\alpha$. So the magnitude of the transverse magnetization after the $n$th pulse is $\sin\alpha\cos^n\alpha$, which forms a new configuration. Once this configuration exists, each following pulse decreases its transverse magnetization with a factor $^1/_2(1+\cos\alpha) = \cos^2(\alpha/2)$ – see the (1,1) element of the rotation matrix (8.3a). So after $n$ pulses the magnitude of the mirror echoes can be roughly estimated to be $A_n$, with

$$A_n = M_0 \sin\alpha \cos^n\alpha \left(\cos^2\left(\frac{\alpha}{2}\right)\right)^{N-n} \tag{8.20}$$

In Fig. 8.10a the amplitude of the echoes is shown as a function of the echo number $n$ for a burst of eight excitation pulses and for flip angles equal to 6° (lower line), 12° (middle line) and 20° (upper line). It shows that for small flip angles (certainly up to ($\alpha \leq 90/N$, which is about 11° for $N = 8$) the amplitudes are proportional to $\alpha$ and nearly constant as a function of $n$, but for higher flip angles the value of the amplitude is no longer proportional to $\alpha$ and decreases as a function of echo number.

In Fig. 8.10b the average magnitude of the echoes, based on (8.20), is shown as a function of the flip angle for an 8-pulse BURST excitation. The figure suggests that there is an optimum flip angle, and for larger flip angles the average echo amplitude decreases again. It should be remembered, however, that the theory presented here is only a very crude approximation, which may only be expected to indicate a likely trend.

More exact calculations, based on the rotation and precession matrices and given in (8.9) and (8.10), show that for flip angles $\alpha$ larger than $(90/N)$ the amplitudes of the echoes as a function of the echo number, $A_n$, show a completely different behaviour and are much smaller than anticipated from (8.20). This is assumed due to destructive interference, which is clearly illus-

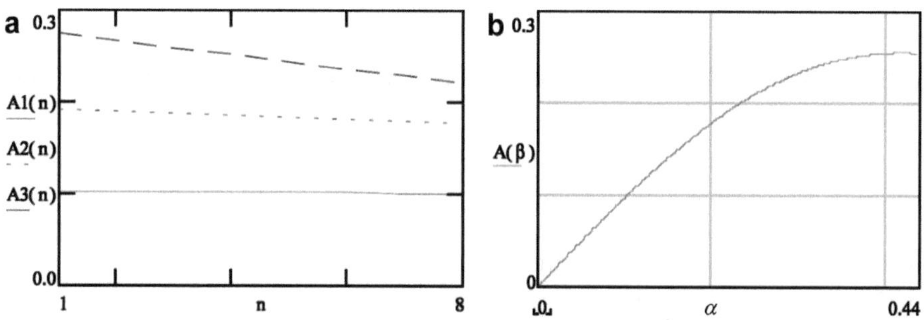


**Fig. 8.10. a** Relative amplitude of the mirror echoes as a function of echo number. Flip angles are 6° (*lower trace*), 12° (*middle trace*) and 20° (*upper trace*). **b** Average relative amplitude of the mirror echoes as a function of the flip angles (in radians)

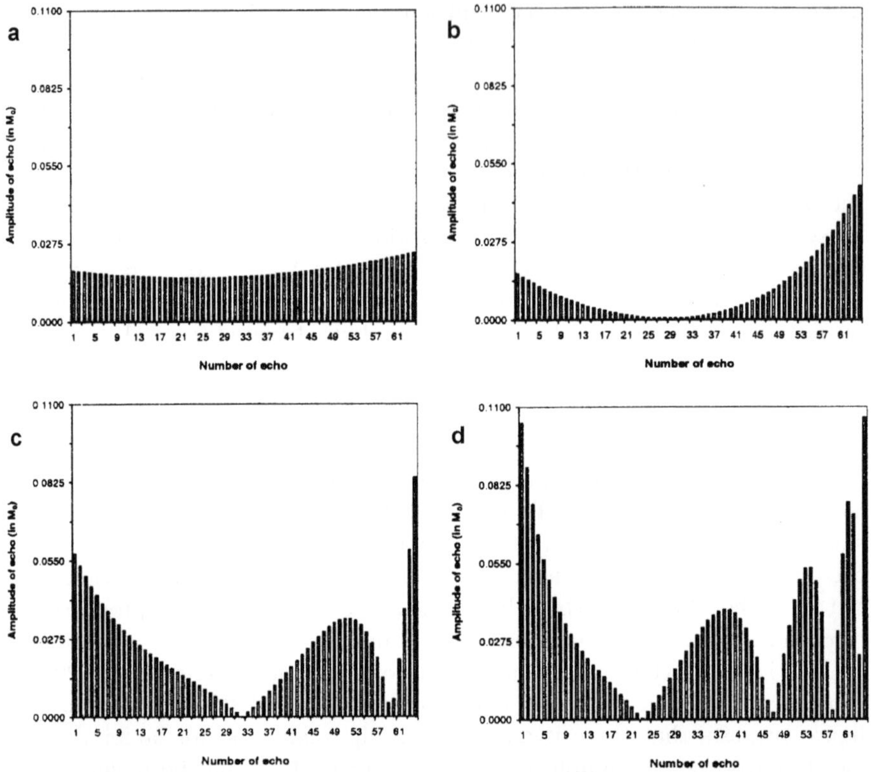


**Fig. 8.11.** Echo Amplitude as a function of the echo number for a 64-pulse burst for four values of the flip angle. **a**: $\alpha_1 = 90°/N$, **b**: $\alpha = 2\alpha_1$, **c**: $\alpha = 3\alpha_1$, **d**: $\alpha = 4\alpha_1$

trated in Fig. 8.11 (see also [12]). The flip angle in this study was varied as $\alpha_1(= 90/N), 2\alpha_1, 3\alpha_1$ and $4\alpha_1$.

A special study of the magnitude of the echoes, based on the application of the complete rotation and precession matrices to the configurations, can be found in [18] and [20].

#### 8.3.1.1. Excitation Profile for BURST with Single-Phase Excitation

In this section, methods will be discussed to make BURST imaging more efficient by exciting a more homogeneous excitation profile. There are several practical suggestions for dealing with this problem. Generally these suggestions are based on phase or frequency modulation of the RF pulse burst. Before we discuss these methods, however, we must discuss the situation with single-phase RF pulses (single-phase BURST) in more detail, so as to set the stage for discussing RF phase modulation schemes.

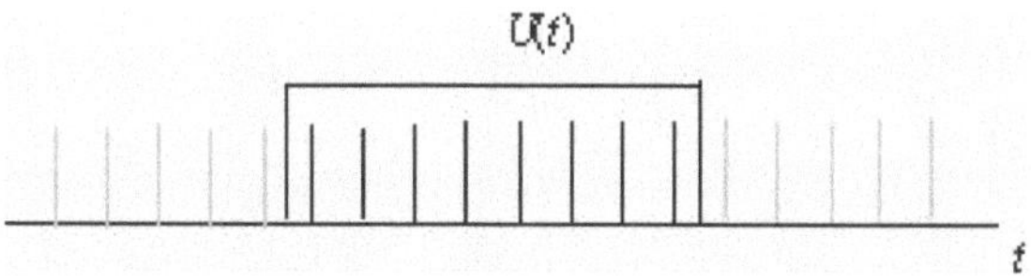


**Fig. 8.12.** The excitation in the time domain as a product of an infinite series of $\delta$ functions and a finite duration function, $U(t)$

In Sect. 2.3.2 it has been shown that, apart from a phase factor, the excitation profile of an RF pulse is the Fourier transform of the RF field in its time domain, divided by the gradient field strength – see (2.16) and (2.17). In this current section, this theory will be applied to a burst of RF excitation pulses.

The RF field strength as a function of time can be described as a infinite series of $\delta$ functions with spacing $\tau$, multiplied by a finite duration function $U(t)$, see (Fig. 8.12). The Fourier transform of an infinite series of $\delta$ functions in time is well-known to be another infinite series of $\delta$ functions in frequency space – see (2.31a) – and in MRI the position within the object in a gradient field is defined by the frequency of the signal (see [1.12]). We thus have

$$\sum_{-\infty}^{\infty} \delta(t - n\tau) \circ\!\!\stackrel{\mathfrak{F}}{\longrightarrow}\!\!\circ \frac{2\pi}{\tau} \sum_{-\infty}^{\infty} \delta(\omega - n\Delta\omega), \tag{8.21}$$

where $\Delta\omega = 2\pi/\tau$. In the object space the distance between the excited stripes, $\Delta x$, is given by $\Delta x = \Delta\omega/\gamma G_x = 2\pi(\gamma\tau G_x)^{-1}$, as shown already in (8.19). The finite duration function is equal to the function $U(t)$, used in (2.35), where $U(t) = 1$ if $-N\tau/2 < t < N\tau/2$, and $U(t) = 0$ otherwise, given by:

$$U(t) \circ\!\!\stackrel{\mathfrak{F}}{\longrightarrow}\!\!\circ N\tau \frac{\sin(\gamma G_x x N\tau/2)}{\gamma G_x x N\tau/2}. \tag{8.22}$$

The half-width of this function, $\delta x$, is given by $\delta x = 2\pi/\gamma G_x \tau N$. Therefore

$$\Delta x = N\delta x, \tag{8.20a}$$

which means that only about $(1/N)$th of the available spins is excited. As is well-known, the resulting excitation profile is given by the convolution of the Fourier transforms of both functions in Fig. 8.12, and its functional dependence can be written as

$$M_T(x) \div C \sum_{-\infty}^{\infty} \left( \frac{\sin[\gamma G_x \tau (x - m\Delta x) N/2}{\gamma G_x \tau (x - m\Delta x) N/2} \right), \tag{8.22}$$

where $C$ is a constant dependent on $M_0$, $B_{\mathrm{RF}}$, and $G_x$.

This magnetization profile is shown in Fig. 8.13. It follows from this figure that we can obtain a periodic excitation profile and, between the maxima, we can in principle distinguish $N-1$ (sub)stripes, of which only one is completely

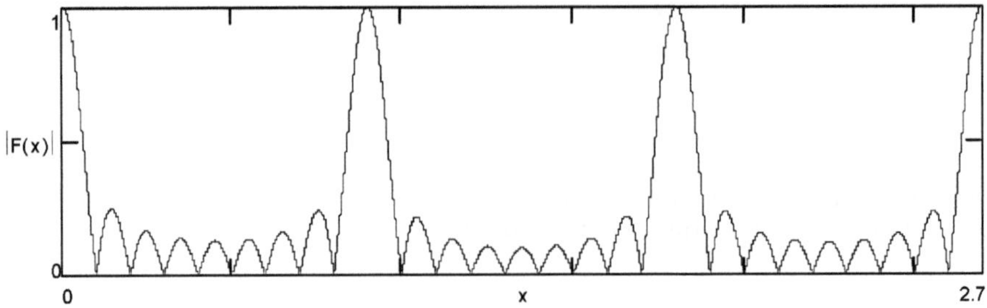


**Fig. 8.13.** Excitation profile $|F(x)|$ for a single phase 8-pulse BURST excitation in a homogeneous object. For an actual object, this function must be multiplied with $M_0(x)$

excited. This results in a low signal, and so also the SNR is decreased by a factor of $N$ with respect to homogeneous excitation.

#### 8.3.1.2. Optimized BURST Excitation Using Phase Modulation

A solution to the problem of inhomogeneous excitation can be found by varying the phase of the RF pulses. We first give an explanation based on simple non-mathematical reasoning. We consider a burst of $N = M^2 = 64$ pulses. The first $M = 8$ pulses have equal phase, which yields an excitation profile as shown in Fig. 8.13. Assume that in the second group of eight pulses each following pulse is given a phase advance of $\Delta\phi = 2\pi/M$ radians with respect to the previous pulse [12]. So the $m$th pulse in the second group has a phase angle equal to $2\pi m/M$. That means that the rotation axis of the RF pulses in the $x' - y'$ plane is each time advanced by an angle $\Delta\phi$ – see for example Sect. 4.2.4.

Spins on locations, where they experience the same phase advance $\Delta\phi$ during the precession periods, will be optimally rotated away from the $z$ axis by the series of RF pulses. The spins that undergo the correct precession angle are located at $\delta x = \Delta x/M$ away from the point where the single phase burst has its maximum, i.e. at the first sub-maximum beside the main maximum in Fig. 8.13. For the next $M$ pulses, we can make the phase advance $2\Delta\phi$ between the pulses and thus obtain a maximum at $2\delta x$ away from the original maximum in Fig. 8.13. This process can be repeated $M-1$ times, which means that $M-1$ (sub)stripes are excited at a distance $m\delta x$ ($m = 0, 1 \ldots M-1$), from the original stripes for a single phase RF pulse burst. If we count the number of groups of $M$ pulses with $n$, and the members within each group with $m$, the phase for the $m$th pulse in the $n$th group is equal to $\phi_{n,m} = nm(2\pi/\sqrt{N})$ radians.

This result shows that all spins in the excited volume are used, and there are two consequences:

1. The flip angle of the burst pulses can be taken to be $90°/\sqrt{N}$, significantly bigger in most cases than $90°/N$, since each single stripe experiences only $\sqrt{N}$ pulses.
2. The SNR of the BURST sequence is improved by a factor $\sqrt{N}$ in comparison with a single-phase burst, and this is due to the larger permitted flip angle.

After these intuitive arguments we can embark on a somewhat more exact theory, based on the theory presented in Sect. 2.3.2. The conclusion of this section, that the slice profile of a low flip-angle slice-selective pulse is proportional to its Fourier transform, also holds for a burst of (small) pulses. A burst of $N$ excitation pulses can be seen as an infinite series of pulses multiplied by a finite duration function, as shown in Fig. 8.12. This (formally) gives an infinite number of stripes with distance $\Delta x$ as shown in Fig. 8.13. The Fourier transform of this excitation profile (transforming again from position (frequency) to time) gives the resulting echoes.

We now consider a change of the relative phases of the excitation pulses. For our discussion, the finite duration function that restricts the infinite series to a finite one does not need to be taken into account, since it does not interfere with the phase. So we now consider an excitation given by:

$$P(t) = \sum_{n=-\infty}^{\infty} \delta(t - n\tau)\exp(j\phi_n), \tag{8.23}$$

where the phase $\phi_n$ is repeated after each group of $M$ pulses. This makes $P(t)$ a periodic function with $M$ pulses per period, $M\tau$. Within a single period, the RF pulses are numbered as $m$, with $0 \le m \le M-1$. Therefore this function can now be transformed as a Fourier series with fundamental frequency $2\pi/M\tau$, to yield the amplitudes of the excited stripes, $a_n$, in accordance with

$$P(t) = \sum_{-\infty}^{\infty} a_n \exp\left(j\frac{2\pi nt}{M\tau}\right), \tag{8.24}$$

where

$$\begin{aligned} a_n &= \int_0^{M\tau} (\delta(t - m\tau)\exp\left(-j\frac{2\pi nt}{M\tau} + j\phi_m\right) \mathrm{d}t \\ &= \sum_{m=0}^{M-1} \exp\left(-j\frac{2\pi nm}{M} + j\phi_m\right). \end{aligned}$$

We shall consider this result for special cases of the phase angles.

#### *8.3.1.2.1. Phase Angle Constant*

When the phase angle $\phi_m$ is constant, say zero, we find that $a_n = M$ when $n = 0$, $M$, $2M$, etc., and $a_n = 0$ otherwise. We observe that this is the same

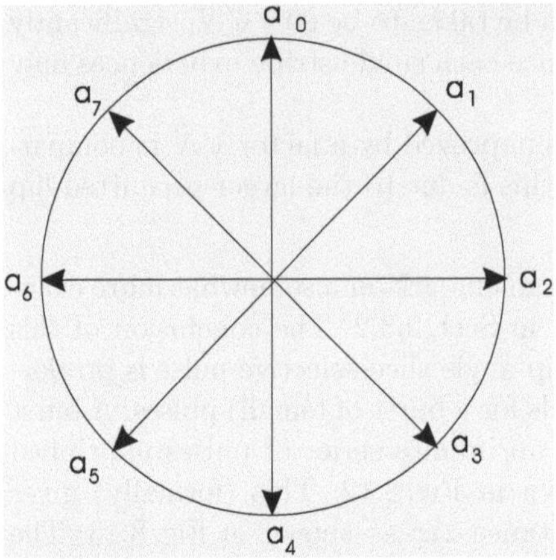


**Fig. 8.14.** Complex coefficients $a_n$ compensate for each other in the addition given by (8.24)

transformation as that of (8.21), without the convolution with the finite-duration function. The fundamental frequency of the Fourier series is in this case $2\pi/\tau$, instead of $2\pi/M\tau$, and the distance of the stripes is $\Delta x$, as should be expected from (8.19).

In Fig. 8.14 the terms of the last sum in (8.24) are shown to compensate for each other for $n \neq 0,\, M,\, 2M,\, \ldots$.

*8.3.1.2.2. Phase Angle* $= 2\pi m/M$

The next case is $\phi_m = 2\pi m/M$, where the phase angle is proportional to $m$. In that case, we can write:

$$a_n = \sum_{n=0}^{M-1} \mathrm{e}^{-j\frac{2\pi n m}{M}}\, \mathrm{e}^{j\frac{2\pi m}{M}} = \sum_{n=0}^{M-1} \mathrm{e}^{-j\frac{2\pi (n-1)m}{M}}, \tag{8.25}$$

which means that $a_n = M$ for $n = 1, M+1, \ldots$, and $a_n = 0$ otherwise. In this case, only single stripes are excited at a distance $\delta x$ to the right of the original maximum for $\phi_m = 0$. The conclusion is that a linear phase increment as a function of the pulse number does not help us to excite the spins more efficiently.

Of course a series of $M$ pulses can be repeated by $N$ further groups of $M$ pulses with $\phi_{n,m} = 2\pi nm/M$, $(0 \leq n \leq N)$, and so on, yielding a more efficient use of the available spins. This is because each substripe is excited only $M$ times in a sequence of $N = M^2$ pulses, describing the situation explained at the beginning of this section.

*8.3.1.2.3. Phase Angle with Quadratic Dependence*

A quadratic dependence of the phase as a function of the pulse number can also be chosen: for example $\phi_m = 1/2\phi_\mathrm{a} m^2$, where $\phi_\mathrm{a}$ is a constant. This dependence means that the phase advance between the pulses $\mathrm{d}\phi_m/\mathrm{d}m$ is linear according to the relation $\phi_a m$. This relationship reminds us of the theory of phase cycling, used for $T_1$-FFE and described in Sect. 4.2.7, where it has been shown that a steady state is only obtained when the phase advance between pulses is: $\phi(m) = \mathrm{d}\phi_m/\mathrm{d}m = m\phi_a + \phi_b$. In [11] it has been shown that, for $\phi_a = 2\pi/M$, the excited stripes all have equal amplitude:

$$a_n = |a_n|\mathrm{e}^{j\varphi_n} = \sum_{m=0}^{M-1} \mathrm{e}^{-j\frac{2\pi mn}{M}} \mathrm{e}^{j\frac{2\pi m^2}{2M}} = \sum_{m=0}^{M-1} \mathrm{e}^{-j\frac{2\pi(nm - \frac{1}{2}m^2)}{M}} \tag{8.26}$$

We shall not reproduce the formal (complicated) theory for the foregoing. However, for $M = 4$ and $m = 0, 1, 2, 3$ this result is easily checked and shows that $|a_n| = 2$ (so equal to $\sqrt{N}$) and the phase of the magnetization in the strips is $\phi_0 = -\pi/4$, $\phi_1 = 0$, $\phi_2 = 3\pi/4$ and $\phi_3 = 0$.

The most general functional dependence that leads to homogeneous excitation is $\phi_m = m^2\phi_a + m\phi_b + \phi_c$, yielding a phase increment $\phi(m) = m\phi_a + \phi_b$, just as is the case with RF spoiling (see Sect. 4.2.7 and also Appendix B of [11]).

One disadvantage of multi-phase BURST sequences is that the echoes have different phases, which must be accounted for in the reconstruction. Therefore in [12] it has been suggested that only the phase angles 0 and $\pi$ are used. Although it looks as though the number of degrees of freedom is limited, a good optimization is still possible, without the disadvantage of phase corrections in the reconstruction process.

#### 8.3.1.3. Combination of BURST with TSE

BURST excitation can in principle be combined with conventional sequences like TSE, as proposed in [13]. Some of the advantages of TSE – for example the insensitivity to $B_0$ inhomogeneity and susceptibility effects – are inherited by such a combination.

In order to obtain a slice-selective excitation, a selection gradient is switched on during each pulse of the pulse burst, which is refocussed at the next pulse. This can be done by inverting the selection gradient between the RF pulses and switching off the dephasing gradient in the read direction during the pulses. In a MRI scanner with maximum gradient 10 mT/m, the inter-pulse interval can be made as small as 2 ms. The sequence is sketched in Fig. 8.15. The BURST excitation contains six pulses. The pulses are phase modulated, as described in the previous section, and have a flip angle slightly below the optimum flip angle of $\pi/(2\sqrt{N})$ rad. In a multi-slice scan, when slices away from the isocenter are scanned, this means that the RF frequency must be appropriately matched for each slice. During the burst excitation the phase-encode gradient is blipped to give the echoes different phase encoding for obtaining adjacent profiles in the $\vec{k}$ plane.

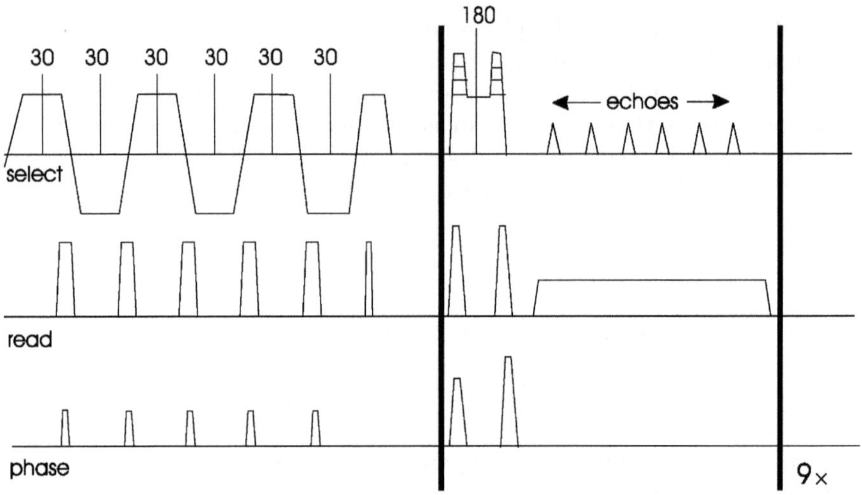


**Fig. 8.15.** Combination of BURST excitation and TSE. The part between the vertical bars is repeated nine times. The phase gradient after a refocussing pulse is refocussed before the next refocussing pulse

Also in Fig. 8.15, the part of the sequence between vertical bars is repeated nine times so that, after a single excitation BURST with six pulses, 54 profiles can be acquired. The refocussing pulses of the sequence have an interval of 26 ms. To avoid unwanted FIDs, crusher gradient pulses are placed around the 180° pulses in all directions. The acquisition time for a single slice is therefore about 234 ms, and for a FOV of 20 cm the resolution is 4 mm. With the phase-encode gradient (which is rewound before the next refocussing pulse), the $T_2$-sensitivity can be chosen by deciding in which refocussing interval the $k = 0$ profiles are acquired.

Dynamic multi-slice acquisition ($n$ slices) is possible when TR is chosen to be $260n$ ms. The interesting property of this sequence is that it is insensitive to susceptibility effects. The SNR of the present scan method is somewhat lower than that of a comparable GRASE sequence [13].

### 8.3.1.4. Gradient Recalled BURST Sequences

Instead of inversion by a 180° pulse, refocussing in a BURST sequence can also be initiated by gradient reversal [14]. Since there is no means to select a slice, a complete volume is excited by the very short BURST pulses. Therefore these gradient-recalled methods are used in combination with 3D acquisition.

Some of the potential applications of these very-fast-gradient echo-recalled BURST sequences are: contrast-agent bolus tracking; cine studies of heart and joint motion; and (possibly) functional imaging of the brain. The phase diagram of a GR-BURST sequence is shown in Fig. 8.16. The gradient is

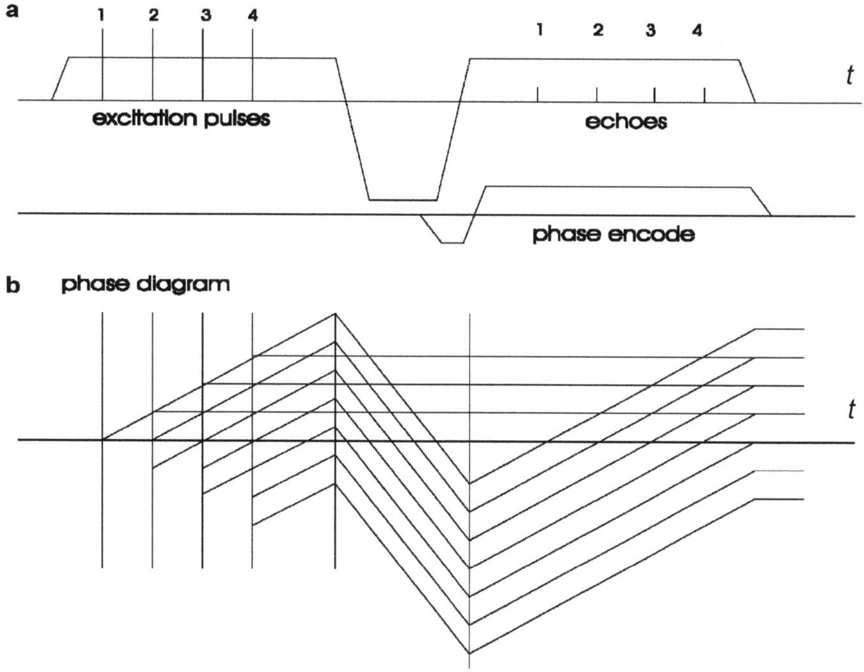


**Fig. 8.16.** Sequence **a** and phase diagram **b** of a four-pulse gradient-recalled BURST sequence

reversed twice to make the time order of the echoes equal to the time order of the excitation pulses, so that all echoes suffer equal $T_2^*$ decay.

For 3D BURST to be acceptably fast, a short TR must be chosen. This results in a dynamic equilibrium situation, named "steady state" or "saturation" (see Sects. 4.2.5 and 4.4), which causes a low signal, similar to the steady state of FFE sequences. Therefore single-phase BURST is applied in which in the consecutive TR intervals the frequency is shifted so as to excite an adjacent stripe. When eight pulses are used, eight stripes are located within one voxel, and so only after eight TR intervals is the same stripe excited again, resulting in a decreased saturation effect and therefore a larger signal. This method is called Frequency Shifted BURST (FS-BURST). The lowering of the SNR, due to the use of only part of the spins per single phase pulse burst, is more than compensated for by the 3D scan method [15].

Field inhomogeneities are not compensated for in a gradient recalled echo sequence, and so one gets $T_2^*$-weighted images. However, for the study of susceptibility changes after intravenous bolus injection, this is an advantage and the sequence could be useful for brain perfusion studies [19].

The steady state for gradient recalled BURST sequences with short TR has been studied in [20] in a way similar to the calculations of steady state in Sect. 4.4, but here all pulses and precession intervals have to be accounted for, which makes the situation very complicated. It appears that strong spoiling by a spoiler gradient added at the end of the sequence in each TR interval is necessary to obtain near-constant echo amplitudes.

### 8.3.1.5. QUEST and PREVIEW

The BURST sequences described up till now do not result in the maximum possible numbers of echoes. In Sect. 8.3 it has been stated that after $N$ pulses there are $3^{N-1}$ configurations. Two-thirds of these configurations are modes with transverse magnetization, which can be refocused using a 180° pulse. This is a much larger number of echoes than in a BURST sequence, where the number of echoes is equal to the number of RF pulses and the different configurations already interfere during the excitation period.

It is therefore possible to generate many more echoes (after phase-encoding: profiles in the $\vec{k}$ plane) per excitation pulse. A sequence in which an optimum number of echoes is generated is the QUEST (QUick Echo Split) sequence [16]. The phase diagram of this technique is shown in Fig. 8.17. It is shown in this figure that four pulses give rise to $(2/3)3^3 = 18$ echoes after

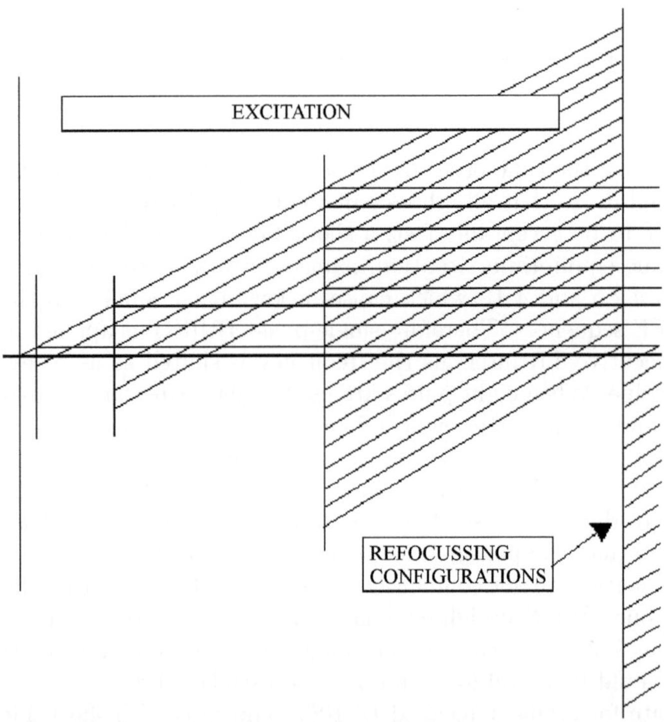


**Fig. 8.17.** Phase diagram of QUEST. The first pulse has a flip angle of 90°, the three following pulses have arbitrary flip angles, and the fifth pulse is the refocussing pulse, generating 18 refocussing configurations

refocussing with a 180° pulse. The scan time is about 40 ms, which makes this sequence one of the fastest known so far.

It is also shown that for this sequence there is almost no interference between the configurations, except for the longitudinal configurations interfering with the transverse configurations at the second and third excitation pulses. Notwithstanding this, the resulting echoes show a wide variation in amplitude and phase [18]. The optimum flip angles of the splitting excitation pulses are 90°, 75°, 75° and 90°. The result, however, is that the ratio between the largest and the smallest echo amplitude is still 7, requiring correction in post processing and increased SNR. The average value of the echo amplitudes is $0.08M_0$ (maximum is $0.1M_0$) and the relative standard deviation is 52.5%.

When the first excitation pulse has a flip angle smaller than 90°, at the time of the second and later pulses there is still longitudinal magnetization left, so that the excitation pulses also cause new FIDs, forming even more configurations and therefore more echoes. This sequence is named PREVIEW [14]. However, PREVIEW has the same difficulties as QUEST: a large relative standard deviation of the echo amplitudes, being a source of artifacts [18]. PREVIEW yields predominantly $T_2$-to-proton density contrast. Other contrast is, of course, possible using magnetization preparation (see Sect. 5.3).

The properties of PREVIEW are studied theoretically in [20] on the basis of the rotation and precession matrices for the configurations. In the same reference it is shown that for an OUFIS sequence the average value is $0.1M_0$ and the relative standard deviation is only 7.7%, as already described in Sect. 8.4.1.

## 8.4 Theory of Configurations and Well-Known Fast-Imaging Sequences

In this section we shall show that the theory of configurations is also applicable to numerical inspection of all scan methods that contain a number of RF pulses at regular distances. The theory can shed new light on some of the properties of well-known scan methods. We have chosen two examples, namely single shot TSE and FFE sequences.

### 8.4.1 Application to TSE

Turbo Spin-Echo (TSE) sequences are discussed in Sect. 3.2. In practice, it is impossible to make ideal refocussing pulses with a flip angle of exactly 180°. The consequences of these non-ideal refocussing pulses are described in Sect. 3.2.2. In Sect. 8.2.4 it is shown that the theory of configurations can also be used to gain more insight into the properties of the echoes in TSE sequences with imperfect refocussing pulses. For numerical calculations, a slight modification of the theory is necessary: the TR intervals between the pulses are taken equal to the time between the excitation pulse and the first refocussing pulse. Then the time between two refocussing pulses is divided into two TR intervals, separated by a virtual pulse with zero flip angle.

At every (non-ideal) refocussing pulse the incoming configurations are split up into three configurations: one defocussing configuration, one refocussing configuration, and one longitudinal configuration (see Sects. 8.2.4 and 4.2.7). As a consequence, it may be expected that both spin echoes (or Eight-Ball echoes) and stimulated echoes are generated. Also, at every imperfect refocussing pulse a new FID is generated. So each echo in the echo train of a TSE sequence is built up from different configurations with a different history, similar to what is discussed in Sect. 4.2.7 and shown in Figs. 4.11 and 8.7. When the configurations, forming together a certain echo, have experienced different phase-encoding gradient surfaces, a single voxel is given different positions in the $y$ direction, leading to serious artifacts. Therefore it is necessary to compensate the phase-encode gradient in each interval between two refocussing pulses by a "rewinder", so that all configurations forming an echo have the same phase encoding.

A recent development is the application of single-shot TSE to Magnetic Resonance Cholangiopancreatography (MRCP). The TSE takes advantage of the very long $T_2$ relaxation time of the fluids involved (bile and pancreatic secretions), which are surrounded by tissue with much shorter $T_2$ relaxation. This allows use of TSE with a turbo factor of (for instance) 128 and a long effective TE, yielding a high signal intensity of the fluids and low intensity of the surrounding tissue [21, 22], see image set VIII-1. Depending on the practical situation we might want to have a maximum echo signal around the time when the lowest $k$-lines are sampled, or a constant echo signal to obtain a narrow point spread function (with high resolution). The amplitudes of the echoes can be influenced by varying the amplitudes of the refocussing pulses.

As an example, we have calculated for a train of 64 echoes the required sequence of refocussing pulses (which now deviate from 180°) necessary to obtain almost constant echo amplitudes on the basis of the theory of configurations. The idea is that, at the beginning of the echo train, longitudinal configurations (position-encoded longitudinal magnetization – see Sect. 4.2.3) are generated. These longitudinal configurations can be converted into transverse configurations much later in the pulse train to form stimulated echoes, so as to increase the total echo amplitude late in the echo train. The result of this study, based on a numerical solution of (8.9) and (8.10), is shown in Fig. 8.18.

In Fig. 8.18a the amplitudes of the refocussing pulses and the echo train of a TSE sequence with constant 160° pulses is shown, and this sequence results in almost pure $T_2$ decay of the echo train. In Fig. 8.18b the flip angles of the refocussing pulses, which yield an echo train with more or less constant amplitudes of the echoes, are shown. The flip angles of the refocussing pulses start at about 100° for the first refocussing pulse and go through a minimum

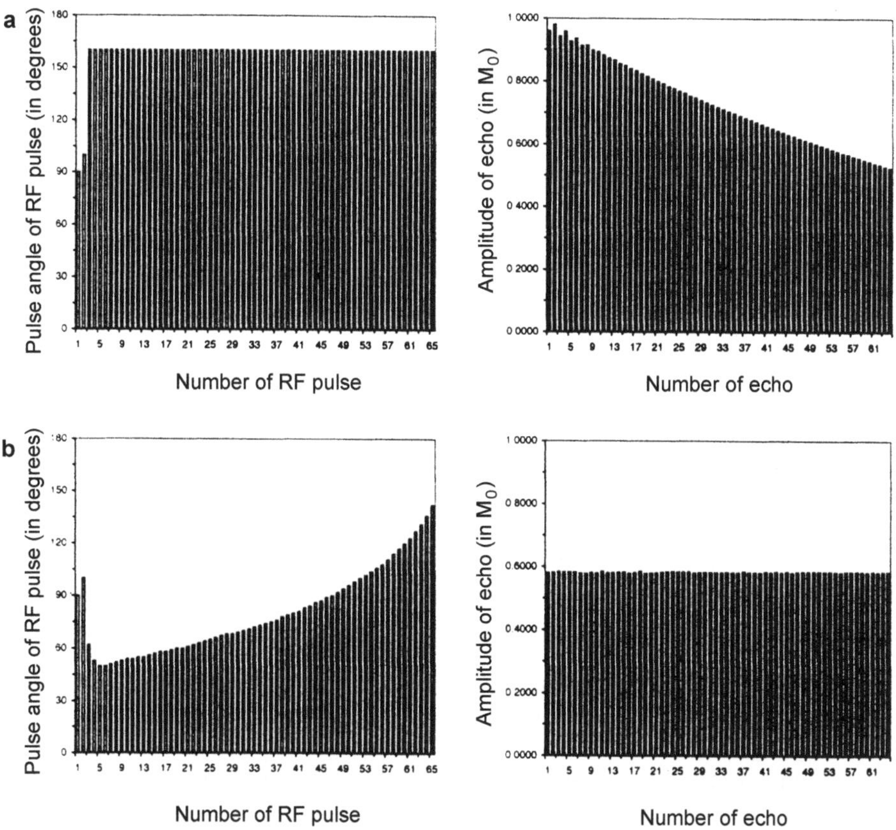


**Fig. 8.18.** Excitation pulse amplitudes and echo amplitudes of a TSE with turbo factor 64 for constant refocussing pulse amplitude **a** and for varying pulse amplitude **b**. $T_1 = 4\,\mathrm{s}$, $T_2 = 1\,\mathrm{s}$. Time between echoes is 10 ms

at the sixth pulse with a flip angle of 48° and then gradually increase to 150°. It is concluded from Fig. 8.18 that asking for constant echo values requires a sacrifice in echo amplitude. Alternatively, in single-shot TSE we may decide not to use the first echoes for acquisition to reduce the artifacts. However, when the echo train is constant, we can start immediately. It depends on the practical application, which optimizing criterion must be chosen.

In conclusion, we can state that it has been demonstrated that it is possible to use the theory of configurations for the optimization of single-shot TSE sequences for the different applications and scan methods used in MRCP. It will be clear that the considerations given above can also be applied to segmented TSE for imaging tissues with shorter $T_2$. For sequences with short or long effective echo time, $\mathrm{TE_{eff}}$ (see Sect. 3.2.1) or when a half-echo (HASTE [21]) is acquired, different types of optimization are possible.

### 8.4.2 Application to FFE

Gradient Echo sequences (FFE or GRE) have been extensively discussed in Chap. 4. All special effects, such as dynamic equilibrium (steady-state) and the banding artifact, have been treated on the basis of the total magnetization vector. However, since it is in principle a multi-pulse sequence, FFE can also be considered in relation to the present theory of configurations, due to its periodic excitation pulses and equal-phase increment in the inter-pulse precession periods. We then have to repeat the theory of Sects. 4.3 and 4.4, but now with precession and rotation matrices in a form suitable to transform the equal-order configurations, as per (8.9). The equation for the steady state, (4.10a), then becomes:

$$\vec{M}_R(t_{n+1}) \equiv \vec{M}_R(t_n) = \stackrel{\Rightarrow}{\mathfrak{R}}_{\phi',\alpha} \stackrel{\Rightarrow}{P}_{\mathrm{TR}} \vec{M}_R(t_n) + \stackrel{\Rightarrow}{\mathfrak{R}}_{\phi',\alpha}(1-E_1)\vec{M}_0,$$

where the rotation matrix $\stackrel{\Rightarrow}{\mathfrak{R}}_{\phi',\alpha}$ is given by (8.5) and the precession matrix $P_{\mathrm{TR}}$ is given by:

$$\stackrel{\Rightarrow}{P}_{\mathrm{TR}} = \begin{pmatrix} \mathrm{e}^{i\theta-\frac{\mathrm{TR}}{T_2}} & 0 & 0 \\ 0 & \mathrm{e}^{-i\theta-\frac{\mathrm{TR}}{T_2}} & 0 \\ 0 & 0 & \mathrm{e}^{-\frac{\mathrm{TR}}{T_1}} \end{pmatrix}.$$

The second term on the right-hand side is due to the fact that $T_1$ relaxation always restores the equilibrium magnetization $M_0$, which is in the $z$ direction.

The configurations can now be introduced, as shown in Sect. 8.2.3, resulting in an equation for the steady state for the configurations:

$$\begin{pmatrix} F_k^+ \\ F_{-k}^{*+} \\ M_{z,k} \end{pmatrix} = \stackrel{\Rightarrow}{\mathfrak{R}}_{\phi',\alpha} \stackrel{\Rightarrow}{P}_{\mathrm{TR}} \begin{pmatrix} F_k^+ \\ F_{-k}^{*+} \\ M_{z,k} \end{pmatrix} + \delta(k) \stackrel{\Rightarrow}{\mathfrak{R}}_{\phi',\alpha}(1-E_1) \begin{pmatrix} 0 \\ 0 \\ M_0 \end{pmatrix}, \quad (8.27)$$

where $\delta(k) = 0$ for $k \neq 0$ and $\delta(k) = 1$ for $k = 0$. This equation is discussed in Appendix A of [5], and we shall not discuss it further here since most of the results obtained with these equations are already known from the discussions in Chap. 4 herein. Instead, we shall have a look at the phase diagrams of the FFE sequences, because these diagrams will give another view on the different FFE sequences (see Fig. 4.17).

A phase diagram for FFE has already been shown in Fig. 4.11 and used for the explanation of RF spoiling. Also, Fig. 8.7 can be seen as a phase diagram for FFE when the number of excitation pulses is made very large. In the phase diagrams mentioned, only the overall phase increment in a TR interval is shown, and the order of the transverse configurations increases linearly between two RF pulses for non-rephased FFE sequences. The different FFE sequences, however, rely on different time behaviour of the gradients within a TR interval.

In Fig. 8.19 the phase diagrams of N-FFE is shown. The zero-th order configuration is named the FID, which is the sum of the ECHO, generated

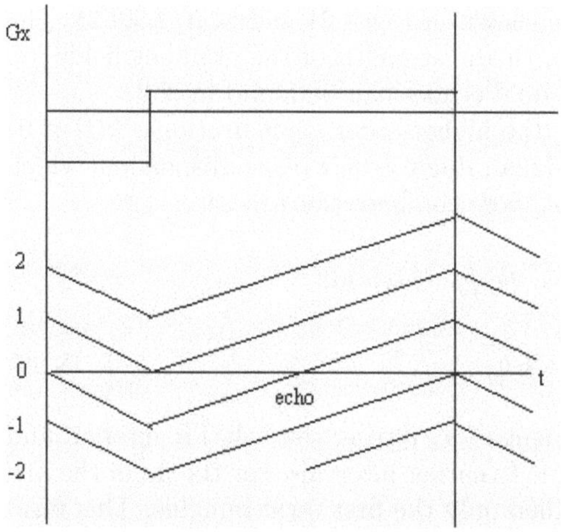


**Fig. 8.19.** Schematic phase diagram of a N-FFE or $T_1$-FFE sequence. The zero-th order mode is used for generating an echo

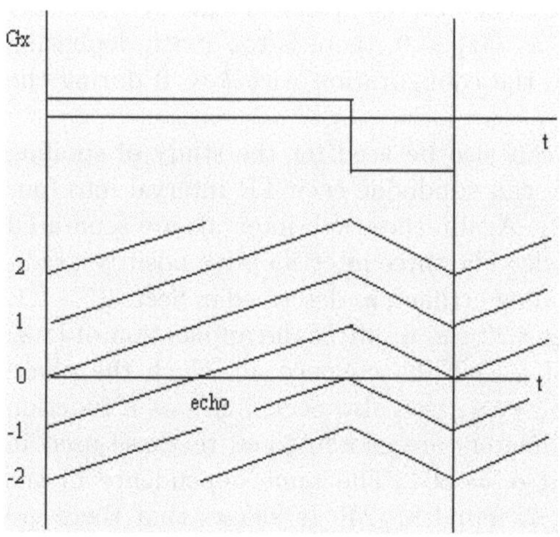


**Fig. 8.20.** Schematic phase diagram of $T_2$-FFE, here the $-1$ configuration forms the echo

by earlier pulses, and the fid of the $n$th pulse. This fid generates the echo (see Sects. 4.2 and 4.5.3.1.1). Note that the same phase diagram describes $T_1$-FFE, where the ECHO is spoiled by phase cycling so that the zero-th order configuration consists only of the fid of the $n$th pulse.

Another member of the FFE family is the $T_2$-FFE. In this case the $-1$ mode generates an echo between two RF pulses, as is shown in Fig. 8.20.

We leave it to the reader to show that more than one configuration can form an echo during a single TR interval, for example by starting with a negative gradient, followed by a positive read out gradient and again a negative gradient before the next pulse (as shown in Fig. 4.21 and Sect. 4.5.3.2). The number of echoes can be varied with the strength of the gradient fields [5]. A practical example of the use of dual echoes can be found in [23].

During acquisition of an echo, the higher-order configurations ($k(t) \neq 0$) have very little influence on the signal, due to their phase dispersion, which reduces the total magnetization of these configurations to zero:

$$\begin{aligned} k(t) &= \gamma \int_{n\mathrm{TR}}^{n\mathrm{TR}+t} G_x(t)\mathrm{d}t' + \gamma k \int_0^{\mathrm{TR}} G_x(t)\mathrm{d}t' \\ &= \gamma \int_{n\mathrm{TR}}^{n\mathrm{TR}+t} G_x(t)\mathrm{d}t + k\, k_0, \end{aligned} \tag{8.28}$$

where the first term is the increment of $k_x$ during the $n$th TR interval and the last term is the value of $k(t)$ due to earlier intervals. For the fid of the $n$th pulse the value of $k$ is zero, and then only the first term remains. This term describes dephasing, and unless the integral (zero-order moment) is zero the fid gives no signal. The ECHO signal at the $n$th pulse is due to configurations generated by earlier pulses. These configurations may have many values of $k$ – for example earlier dephasing intervals are compensated for by rephasing intervals after inversion. As soon as $k(t) \neq 0$, there is too much dephasing to yield a detectable signal. Only the configuration with $k = 0$ during the acquisition window yields an echo.

The theory of configurations can also be used for the study of spoiling under different circumstances. We can subdivide each TR interval into four sub-intervals, as shown in Fig. 8.21. Again, these sub-intervals are separated by virtual pulses with zero flip angle. The three intervals with positive gradients are necessary to avoid the banding artifact, as described in Sect. 4.5.3.1.1.

As an example we show in Fig. 8.22 the result of the application of (8.9) and (8.10) to the steady state of a $T_1$-FFE sequence, in which the phase is varied according to $\phi(n) = n\phi_\mathrm{a} + \phi_\mathrm{b}$, (see also Sect. 4.2.7) as a function of $\phi_\mathrm{a}$ and with $\phi_\mathrm{b} = 0$. The parameters are chosen equal to those used in [24]: $T_1/\mathrm{TR} = T_2/\mathrm{TR} = 20$ and $\alpha = 30°$. The same dependence of the steady-state magnetization on $\phi_\mathrm{a}$ is found in [24]. It follows that there are a number of values of the phase increment $\phi_\mathrm{a}$ that satisfy the equation of the magnetization in an ideally spoiled $T_1$-FFE sequence, as per (4.16). The value $\phi_\mathrm{a} = 117°$ is often used in practice. It is, however, clear that small deviations from the correct values of $\phi_\mathrm{a}$ may cause large errors.

A more practical set of parameters is $T_1/T_2 = 10$ and $T_1/\mathrm{TR} = 160$, 80, 40 and 20, and $\alpha = 40°$. The resulting steady-state magnetization as a function of the phase increment $\phi_\mathrm{a}$ is shown in Fig. 8.23. The general aspect of the curves is similar to that of Fig. 8.22; however, the difference between the maxima and minima is much smaller. This might mean that the exact choice of $\phi_\mathrm{a}$ is not very relevant for large TR and that much smaller values of $\phi_\mathrm{a}$ can be used.

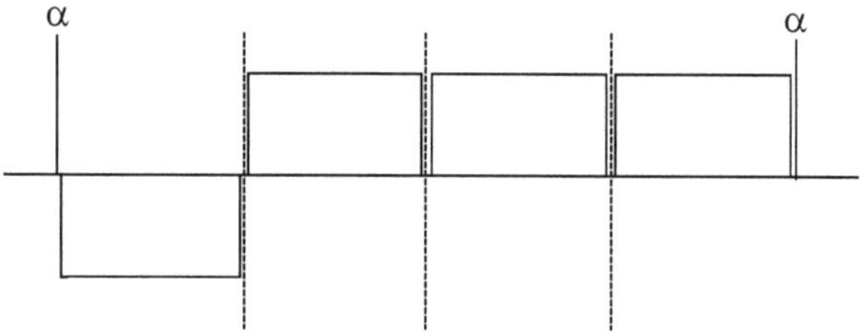


**Fig. 8.21.** Subdivision of one TR interval into four sub-intervals separated by virtual pulses with zero flip angle, to allow the application of the theory of configurations

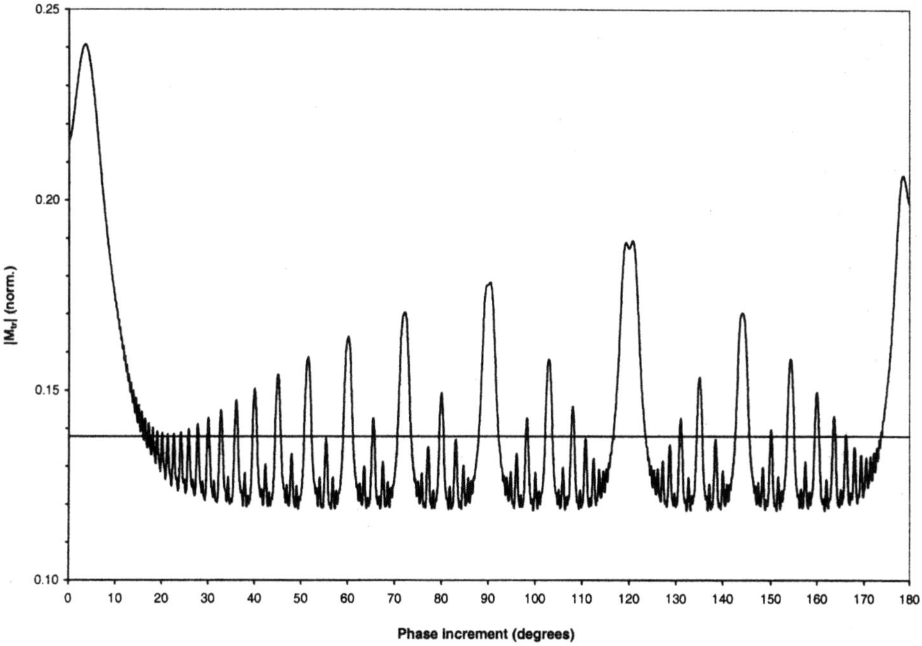


**Fig. 8.22.** Steady-state magnetization of a $T_1$-FFE sequence as a function of the phase angle $\phi_a$. The horizontal line shows the steady-state value given by (4.16)

The examples shown in this section prove that the theory of configurations is a strong tool for the study of all types of multi-pulse sequences.

A beautiful example of the application of configuration theory to gradient echo sequences can be found in [15] of Chap. 5.

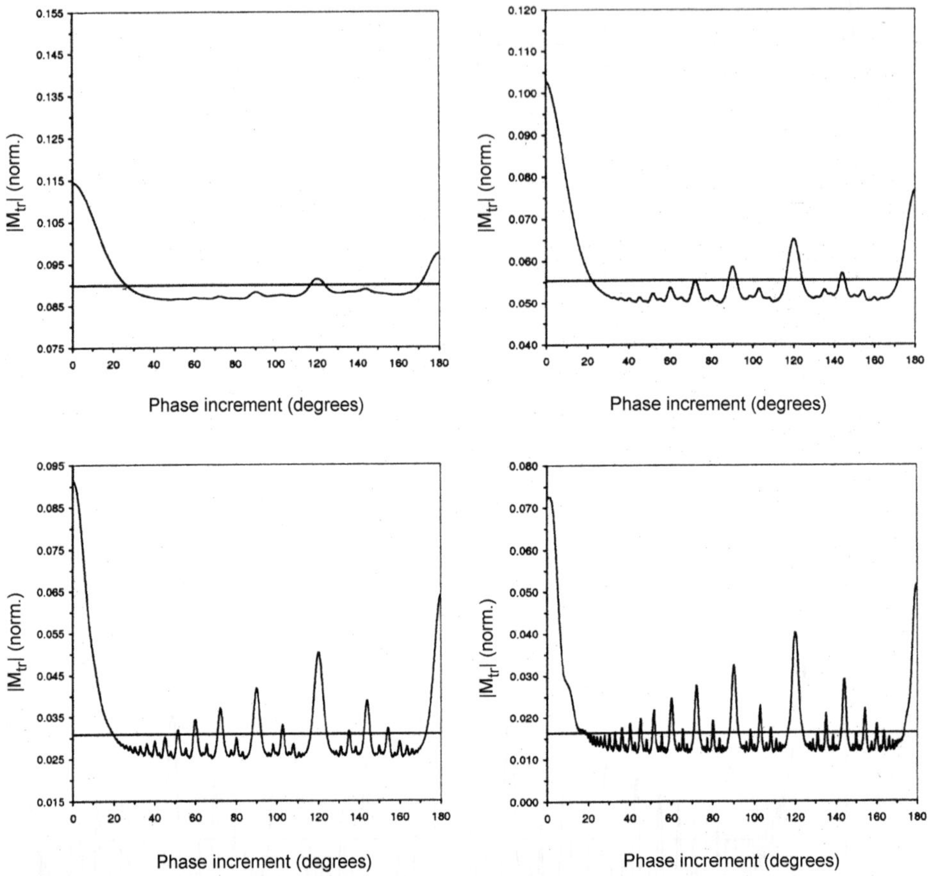


**Fig. 8.23.** Steady-state magnetization of a $T_1$-FFE sequence as a function of $\phi_a$, with $T_1 = 10T_2$, $\mathrm{TE} = 1/2\mathrm{TR}$. **a** $T_1/\mathrm{TR} = 20$, **b** $T_1/\mathrm{TR} = 40$; **c** $T_1/\mathrm{TR} = 80$; and **d** $T_1/\mathrm{TR} = 160$. The peaks in the curves are at $(360/n)^\circ$

## 8.5 Rotation and Precession Matrices and RF Pulse Design

Although the design of RF pulses is beyond the scope of this book, it is interesting to realize that the rotation and precession matrices, which are the basis of the theory of configurations, can also be used for RF pulse design. Due to the non-linearity of the Bloch equations, the design of a large-flip-angle RF pulse, exciting a required-slice profile, has always been a formidable problem, which can only be solved with iterative methods or complicated analytical methods [25].

In Sect. 2.3.1 it was shown that a selective, low-flip-angle pulse excites a slice profile that is the Fourier transform of the time dependence of the envelope of the RF field. For selective large-flip-angle RF pulses, a more complicated relation exists between the slice profile and the amplitude-vs-time and phase-vs-time functions of the RF envelope, and this is due to the non-linearity of the Bloch equations.

This relation can be calculated assuming that the RF pulse may be divided in short time intervals of equal length $\delta t$ with precession, separated by hard RF pulses with a strength proportional to the momentaneous value of the RF field times $\delta t$ (see Fig. 8.21). This method is named the forward Shinnar–le Roux (SLR) Transformation, after its inventors [26–27]. Even more exciting is that they proved that the inverse process is also possible by simply inverting the rotation and precession matrices of the sub-pulses: given the slice-magnetization profile in the form of a discrete Fourier series, it is possible to find the exact form of the RF pulse that caused the slice profile. This is called the Inverse Shinnar–le Roux (ISLR) Transformation. This theory therefore includes the inherent non-linearity of the Bloch equation.

### 8.5.1 Shinnar–Le Roux (SLR) Transformation

A selective RF pulse (a soft pulse) can be subdivided in $N$ subpulses with equal length $\delta t$, as shown in Fig. 8.24. Each subpulse $n$ is subdivided in a part with rotation through angle $\alpha_n = \gamma|B_n|\delta t$ around an axis having an angle $\phi_n$ with the $x'$ axis in the $x' - y'$ plane, and a precession part with a precession angle $\theta = \gamma G z \delta t$, where $G$ is the gradient strength during the RF pulse and $z$ is the selection direction. For sufficiently small $\delta t$, this separation in rotation and precession is a valid representation of reality. We can thus apply the equations for the calculation of rotation, (8.4), and precession, (8.5), with $\phi = \theta$, from the theory of configurations, to each time slot $\delta t$.

The combined effect of precession followed by rotation within one subpulse can be calculated from the product of (8.4) and (8.5):

$$\overset{\Rightarrow}{\mathfrak{R}}_{\phi_n,\alpha_n} \overset{\Rightarrow}{P} = \begin{pmatrix} \frac{1}{2}(1+\cos\alpha_n)\mathrm{e}^{j\theta} & \frac{1}{2}(1-\cos\alpha_n)\mathrm{e}^{-j(\theta-2\phi_n)} & -j\sin\alpha_{nj}\mathrm{e}^{j\phi_n} \\ \frac{1}{2}(1-\cos\alpha_n)\mathrm{e}^{j(\theta-2\phi_n)} & \frac{1}{2}(1+\cos\alpha_n)\mathrm{e}^{-j\theta} & j\sin\alpha_n\mathrm{e}^{-j\phi_n} \\ \frac{-j}{2}\sin\alpha_n\mathrm{e}^{j(\theta-\phi_n)} & \frac{j}{2}\sin\alpha_n\mathrm{e}^{-j(\theta-\phi_n)}. & \cos\alpha_n \end{pmatrix}. \tag{8.29}$$

Note the relation between the matrix elements, which can be used to simplify the calculations. We shall introduce here a simplified notation and write the matrix in (8.29) as:

$$\overset{\Rightarrow}{\mathfrak{R}}_{\phi_n,\alpha_n} \overset{\Rightarrow}{P} = \begin{pmatrix} r_{11}(n)\mathrm{e}^{j\theta} & r_{12}(n)\mathrm{e}^{-j\theta} & r_{13}(n) \\ r_{21}(n)\mathrm{e}^{j\theta} & r_{22}(n)\mathrm{e}^{-j\theta} & r_{23}(n) \\ r_{31}(n)\mathrm{e}^{j\theta} & r_{23}(n)\mathrm{e}^{-j\theta} & r_{33}(n) \end{pmatrix}. \tag{8.29a}$$

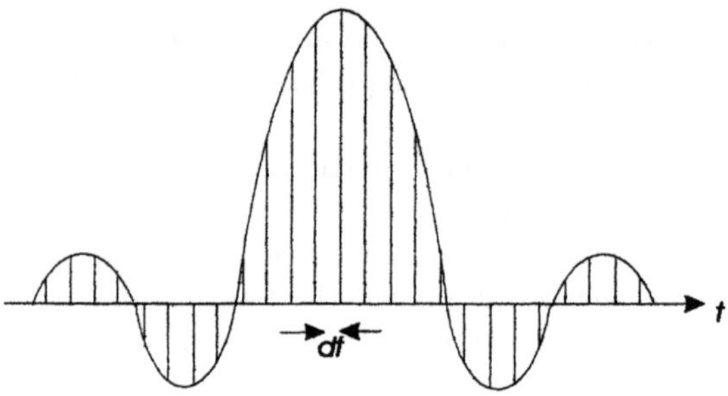


**Fig. 8.24.** Subdivision of an RF pulse in a number of square subpulses

Starting from a predefined magnetization state (for example, the equilibrium magnetization in the $z$ direction), with this matrix the rotation and precession of the magnetization at the end of the first subpulse become (see also (8.1))

$$\vec{M}(1) = \begin{pmatrix} r_{13}(1)M_0 \\ r_{23}(1)M_0 \\ r_{33}(1)M_0 \end{pmatrix} = \begin{pmatrix} M_T(1) \\ M_T^*(1) \\ M_z(1) \end{pmatrix}, \tag{8.30}$$

from which we obtain $M_{\mathrm{T}}$ and $M_z$. The effect of the second subpulse is therefore:

$$\begin{pmatrix} M_T(2,\theta) \\ M_T^*(2,\theta) \\ M_z(2,\theta) \end{pmatrix} = \begin{pmatrix} r_{11}(2)\mathrm{e}^{j\theta} & r_{12}(2)\mathrm{e}^{-j\theta} & r_{13}(2) \\ r_{21}(2)\mathrm{e}^{j\theta} & r_{22}(2)\mathrm{e}^{-j\theta} & r_{23}(2) \\ r_{31}(2)\mathrm{e}^{j\theta} & r_{23}(2)\mathrm{e}^{-j\theta} & r_{33}(2) \end{pmatrix} \begin{pmatrix} M_T(1) \\ M_T^*(1) \\ M_z(1) \end{pmatrix}$$

$$= \begin{pmatrix} r_{11}(2)r_{13}(1)\mathrm{e}^{j\theta} + r_{12}(2)r_{23}(1)\mathrm{e}^{-j\theta} + r_{13}(2)r_{33}(1) \\ r_{21}(2)r_{13}(1)\mathrm{e}^{j\theta} + r_{22}(2)r_{23}(1)\mathrm{e}^{-j\theta} + r_{23}(2)r_{33}(1) \\ r_{31}(2)r_{13}(1)\mathrm{e}^{j\theta} + r_{23}(2)r_{23}(1)\mathrm{e}^{-j\theta} + r_{33}(2)r_{33}(1) \end{pmatrix} M_0. \tag{8.31}$$

After the second pulse, terms with $\exp(j\theta)$ and $\exp(-j\theta)$ appear, which is the beginning of a discrete Fourier series. The third subpulse will introduce terms with $\exp(2j\theta)$ and $\exp(-2j\theta)$, etc. and after the $N$th subpulse the highest-order terms have a $\theta$ dependence of $\exp\{j(N-1)\theta\}$ and $\exp\{-j(N-1)\theta\}$.

This repeated application of the precession–rotation matrix, (8.29), is named the forward Shinnar–Le Roux (SLR) transformation. Since $\theta = \gamma Bz\delta t$, where $\delta t$ is taken to be constant, the result is:

$$M_T(N,\theta) = \sum_{-(N-1)}^{N-1} F_n \mathrm{e}^{jk\theta}, \quad \text{and} \quad M_z(N,\theta) = \sum_{-(N-1)}^{N-1} M_{z,k}\mathrm{e}^{jk\theta}. \tag{8.32}$$

This result describes the slice profile and the phase of the resulting transverse magnetization after the pulse.

The longitudinal magnetization is always a real quantity, and so $M_{z,k} = M^*_{z,-k}$. The transverse magnetization, however, may be complex, describing its direction in the transverse plane (the "phase"), and so $F_k \neq F^*_{-k}$. This latter fact is also found in the analytic theory of low-flip-angle pulses (see Sect. 2.3.2), where the phase of the transverse magnetization that results from a sinc pulse is shown to be a linear function of $z$, the position in the selection direction, which must be corrected for by applying a refocussing gradient – see (2.16). A completely different phase behaviour is also possible, as is shown in [28], where it is shown that inherently refocussed RF pulses have a completely different phase-versus-position profile in the selection direction.

For a purely AM pulse, $\phi_n$ can be chosen to be constant. When the phase $\phi_n$ is changed as a function of the sub-pulse number, the slice profile of FM pulses can also be calculated.

#### 8.5.1.1. The Inverse Shinnar–Le Roux Transformation (ISLR Transformation)

The process described in the previous section can be inverted. First we can express the required longitudinal magnetization profile in a discrete Fourier series of $N$ terms. Actually we start with $L$ terms ($L > 2N$, in view of the Nyquist theorem) and then retain the first $N$ terms of this Fourier series to avoid backfolding. This yields the Fourier series development of $M_z(N,\theta)$ and $M_{\mathrm{T}}(N,\theta)$ as shown in (8.32). The transverse magnetization profile at the end of the last subpulse can be calculated from $M_0^2 = |M_{\mathrm{T}}|^2 + M_z^2$, except for its phase. The transverse magnetization can point in any direction in the $x' - y'$ plane, which makes it impossible to determine uniquely its phase beforehand.

However, much is known about small-flip-angle RF pulses (see Sect. 2.3.2) and about moderate-flip-angle pulses and their slice profiles [28]. These results can be used as a first estimate of the slice profile due to the pulse, including the phase of the transverse magnetization. Another possibility is to assume a slice profile with position (frequency) and phase properties similar to the frequency and phase response of a FIR filter [29]. These assumptions result in a slice profile that satisfies requirements with respect to ripple inside and outside the slice and to the slope of the magnetization profile of the slice. However, the resulting pulses are not optimized for low power dissipation, which is a major requirement in MRI. Even so, several very reasonable estimates can be made about the phase of the magnetization profile caused by an RF pulse of a certain shape.

The required magnetization profile after the $N$th (last) subpulse of an RF pulse can be developed in Fourier series, as per (8.32). This Fourier series will be the starting point of the inverse Shinnar–le Roux transformation. First we have to find the inverse transformation matrix of a sub-section.

The inverse transformation of the magnetization by the hard pulse of the $n$th segment can be found by inverting the rotation matrix of the $n$th segment, as per (8.5), which is done by applying (8.5) to a rotation $-\alpha_n$ around the same axis, which is characterized by $\phi_n$:

$$\overset{\Rightarrow}{\mathfrak{R}}{}^{-1}_{\phi_n,\alpha_n} = \begin{pmatrix} \frac{1}{2}(1+\cos\alpha_n) & \frac{1}{2}(1-\cos\alpha_n)e^{j2\phi_n} & j\sin\alpha_n e^{j\phi_n} \\ \frac{1}{2}(1-\cos\alpha_n)e^{-j2\phi_n} & \frac{1}{2}(1+\cos\alpha_n) & j\sin\alpha_n e^{-j\phi_n} \\ \frac{j}{2}\sin\alpha_n e^{j\phi_n} & \frac{-j}{2}\sin\alpha_n e^{j\phi_n} & \cos\alpha_n \end{pmatrix}. \quad (8.33)$$

This result is easily checked by multiplying the matrices of (8.33) and (8.5) for $\alpha_n$. The inverse of the precession matrix, (8.4), is then:

$$\overset{\Rightarrow}{P}{}^{-1} = \begin{pmatrix} e^{-j\theta} & 0 & 0 \\ 0 & e^{j\theta} & 0 \\ 0 & 0 & 1 \end{pmatrix}. \quad (8.33a)$$

The result of the inversion of the effect of a complete sub-section is therefore:

$$\overset{\Rightarrow}{P}{}^{-1}\overset{\Rightarrow}{\mathfrak{R}}{}^{-1}_{\phi_n,\alpha_n} = \begin{pmatrix} \frac{1}{2}(1+\cos\alpha_n)e^{-j\theta} & -\frac{1}{2}(1-\cos\alpha_n)e^{j(2\phi_n-\theta)} & j\sin\alpha_n e^{j(\phi_n-\theta)} \\ \frac{1}{2}(1-\cos\alpha_n)e^{-j(2\phi_n-\theta)} & \frac{1}{2}(1+\cos\alpha_n)e^{j\theta} & -j\sin\alpha_n e^{-j(\phi_n-\theta)} \\ \frac{j}{2}\sin\alpha_n e^{-j\phi_n} & \frac{-j}{2}\sin\alpha_n e^{j\phi_n} & \cos\alpha_n \end{pmatrix}. \quad (8.34)$$

The matrix $\overset{\Rightarrow}{P}{}^{-1}\overset{\Rightarrow}{\mathfrak{R}}{}^{-1}$ has in the first row the term $\exp(-j\theta)$, in the second row the term $\exp(j\theta)$, and in the last row no dependence on $\theta$ appears. The transformation matrix is first applied to the last sub-section (section $N$) and therefore applied to the known Fourier series of the profile of the slice magnetization, as per (8.32). Thus:

$$\begin{pmatrix} M_T(N-1) \\ M_T^*(N-1) \\ M_z(N-1) \end{pmatrix} = \begin{pmatrix} R_{11}(N)e^{-i.\theta} & R_{12}(N)e^{-i\theta} & R_{13}(N)e^{-i\theta} \\ R_{21}(N)e^{i\theta} & R_{22}(N)e^{i\theta} & R_{23}(N)e^{i\theta} \\ R_{31}(N) & R_{32}(N) & R_{33}(N) \end{pmatrix} \begin{pmatrix} M_T(N) \\ M_T^*(N) \\ M_z(N) \end{pmatrix}, \quad (8.35)$$

where $R_{i,j}$ are the elements of the inverse rotation matrix $\mathfrak{R}$,without any $\theta$ dependence. Now we can calculate $\alpha_N$ and $\phi_N$ from the fact that the highest-order term of the Fourier series of $M_z(N)$ has a $\theta$ dependence equal to $\exp\{j(N-1)\theta\}$, while the highest-order term of $M_z(N-1)$ contains

$\exp\{j(N-2)\theta\}$. Therefore multiplying the third row by the magnetization vector in the right-hand side of (8.35) yields:

$$R_{31}(N)F_{N-1}(N) + R_{32}(n)F^*_{-(N-1)}(N) + R_{33}(N)M_z(N) = 0$$

or

$$j\sin\alpha_N\left(F_{N-1}(N)\mathrm{e}^{j\phi_N} - F^*_{-(N-1)}(N)\mathrm{e}^{-j\phi_N}\right) \\ +2\cos\alpha_N M_{z,N-1}(N) = 0. \qquad (8.36)$$

From this complex equation $\alpha_N$ and $\phi_N$ can be calculated. Then also the terms of the precession – rotation matrix, (8.34), can also be calculated. With this knowledge, all Fourier coefficients of $M_{\mathrm{T}}(N-1)$ and $M_z(N-1)$ can be calculated, and so the backward transformation can be performed. This process can be repeated for all sub-sections, reducing each time the length of the Fourier series.

It is not our aim to perform this transformation, which is not practically feasible without numerical calculations, but it is clear that a solution of the Bloch equation is possible by developing the magnetization in a discrete Fourier series. The SLR transformation can also be formulated on the basis of the Cayley–Klein parameters [30], which is done in [29].

Image Sets
Chapter 8

# VIII-1 Magnetic Resonance Cholangio-Pancreatography (MRCP) and Influence of Flip Angle of the Refocussing Pulses

Images obtained with help of A. van der Molen and J.M.A. van Engelshoven, University Hospital Maastricht, The Netherlands

(1) The liquid content of the bile and pancreatic ducts is free of cells and macromolecules so that its $T_1$, as well as its $T_2$ are high. This allows the visualization of this tissue in Turbo Spin Echo with very long TE. The strong $T_2$ weighting is sufficient to suppress nearly completely the signal from all other tissue, so that the data can be used immediately for a maximum intensity projection image of the duct system (3D acquisition), or as a projective image (single-slice acquisition) with very similar information. The attractiveness of these types of scan for the study of the duct system has called the attention of clinical groups and the abbreviation MRCP (for magnetic resonance cholangio-pancreatography) has become customary. Single-slice MRCP can, for instance, be obtained within a single breathhold and employed to obtain a functional study of the pancreas by running a dynamic series after administration to the patient of a pancreas stimulation drug [31].

(2) The details of the single-slice TSE scan used for MRCP scans are rather exacting. Fat (with a long $T_2$) has to be suppressed by a separate pulse. The first echoes of each echo train still carry high signals from other tissue and are not used for imaging (dummy echoes). To suppress the signal of blood, $TE_{eff}$ should be larger than about 600 ms. High spatial resolution is needed because the bile and pancreatic ducts can be narrow and this calls for long echo trains.

Long echo trains can be reached by allowing values of $TE_{eff}$ above 1 s, but then already too much of the signal of the choleic and pancreatic fluids is lost because of their $T_2$ decay. So the effective echo time $TE_{eff}$ and the duration of the echo train TT are compromises. The sequence is often used with a half scan technique in which the $k_y$ order increases linearly and $TE_{eff}$ is less than half TT [21], but a very asymmetric half scan and hence a long TT effectively lowers the resolution because of signal loss from $T_2$ decay.

With the short echo spacings possible in modern systems, a single echo train can be used that contains all $k_y$ values for a complete projection image (single-shot TSE). Single-shot TSE is attractive because of its robust image quality, free of motion artifacts. Moreover, when sticking to the concept of imaging in a single breathhold (10 s), multi-shot TSE has little or no advantage. Additional excitations have to be separated by a large TR because of the long $T_1$ of the tissue of interest. Moreover, during the long echo train the longitudinal magnetization does not recover. As a result TR should be at least 6 s and only two excitations can be given within one breathhold period.

So, compared to single-shot TSE, the potential gain in SNR and/or resolution from multi-shot single breathhold scanning is limited.

Although short $\Delta$TE is desirable to combine adequate $TE_{eff}$ and resolution, its value should not be shorter than needed so that a narrow bandwidth can be kept (e.g. water–fat shift $> 1$ pixel).

(3) A final argument is the RF power dissipated in the patient during the TSE sequence. When used in a dynamic series, the specific absorbed rate (SAR) from this type of scan can be close to the limit allowed for reasons of safety. This is because of the large number of refocusing pulses given per unit time. Per pulse, the energy dissipated is proportional to the RF field strength and to the square of the flip angle used, so that it can be interesting to reduce this angle, especially in strong modern high-powered systems. In this image set, single-shot MRCP scans are compared with a varying value of the refocusing flip angle.

Parameters: $B_0 = 1.5\,\mathrm{T}$, surface coil; matrix $= 256\times512$; $TE_{eff} = 850\,\mathrm{ms}$; first 20 echoes not used for reconstruction; echo train length 153; half scan factor 0.75; $\Delta TE = 12.8\,\mathrm{ms}$; water–fat shift 1.9 pixels; $TT = 2214\,\mathrm{ms}$; $d = 60\,\mathrm{mm}$; coronal images of normal volunteer. Per scan the refocus flip angles had a constant value, except for the first refocussing pulse which always was 180°.

| image no. | 1 | 2 | 3 | 4 | 5 |
|---|---|---|---|---|---|
| flip angle of the refocusing pulse (degrees) | 160 | 140 | 120 | 100 | 90 |
| theoretical relative amplitude of echo 32 | 0.99 | 0.95 | 0.90 | 0.80 | 0.74 |
| experimental SNR (pancreatic duct) | 19.7 | 15.1 | 15.0 | 12.5 | 9.6 |
| experimental SNR (bile duct) | 47.8 | 42.6 | 44.6 | 45.0 | 39.7 |

(4) The amplitude of echo 32, compared to the case of 180° refocusing pulses, was calculated using the theory of Sect. 8.4.1 (assuming $T_1 = 4\,\mathrm{s}$, $T_2 = 1\,\mathrm{s}$ and $\Delta TE = 10\,\mathrm{ms}$). In these calculations, all echo amplitudes were obtained. After the first few oscillating values, the calculated amplitudes decay smoothly at a rate that is similar for all flip angles (see e.g. Fig. 8.18), so that the amplitude of echo 32 directly provides a measure of the signal-to-noise ratio SNR of the TSE image. So for flip angles that are not too small the theory predicts the possibility to obtain artifact-free TSE images with reduced flip angle refocussing pulses as well as a moderate loss in SNR with decreasing flip angle.
Comparison between the images shows that adequate background suppression is maintained for all flip angles used. SNR was estimated from the images for the pancreatic duct as well as the bile duct by taking the ratio of its brightness to the noise in the image background. In the pancreatic duct the observed decrease of SNR is more pronounced than that in the bile duct. The decrease with flip angle of the predicted SNR is between that of the inspected liquids and confirms that $T_2$ decay is the main cause of signal loss in the echo train of the TSE sequence even when using reduced flip angles

for the refocusing pulses. This suggests a relatively small $T_2$ of the pancreatic fluid.

(5) The image set shows that the use of flip angles below 180° is possible and can have practical value to reduce SAR in dynamic single shot TSE at acceptable loss of SNR, especially in modern high-powered MR systems.

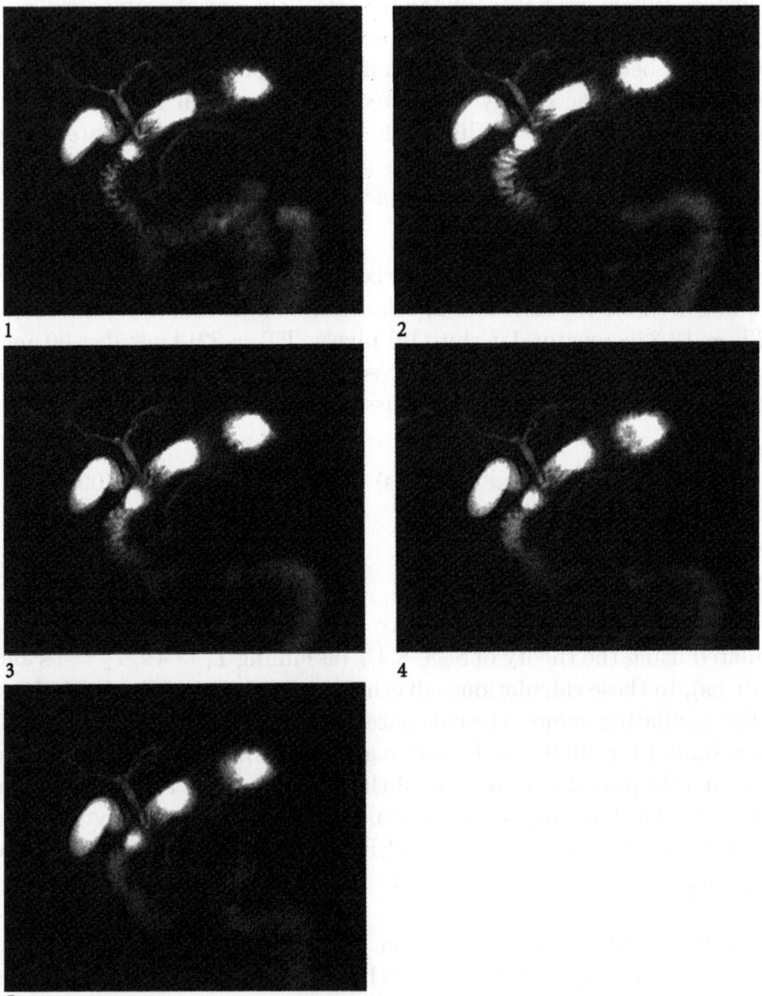


**Image Set VIII-1**

# Appendix

## Descriptive Terms of MR Scan Methods

In our discussions we have met examples of most of the existing basic acquisition methods used in MRI. Many other methods exist, but they are generally derivations or combinations of the acquisition methods mentioned so far. The confounding variety of terms and abbreviations used in the description of such MR acquisition methods is a burden to the communication between MR scientists, especially for new-comers in the field. Yet after writing this book we have the impression that a step towards a more coherent nomenclature is possible. This step should be helpful in offering in a short hand format the essential information for each possible acquisition method. By mutual agreement between authors, a similar proposal is given in another recent textbook on MRI:

Magnetic Resonance Imaging by E.M. Haacke, R.W. Brown, M. Thompson and R. Venkatesan, Wiley, 1999.

### Definitions

Some terms in this nomenclature need to be defined first.

**Scanning** [Performing an imaging experiment]. Scanning in an MR system is the sum of all actions needed to obtain an image: a scan is specified by the system initiation phase, its sequence structure, its imaging parameters (which are adjusted initially), the physiological conditions, and the method of reconstruction of the data.

**Sequence Structure.** The sequence structure describes the entire set of events generated by the MR system to obtain the data necessary for reconstruction. This (data acquisition) process is subdivided in a series of periods [cycles] in which all events are repeated, but not necessarily with the same amplitude (such as the phase encode gradient strength). The combined RF and gradient waveforms in such a period are referred to as an MR sequence [or sequence cycle].

**Sequence** [Sequence Cycle]. A sequence always starts with an excitation pulse. In a conventional scan (Chap. 2) the magnetization profile of a single line in the $\vec{k}$ plane is acquired per sequence [cycle]. In a TSE or multi shot EPI or a segmented spiral scan (Chap. 3) in a single sequence [cycle] the profiles of several lines (or a spiral arm) in the $\vec{k}$ plane are acquired. Sometimes all necessary profiles are acquired in one sequence (a single shot acquisition), in that case the sequence structure and the sequence [cycle] are identical. The repetition time of the sequence is called TR. If TR $< T_1$ the sequences for the actual data acquisition may be preceded by one or more run in the sequences [magnetization preparation] so that a dynamic equilibrium is established (see Chap. 4).

## Characteristics

The nomenclature for each scan method is based on the characteristics of the sequence used, as will be shown below. More than the sequence structure and the scan initialization are, the sequence is the main descriptive element of the scan method.

**Echoes.** The excitation pulse with which a sequence starts generates a FID (see Sect. 1.1.5). Normally this FID is not measured directly (except in sequences such as spiral imaging, Sect. 3.6.1), but is transformed into one or more spin echoes (Sect. 2.4) and/or gradient field echoes (Sect. 2.5). We then speak of either Spin Echo methods or Gradient Echo methods, respectively (in our text we frequently used "Field Echo sequences" adhering to the local convention of the MR systems used for our image sets), although a mixture of these is also possible.

**Repeated Echoes.** In some sequences save acquisition time, several echoes are formed and phase encoded after a single excitation pulse. For example in TSE (or RARE, see Sect. 3.2) the spin echo is repeated, so a more informative name for TSE would be Repeated Spin Echo (RSE). In EPI repeated gradient echoes are acquired (see Sect. 3.3). This sequence could similarly be named Repeated Gradient Echo (RGE). In GRASE imaging (Sect. 3.4) the sequence naming could be Repeated Gradient and Spin Echo.

**Magnetization State.** The magnetization state describes the magnetization at the start of the sequence [cycle] (just before the excitation pulse).

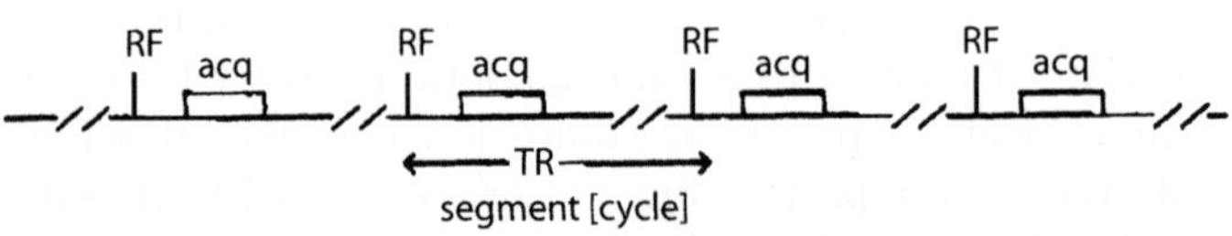


Segment structure of a scan. Each segment [cycle] consists of excitation RF and acquisition. No gradient waveforms are shown.

a. Steady-state methods have an equal magnetization state at the start of each sequence. Conventional $T_2$-weighted Spin Echo or Gradient Echo scan methods with TR > $T_1$ are, according to this definition, steady-state methods because each sequence starts from the equilibrium magnetization, $M_0$. When TR < $T_1$, the magnetization state at the start of each sequence is still equal (after the run-in sequences in which dynamic equilibrium is established), but smaller than $M_0$ (see Sects. 4.4 and 4.5), so we still speak of steady state. In addition the magnetization state may be influenced by preparation pulses of various design (prepared magnetization state, see below).

b. Transient-state methods are all methods in which the magnetization state at the start of each sequence changes, in phase and/or in magnitude. For example, when in an FFE scan method (with very short TR) the run in sequences are left out, so that the acquisition sequences run during the transient between the initial magnetization state and the dynamic equilibrium, we speak of a Transient Field Echo method (TFE, see Chap. 5). The initial state may be the equilibrium state, in which at the start of each sequence the longitudinal magnetization is equal to the equilibrium magnetization $M_0$ or it may be influenced by previous eyents, such as preparation pulses.

**Prepared Magnetization State.** In Magnetization-Prepared (MP) methods each sequence [cycle] or group of sequences is preceded by one or more RF pulses and gradient lobes to influence the magnetization state at the start of the sequence (see Sect. 5.3). Magnetization Preparation is used to influence the weighting of the resulting images and can make $T_1$ or $T_2$ weighting more dominant. Also other weighting parameters may be influenced (for example diffusion weighting).

**Spoiling.** Spoiling (such as by RF phase cycling of the excitation pulses or by gradients) may be applied to minimize in the acquired signal the observed coherence between the contribution of the FID and the transverse magnetization resulting from earlier excitations at the start of each sequence (see Sects. 4.2.7 and 4.5.3.3). This technique plays an important role in some steady-state FFE methods. In transient-state methods, spoiling may also be applied, but its effect is less clear.

**Conclusion.** The features described above are summarized in Table 1.
For a scan method one always has a choice for each of the features given in Table 1. The features that have been presented are orthogonal and the nomenclature that in conjunction with the authors of the aforementioned text, we propose is based on specifying these choices.

**Table 1.** Characterization of scan methods

| Magnetization State | | | Echo | |
|---|---|---|---|---|
| | | | number | type |
| unprepared<br>prepared | unspoiled<br>spoiled | steady<br>transient | single<br>repeated | spin echo<br>gradient echo |

## Examples

We shall finish this first (still incomplete) attempt of the systematic naming of scan methods with a few examples.
*N-FFE or GRASS.* This scan method is built up of unprepared, unspoiled, steady-state, single-gradient echo sequences. Here it becomes visible that not always all choices need to be mentioned. So we propose a "grammar rule" to be used in connection with Table 1 that when a feature is not mentioned always the upper choice in the table is implied (except for the last column: the choice between a spin echo or a gradient echo is always mentioned separately). With this "grammar rule" the N-FFE method can simply be named a gradient echo method.
*EPI.* A scan method consisting of a set of one or more repeated field-echo sequences, or a repeated gradient-echo method.
*MP-$T_1$-TFE or MP-RAGE.* A scan method consisting of prepared, spoiled, transient-state, single gradient-echo sequences, or a prepared, spoiled, transient gradient-echo method.
*SE.* Usually this name is used for a method using an unprepared, unspoiled steady-state, single spin-echo sequence. With the grammar rule mentioned above it is: a spin-echo method.

# References

## Historical Introduction


1. Nuclear induction, F. Bloch, W.W. Hanson, M.E. Packard. *Phys. Rev.*, **69**, p. 127, 1946
2. Resonance absorption by nuclear magnetic moments in solid, E.M. Purcell, H.C. Torrey, R.V. Pound, *Phys. Rev.*, **69**, p. 37, 1946
3. Mesure de temps de relaxation T2 en présence d'une inhomogeneité de champs magnétique supérieur à la largeur de raie, R. Gabillard, *CR Acad. Sci. Paris*, **232**, 1951
4. Tumor detection by nuclear magnetic resonance, R. Damadian, *Science*, **171**, p. 1151, 1971
5. Image formation by induced local interactions: examples of employing nuclear magnetic resonance, P.C. Lauterbur, *Nature*, **242**, p. 190, 1973
6. Tumor imaging in a live animal by field focussing NMR (FONAR), R. Damadian et al., *Physiol. Chem. Phys.*, **8**, p. 61, 1976
7. Apparatus and method for detecting cancer in tissue. R. Damadian, US Patent No 3789823 filed 17 March 1972
8. *Magnetic Resonance Zeugmatography*, P.C. Lauterbur et al., Proc XVIII Ampere Congress, Nottingham (Amsterdam, North Holland, 1974) pp. 27–29
9. Spin mapping: the application of moving gradients to NMR, W.S. Hinshaw, *Phys. Letters* **48A**, p. 78, 1974
10. Image formation by nuclear magnetic resonance: the sensitive point method, W.S. Hinshaw, *J. Appl. Phys.*, **47**, p. 3709, 1976
11. Radiographic thin-section image of the wrist by nuclear magnetic resonance, W.S. Hinshaw, P.A. Bottomley, G.N. Holland, *Nature (London)*, **270**, p. 723, 1977
12. Display of cross sectional anatomy by nuclear magnetic resonance imaging, W.S. Hinshaw, E.R. Andrew, P.A. Bottomley et al., *Br. J. Radiol.* **51**, p. 273, 1980
13. Nuclear magnetic resonance tomography of the brain: A preliminary clinical assessment with demonstration of pathology, R.C. Hawkes, G.N. Holland, W.S. Moore et al., *J. Comp. Assist. Tomography*, **4(5)**, p. 577, 1980
14. *Imaging by Nuclear Magnetic Resonance*, J.M.S. Hutchison, Proc. 7th LH Gray Conf., Leeds (Wiley, Chichester, 1976) pp. 135–141
15. Image formation in NMR by a selective irradiative process, A.N. Garroway, P.K. Grannell, P. Mansfield, *J. Phys. C*, **7**, p. 457, 1974
16. Line scan proton spin imaging in biological structures by NMR, P. Mansfield, A.A. Maudsley, *Phys. Med. Biol.*, **23**, p. 847, 1976
17. Human whole body line scan imaging by NMR, P. Mansfield, I.L. Pykett, P.G. Morris et al., *Br. J. Radiol.*, **52**, p. 242, 1979
18. NMR Fourier Zeugmatography, Kumar, D. Welti, R.R. Ernst, *J. Magn. Res.*, **18**, p. 69, 1975

19. Spin warp NMR imaging and applications to human whole-body imaging, W.A. Edelstein, J.M.S. Hutchison et al., *Phys. Med. Biol.*, **25**, p. 571, 1980
20. NMR whole body imager operating at 3.5 kGauss, L.E. Crooks, J.C. Hoenninger, M. Arakawa et al., *Radiology*, **143**, p. 169, 1982
21. Blood flow rates by NMR measurements. J.R. Singer, *Science*, **130**, p. 1652, 1959
22. Direct cardiac NMR imaging of the heart wall and blood flow velocity. P. van Dijk, *J. Comp. Assist. Tomogr.*, **429**, 1984
23. Measurement of flow with NMR imaging using a gradient pulse and phase difference technique, D.J. Bryant et al., *J. Comp. Assist. Tomogr.*, **8**, p. 588, 1984
24. *Three-dimensional display of blood vessels in MRI*, S. Rossnick, G. Laub, R. Braekle et al., Proc. IEEE Computers in Cardiology Conf. New York 1986, p. 193
25. NMR angiography based on inflow, J.P. Groen, R.G. de Graaf, P. van Dijk, Soc. Magn. Res. Imaging Med., 6th Annual meeting, August 20–26, 1988 San Francisco
26. Three dimensional phase contrast angiography, C.L. Dumoulin, S.P. Souza et al., *Magn. Res. Med.*, **9**, p. 139, 1989
27. The k-trajectory formulation of the NMR imaging process with application in analysis and synthesis of imaging methods. D.B. Twieg *Med. Phys.*, **10**, p. 610, 1983
28. Very fast MR imaging by field echoes and small angle excitation, P. Van der Meulen, J.P. Groen, J.J.M. Cuppen, *Magn. Res. Imag.*, **3**, p. 297, 1985
29. FLASH imaging. Rapid NMR imaging low flip angle pulses, A. Haase, J. Frahm, D. Matthaei et al., *J. Magn. Res.*, **67**, p. 258, 1986
30. RARE imaging: A fast imaging method for clinical MR, J. Hennig, A. Nauerth, H. Friedburg, *Magn. Res. Med.*, **3**, p. 823, 1986
31. Multiplanar image formation using NMR spin echoes, P. Mansfield, *J. Phys. C: Solid State Phys.*, **10**, L55, 1977

## Chapter 1

1. *The Principles of Nuclear Magnetism* A. Abragam (Oxford University Press, Oxford 1978) ISBN 019-851236-8
2. *NMR Imaging in Biomedicine* P. Mansfield and P.G. Morris (Academic Press, New York 1982) ISBN 012-025562-6
3. *Nuclear Magnetic Resonance Imaging in Medicine and Biology* P. Morris (The Clarendon Press, Oxford 1986) ISBN 019-855155-X
4. *Manual of Clinical Magnetic Resonance Imaging* Second Edition J.P. Heiken and J.J. Brown (Raven Press, New York 1991) ISBN 088-167744-2
5. *MRI Workbook for Technologists* Carolyn Kaut (Raven Press, New York 1992) ISBN 088-167876-7
6. *Magnetic Resonance Imaging* D. Stark and W.G. Bradley (Mosby Year Book, St. Louis 1992) ISBN 0-8016-4930-7
7. *Magnetic Resonance in Medicine* P. Rinck (Blackwell Scientific Publications, Oxford 1993) ISBN 0-632-03789-4
8. *Magnetic Resonance Imaging, Physical Principles and Sequence Design* E.M. Haacke, R.W. Brown, M.R. Thompson, R. Venkatesan (Wiley-Liss, New York 1999) ISBN 0-471-35128-8

9. Spin Echoes, E.L. Hahn, *Phys. Rev.* **20(4)**, p. 580, 1950
10. *Magnetic Resonance Imaging Techniques* A.M. Parikh (Elsevier, New York 1991) ISBN 0-444-01634-1
11. *Biomedical Magnetic Resonance Imaging* C.N. Chen and D.I. Hoult (Adam Hilger, Bristol 1989) ISBN 0-85274-118-9
12. *The Fourier Transform and its Applications* R. Bracewell (McGraw Hill, New York 1965)
13. *The Fast Fourier Transform* (Prentice Hall, Englewood Cliff, NS 1974) ISBN 013-307496-X.
14. *Digital Image Processing* R.C. Gonzalez, P. Wintz (Addison Press, Reading 1977) ISBN 0-201-11026-1
15. Permanent Magnetic Systems for NMR Tomography, H. Zijlstra, *Philips Journal of Research*, **40**, pp. 259–288, 1985
16. *Handbook of Mathematical Functions* M. Abramowitz and I. Stegun (Dover Publications, New York 1965)
17. Gradient Coil Design, a Review of Methods, R. Turner, *Magn. Res. Im.*, **11**, pp. 903–920, 1993
18. Parallelable PWM Amplifier, R.S. Burwen, *IEEE Transactions on Instrumentation and Measurement*, **36(4)**, pp. 1001–1003, 1987
19. Design and Evaluation of Shielded Gradient Coils, J.W. Carlson, KA. Derby, K.C. Hawryszko and M. Weideman, *Magn. Res. in Med.*, **26**, pp. 191–206, 1992
20. Highly Selective $\pi/2$ and $\pi$ Pulse Generation, M.S. Silver and D.I. Hoult, *J. Mag. Res.*, **59**, pp. 347–351, 1984
21. Local Intensity Shift Artifact, P.H. Wardenier, *SMRM Book of Abstracts* 1989
22. Calculation of the Quadrupole Intensity Artifact in MRI, A.M.J. van Amelsfoort, T. Scharten and P. Wardenier, *SMRM Book of Abstracts* 1988, p. 119
23. The NMR Phased Array, P.B. Roemer, W.A. Edelstein, C.E. Hayes, S.P. Souza and O. Müller, *Mag. Res. in Medicine*, **16**, pp. 192–225, 1990
24. *Electrocardiography: A Physiologic Approach* D. Mirvis (Mosby Year Book, St Louis 1993) ISBN 08-01674794
25. *ABC of the ECG, A Guide to Electrocardiography* J. Boutkan (Philips Press, 1968)
26. Understanding and using DICOM, the data exchange standard for biomedical imaging, W.D. Bidgood Jr, S.C. Horii, F.W. Prior, D.E. Van Syckle, *J. Am. Med. Inform. Assoc.* **4**, pp 199–212, 1997


## Chapter 2


1. Nuclear Induction, F. Bloch, *Phys. Rev.*, **70**, p. 460, 1946
2. *Magnetic Resonance Imaging* D. Stark and W.G. Bradley, Mosby Year Book, St Louis, 1992, Chapter 4
3. A *k*-space Analysis of Small Tip Angle Excitation, J. Pauly, D. Nishimura, A. Mackovski *J. Magn. Res.*, **81**, pp. 43–56, 1989
4. A Linear Class of Large Tip Angle Selective Excitation Pulses, J. Pauly, D. Nishimura, A. Mackovski, *J. Magn. Res.*, **82**, pp. 571–587, 1989
5. The Art of Pulse Crafting, W.S. Warren, MS. Silver, *Advances in Magnetic Resonance*, Volume 12, Academic Press New York, 1988, pp. 247–388
6. Parameter Relations for the Shinnar–Le Roux Selective Pulse Design Algorithm, P. le Roux, D. Nishimura, A. Mackowsky *IEEE Trans. on Med. Imaging*, **10**, pp. 53–65, 1991

7. Variable Rate Selective Excitation, S. Conolly, D. Nishimura, A. Mackovsky *J. Magn. Res.*, **78**, pp. 440–458, 1988
8. IEC 601-1, 1988 and IEC 601-1-1, 1992, Part 2 *Particular Requirements for Safety of Nuclear Resonance Equipment*
9. E.L. Hahn, Spin Echoes, *Phys. Rev.*, **20(4)**, p. 580, 1950
10. Proton NMR Tomography, P.R. Locher, *Philips Technical Review*, **41**, pp. 73–88, 1983
11. Application of Reduced Encoding Imaging with Generalized-Series Reconstruction (RIGR) in Dynamic MR Imaging, S. Chandra, Z.-P. Liang, A.Webb, H. Lee, H. Douglas Morris, P.C. Lauterbur, *J. Magn. Res. Im.*, **6**, 783–797, 1996
12. "Keyhole" Method for Accelerating Imaging Contrast Agent Uptake, J.J. v. Vaals, M.E. Brummer, W.T. Dixon, H.H. Tuithof, H. Engels, R.C. Nelson, B.M. Gerity, J.L. Chezmar, J.A. den Boer, *J. Magn. Res. Im.*, **3**, 671–675, 1993
13. Rapid Images and MR Movies, A. Haase, J. Frahm, O. Matthaei, K.D. Merboldt, W. Heanike *SMRM Book of Abstracts*, 1985, pp. 980–981
14. Very Fast MR Imaging by Field Echos and Small Angle Excitation, P.v.d. Meulen, J.P. Groen, J.M. Cuppen *Magn. Res. Im.*, **3**, pp. 297–299, 1985
15. Artifacts in Magnetic Resonance Imaging, R.M. Henkelman, M.J. Bronskill *Reviews of Magnetic Resonance in Medicine*, **2(1)**, 1987
16. Analysis of $T_2$ Limitations and Off Resonance Effects on Spatial Resolution and Artifacts in Echo Planar Imaging, F. Farzaneh, S.J. Riederer, N.J. Pelc *Magn. Res. in Med.*, **14**, pp. 123–139, 1990
17. Short TI Inversion Recovery Sequence: Analyses and Initial Experience in Cancer Imaging, A.J. Dwyer et al. *Radiology*, **169**, pp. 827–836, 1988
18. $^1$H NMR Chemical Shift Imaging, A. Haase, J. Frahm, W. Hänicke, D. Matthaei *Phys. Med. Biology*, **30(4)**, pp. 341–344, 1985
19. Proton Spin Relaxation Studies of Fatty Tissue and Cerebral White Matter, R.L. Kamman, K.G. Go, A.J. Muskiet, G.P. Stomp, P.v. Dijk, H.J.C. Berendsen *Magn. Res. In. Med.*, 2, pp. 211–220, 1984
20. Contrast between White and Grey Matter: MRI Appearance with Aging, S. Magnaldi, M. Ukmar, R. Longo, R.S. Pozzi-Mocelli *Eur. Rad.*, **3**, pp. 513–319, 1993
21. Sensitivity Encoding for Fast MRI, K.P. Preussmann, M. Weiger, M.B. Scheidegger, P. Boesiger, *Magn. Res. in Med.*, **42**(5), pp. 952–962, 1999
22. Simultaneous Acquisition of Spatial Harmonics (SMASH): Ultra Fast Imaging with Radiofrequency Coil Arrays, D.K. Sodickson, W.J. Manning, *Magn. Res. in Med.*, **38**, 591–603, 1997
23. *MRI Scan Time Reduction through Non-Uniform Sampling*, G.J. Marseille, Doctoral Thesis, Technical University of Delft, The Netherlands, 1997
24. Driven Equilibrium Fourier Transform Spectroscopy. A New Method for Nuclear Magnetic Resonance Signal Enhancement. E.D. Becker, J.A. Ferrati, T.C. Farrar. *J. Am. Chem. Soc.*, **91**, pp. 7784–7785, 1969
25. Black Blood $T_2$ weighted Inversion Recovery MR Imaging of the Heart. O.P. Simonetti, J.P. Finn, R.D. White, G. Laub, D.A. Henry. *Radiology*, **19**, pp. 49–57, 1996

## Chapter 3

1. RARE Imaging, A Fast Imaging Method for Clinical MR, J. Hennig, A. Nanert, H. Friedburg, *Magn. Res. in Med.*, **3**, pp. 823–833, 1986
2. P. Mansfield, I.L. Pykett, *J. Magn. Res.*, **29**, p. 355, 1978

3. GRASE (Gradient and Spin Echo) Imaging, A Novel Fast Imaging Technique, K. Oshio, D.A. Feinberg, *Magn. Res. in Med.*, **20**, pp. 344–349, 1991
4. Radial Turbo Spin Echo Imaging, V. Rascke, D. Holz, W. Schepper, *Magn. Res. in Med.*, **32**, pp. 629–638, 1994
5. Fast Spiral Coronary Imaging, C.H. Meyer, B. Hu, D.G. Nishimura, A. Mackovski, *Magn. Res. in Med.*, **28**, pp. 202–213, 1992
6. Square Spiral Fast Imaging, C.H. Meyer, A. Mackovski, *SMRM Book of Abstracts*, 1989, p. 362
7. Phase Encode Order and its Effect on Contrast and Artifact in Single Shot RARE Sequences, R.V. Mulkern, P.S. Melke, P. Jahab, N. Higushi, FA. Jolesz, *Med. Phys.*, **18(5)**, pp. 1032–1037, 1991
8. On the Application of Ultra Fast RARE Experiments, D.G. Norris, P. Boerner, T. Reese, D. Leibfritz *Magn. Res. in Med.*, **27**, pp. 142–164, 1992
9. $T_2$ Weighted Thin Section Imaging with Multislab Three Dimensional RARE Techniques, K. Oshio, F.A. Jolesz, P.S. Melki, R.V. Mulkern *J. Magn. Res. Im.*, **1**, pp. 695– 700, 1991
10. Interleaved Echo Planar Imaging on a Standard MRI System, K. Buts, S.J. Riederer, R.L. Ehman, R.M. Tomson, CR. Jack *Magn. Res. in Med.*, **31**, pp. 67–72, 1994
11. Ultrafast Interleaved Gradient Echo Planar Imaging on a Standard Whole Body System, G.C. McKinnon *Magn. Res. in Med.*, **30**, pp. 609–616, 1993
12. Analysis of $T_2$ Limitations and Off Resonance Effects on Spatial Resolution and Artifacts in Echo Planar Imaging, F.E. Farzaneh, S.J. Riederer, N. Pelc *Magn. Res. in Med.*, **14**, pp. 123–139, 1990
13. Limits to Neuro-Stimulation in Echo Planar Imaging, P. Mansfield, P.R. Harvey *Magn. Res. in Med.*, **29**, pp. 746–758, 1993
14. Non-axial Whole Body Instant Imaging, R.M. Weisskopf, M.C. Cohen, R.R. Rzedzian *Magn. Res. in Med.*, **29**, pp. 276–303, 1993
15. Single Shot GRASE Imaging without Fast Gradients, K. Oshio, D.A. Feinberg *Magn. Res. in Med.*, **26**, pp. 355–360, 1992
16. *Basic Principles of MR Imaging* A Philips Publication, product code 4522 984 30501, p. 79, Fig. 63
17. Gradient Echo Shifting in Fast MRI Techniques (GRASE Imaging) for Correction of Field Inhomogeneity Errors, D.A. Feinberg, K. Oshio *J. Magn. Res.*, **97**, pp. 177–183, 1992
18. *Spiral Scanning, Simulation of "Square" Spiral Imaging* J. Brown, J. Larson, M.T. Vlaardingerbroek, S. Wengi., HP internal Report
19. A Method of Measuring Field Modulation Shapes. Application to High-Speed NMR Spectroscopic Imaging, T. Onodero, S. Matsui, K. Sekihara *J. Phys. E, Scientific Instruments*, **20**, pp. 416–419, 1987
20. Selection of a Convolution Function for Fourier Inversion Gridding, J.I. Jackson, C.H. Meyer, D.G. Nishimura, A. Mackovski *IEEE Trans. MI.*, **10**, pp. 473–478, 1991
21. Deblurring for Non 2D Fourier Transform MRI, D.C. Noll, J. Pauly, C.H. Meyer, D.G. Nishimura, A. Mackovski *Magn. Res. in Med.*, **25**, pp. 319–333, 1992
22. Fast Magnetic Resonance Imaging with Simultaneously Oscillating and Rotating Gradients, S.J. Norton *IEEE Trans. Med. Im.*, **6(1)**, pp. 21–31, 1987
23. Continuous Radial Data Acquisition for Dynamic MRI, V. Rasche, R.W. de Boer, D. Holz, R. Proksa *Magn. Res. in Med.*, **34**, 754–761, 1995
24. Radial Turbo Spin Echo Imaging, V. Rasche, D. Holz, W. Schepper *Magn. Res. in Med.*, **32**, pp. 629–638, 1994

25. A $k$-Space Analysis of Small Tip Angle Excitation, J. Pauly, D.G. Nishimura, A. Mackovski *J. Magn. Res.*, **81**, pp. 43–56, 1989
26. Simultaneous Spatial and Spectral Selective Excitation, C.H. Meyer, J. Pauly, A. Mackovski, D.G. Nishimura, *Magn. Res. in Med.*, **15**, pp. 287–304, 1990
27. Magnetic Resonance Imaging of Brain Iron, B. Drayer, P. Burger, R. Darwin, S. Riederer, R. Herfkens, G.A. Johnson *Am. J. of Neuroradiology*, **7**, pp. 373–380, 1986
28. Why Fat is Bright in RARE and Fast Spin Echo Imaging, R.M. Henkelman, P.A. Hardy, J.E. Bischop, C.S. Moon, D.B. Plewes *J. of Magn. Res. Im.*, **2**, pp. 533–540, 1993
29. In Vivo and in Vitro MR Imaging of Hyaline Cartilage: Zonal Anatomy, Imaging Pittfalls and Pathologic Conditions, J.G. Waldschmidt, R.J. Rilling, A.A. Kajdacsy-Balla, M.D. Boynton, S.J. Erickson *Radiographics*, **17**, 1387–1402, 1997

## Chapter 4

1. Phase and Intensity Anomalies in Fourier Transform NMR, R. Freeman and H.D.W. Hill *J. Magn. Res.*, **4**, p. 366, 1971
2. Image Formation by Nuclear Magnetic Resonance, W.S. Hinshaw *J. Appl. Phys.*, **47**, p. 3907, 1976
3. Very Fast MR Imaging by Field Echos and Small Angle Excitation, P.v.d. Meulen, J.P. Groen and J.J.M. Cuppen *Magn. Res. Im.*, **3**, p. 297, 1985
4. Rapid Fourier Imaging using Steady State Free Precession, R.C. Hawkes and S. Patz *Magn. Res. in Med.*, **4**, p. 9, 1987
5. Fast Field Echo Imaging, P.v.d. Meulen, J.P. Groen, A.M.C. Tinus and G. Brunting *Magn. Res. Im.*, **6**, p. 355, 1988
6. An Analysis of Fast Imaging Sequences with Transverse Magnetisation Refocussing, Y. Zur, S. Stokar and P. Bendel *Magn. Res. in Med.*, **6**, p. 174, 1988
7. Editorial *Magn. Res. Im.*, **6**, p. 353, 1988
8. Recent progress in Fast MR Imaging, J.J.v. Vaals, J.P. Groen and G.H.v. Yperen *Medica Mundi*, **36(2)**, p. 152, 1991
9. Spin Echo's, E.L. Hahn *Phys. Rev.*, **80(4)**, p. 580, 1950
10. Errors in the Measurement of T2 Using Multiple Echo MRI Techniques, S. Majumdar, S.C. Orphanoudakis, A. Gmitro, M. O'Donnel and J.C. Gore *Magn. Res. in Med.*, **3**, p. 397, 1986
11. Errors in T2 Estimation using Multislice Multiple Echo Imaging, A.P. Crawly and R.M. Henkelman *Magn. Res. Im.*, **4**, p. 34, 1987
12. Spoiling of Transverse Magnetization in Steady State Sequences, Y. Zur, M.L. Wood and L.J. Neuringer *Magn. Res. in Med.*, **21**, p. 251, 1991
13. Motion-Insensitive, Steady-State Free Precession Imaging, Y. Zur, M.L. Wood and L.J. Neuringer *Magn. Res. in Med.*, **16**, p. 444, 1990
14. Elimination of Transverse Coherences in FLASH MRI, A.P. Crawley, M.L. Wood and R.M. Henkelman *Magn. Res. in Med.*, **8**, p. 248, 1988
15. *Rapid Images and NMR Movies, Book of Abstracts 4th SMRM* London 1985, page 980, A. Haase, J. Frahm, D. Mathei, K.-D. Merbold and W. Haenecke
16. FLASH Imaging, Rapid Imaging using Low Flip Angle Pulses, A. Haase, J. Frahm, D. Mathei, W. Haenicke and K.-D. Merboldt *J. Magn. Res.*, **67**, p. 256, 1986

17. The Application of SSFP in Rapid 2 DFT NMR Imaging, FAST and CE-FAST sequences. M.L. Gyngell *Magn. Res. Im.*, **6**, p. 415, 1988
18. FISP – a new Fast MR Sequence, A. Oppelt, R. Graumann, H. Barfuss, H. Fisher and W. Hartl *Electromedica*, **54**, p. 15, 1986
19. P. Mansfield and P.G. Morris *NMR Imaging in Biomedicine*, Academic Press, New York, 1982
20. Proton NMR Tomography, P.R. Locher *Philips Technical Review*, **41(3)**, p. 73, 1983/84
21. FADE – A New Fast Imaging Sequence, T.W. Redpath and R.A. Jones *Magn. Res. in Med.*, **6**, p. 224, 1988
22. *Multi Echo True FISP Imaging*, O. Heid, M. Deimling, SMR Book of Abstracts, 1995, p. 481.
23. jiTR Spoiling, Transverse Magnetization Spoiling in SSFP Sequences using a Random Time Jittered TR *Abstracts SMRM* 1990, p. 1159
24. MR Angiography Based on Inflow, J.P. Groen, R. de Graaf and P.v. Dijk *Abstracts SMRM*, 1988 p. 906
25. Optimized articular cartilage-fluid contrast in high-resolution orthopedic 3D imaging using a single echo sequence, R. Springorum, T. Rozijn, J.v.d. Brink, T. Schäffter, J. Groen, Proceedings ESMRMB 1998


## Chapter 5


1. Advances in cardiac application in sub-second FLASH MRI, D. Chien, K.-D. Merbold, W. Hanecke, H. Bruhn, M.L. Gyngell and J. Frahm *Magn. Res. Im.*, **8**, p. 829, 1990
2. MR Mammography, W. Kaiser *Medica Mundi*, **36(2)**, p. 168, 1991
3. Snapshot FLASH MRI applications to $T_1$, $T_2$ and chemical shift Imaging, A. Haase *Magn. Res. in Med.*, **13**, p. 77, 1990
4. Approach to Equilibrium in Snapshot Imaging, R.A. Jones and P.A. Rinck *Magn. Res. Im.*, **8**, p. 797, 1990
5. Signal Strength in Sub-second FLASH MRI Sequences, The Dynamic Approach to Steady State, W. Hanecke, K.-D. Merbold, D. Chien, M.L. Gyngell, H. Bruhm, J. Frahm *Abstracts SMRM*, 1990, p. 458
6. Variable Flip Angle Snapshot GRASS Imaging, A.E. Holsinger and S.J. Riederer *Abstracts SMRM*, 1990, p. 453
7. Optimized (incremented) RF angle Gradient Echo imaging, M.K. Stehling *Abstracts SMRM*, 1990, p. 459
8. Rapid Three Dimensional $T_1$ Weighted MR Imaging with the MP-Rage Sequence, J.P. Muggler, J.R. Brookman *Magn. Res. Im.*, **1**, p. 561, 1991
9. $T_2$ weighted Three Dimensional MP-Rage Imaging, J.P. Muggler, T.A. Spraggins, J.R. Brookman *J. Magn. Res. Im.*, **1**, p. 731, 1991
10. DPSF: Snapshot Flash Diffusion/Perfusion Imaging, W.H. Perman, M. Gado, J.C. Sandstrom *Abstracts SMRM*, 1990, p. 309
11. Imaging of Diffusion and Micro-Circulation with Gradient Sensitization: Design, Strategy and Significance, D. le Bihan, R. Turner, C.T.W. Moonen and J. Pekar *J. Magn. Res. Im.*, **1**, p. 7, 1991
12. Excitation Angle Optimization for Snapshot FLASH and a Signal Comparison with EPI, D.G. Norris *J. Magn. Res.*, **91**, p. 190, 1991
13. The Importance of Phase-Encoding Order in Ultra-Short TR Snapshot Imaging, A.E. Holsinger and S.J. Riederer *Magn. Res. Im.*, **16**, p. 481, 1990

14. Strategies to Improve Contrast in TurboFLash Imaging: Reordered Phase Encoding and K-Space Segmentation, D. Chien, D.A. Atkinson and R.A. Edelman *J. Magn. Res. Im.*, **1**, p. 63, 1991
15. Steady State preparation for Spoiled Gradient Echo Imaging, R.B. Busse, S.J. Riederer, *Magn. Res in Med.* **45**, 653–661, 2001
16. Magnetization Prepared True FISP Imaging. M. Deimling, O. Heid. *Proc. 2nd Ann. Meeting ISMRM*, San Fransisco, 1994, p. 495
17. Magnetization Preparation during the Steady State: Fat Saturated 3D True FISP, K. Scheffler, O. Heid, J. Hennig, *Magn. Res. in Med.*, **45**, pp. 1073–1080, 2001.

## Chapter 6

1. Relaxation Effects in Nuclear Magnetic Resonance Absorption, Bloembergen, Pound, Purcel *Phys. Rev.*, **73**, p. 679, 1948
2. *Magnetic Resonance Imaging*, D. Stark and W.G. Bradley (editors), Mosby Year Book, St. Louis, 1992
3. Simultaneous Measurement of Regional Blood Volume and Capillary Water Permeability with Intravascular MR Contrast Agents, C. Schwarzbauer, S.P. Morissey, R. Deichmann, H. Adolf, U. Noth, K.V. Toyka, A. Haase, *Proc. ISMRM*, New York 1996, p. 1577
4. Design and Implementation of Magnetization Transfer Sequences for Clinical Use, J.V. Hajnal, C.J. Bandom, A. Oatridge, I.R. Young, G.M. Bydder *J. of Computer Assisted Tomography*, **16**, pp. 7–18, 1992
5. Turbo-Mix $T_1$ Measurements and MTC exchange Rate $K_{for}$ Calculations, R.W. de Boer, A. Eleveld *SMRM Abstracts*, 1993, p. 175
6. Quantitative $^1$H Magnetization Transfer Imaging *in Vivo*, J. Eng, T.L. Ceckler, R.S, Balaban, *Magn. Res. in Med.*, **17**, 304–314, 1991
7. Magnetization Transfer Contrast with Periodic Pulsed Saturation, H.N. Yeung, A.M. Aisen, *Radiology*, **183**, 209–214, 1992
8. Magnetization Transfer Contrast, R.W. de Boer, *Medica Mundi*, **40/2**, 64–83, 1992
9. Improved Time of Flight Angiography of the Brain with Magnetization Transfer Contrast, R.E. Edelman, S.S. Ahn, D. Chien, Wei Li, A. Goldman, M. Mantello, J. Kramer, J. Kleefield, *Radiology*, **184**, 395–399, 1992
10. MR enhancement of Brain Lesions, Increased Contrast Dose Compared with Magnetization Transfer, M. Knauth, M. Forsting, M. Hartmann, S. Heiland, T. Bolder, K. Sartor, *AJNR*, **17**, 1853–1859, 1996
11. Magnetization Transfer Contrast in Multiple Sclerosis, R.I. Grossman, *Ann. Neurology*,**36**, Suppl: S97–99, 1994
12. Use of Magnetization Transfer for Improved Contrast on Gradient Echo MR Images of the Cervical Spine, D.A. Finelli, G.C. Hurst, B.A. Karaman, J.E. Simon, J.L. Duerk, E.M. Bellon *Radiology*, **193**, 165–171, 1994
13. Analysis of Water-Macromolecule Proton Magnetization Transfer in Articular Cartilage, D.K. Kim, T.L. Ceckler, V.C. Hascall, A. Calabro, R.S. Balaban, *Magn. Res. in Med.*, **29(2)**, 211–216, 1993
14. *Magnetic Resonance Imaging*, D. Stark and W.G. Bradley (editors), Mosby Year Book, St. Louis, 1992, Chapter 14
15. Basic Physics at MR Contrast Agents and Maximization of Image Contrast, R.E. Hendrick, E.M. Haacke *J. Magn. Res. Im.*,**3**, pp. 137–148, 1993

16. The Signal-to-Noise Ratio of the Nuclear Magnetic Resonance Experiment, D.I. Hoult, R.E. Richards *J. Magn. Res.*, **24**, p. 71, 1976
17. Resolution and Signal-to-Noise Relationships in NMR Imaging in the Human Body, J.M. Libove, J.R. Singer *J. Phys. E: Scientific Instruments*, **13**, pp. 38–44, 1980
18. Magnetic Resonance Imaging, Effects of Magnetic Field Strength, J. Hoenninger, B. McCasten, J. Watts, L. Kaufmann *Radiology*, **151**, pp. 127–133, 1984
19. Improvement of SNR at low Fieldstrength using Mutually Decoupled Coils for Simultaneous NMR Imaging, C. Leussler and D. Holz *SMRM Abstracts*, 1991, p. 724
20. Multifrequency Selective RF pulses for Multislice MR Imaging, S. Muller *Magn. Res. in Med.*, **6**, pp. 364–371, 1988, and **10**, pp. 145–155, 1989
21. $T_1$-Calculations, Combining Ratios and Least Squares, J.J.E. in den Kleef, J.J.M. Cuppen *Magn. Res. in Med.*, **5**, pp. 513–524, 1987
22. Protocols and Test Objects for the assessment of MRI Equipment, R.A. Lerski, D.W. McRobbie, J.D. Certaines *Magn. Res. Im.*, **6**, pp. 195–199, 1988
23. Age-Related Changes in Proton $T_1$ values of Normal Human Brain, R.G. Steen, S.A. Gronemeyer, J.S. Taylor *J. of Magn. Res. Im.*, **5**, pp. 43–48, 1995
24. Use of Fluid Attenuated Inversion Recovery (FLAIR) Pulse Sequences in MRI of the Brain, J.V. Hajnal, D.J. Bryant, L. Kosuboski, I.M. Pattany, B. de Ceane, P.D. Lewis, J.M. Pennock, A. Oatridge, I.R. Young, G.M. Bydder, *J. Computer Aided Tomography*, **16**, pp. 841–844, 1992
25. MR Imaging of the Breast; Fast Imaging Sequences with and without the use of Gd-DPTA, W.A. Kaiser, E. Zeitler *Radiology*, **170**, pp. 681–686, 1989
26. Pharmacokinetic Analysis of Gd-DTPA Enhancement in Dynamic Three-Dimensional MRI of Breast Lesions, J.A. den Boer, R.K.K.M. Maenderop, J. Smink, G. Dornseiffen, P.W.A.A. Koch, J.H. Mulder, C.H. Slump, E.D.P. Volker, R.A.I. de Vos *J. of Magn. Res. Im.*, **7**, pp. 702–715, 1997
27. Magnetization Transfer Contrast (MTC) and Tissue Water Proton Relaxation in Vivo, S.D. Wolf, R.S. Balaban *Magn. Res. in Med.*, **10**, pp. 135–144, 1989

## Chapter 7

1. Artifacts in Magnetic Resonance Imaging, R.M. Henkelman, M.J. Bronskill *Reviews of Magn. Res. Im.*, **2(1)**, 1987
2. *Magnetic Resonance Imaging: Cardiovascular System* C.G. Blackwell, G.B. Cranney, G.M. Prohost, Gower Medical Publishing, New York, 1992. ISBN 1-56375-000-7
3. *Magnetic Resonance Angiography, Concepts and Applications* E.J. Potchen, E.M. Haacke, J.E. Siebert, A. Godschalk, Mosby, St Louis, 1993. ISBN 1-55664-270-9
4. Motion Triggered Cine SIR Imaging of Active Joint Movement, U.H. Melchert, C. Schröder, J. Brossmann, C. Muhle *Magn. Res. Im.*, **10**, pp. 457–460, 1992
5. "Functional MRI" of the Patellofemoral Joint: Comparison of Ultra Fast MRI, Motion Triggered MRI and Static MRI, C. Muhle, J. Brossmann, U. Melchert, C. Schröder, R. de Boer *European Radiology*, 1995
6. Adaptive Technique for High Resolution MR Imaging of Moving Structures Navigator Echoes, R.L. Edelmann, J.P. Felmlee *Radiology*, **173**, pp. 255–263, 1989

7. Optimum Electrocardiographic Electrode Placement for Cardiac Gated MRI, R.N. Dunich, L.W. Hedlund, R.J. Herfkens, E.K. Fram, J. Utz *Investigative Radiology*, **22**, pp. 17–22, 1987
8. Retrospective Cardiac Gating: a Review of Technical Aspects and Future Directions, G.W. Lenz, E.M. Haacke, RD. White *Magn. Res. Im.*, **7**, pp. 445–455, 1989
9. MR Angiography with Pulsatile Flow, R.G. de Graaf, J.P. Groen *Magn. Res. Im.*, **10**, pp. 23–34, 1992
10. Coronary Arteries Breath-hold MR Angiography, R.R. Edelman, W.J. Manning, D. Bursten, S. Pauli *Cardiac Radiology*, **181**, pp. 641–643, 1991
11. Three-dimensional MR Imaging of the Coronary Arteries: Preliminary Clinical Experience, C.B. Parschal, E.M. Haacke, L.P. Adler *J. Magn. Res. Im.*, **3**, pp. 491–500, 1993
12. Fast Spiral Coronal Imaging, C.H. Meyer, B.S. Hu, D.G. Nishimura, A. Mackovski *Magn. Res. in Med.*, **28**, pp. 202–213, 1992
13. Suppression of Respiratory Motion Artifacts in MRI, M.L. Wood, R.M. Henkelman *Med. Phys.*, **13(6)**, pp. 794–805, 1986
14. Respiratory Motion of the Heart, Kinematics and the Implications for the Spatial Resolution in Coronary Imaging, Yi Wang, S.J. Riederer, R.L. Ehman *Magn. Res. in Med.*, **33**, 713–719, 1995
15. Two Dimensional Coronary MR Angiography without Breathholding, J.N. Oshinski, L.H. Hofland, S. Mukumdan, W.T. Dixon, W.J. Parks, R.I. Pettigrew *Radiology*, **201**, 737–743, 1996
16. Prospective Adaptive Navigator Correction for Non Breathhold MR Coronary Angiography, M.V. McConnel, V.C. Khasgiwala, B.J. Savord, M.H. Chen, M.L. Chuang, R.R. Edelman, W.J. Manning *Magn. Res. in Med*, **37**, 148–152, 1997
17. True Myocardial Motion Tracking, S.F. Fischer, G.C. McKinnon, SIB. Scheidegger, W. Prins, D. Meier, P. Boesiger, *Magn. Res. in Med.*, **31**, 401–413, 1994
18. Magnetic Resonance Angiography, D.G. Nishimura, A. Macovski, J. Pauly *IEEE Trans. MI*, **5(3)**, pp. 140–151, 1986
19. Motion Induced Phase Shifts in MR, M. Kouwenhoven, M.B.M. Hofman, M. Sprenger *Magn. Res. in Med.*, **33**, 766–777, 1995
20. *Vascular Diagnostics* P. Lanser, J. Rösch, Springer Verlag, Heidelberg 1994, pp. 375–400. ISBN 3-540-57939-7
21. Physical Principles and Applications of MRA, J.E. Siebert, E.J. Potchen *Seminars in US, CT and MRI*, **13(4)**, pp. 227–245, 1992
22. Quantitative NMR Imaging of Flow, J.M. Pope and S. Yao *Concepts in Magnetic Resonance*, **5**, pp. 281–302, 1993
23. Encoding Strategies for Three Direction Phase Contrast MR Imaging of Flow, N.J. Pelc, M.A. Bernstein, A. Shimakawa, G.H. Glover, *J. Magn. Res. Im.*, **1**, pp. 405–413, 1991
24. 3D Flow Visualization in Phase Contrast Angiography, J.F.L. De Becker, M. Fuderer, M. Kouwenhoven *SMR Abstracts*, 1993, p. 450
25. Flow Velocity Quantification in Human Coronary Arteries with Fast Breath-Hold MR Angiography, R.R. Edelman, W.J. Manning, E. Cervino, W. Li *J. Magn. Res. Im.*, **3**, pp. 699–703, 1993
26. The Application of Breath Hold Phase Velocity Mapping Techniques to the measurement of Coronary Artery Blood Flow Velocity, J. Keegan, D. Firmin, P. Gatehouse, D. Longmore *Magn. Res. in Med.*, **31**, pp. 526–536, 1994
27. Coronary Artery Imaging in Multiple 1-sec. Breath Holds, M. Doyle, M.B. Scheidegger, R.G. de Graaf, J. Vermeulen, G.M. Pohost *Magn. Res. Im.*, **11**, pp. 3–6, 1993

28. Contrast enhanced MR Angiography, Methods, Limitations and Possibilities, M. Kouwenhoven, *Acta Radiologica*, **38**, 1997. ISBN 87-16-15614-5
29. Time Resolved Contrast Enhanced 3D MR Angiography, F.R. Korosec, R. Frayne, T.M. Grist, C.A. Mistretta *Magn. Res. in Med.*, **36**, pp. 345–351, 1996
30. Dynamic Gd:DTPA Enhanced 3DFT Abdominal MR Angiography, M.R. Prince, E.K. Yucel, J.A. Kaufman, A.C. Waltman *J. Magn. Reson. Imaging*, **3**, p. 877 (1993)
31. Automated Detection of Bolus Arrival and Initiation of Data Acquisition in Fast, Three-Dimensional, Gadolinium-enhanced MR Angiography, T.K.F. Foo, M. Saranathan, M.R. Prince, T.L. Chenevert *Radiology*, **203**, pp. 275–280 (1997)
32. Improved MTC Angiography with Spatially Varying Off-Resonance Frequency, M. Kouwenhoven, L. Hofland, R.W. de Boer, J.J. van Vaals, *Proceedings 12th Annual Meeting SMRM*, p. 383, 1993
33. Perfusion Imaging, D.A. Detre, J.S. Liegh, D.S. Williams, A.P. Koretsky *Magn. Res. Med.*, **23**, pp. 37–45, 1991
34. Multiple Readout Selective Inversion Recovery Angiography, S.J. Wang, D.G. Nishimura, A. Mackovski, *Magn. Res. in Med.*, **17**, pp. 244–251, 1991
35. Breath-Hold 3D STAR Angiography of the renal Arteries Using Segmented Echo Planar Imaging, P.A. Wielopolski, M. Adamis, P. Prasad, J. Gaa, R.R. Edelman, *Magn. Res. in Med.*, **33**, pp. 432–438, 1995
36. Fast Selective Black Blood MR Imaging, R.R. Edelman, D. Chien, D. Kim, *Radiology* **181**, pp. 655–660, 1991
37. A Method for $T_1\rho$ imaging, R.E. Sepponen, J.A. Pohjonen, J.T. Sipponen, J.I. Tanntu, *J. Comput. Assist. Tomogr.*, **9**, pp. 1007–1011, 1995
38. Myocardial Suppression *in vivo* by Spin Locking with Composite Pulses, W.T. Dixon, J.N. Oshinski, J.D. Trudeau, B.C. Arnold, R.I. Pettigrew, *Magn. Res. in Med.*, **36**, pp. 90–94, 1996
39. Comparison of four Magnetization Preparation Schemes to Improve Blood-Wall Contrast in Cine Short Axis Cardiac Imaging, S.I.K. Semple, R.W. Redpath, F.I. McKiddie, G.D. Waiter, *Magn. Res. in Med.*, **39**, pp. 291–199, 1998
40. Flow-Independent Magnetic Resonance Projection Angiography, G.A. Wright, D.G. Nishimura, A. Mackovski, *Magn. Res. in Med.*, **17**, pp. 126–140, 1991
41. Three-Dimensional Flow-Independent Peripheral Angiography, J.H. Brittain, E.W. Olcott, A. Szuba, G.E. Gold, G.A. Wright, P. Irarrasaval, D.G. Nishimura, *Magn. Res. in Med.*, **38**, pp. 343–354, 1997
42. Black Blood Imaging, W. Lin, E.M. Haacke, R. Edelman, pp. 160–172 of Ref. [7.3]
43. Private communication, S.M.J.J.G. Nijsten and D. Kaandorp, Technical University Eindhoven
44. The Effects of Time Varying Intravascular Signal Intensity and $k$-Space Acquisition Order on Three-Dimensional MR Angiography Image Quality. J.H. Maki, M.R. Prince, F.J. Londy, T.L. Chenevert, *J. Magn. Reson. Im.*, **6**, pp. 642–651, 1996
45. Measurement of Coronary Flow using Time of Flight echo Planar MRI, B. Poncelet, R.M. Weisskoff, V.J. Wedeen, T.J. Brady, H. Kantor *Proc SMRM* 1993
46. *Tracer Kinetic Methods in Medical Physiology* N.A. Lassen, W. Perl (Raven Press, New York 1979)
47. Assessment of Cerebral Blood Volume with Dynamic Susceptibility Contrast Enhanced Gradient Echo Imaging, F. Gückel, G. Brix, K. Rempp, M. Deimling, J. Rother, M. Georgi *J. Comp. Aided Tomography*, **18**, pp. 344–351, 1994

48. High-Resolution Measurement of Cerebral Blood Flow using Intravascular Tracer Bolus Passage. Part I: Mathematical Approach and Statistical Analysis; Part II: Experimental Comparison and Preliminary Results, L. Ostergaard, A.G. Sörensen, K.K. Kwong, R.M. Weisskoff, C. Gyldenstad, BR Rosen *Mag. Res. Med.*, **36**, pp. 715–736, 1996
49. *Functional Brain Imaging* W.W. Orrison, J.D. Lewine, A. Sanders, M.E. Hartshorne, Editors (Mosby, St Louis 1995), ISBN 0-8151-6509-9, p. 270
50. Susceptibility Contrast Imaging of Cerebral Blood Volume: Human Experience. B.R. Rosen, J.W. Belliveau, H.J. Aronen, D. Kennedy; B.R. Buchbinder, A. Fischman, M. Gruber, J. Glas, R.M. Weisskoff, M.S. Cohen, et al., *Magn. Reson. Med.*, **22**, pp. 293–299, 1991
51. MR Contrast Due to Intravascular Magnetic Susceptibility Perturbations. J.L. Boxerman, L.M. Hamberg, B. Rosen, R.M. Weiskoff, *Mag. Res. Med.*, **34**, pp. 555–566, 1995
52. Clinical MR Diffusion/Perfusion Protocol for Hyperacute Stroke, M. Moseley, A. De Crespigny, D. Tong, M. O'Brien, K. Butts, G. Albers, M. Marks *Proc. ISMRM 4th scientific meeting*, p. 567, 1997
53. Magnetic Resonance Imaging of Perfusion Using Spin Inversion of Arterial Water. D.S. Williams, J.A. Detre, J.S. Leigh, A.P. Koretsky, *Proc. Natl. Acad. Science, USA*, **89**, pp. 212–216, 1992
54. Tissue Specific Perfusion Imaging Using Arterial Spin Labelling, J.A. Detre, W. Zhang, D.A. Roberts, A.C. Silva, D.S. Williams, D.J. Grandis, A.P. Koretsky, J.S. Leigh *NMR Biomed.*, **7**, pp. 75–82, 1994
55. Qualitative Mapping of Cerebral Blood Flow and Functional Localisation with Echo-Planar MR Imaging and Signal Targeting with Alternating Radio Frequency, R.R. Edelman, B. Siewert, D.G. Darby, V. Thangaraj, A.C. Nobre, M.M. Mesulam, S. Warach *Radiology*, **192**, pp. 513–520, 1994
56. Slice Profile Effects in Adiabatic Inversion: Application to Multi-Slice Perfusion Imaging, L.R. Frank, E.C. Wong, R.B. Buxton, *Magn. Res. in Med.*, **38**, pp. 558–564, 1997
57. Dynamic Magnetic Resonance Imaging of the Human Brain Activity During Primary Sensory Stimulation, K.K. Kwong, J.W. Belliveau, D.A. Chesler, I.E. Goldberg, R.M. Weisskopf, D. Poncelet, B.E. Hoppel et al. *Proc. Natl Acad. Science (USA)*, **89**, pp. 5675–5679, 1992
58. Correction for Vascular Artifacts in Cerebral Blood Flow Values Measured by using Arterial Spin Tagging Techniques, F.Q. Ye, V.S. Mattay, P. Jezzard, J.A. Frank, D.R. Weinberger, A.C. McLaughlin *Mag. Res. in Med.*, **37**, pp. 226–235, 1997
59. Spin Diffusion Measurements: Spin Echoes in the Presence of a Time Dependent Field Gradient, E.O. Stejskal and J.E. Tanner *Journal of Chemical Physics*, **42(1)**, pp. 288–292, 1965
60. Molecular Diffusion Nuclear Magnetic Resonance Imaging, D. le Bihan *M.R. Quarterly*, **7(1)**, pp. 1–30, 1991
61. Imaging of Diffusion and Microcirculation with Gradient Sensitization, Design, Strategy and Significance, D. le Bihan, R. Turner, C.T.W. Moonen, J. Pekar *J. Magn. Res. Im.*, **1**, pp. 7–28, 1991
62. Isotropic Diffusion-Weighted and Spiral Navigated Interleaved EPI for Routine Imaging of Acute Stroke, K. Butts, J. Pauly, A. de Crespigny, M. Moseley *Magn. Res. in Med.*, **38**, pp. 741–749, 1997
63. Optimized Isotropic Diffusion Weighting, E.C. Wong, R.W. Cox, A.,W. Song *Magn. Res. in Med.*, **34**, pp. 139–143, 1995
64. MR Imaging of Motion with Spatial Modulation of Magnetization, L. Axel, L. Dougherty, *Radiology*, **171**, pp. 841–845, 1989

65. Improved Myocardial Tagging Contrast, S.E. Fisher, G.C. McKinnon, S.E. Maier, P. Boesiger, *Magn. Res. in Med.*, **30**, pp. 191–200, 1993
66. Peripheral vascular tree stenoses: evaluation with moving-bed infusion-tracking MR angiography, K.Y. Ho, T. Leiner, M.W. de Haan, A.G. Kessels, P.J. Kitslaar, J.M. van Engelshoven, *Radiology* **206**, pp. 683–692, 1998
67. Three-Dimensional Flow Independent Angiography of Aortic Aneurisms using Standard Fast Spin Echo, D.W. Kaandorp, P.F.F. Wijn, K. Kopinga, *Proc. ISMRM*, Sydney, 1998, p. 792
68. MR Perfusion and Diffusion Imaging in Ischaemic Brain Disease, J.A. den Boer, P. Folkers, *Medica Mundi*, **41**, pp. 21–35, 1997
69. Multislice Functional Perfusion Imaging Using a Magnetisation Transfer Insensitive Labeling Technique, X.Golay, M. Stuber, K. Pruessmann, R. Luechinger, O.M. Weber, M.B. Scheidegger, D. Meier, P. Boesiger, Proc. 6th Scientific Meeting ISMRM, Sydney, 1998; p. 1206
70. Transfer Insensitive Labelling Technique (TILT). X. Golay, M Stuber, K.P. Preussmann, D. Meier, P. Boesiger, *Mag. Res. in Med*, to be published
71. Transfer Insensitive Labelling Technique (TILT): Application to Multislice Functional Perfusion Imaging. X. Golay, M. Stuber, K.P Preussmann, D. Meier, P. Boesiger, *Journ. of Magn. Res Im.*, **9**, pp. 454–462, 1999
72. Diffusion Tensor MR Imaging of the Brain. C. Pierpaoli, J.P. Jezzard, P.J. Basser, A. Barnett, G. DiChiro, *Radiology*, **201**, pp. 637–648, 1996
73. Simultaneous Tagging and Through-Plane Velocity Quantitation: a 3D Myocardial Tracking Algorithm, J.P.A. Kuyer, J.T. Marcus, M.J.W. Goette, A C. van Rossum, R.M. Heethaar, *Journ. of Magn Res. Im.*, **9**, pp. 409–419, 1999
74. Measurement of Rat Brain Perfusion by NMR using Spin Labelling of Arterial Water: In-Vivo Determination of the Degree of Spin Labelling, W. Zhang, D.S. Williams, A.P. Koretski, *Magn. Res in Med.*, **29**, pp. 416–421, 1993
75. Variations in Blood Flow Waveforms in Stenotic Renal Arteries by 2D Phase Contrast Cine MRI, J.J.M. Westenberg, M.N.J.M. Wasser, R.J. v.d. Geest, P.M.T. Pattynama, A. de Roos, J. Vanderschoot, J.H.C. Reiber, *Journ. of Magn. Res. Im.*, **8**, pp. 590–597, 1998
76. MR Angiography of Occlusive Disease of the Arteries in Head and Neck: Current Concepts, B.C. Bouwen, R.M. Quencer, P. Margosian, P.M. Pattany, *Am. Journ. of Radiology*, **162**, pp. 9–18, 1994
77. MR Angiography of Normal Pelvic Arteries: Comparison of Signal Intensity and Contrast to Noise Ratio for Three Different Inflow Techniques, K. Yucel, M.S. Silver, A P. Carter, *Am. Journ. of Radiology*, **163**, pp. 197–201, 1994
78. Selective Projection Imaging: Application to Radiography and NMR, A. Mackovski, *IEEE Trans. on Med. Im.*, **MI-1**, pp. 42–47, 1982
79. Pulsed Field Gradient Nuclear Magnetic Resonance as a tool for Studying Translational Diffusion, Part I, Basic Theory. W.C. Price, *Concepts Magn. Resonance*, **9**, pp. 299–336, 1997
80. Restricted Self Diffusion of Protons in Colloidal Systems by the Pulsed-Gradient, Spin Echo Method, J.E. Tanner and E.O. Stejskal, *Journal of Chem. Physics*, **49-1**, pp. 1768–1777, 1968
81. Principles and Application of Self Diffusion Measurements by Nuclear Magnetic Resonance, J. Kärger, H. Pfeifer, W. Heink, *Adv. Magnetic Resonance*, **12**, pp. 1–89, 1988
82. Restricted Diffusion and Exchange of Intracellular Water: Theoretical Modeling and Diffusion Time Dependence of 1H NMR Measurements on Perfused Glial Cells, J. Pfeuffer, U. Flögel, W. Dreher, D. Leibfritz, *NMR in Biomedicine*, **11**, pp. 19–31, 1998

83. *NMR Imaging of Materials* (Monographs on the Physics and Chemistry of Materials), B. Blümich, (Oxford Science Publications, Oxford), 2000. ISBN 0 19 850683 X
84. *Principles of Nuclear Magnetic Resonance Microscopy*, P.T. Callaghan (Clarendon Press, Oxford, 1991)
85. *NMR Tomography, Diffusometry, Relaxometry*, R. Kimmich (Springer Verlag, Berlin, Heidelberg, New York, 1997)
86. Emphysema: Hyperpolarized Helium 3 Diffusion Imaging of the Lungs, Compared with Spirometric Indices, Initial Experience, M. Salermo, E.E. de Lange, T.A. Altes, J.D. Truwitt, J.R. Brookman, J.P. Muggler III, *Radiology*, **22**, p. 252, 2001
87. A General Kinetic Model for Quantitative Perfusion Measurement with Arterial Spin Labeling, R.B. Buxton, L.R. Frank, E.C. Wong, B. Siewart, S. Warach, R.R. Edelman, *Magn. Res. in Med.*, **40**, pp. 383–396, 1998
88. Novel Real-Time R-Wave Detection Algorithm Based on the Vector Cardiogram for Accurate Gated Magnetic Resonance Acquisition, S.E. Fisher, S.A. Wickline, C.H. Lorentz, *Magn. Res. in Med.*, **42**, p. 361, 1999
89. High b-Value Diffusion Weighted MRI of Normal Brain, J.H. Burdette, D.D. Durden, A.D. Elster, Y.F. Yen, *J. Comp. Ass. Tomography*, **25**, p. 515, 2001
90. The Value of $b$ Required to Avoid $T_2$ Shine Through from Old Lacunar Infarcts in Diffusion Weighted Imaging, B. Geijer, P.C. Sundgren, A. Lindgren, S. Brocksted, F. Stahlberg, S. Holtas, *Neuroradiology*, **43**, p. 511, 2001


## Chapter 8


1. Effects of Diffusion in Nuclear Magnetic Resonance Split Echo Experiments, D.E. Woessner, *J. Chem. Phys.*, **34**, pp. 2057–2061, 1961
2. Diffusion and Field Gradient Effects in NMR Fourier Spectroscopy, R. Kaiser, E. Bartholdy and R.R. Ernst, *J. Chem. Phys.*, **60-8**, pp. 1966–2979, 1974
3. Echoes, How to Generate, Recognise, Use or Avoid Them in MR Imaging Sequences, J. Hennig, *Concepts in Magn. Res.*, **3**,125–143, 1991
4. Burst Imaging, J. Hennig and M. Hodapp, *MAGMA*, **1**, pp. 39–48, 1993
5. On the Stationary State of Gradient Echo Imaging, W.T. Sobol and D.M. Gauntt, *Journ. of Magn. Res. Im.*, **6**, pp. 384–398, 1996
6. RARE Imaging: a Fast Imaging Method for Clinical MR, J. Hennig, A. Nauert, H. Friedburg, *Magn. Res. in Med.*, **3**, pp. 823–833, 1986
7. The Sensitivity of Low Flip Angle RARE Imaging, D.A. Alsop, *Magn. Res. in Med.*, **37**, pp. 176–184, 1997
8. Effects of Diffusion on Free Precession on Nuclear Magnetic Resonance Experiments, H.Y. Carr, E.M. Purcel, *Phys. Rev.*, **94**, pp. 630–638, 1954
9. Modified Spin Echo Method for Measuring Nuclear Relaxation Times, S. Meibohm, D.Gill, *Rev. Scientific Instruments*, **29**, pp. 688–691, 1958
10. DANTE Ultrafast Imaging Sequence (DUFIS), I.J. Lowe and R.E. Wysong, *J. Magn. Res.*, **B 101**, pp. 106–109, 1993
11. Analytical Solution for Phase Modulation in BURST Imaging with Optimum Sensitivity, P.v. Gelderen, Ch.T. Moonen, J.H. Duyn, *J. Magn. Res.*, **B 107**, pp. 78–82, 1995
12. Optimised Ultra Fast Imaging Sequence (OUFIS), L. Zha, I.J. Lowe, *Magn. Res. in Med.*, **33**, pp. 377–395, 1995

13. Susceptibility Insensitive Single Shot MRI Combining BURST and Multiple Spin Echoes, P.v. Gelderen, Ch.T. Moonen, J.H. Duyn, *Magn. Res. in Med.*, **3**, pp. 439–442, 1995
14. Ultra Rapid Gradient Echo Imaging, O. Heid, M. Deimling, W.J. Huk, *Magn. Res. in Med.*, **33**, pp. 143–149, 1995
15. Fast Volume Scanning with Frequency Shifted BURST MRI, J.H. Duyn, P.v. Gelderen, G. Liu, Ch. T.W. Moonen, *Magn. Res. in Med.*, **32**, pp. 429–432, 1994
16. QUEST, A Quick Echo Split NMR Imaging Technique, O. Heid, M. Deimling, W. Huk, *Magn. Res. in Med.*, **29**, 280–283, 1993
17. PREVIEW: a New Ultrafast Imaging Sequence Regarding Minimal Gradient Switching, C.J. Counsell, *Magn. Res. Imaging*, **11**, pp. 603–616 1993
18. Analysis and Evaluation of Ultra-Fast MR Sequences, J. Kürsch, A.R. Brenner, T.G. Noll, *Proc. ISMRI*, 1996
19. 3D Bolus Tracking with Frequency Shifted BURST MRI, J.H. Duyn, P.v. Gelderen, P. Barker, J.A. Frank, V.S. Mattay, Ch.T.W. Moonen, *J. of Comp. Assisted Tomography*, **18 (5)**, pp. 680–687, 1994
20. Optimisation of Ultrafast Multi-Pulse Sequences for Dynamic MR Imaging, A.R. Brenner, A. Glowinski, J. Kürsch, M. Drobnitzki, T.G. Noll, R.W. Gunther, *Proc. ISMRI*, 1996
21. MR Cholangiopancreatography Using HASTE Sequences, T. Myazaki, Y. Yamashita, T. Tsuchigame, H. Yamamoto, J. Orata, M. Takahashi, *Am. J. of Radiology*, **166**, pp. 1297–1303, 1996
22. Current Status of Cholangiopancreatography, C. Reinhold, P.M. Bret, *Am. J. of Radiology*, **166**, pp. 1285–1295, 1996
23. Optimization of a Dual Echo in the Steady State (DESS) Free-Precession Sequence for Imaging Cartilage, P.A. Hardy, M.P. Recht, D. Piraino, D. Thomasson, *J. of Magn. Res. Im.*, **6**, pp. 329–335, 1996
24. Steady-State Effects in Fast Gradient Echo Magnetic Resonance Imaging, J.H. Duyn, *Magn. Res. in Med.*, **37**, pp. 559–568, 1997
25. Selective Pulse Creation by Inverse Solution of the Bloch–Ricatti Equation, M. Silver, R. Joseph, D. Hoult, *J. Magn. Res.*, **59**, p. 347, 1984
26. The Synthesis of Pulse Sequences Yielding Arbitrary Magnetization Vectors, M. Shinnar, S. Eleff, H. Subramanian, J.S. Leigh, *Magn. Res. in Med.*, **12**, pp. 74–80, 1989 (See also the three following papers in the same Journal)
27. P. le Roux, French Patent 8610179, 1986
28. A Linear Class of Large Tip Angle Selective Excitation Pulses, J. Pauly, D. Nishimura, A. Mackovski, *J. Magn. Res.*, **82**, pp. 571–587, 1989
29. Parameter Relations for the Shinnar–le Roux Selective Excitation Pulse Design Algorithm, J. Pauly, P. le Roux, D. Nishimura, A. Mackovski, *IEEE Trans. Med. Im.*, **10**, pp. 53–65, 1991
30. *Classical Mechanics*, H. Goldstein (Addison-Wesley Publ. Comp., Reading Inc, Mass. 1950) Libr. of Congress Catalog No 50-7669
31. Pancreatic Duct, Morphologic and Functional Evaluation with Dynamic Pancreatography after Secretin Stimulation, C. Matos, T. Metens, J. Deviere, N. Nicaise, P. Braude, G.v. Yperen, M. Cremer, J. Struiven, *Radiology*, **203**, pp. 435–441, 1997

# Index of Abbreviated Terms

Full terms are identical to terms in the index

| | |
|---|---|
| IR | Inversion Recovery sequence 89, 180 |
| ISLR | Inverse Shinnar–Le Roux transformation 465 |
| MIP | Maximum Intensity Projection 341 |
| MOTSA | Multi-Overlapping Thin Slab Acquisition 346 |
| MRA | Magnetic Resonance Angiography 4, 340, 344, 346 |
| MRCP | Magnetic Resonance Cholangio Pancreatography 456, 470 |
| MRI | Magnetic Resonance Imaging 2, 9 |
| MT | Magnetization Transfer 277, 310, 318, 349 |
| MTC | Magnetization Transfer Contrast 280 |
| N-FFE | Normal Fast Field Echo sequence 219, 229, 237 |
| NMR | Nuclear Magnetic Resonance 1 |
| NSA | Number of Samples Acquired 289 |
| OUFIS | Optimized Ultra-Fast Imaging Sequence 442 |
| PCA | Phase Contrast Angiography 4, 340, 399 |
| PEAR | Phase Encode Artifact Reduction 324, 376, 388 |
| PSIF | reversed FISP sequence 219 |
| Q | Quality of the RF coil 63, 288 |
| QUEST | QUick Echo Split Technique 454 |
| RARE | Rapid Acquisition by Repeated Echo sequence 135 |
| REST | REgional SaTuration 345 |
| rCBV | relative Cerebral Blood Volume 360, 414 |
| R-FFE | Refocused Fast Field Echo sequence 209, 219, 224, 229 |
| ROPE | ReOrdered Phase Encoding 324 |
| SE | Spin Echo sequence 64, 96, 99, 197 |
| SENSE | SENSitivity Encoding 129, 301 |
| SNR | Signal to Noise Ratio 283, 297, 312, 314 |
| SPAMM | SPAtial Modulation of Magnetization 396 |
| SPGR | SPoiled GRadient sequence 219 |
| SPIR | Spectral Presaturation by Inversion Recovery sequence 92 |
| SSFP | Steady State Free Precession sequence 2, 219 |
| STAR | Signal Targeting with Alternating Radiofrequency 363 |
| STIR | Short Tau Inversion Recovery sequence 90, 107 |
| $T_1$-FFE | $T_1$-enhanced Fast Field Echo sequence 219, 231, 237 |
| $T_2$-FFE | $T_2$-enhanced Fast Field Echo sequence 209, 219, 222, 229 |
| TE | Echo Time 17, 99 |
| TFE | Transient Field Echo sequence 248, 264, 264, 475 |
| TF | Turbo Factor 137, 172 |
| TI | Inversion Time 89 |
| TILT | Transfer Insensitive Labelling Technique 416 |
| TONE | Tilted Optimized Non saturating Excitation 345 |
| TR | Repetition Time 19, 96, 193 |
| TSE | Turbo Spin Echo sequence 137, 172, 175, 180, 455 |
| URGE | Ultra Rapid Gradient Echo sequence 442 |
| VENC | Velocity ENCoding 331, 399 |
| VERSE | Variable Rate Selective Excitation 62 |

# Index

The New Springer Global Website
Be the first to know
Benefit from new practice-driven features.
Search all books and journals – now faster and easier than ever before.
Enjoy big savings through online sales.
springeronline.com – the innovative website with you in focus.
springeronline.com
The interactive website for all Springer books and journals
Springer
010048x

MIX
Papier aus verantwortungsvollen Quellen
Paper from responsible sources
FSC® C105338

If you have any concerns about our products,
you can contact us on
**ProductSafety@springernature.com**

In case Publisher is established outside the EU,
the EU authorized representative is:
**Springer Nature Customer Service Center GmbH**
**Europaplatz 3, 69115 Heidelberg, Germany**

Printed by Libri Plureos GmbH
in Hamburg, Germany